Internet and Mobile Phone Addiction

Internet and Mobile Phone Addiction

Health and Educational Effects

Special Issue Editor

Olatz Lopez-Fernandez

MDPI • Basel • Beijing • Wuhan • Barcelona • Belgrade

Special Issue Editor
Olatz Lopez-Fernandez
Monash University
Australia

Editorial Office
MDPI
St. Alban-Anlage 66
4052 Basel, Switzerland

This is a reprint of articles from the Special Issue published online in the open access journal *International Journal of Environmental Research and Public Health* (ISSN 1660-4601) from 2017 to 2018 (available at: https://www.mdpi.com/journal/ijerph/special_issues/internet_addiction)

For citation purposes, cite each article independently as indicated on the article page online and as indicated below:

LastName, A.A.; LastName, B.B.; LastName, C.C. Article Title. *Journal Name* **Year**, *Article Number*, Page Range.

ISBN 978-3-03897-604-2 (Pbk)
ISBN 978-3-03897-605-9 (PDF)

Contents

About the Special Issue Editor

Olatz Lopez-Fernandez is a Senior Lecturer in the Turning Point, Eastern Health Clinical School at Monash University, and Course Coordinator of the Master in Addictive Behaviours. Her career has spanned positions in international universities including University of Barcelona, Catholic University of Louvain, and Nottingham Trent University. She is a social, educational, and health sciences academic with research interests in Internet use-related addiction problems, and educational technology in higher education.

Preface to "Internet and Mobile Phone Addiction"

Internet use-related addiction problems such as Internet addiction or problematic mobile phone use (including other technological behavioural addictions: cybersex, problem social networking, and gaming disorder) have been defined and conceptualised in multiple ways. However, almost all definitions have similar core elements describing the addictive symptomatology presented by individuals who behave addictively using technologies, which include using the Internet excessively and problematically. In other words, they use fixed or mobile Information and Communication Technologies (ICT) and their applications (e.g., video games, social networks, and gambling sites) in a way that constitutes the most critical activity in their daily life. These cause them distress and functional impairment, which is perceived as a loss of control impacting on their health and wellbeing. These problems are not restricted to the Internet in general and can be applied to a full range of specific online activities that potentially could affect users' health (personal, social, and environmental). Indeed, they constitute a biopsychosocial phenomenon.

To date, most published Internet use-related addiction problems literacy research has focused on assessing, through psychometric and epidemiological studies, the different problems detected since 1996, starting with the concept of 'Internet addiction'. Nevertheless, 'video game addiction' was simultaneously researched and recently, included as a mental disease in the 11th Revision of the International Classification of Diseases (ICD-11) by the World Health Organization (WHO). Previously, the American Psychiatric Association (APA), in 2013, requested additional research on 'Internet Gaming Disorder' (IGD) as a "Condition for Further Study" in the third appendix of their fifth Diagnostic And Statistical Manual Of Mental Disorders (DSM-5).

Thus, a better understanding has emerged of the extent to which these addictive problems related to the use of fixed or mobile ICTs are mediated by personal, technological, and environmental demands, which have emerged during the 21st century. It may be argued this phenomenon, at present, could be considered a contemporary problem, classed as a psychopathological disease in the case of Gaming and Gaming Disorders by the WHO. Indeed, the complexity of the phenomena under the umbrella term of 'Internet addiction' is considerable, and requires a critical understanding of its scientific and clinical evidence. The context in which Internet use-related addiction problems literacy has been developed and been applied is international, beginning in the mid-nineties in the Anglo-Saxon countries. This expansion has led to much greater attention being given to ways of reducing the individual demands and complexity of individuals who suffer one or more of these health problems which are emerging worldwide. A range of tools and treatments are to help create and develop health programs in organisations from Asia, Europe, and America. These proposed strategies seek to reduce the individual demands on people developing these addictive problems. However, more knowledge and dissemination is needed to know how to prevent these potential health issues, and education on them could be a protective factor.

A Special Issue on health and educational effects due to excessive Internet or mobile phone use has been tackled in the International Journal of Environmental Research and Public Health (IJERPH) titled: "Internet and Mobile Phone Addiction: Health and Educational Effects", which was developed between 2017 and 2018. The recognition of these Internet use-related addiction problems were also being internationally debated. These problems have finally started to be accredited by Public Health organisations such as the WHO. For detailed information on the Special Issue referred to, you can visit https://www.mdpi.com/journal/ijerph/special issues/internet addiction.

In summary, the APA and theWHOhave been considering the inclusion of other sub-syndromal behavioural addictions (namely, ''Disorders Due to Addictive Behaviours" with online and offline settings). The number of these ICTs, such as the Internet and mobile technologies, is expected to rise in the next years according to the International Telecommunication Union, an agency of the United Nations. However, we still do not know the phenomenology, the developing course, and the potential number of people that already have or could develop these potential behavioural addictions through technologies per country. The risk of these addictive problems has received increasing attention. As a result, disordered addictive behaviours supported by technologies, such as the Internet and mobile phones, have drawn interest internationally from different sectors. Researchers and practitioners in this field have been looking at advances in these potential behavioural addictions concerning care, management, and prevention, including diagnosis, treatment, and co-morbidity. Developing updated knowledge and strategies for reducing risk factors that predispose populations and gaining an understanding of the nature of these problems regarding their public health, social, and educational impact is needed. In conclusion, this Special Issue is dedicated to the subject area of these potential addictive problems for its detection, treatment, and prevention with a cross-cultural approach.

Olatz Lopez-Fernandez
Special Issue Editor

International Journal of
Environmental Research and Public Health

MDPI

Article

Commuting, Life-Satisfaction and Internet Addiction

Bernd Lachmann [1,*], Rayna Sariyska [1], Christopher Kannen [2], Maria Stavrou [3] and Christian Montag [1,4]

1 Molecular Psychology, Institute of Psychology and Education, Ulm University, Helmholtzstrasse 8/1, 89081 Ulm, Germany; rayna.sariyska@uni-ulm.de (R.S.); christian.montag@uni-ulm.de (C.M.)
2 Department of Informatics, University of Bonn, 53012 Bonn, Germany; info@ckannen.com
3 Department of Psychology, Goldsmiths, University of London, London SE14 6NW, UK; mstav010@gold.ac.uk
4 Key Laboratory for NeuroInformation/Center for Information in Medicine, School of Life Science and Technology, University of Electronic Science and Technology of China, Chengdu 611731, China
* Correspondence: bernd.lachmann@uni-ulm.de; Tel.: +49-731-502-6550; Fax: +49-731-503-2759

Received: 13 September 2017; Accepted: 2 October 2017; Published: 5 October 2017

Abstract: The focus of the present work was on the association between commuting (business and private), life satisfaction, stress, and (over-) use of the Internet. Considering that digital devices are omnipresent in buses and trains, no study has yet investigated if commuting contributes to the development of Internet addiction. Overall, $N = 5039$ participants ($N = 3477$ females, age $M = 26.79$, $SD = 10.68$) took part in an online survey providing information regarding their commuting behavior, Internet addiction, personality, life satisfaction, and stress perception. Our findings are as follows: Personality seems to be less suitable to differentiate between commuter and non-commuter groups, which is possibly due to commuters often not having a choice but simply must accept offered job opportunities at distant locations. Second, the highest levels of satisfaction were found with income and lodging in the group commuting for business purposes. This might be related to the fact that commuting results in higher salaries (hence also better and more expensive housing style) due to having a job in another city which might exceed job opportunities at one's own living location. Third, within the business-commuters as well as in the private-commuter groups, females had significantly higher levels of stress than males. This association was not present in the non-commuter group. For females, commuting seems to be a higher burden and more stressful than for males, regardless of whether they commute for business or private reasons. Finally, we observed an association between higher stress perception (more negative attitude towards commuting) and Internet addiction. This finding suggests that some commuters try to compensate their perceived stress with increased Internet use.

Keywords: commuting; well-being; personality; gender; stress; Internet addiction

---

1. Introduction

Commuting is a wide spread occurrence with millions of people afflicted around the globe [1]. Therefore, commuting behavior has the potential to influence both well-being and life satisfaction of a great part of the population either positive [2], where people regard travelling itself as enjoyable or negative [3], where they see it as a burden. The reasons to accept extended travel times are diverse. Some individuals commute to facilitate a better housing situation or to combine family and vocational goals [4]. Other reasons to commute might include promising career perspectives or financial incentives [5].

From an economical point of view, it seems logical that the costs and benefits of commuting should be at equilibrium to achieving decent levels of life satisfaction [3]. This means that increased costs of commuting (e.g., in terms of higher stress or lower well-being) should be compensated in some way, by higher benefits of the provided job opportunity, for example. Interestingly, this equilibrium is not always met and commuters are often willing to carry higher burdens than non-commuters. This has been coined as the 'Commuting Paradox' in the literature [5].

In the commuting literature physiological issues have also been investigated. A recent study examined the relation between commuting distance, cardiorespiratory fitness, and metabolic risk factors [6]. They found a negative association between commuting distance and physical activity (doing sports on a regular basis) as well as cardiorespiratory fitness and a positive association between commuting distance and Body-Mass-Index, waist circumference, and systolic/diastolic blood pressure. Another study from Norway [7] explored the association between long commutes and subjective health complaints. Again, those who reported longer travel times had more musculoskeletal pain and gastrointestinal problems. Moreover, commuters with a travel history of more than 10 years reported significantly more health complaints than those commuting for two years or less. The perceived level of stress should also be considered [8], since the commuting situation is often difficult to control (e.g., traffic jams, delays in public transport, bad weather conditions). This in turn has the potential to contribute to higher stress levels and frustration. Taking into account the link between health and well-being [9,10] the commuting situation should be associated with well-being. Interestingly, there is evidence that females suffer more under the burden of commuting than males. A study analyzing the effects of commuting on health with regard to gender [11] revealed that females with longer commuting times did seek medical advice and called in sick more often compared to males. An association between commuting and perceived higher stress levels could only be observed in females [12,13]. Presumably, the commuting situation, ceteris paribus, has a more adverse effect on females than males.

Aside from the relation between commuting and health/perceived stress, several studies investigated the direct link between commuting and life satisfaction. Here, the findings are mixed. An early study by Stutzer & Frey [3] concluded that commuting and life satisfaction are negatively related. Moreover, they demonstrated that commuting was characterized by rather low levels of positive effect, with a simultaneous fairly high negative effect. Similar findings were observed in another study investigating the commuting situation, salary, and life satisfaction [4]. In this work, commuting was negatively associated with overall life satisfaction but had no effect on the domains of life satisfaction, work and family. Furthermore, Lyons & Chatterjee [14] reported detrimental effects of commuting on stress, fatigue, and overall dissatisfaction. However, according to the authors, overall life satisfaction could even be increased if commuters can make their own decisions on how to use their travel time. If commuters experience these choices as worthwhile, a positive impact on life satisfaction could be the result [15]. In contrast to the studies mentioned so far, a positive relationship between commuting time and life satisfaction was found by Morris [16]. The authors observed such an association between commuting time and life satisfaction more strongly in rural areas and small cities. This association was visible only to a much weaker extent in large cities, most probably due to the higher degree of traffic congestions. Usually, traffic congestions are not controllable and therefore limit the individual's self-determination. Subsequently, this could lead to a change in life satisfaction [17]. One study examined the association between life satisfaction and work commute [18]. The participants of this study reported mostly positive or neutral feelings during work commute and, consequently, a higher level of happiness. As possible reasons for this, the authors postulate that short work commutes could provide a buffer between work and private sphere, which in turn contributes to an increased level of well-being. For longer commutes, social and entertainment activities could counteract stress or boredom, as well as increase positive effects.

It is worth noting that one possible reaction to stress might be the increased use of the Internet [19]. Since smartphones offer Internet access and are more often than not in people's possession [20] it seems worthwhile to investigate the association between perceived stress (also in terms of attitude towards commuting) and Internet use [21] while commuting. If commuters show a negative attitude towards commuting and/or a high stress level one way to compensate for this could be an increased use of the Internet. As a growing number of researchers around the globe are currently investigating if problematic Internet use represents a societal problem (for an overview see Montag & Reuter [22]; Brand et al. [23]: I-PACE model; Petry & O'Brien [24]: Inclusion of Internet Gaming Disorder in Section III of DSM-5), the question arises if commuting itself also represents a vulnerability factor in becoming addicted to the Internet. Long commutes, in particular, could lead to excessive usage of digital channels. On the other hand, it is imaginable that commuting leads to less time spent on the Internet aside from commuting, because much of what needs to be done online has been already finished during the commute. In short, there is little awareness of studies investigating associations between Internet addiction and commuting. That is the reason why we also address this topic in this work. Internet addiction has been investigated for more than 20 years now [25,26] and much progress has been made to understand problematic Internet use. Although no consensus has been reached with respect to necessary symptoms of Internet addiction, symptoms such as preoccupation with the Internet, withdrawal symptoms when not being online, loss of control and problems in social/work life due to the overuse are of importance [27] Prevalence rates differ across the world, but in Germany (where the current investigated sample has been recruited) about 1% of the population is afflicted [28]. In the context of Internet addiction, it is necessary to distinguish between generalized (generalized pathological Internet use) and specific forms (e.g., excessive online gambling, shopping or social network use) of Internet addiction [29–31]. The present work focuses on unspecific tendencies towards Internet addiction in the realm of commuting.

Overall, an association between commuting and life satisfaction was found in many studies; however, the direction of this association is not trivial to understand. Moreover, it is noticeable that the reviewed research focuses on non-commuters vs. commuters, but they do not ask the question of whether people commute because of private (e.g., to see one's own partner on the weekend) or business reasons (going to work and return back home). Reasons to commute could be an important factor because commuting may not always be considered as a burden [5], but as a fulfilling activity [2]. It is, therefore, conceivable that the motivation behind commuting or personality factors have an impact on the well-being of the commuters. Furthermore, there seems to be a gender specific effect in commuting, with females showing higher stress levels than males [13]. Therefore, in the present study (i) we investigated the association between commuting and life satisfaction in three separate groups: non-commuters, business commuters, and private commuters. In this context, we also examined the underlying personality structure in these groups since several studies reported associations between personality and life satisfaction [32–35] (see Supplementary Material for analyses on the associations of personality). Moreover (ii), we intended to replicate earlier findings about higher stress perception of females in commuting situations [11–13]. This time we also wanted to extend this in the realm of being a non-commuter, business-commuters, and private commuter. Finally (iii), we investigated the association between Internet addiction, life satisfaction, and stress (in relation to commuting). We expected a positive association between a negative attitude toward commuting (and high stress perception) and high Internet addiction.

2. Materials and Methods

For the present study, we asked participants to provide information via a specific designed online questionnaire covering various aspects of commuting. The online questionnaire could be filled in using any suitable device (e.g., tablet, smartphone, personal computer) with access to the Internet.

2.1. Participants

Overall N = 5039 participants (N = 3477 females) answered the online questionnaire and provided socio-demographic information, information on personality data and life satisfaction, as well as data concerning their commuting behavior. The mean age of the sample was 26.79 (SD = 10.68) ranging from 11 years to 98 years. Concerning educational training within the sample, the number of school leaving certificates was distributed as follows: a total of 31.8% had no school leaving certificate, 30.8% had a secondary school leaving certificate, 14.9% had a Baccalaureate-Diploma, and 22.5% had a university degree. Participation was voluntarily and completely anonymous. There was no monetary incentive, but upon completion of the questionnaire, all participants got a brief individual feedback on their personality profiles, life satisfaction, and Internet use (Internet addiction) based on the data provided. The local ethics committee of the Ulm University, Ulm, Germany approved the study, and all participants gave electronic consent prior to participation.

2.2. Materials

The data for the present study were gathered by means of an online questionnaire. In addition to collecting data on demographics, we requested information on personality (see Supplementary Material. For further information on the used personality questionnaire please refer to Rammstedt et al. [36] and John et al. [37]), life satisfaction, and Internet addiction. Furthermore, the participants gave information on their commuting status (none, business, private) and their (emotional) attitude towards commuting, particularly stress. To assess the overall attitude towards commuting, we asked the participants four questions ("Commuting does not matter to me" (item 1, inverse coding), "Commuting deteriorates my mood" (item 2), "Commuting deteriorates my quality of life" (item 3), and "Commuting stresses me" (item 4)). All items could be rated from 1 ("I do not agree at all") to 5 ("I totally agree"). For the analyses, item 4 was analyzed both as a single item (to assess the participants stress level in relation to commuting) and also combined in the short four-item scale described above (to assess the overall 'attitude towards commuting'; ATC). The scores of the four items are simply added up, after reversing the score of item 1. Cronbach's alpha for the ATC scale was alpha = 0.85.

Life satisfaction as one distinct part of subjective well-being, aside from positive and negative effect [38], was measured via questions retrieved from the German Socio-Economic Panel (SOEP) [39]. One section of the panel covers the current life situation within several areas, contributing to overall life satisfaction. For the purpose of this survey, we asked for the degree of satisfaction in the following areas: health, job, income, lodging, leisure, and overall satisfaction with life. Following a recommendation of the SOEP, the question for overall satisfaction with life was presented at the end of the life satisfaction questionnaire. This was done to avoid possible interference with specific domains of life satisfaction. It is important to note that overall life satisfaction is not a simple composite of the various domains of life satisfaction. In fact, all life satisfaction items are considered to be distinct, but also overlap to some extent (e.g., a person more satisfied with his leisure might have, as a consequence, a higher score on overall satisfaction). The items were answered, using a Likert scale, ranging from 0 ("completely dissatisfied") to 10 ("completely satisfied").

To gather data on Internet overuse we administered a short version of the Internet Addiction Test IAT [25], the short Internet Addiction Test (s-IAT) from [40]. This inventory consists of 12 items as opposed to the original version, which contains 20 items. The psychometric quality of the s-IAT has been considered to be of good effect [40]. The Cronbach's alpha in our sample was high (alpha = 0.88). To more effectively assess the association between commuting and Internet use we asked an additional question; "because of commuting I use digital devices more often" (CMD). This could be rated from 1 ("I do not agree at all") to 5 ("I totally agree").

2.3. Procedure

Since the size of the total sample greatly relied upon the publicity of the online questionnaire, the study was introduced during interviews with several national radio and TV stations. This approach was taken to ensure high media coverage throughout Germany, and to avoid biases caused by the restriction of local samples. The audience was given a short introduction to the rationale of the study, combined with information on how to access the online questionnaire. Completed questionnaires were stored on servers and processed for further analyses.

2.4. Statistical Analysis

The statistical analyses were conducted using SPSS 22.0 for windows (IBM SPSS Statistics, Chicago, IL, USA). Differences in life satisfaction and personality variables for non-commuters, business commuters, and private commuters were investigated using ANOVAs, with the additional inclusion of gender effects. The associations between perceived stress, attitude towards commuting, Internet addiction, and life satisfaction were examined within the three groups of commuter type using Pearson correlations, as well as ANOVAs to test for gender effects.

3. Results

3.1. Data Cleaning and Descriptive Statistics

After inspection of the data, 25 participants were discovered to have provided conspicuous age information: A total of 20 participants reported an age of zero. Information on age was given using a slider ranging from 0 to 99 years, where 0 was the default value. Indicating an age of zero means that the participants did not provide any information on their age. Furthermore, we found 5 participants reporting an age between 1 and 9 years. Considering the rationale of the present study, it was uncertain as to whether or not these individuals indeed reported their real age and, if such was the case, were able to complete the questionnaire in a suitable way. Therefore, we decided to exclude these 25 participants, an overall (0.5% of the original sample; 18 commuting either for business or private plus 7 non-commuters) from the sample, which left an overall sample size of $N = 5039$ participants ($N = 5064$ before) for the analyses. Moreover, $N = 409$ participants reported both business-commuting and private-commuting, which forfeited their inclusion into one specific group. There were no further exclusions. We further investigated participants between 11 and 14, because at 15 years, persons are allowed to start working (with a vocational training). A closer inspection of participants with an age less than 15 showed that 28 persons (in our sample a total of 248 participants were younger than 15 years) reported to undertake business-commuting. This seems to be unusual considering the age of these participants. On the other hand, commuting to and from school (97.6% of the participants younger than 15 years reported to be still in school) is obligatory and therefore might be for some participants more connected to the term business rather than private-commuting. For that reason we did not exclude those participants from the sample. All variables were normally distributed and there were no outliers. Table 1 shows the descriptive statistics for the complete sample, and non-commuters (means, standard deviation, observed minimum and maximum values, skew and standard deviation of the skew) were calculated life satisfaction variables, Internet addiction (s-IAT), CSM ("Commuting stresses me"), and ATC (overall "attitude towards commuting").

Table 1. Means and Standard Deviation, Minimum, Maximum and Skewness for life satisfaction variables, and Internet variables for the complete sample (CS) and non-commuters (NC).

Variables	Min	Max	Mean	SD	Skewness	SD
Overall satisfaction	0	10	7.28/7.27	2.19/2.20	−1.06/−1.05	0.03/0.04
Health	0	10	6.62/6.65	2.30/2.30	−0.71/−0.69	0.03/0.04
Job ($N = 4616/3401$)	0	10	6.81/6.82	2.42/2.43	−0.86/−0.87	0.03/0.04
Income	0	10	5.38/5.36	2.96/3.00	−0.28/−0.26	0.03/0.04
Lodging	0	10	7.30/7.33	2.53/2.54	−1.04/−1.05	0.03/0.04
Leisure	0	10	6.55/6.58	2.38/2.37	−0.56/−0.56	0.03/0.04
Family	0	10	7.07/7.10	2.60/2.60	−0.90/−0.90	0.03/0.04
CSM ($N = 1413$)	1	5	2.86/2.86	1.20/1.20	−0.05	0.07
CMD	1	5	2.85/3.11	1.45/1.32	−0.02/−0.27	0.07/0.20
ATC	4	20	10.72/11.29	3.85/3.77	0.16/0.11	0.07/0.20
s-IAT	12	60	27.01/27.06	8.37/8.45	0.74/0.72	0.03/0.04

($N = 5039/3774$, CS/NC), Internet use (s-IAT), CSM ("Commuting stresses me"), ATC (overall "attitude towards commuting"), and CMD ("because of commuting I use digital devices more often").

Please refer to Table 2 for the descriptive statistics of business-commuters and private commuters.

Table 2. Means and Standard Deviation, Minimum, Maximum, and Skewness for life satisfaction variables, and Internet variables for business-commuters (BC) and private commuters (PC).

Variables	Min	Max	Mean	*SD*	Skewness	*SD*
Overall satisfaction	0	10	7.43/6.97	2.08/2.52	−1.17/−1.00	0.09/0.18
Health	0	10	6.55/6.58	2.22/2.36	−0.75/−1.00	0.09/0.18
Job (N = 663/153)	0	10	6.78/6.72	2.38/2.31	−0.80/−0.79	0.10/0.20
Income	0	10	5.75/4.73	2.71/3.05	−0.43/−0.16	0.09/0.18
Lodging	0	10	7.44/6.66	2.39/2.77	−1.19/−0.76	0.09/0.18
Leisure	0	10	6.48/6.51	2.32/2.52	−0.60/−0.72	0.09/0.18
Family	0	10	7.09/6.71	2.55/2.76	−0.91/−0.78	0.09/0.18
CSM	1	5	2.86/2.86	1.19/1.23	1.19/−0.14	0.09/0.18
CMD	1	5	2.66/2.94	1.48/1.40	0.19/−0.11	0.09/0.18
ATC	4	20	10.71/10.61	3.87/3.93	0.19/0.12	0.09/0.18
s-IAT	12	60	26.30/27.77	7.71/8.49	0.75/0.68	0.09/0.18

(N = 676/180, BC/PC). Internet use (IAT), CSM ("Commuting stresses me"), ATC (overall "attitude towards commuting"), and CMD ("because of commuting I use digital devices more often").

3.2. Association between Life Satisfaction and Commuting Status

A one-way ANOVA showed a main effect of commuting status on income ($F_{(2,4627)}$ = 9.53, $p < 0.001$; Levene's test: $F_{(2,4627)}$ = 7.40, p = 0.001), lodging ($F_{(2,4627)}$ = 7.03, p = 0.001; Levene's test: $F_{(2,4627)}$ = 6.00, p = 0.003), and overall LS ($F_{(2,4627)}$ = 3.41, p = 0.033; Levene's test: $F_{(2,4627)}$ = 4.61, p = 0.010). Since Levene's test indicated the presence of heteroscedasticity, we used the Games-Howell test for post hoc analyses as recommended by Field & Miles [41]. The Games-Howell post-hoc analysis revealed significant differences between income scores of the non-commuting and business-commuting groups (p = 0.002; −0.39, 95% CI [−0.66, −0.12]), the non-commuting and private-commuting groups (p = 0.020; 0.63, 95% CI [0.08, 1.18]), and the business-commuting and private-commuting groups ($p < 0.001$; 1.02, 95% CI [0.43, 1.60]). For lodging scores, differences were found between non-commuters and private commuters (p = 0.004; 0.68, 95% CI [0.18, 1.18]) as well as between business-commuters and private-commuters (p = 0.002; 0.79, 95% CI [0.25, 1.32]). No effects on overall life satisfaction were found by post hoc analyses. To visualize our findings, please refer to Figure 1.

We also investigated these associations for gender effects. A one-way ANOVA showed a main effect of commuting status on income ($F_{(2,1420)}$ = 4.32, p = 0.014; Levene's test: $F_{(2,1420)}$ = 9.06, $p < 0.001$) for males. Again, due to a positive Levene's test, we used a Games-Howell test for the post hoc analysis and found a significant difference between non-commuting and business-commuting groups (p = 0.005; −0.57, 95% CI [−1.00, −0.15]) with higher income scores for the business-commuting group. For females, the analysis of variance revealed a main effect of commuting status on income ($F_{(2,3204)}$ = 6.00, p = 0.003; Levene's test: $F_{(2,3204)}$ = 2.41, p = 0.090),), lodging ($F_{(2,3204)}$ = 4.81, p = 0.008; Levene's test: $F_{(2,3204)}$ = 3.47, p = 0.031), and overall LS ($F_{(2,3204)}$ = 3.17, p = 0.042; Levene's test: $F_{(2,3204)}$ = 2.17, p = 0.114). A Games-Howell test revealed a significant difference for females between lodging scores of non-commuting and private-commuting groups (p = 0.015; 0.63, 95% CI [0.09, 1.17]) and of business-commuting and private-commuting groups (p = 0.006; 0.79, 95% CI [0.18, 1.40]). A generalized Tukey 2 post hoc test was used for income and overall life satisfaction since a Levene's test did not indicate unequal variances. Females showed significant differences for income scores of the non-commuting and private-commuting groups (p = 0.007; 0.80, 95% CI [0.17, 1.43]) as well as for the business-commuting and private-commuting groups (p = 0.002; 1.03, 95% CI [0.32, 1.73]). Again, overall life satisfaction showed no significant result in the post hoc analysis. An independent-sample t-test provided significantly lower scores for health (M_{female} = 6.56, SD_{female} = 2.32; M_{male} = 6.89, SD_{male} = 2.25; $t(3772)$ = 3.99, $p < 0.001$) and leisure (M_{female} = 6.50, SD_{female} = 2.37; M_{male} = 6.78, SD_{male} = 2.36; $t(3772)$ = 3.31, p = 0.001) for females compared to males, but only in the non-commuting group.

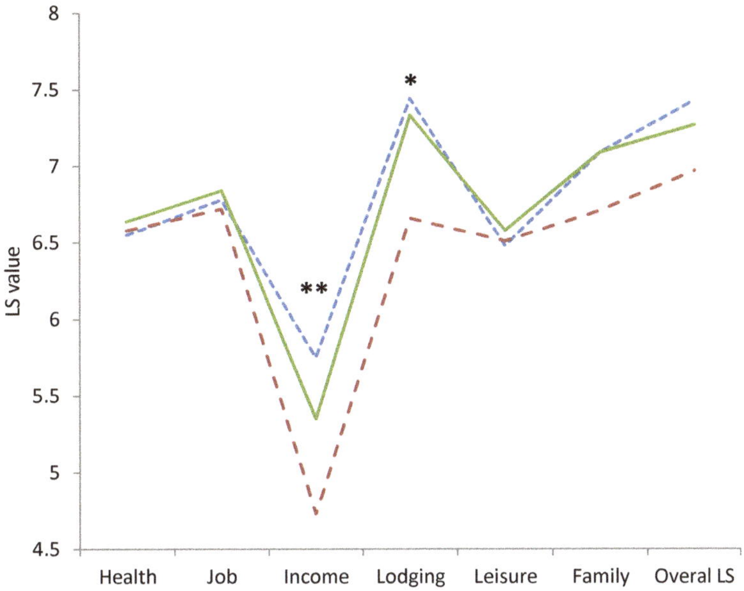

Figure 1. Life satisfaction values depending on commuting status: (BC) business-commuter (dotted blue line), (NC) non-commuters (solid green line), and (PC) private-commuters (dashed red line); * $p = 0.01$, ** ($p < 0.001$) significant differences between all three groups.

The sample sizes used in the analyses for all groups considering commuting status and gender are summarized in Table 3. Age was not associated with commuting status ($r = 0.003$, $p = 0.834$).

Table 3. Sample sizes depending on commuting status and gender.

Sample	B/P *-Commuting	Non-Commuting	Business-Commuting	Private-Commuting
Complete $N = 5039$	409	3774	676	180
Female $N = 3477$	270	2684	391	132
Male $N = 1562$	139	1090	285	48

* B/P: Business and Private.

3.3. Perception of Stress and Overall Attitude Towards Commuting

For the total sample, the CSM and ATC variable ("Commuting stresses me"; "Attitude towards commuting") was negatively correlated with all life satisfaction variables in the business-commuters group but not in the private-commuters group. The same result was found when differentiating for gender; however, for both females and males all correlations between CSM/ATC and life satisfaction variables were negative (or not present) in the business-commuting group and nonexistent for females in the private-commuting group. In the private-commuting group we observed one significant negative correlation for males between CSM and overall LS ($r = -0.39$, $p = 0.006$, after Bonferroni correction for multiple testing; corrected alpha: 0.05/8=0.00625). All correlations are summarized in Table 4.

Table 4. Correlations between CSM ("Commuting stresses me"), ATC ("Attitude towards commuting"; 4 item scale), and life satisfaction variables.

Commuting Status	Health	Job	Income	Lodging	Leisure	Family	Overall LS
BC	−0.11 **/−0.13 **	−0.25 **/−0.28 **	−0.22 **/−0.22 **	−0.16 **/−0.18 **	−0.21 **/−0.26 **	−0.10 **/-0.16 **	−0.14 **/−0.19 **
PC	−0.02/−0.01	−0.06/−0.10	0.01/0.03	−0.10/−0.11	−0.07/−0.09	0.01/0.02	−0.11/−0.11
BC (male)	−0.15 */−0.18 **	−0.28 **/−0.33 **	−0.22 **/−0.23 **	−0.12 */−0.15 *	−0.24 **/−0.34 **	−0.14 */−0.23 **	−0.17 **/−0.25 **
BC (female)	−0.07/−0.09	−0.24 **/−0.26 **	−0.21 **/−0.20 **	−0.19 **/−0.20 **	−0.19 **/−0.20 **	−0.07/−0.12 *	−0.14 **/−0.16 **
PC (male)	−0.14/−0.14	−0.41 */−0.39 *	0.03/0.10	−0.17/−0.03	−0.14/−0.01	−0.03/0.16	−0.39 **/−0.31 *
PC (female)	0.02/0.03	0.03/−0.04	0.04/0.05	−0.08/−0.15	−0.03/−0.11	0.00/−0.05	−0.03/−0.05

BC = business-commuting, PC = private commuting, CSM/ATC; * $p < 0.05$, ** $p < 0.001$.

An ANOVA revealed that CSM scores were significantly higher for females (M_{female} = 3.00, SD_{female} = 1.17) compared to males (M_{male} = 2.68, SD_{male} = 1.20); $F_{(1674)}$ = 12.13, p = 0.001) within the business-commuting group and the private-commuting group (M_{female} = 3.05, SD_{female} = 1.19; M_{male} = 2.33, SD_{male} = 1.19; $F_{(1178)}$ = 12.59, $p < 0.001$). We observed no significant gender differences for the CSM score between business-commuting group, private-commuting group and non-commuting-group. However, ATC scores were significantly higher for females (M_{female} = 11.09, SD_{female} = 3.88) compared to males (M_{male} = 10.19, SD_{male} = 3.80); $F_{(1674)}$ = 9.19, p = 0.003) within the business-commuting group and the private-commuting group (M_{female} = 11.14, SD_{female} = 3.81; M_{male} = 9.13, SD_{male} = 3.92; $F_{(1178)}$ = 9.76, p = 0.002). No differences were observed between the business-, private-, and non-commuting groups (CSM: $F_{(2,1001)}$ = 0.01, p = 0.994; ATC: $F_{(2,1001)}$ = 1.58, p = 0.207).

3.4. Association of Internet Use, Attitude Towards Commuting, Stress, and Life Satisfaction

The s-IAT was positively associated with the attitude towards commuting (ATC) and stress perception (CSM) but only in the business-commuting group. For females, the association between CSM and IAT was r = 0.14, p = 0.005, and between ATC and IAT r = 0.12, p = 0.015. Males showed an association of r = 0.18, p = 0.003 between CSM and IAT score, and r = 0.18, p = 0.002 between ATC and IAT values. A positive correlation between ATC/CSM scores and CMD ("because of commuting I use digital devices more often") was observed, again only in the business-commuter group. For females, the association between ATC and CMD was r = 0.24, $p < 0.001$ (males: r = 0.22, $p < 0.001$), while the association for CSM and CMD was r = 0.21, $p < 0.001$ (males: r = 0.22, $p < 0.001$). For females in the non-commuting group, the correlation between excessive Internet use and overall life satisfaction was $r = -0.17$, $p < 0.001$, while for males the correlation was $r = -0.11$, $p < 0.001$. For business and private commuters, we found significant associations between Internet addiction and life satisfaction only for females (business: $r = -0.13$, p = 0.009; private: $r = -0.23$, p = 0.009).

4. Discussion

The aim of this research was to extend already existing research on the association between personality, life satisfaction, and stress perception as a function of commuting status. No differences for personality were found between non-commuters, business-commuters, and private-commuters (see Supplementary Material as well as Baretta et al. [42] and Ruedl [43]). Significant associations were observed between income and lodging depending on commuting status. The business-commuting group showed the highest scores, while the private-commuting group the lowest scores for these areas of life satisfaction. As expected, stress perception was higher for females in the business-commuting group than males. In the private-commuting group the stress perception of females stayed high, whereas the stress perception of males decreased. However, a significant gender effect in relation to CSM/ATC was found only within the business and private-commuting group, i.e., not between the business- and private-commuting group or in the non-commuting group. A positive association was found between Internet addiction and both ATC and CSM, in the business-commuting group, regardless of gender. Moreover, a negative association between Internet addiction and overall life satisfaction (females and males) was found in the complete sample under investigation, as well as a negative association between Internet addiction and overall life satisfaction in the business and private-commuter groups, but for females only.

The association between life satisfaction and commuting [4,15,16,18] was also investigated. For two areas of life satisfaction, namely income and lodging, significant results were found. Interestingly, private-commuters showed lower and business-commuters higher life satisfaction scores than non-commuters for income and lodging. This finding is somewhat unexpected, as our rationale was that lower self-determination in the business-commuter group would lead to lower life satisfaction scores. However, one can infer that this emphasizes the distinct categorization between business-commuters and private-commuters. We discuss two possible explanations for this finding: First, self-determination might not be as a prominent predictor of life satisfaction as expected (at least in the context of commuting), and second, the presence or absence of a financial compensation in a commuting situation influences life satisfaction (life satisfaction scores for income shows the highest effects in our analyses). According to Erikson [44] and Newman et al. [17] self-determination represents an important factor that promotes higher life satisfaction scores. This, in fact, should lead to higher life satisfaction scores in the private-commuter group since within this group, the freedom of choice whether to commute or not should be higher as in the business-commuter group (as discussed above). The main reason why this is not the case and life satisfaction scores are higher for business-commuters could be due to the distinctiveness of the commuting situation: Business-commuters receive a financial compensation from the job, whereas private-commuters have only costs without any compensation. In this context, we propose that the monetary effect on life satisfaction, especially on income, could be much stronger than the influence of self-determination. Further analyses considering gender effects revealed that significant differences for lodging were present but only for females. Those who commuted for private reasons exhibited not only the lowest scores for income but also the lowest satisfaction scores in lodging. Considering that lodging is more important for females than males [45], it seems comprehensible that satisfaction scores for females in lodging are lowest for private commuters; the money they have spent on their private commute is no longer available to adequately furnish a home. Again, for business-commuting there is, at least, a financial compensation, which potentially enhances life satisfaction in this area.

Our findings regarding stress perception were mostly in accordance with previous findings asserting that females experience more stress in a commuting situation than males [8,46]. Interestingly, the level of perceived stress of females remained almost stationary regardless of whether they commuted for business or private reasons, whereas the stress level of males was much lower in the private-commuting group. Roberts et al. [13] proposed that the larger responsibility of females compared to males for daily household tasks could be the reason for higher stress levels of commuting females. In this case, it would be plausible that the reason for commuting (business or private) would have no influence on the female's perception of stress because the outcome (possible neglect of household tasks due to commuting) is independent from the reason of commuting. The fact that significantly higher stress levels for females compared to males were present within the commuting groups, but not in the non-commuting group, further exacerbates findings concerning the detrimental effects of commuting, especially for females.

Also noticeable is the association between the attitude towards commuting/stress and Internet addiction [47]. Our findings point to a possible compensation of a negative attitude towards commuting by increased excessive Internet use. Interestingly, this association was only present in the business-commuter group and was independent of gender. Moreover, for the business-commuter group, again independent of gender, we observed a positive association between stress perception (attitude towards commuting) and CMD ("because of commuting I use digital devices more often"), which further supports this observation. The findings concerning the association between overall life satisfaction and Internet addiction, where females show more robust and stronger negative associations between Internet addiction and life satisfaction, also support earlier findings [48]. In this study, different thresholds have been suggested for females and males with respect to negative effects on life satisfaction because of high Internet use. Considering the present Internet addiction and commuting findings, one can denote forthcoming benefits from future studies examining what kinds of activities are done while commuting, and at what frequency. For instance, commuting time can be spent on work or learning new skills (e.g., via massive open online courses [49]) and should not be seen as fostering addictive tendencies towards digital technologies, per se. As previously mentioned, the diagnosis of Internet addiction depends on many criteria, but in the current research, it is merely the stressed commuter that is taken into account with regards to having higher Internet addiction tendencies.

The present study has strengths and also limitations worth noting. Firstly, using a cross-sectional design refrain us from causal inferences. It would be desirable to support the results of this study by replicating the main findings in longitudinal and experimental studies. Secondly, despite collecting a rather large sample, some group sizes (particularly the private-commuting group) used in the analyses are relatively small (see results section). Nevertheless, based on our considerations and findings we state that the discrimination between business and private commuters' data can be used in a meaningful way to shed further light on the intriguing association between psychological (and even physiological) variables and commuting behavior. In addition to this, our data collection was accomplished by means of self-report questionnaires, which usually inherits a certain chance for biases. Future studies could minimize this potential shortcoming by using multiple approaches, such as peer reports or data logging to collect data. Furthermore, it should be noted that statements about the generalization of our findings should be done so with caution, since it is difficult to precisely pinpoint the locations of the respondents. On the other hand, the present study was promoted nationwide in Germany and relies on a fairly large sample, which adds to the validity of the findings. In addition, it is important to consider that the ratio of participants (non-commuters, business-commuters, private-commuters) simply reflects the response-rate of the participants from a nationwide promotional campaign. Naturally, this ratio may not be fully representative of the general population. Lastly, a concern of the present study was to learn more about the association between commuting, life satisfaction and Internet addiction. Therefore, we focused on the motivation to commute (private and business) rather than the association between duration/distance and life satisfaction, which has previously been investigated. Since there is evidence that some commuters possibly consider commuting as a fulfilling activity (independent of the duration/distance of the commute) this may be a consideration for further studies to shed light on.

5. Conclusions

In conclusion, the present study underlines the association between life satisfaction, stress perception, and commuting. Extending previous findings in literature, our results show that the reason to commute (business or private) has the potential to explain differences in life satisfaction variables, namely income and lodging. We also replicated previous findings concerning the gender specific role of stress in a commuting situation and its specific impact on females. Furthermore, we demonstrated that this association (in contrast to life satisfaction and commuting) did not vary, regardless of whether the commute was due to business or private reasons. Finally, the current study finds an association between the attitude towards commuting/stress and Internet addiction. Here, a more negative attitude towards commuting was associated with higher tendencies towards excessive use of the Internet. For future research, it would be beneficial to further investigate the association between the frequency and type of commuting behavior with Internet addiction. Moreover, strong insights can be provided by the use of applications to track activity on ubiquitous smartphones, while allowing access to specific questionnaires, and having GPS tracking activated. Essentially, this would mean combining self-reports with objective data directly collected from the smartphone. This research methodology will be complimentary to the emerging research discipline of Psychoinformatics [50]. Furthermore, this design would allow longitudinal studies to be conducted in an accessible and efficient way. Finally, future studies should aim to assess smartphone addiction, as inclinations of an overlap between Internet and smartphone addiction have been shown to exist [51,52], though they are not synonymous. Moreover, smartphones are clearly of high relevance in a commuter's life, and recently, associations between smartphone addiction and lower productivity have also been reported [53].

Supplementary Materials: The following are available online at www.mdpi.com/1660-4601/14/10/1176/s1, Table S1: Means and Standard Deviation, Minimum, Maximum, and Skew of personality variables for for the complete sample (CS) and non-commuters (NC), Table S2: Means and Standard Deviation, Minimum, Maximum, and Skew of personality variables for business-commuters (BC) and private commuters (PC).

Acknowledgments: The present study was funded by the German Research Foundation (MO 2363/2-1). Moreover, the position of Christian Montag is funded by a Heisenberg grant awarded to him by the German Research Foundation (MO 2363/3-2).

Author Contributions: Bernd Lachmann and Christian Montag designed the study. Moreover they developed the attitude towards commuting (ATC) items. Christopher Kannen programmed the online platform and preprocessed the data. Bernd Lachmann performed statistical analyses. Bernd Lachmann wrote the manuscript. Christian Montag, Bernd Lachmann, Rayna Sariyska and Maria Stavrou critically worked on the manuscript. Moreover, Maria Stavrou checked the manuscript for language. All authors contributed substantially to the final version of the paper.

Conflicts of Interest: The authors declare no conflict of interest.

References

1. Becker, R.; Cáceres, R.; Hanson, K.; Isaacman, S.; Loh, J.M.; Martonosi, M.; Rowland, J.; Urbanek, S.; Varshavsky, A.; Volinsky, C. Human mobility characterization from cellular network data. *Commun. ACM* **2013**, *56*, 74–82. [CrossRef]
2. Mokhtarian, P.L.; Salomon, I. Travel for the fun of it. *ACCESS Mag.* **1999**, *15*, 26–31.
3. Stutzer, A.; Frey, B.S. Commuting and life satisfaction in Germany. *Small* **1992**, *44*, 179–189.
4. Pfaff, S. Pendelentfernung, Lebenszufriedenheit und Entlohnung/Commuting Distance, Life Satisfaction, and Wages. *Z. Soziol.* **2014**, *43*, 113–130. [CrossRef]
5. Stutzer, A.; Frey, B.S. Stress that doesn't pay: The commuting paradox. *Scand. J. Econ.* **2008**, *110*, 339–366. [CrossRef]
6. Hoehner, C.M.; Barlow, C.E.; Allen, P.; Schootman, M. Commuting distance, cardiorespiratory fitness, and metabolic risk. *Am. J. Prev. Med.* **2012**, *42*, 571–578. [CrossRef] [PubMed]
7. Urhonen, T.; Lie, A.; Aamodt, G. Associations between long commutes and subjective health complaints among railway workers in Norway. *Prev. Med. Rep.* **2016**, *4*, 490–495. [CrossRef] [PubMed]
8. Gottholmseder, G.; Nowotny, K.; Pruckner, G.J.; Theurl, E. Stress perception and commuting. *Health Econ.* **2009**, *18*, 559–576. [CrossRef] [PubMed]
9. Friedman, H.S.; Kern, M.L. Personality, Well-Being, and Health. *Annu. Rev. Psychol.* **2014**, *65*, 719–742. [CrossRef] [PubMed]
10. Gana, K.; Bailly, N.; Saada, Y.; Joulain, M.; Trouillet, R.; Hervé, C.; Alaphilippe, D. Relationship between life satisfaction and physical health in older adults: A longitudinal test of cross-lagged and simultaneous effects. *Health Psychol.* **2013**, *32*, 896–904. [CrossRef] [PubMed]
11. Künn-Nelen, A. Does commuting affect health? *Health Econ.* **2016**, *25*, 984–1004. [CrossRef] [PubMed]
12. Feng, Z.; Boyle, P. Do long journeys to work have adverse effects on mental health? *Environ. Behav.* **2014**, *46*, 609–625. [CrossRef]
13. Roberts, J.; Hodgson, R.; Dolan, P. "It's driving her mad": Gender differences in the effects of commuting on psychological health. *J. Health Econ.* **2011**, *30*, 1064–1076. [CrossRef] [PubMed]
14. Lyons, G.; Chatterjee, K. A human perspective on the daily commute: Costs, benefits and trade-offs. *Transp. Rev.* **2008**, *28*, 181–198. [CrossRef]
15. Fichter, C. Mobilität: Macht Pendeln unglücklich? Wirtschaftspsychol. Aktuell. 2015, pp. 23–26. Available online: http://www.wirtschaftspsychologie-aktuell.de/files/wirtschaftspsychologie-aktuell-2-2015-fichter.pdf (accessed on 13 September 2017).
16. Morris, E.A. Should we all just stay home? Travel, out-of-home activities, and life satisfaction. *Transp. Res. Part A Policy Pract.* **2015**, *78*, 519–536. [CrossRef]
17. Newman, D.B.; Tay, L.; Diener, E. Leisure and subjective well-being: A model of psychological mechanisms as mediating factors. *J. Happiness Stud.* **2014**, *15*, 555–578. [CrossRef]

18. Olsson, L.E.; Gärling, T.; Ettema, D.; Friman, M.; Fujii, S. Happiness and satisfaction with work commute. *Soc. Indic. Res.* **2013**, *111*, 255–263. [CrossRef] [PubMed]
19. Ostovar, S.; Allahyar, N.; Aminpoor, H.; Moafian, F.; Nor, M.B.M.; Griffiths, M.D. Internet addiction and its psychosocial risks (depression, anxiety, stress and loneliness) among Iranian adolescents and young adults: A structural equation model in a cross-sectional study. *Int. J. Ment. Health Addict.* **2016**, *14*, 257–267. [CrossRef]
20. Lepp, A.; Li, J.; Barkley, J.E.; Salehi-Esfahani, S. Exploring the relationships between college students' cell phone use, personality and leisure. *Comput. Hum. Behav.* **2015**, *43*, 210–219. [CrossRef]
21. Riedl, R.; Kindermann, H.; Auinger, A.; Javor, A. Technostress from a neurobiological perspective. *Bus. Inf. Syst. Eng.* **2012**, *4*, 61–69. [CrossRef]
22. Montag, C.; Reuter, M. *Internet Addiction: Neuroscientific Approaches and Therapeutical Implications Including Smartphone Addiction*; Springer: Berlin, Germany, 2017.
23. Brand, M.; Young, K.S.; Laier, C.; Wölfling, K.; Potenza, M.N. Integrating psychological and neurobiological considerations regarding the development and maintenance of specific Internet-use disorders: An Interaction of Person-Affect-Cognition-Execution (I-PACE) model. *Neurosci. Biobehav. Rev.* **2016**, *71*, 252–266. [CrossRef] [PubMed]
24. Petry, N.M.; O'brien, C.P. Internet gaming disorder and the DSM-5. *Addiction* **2013**, *108*, 1186–1187. [CrossRef] [PubMed]
25. Young, K.S. Internet addiction: The emergence of a new clinical disorder. *Cyberpsychol. Behav.* **1998**, *1*, 237–244. [CrossRef]
26. Young, K.S. Psychology of computer use: XL. Addictive use of the Internet: A case that breaks the stereotype. *Psychol. Rep.* **1996**, *79*, 899–902. [CrossRef] [PubMed]
27. Tao, R.; Huang, X.; Wang, J.; Zhang, H.; Zhang, Y.; Li, M. Proposed diagnostic criteria for internet addiction. *Addiction* **2010**, *105*, 556–564. [CrossRef] [PubMed]
28. Rumpf, H.; Meyer, C.; Kreuzer, A.; John, U.; Merkeerk, G. Prävalenz der Internetabhängigkeit (PINTA). Available online: https://www.bundesgesundheitsministerium.de/fileadmin/Dateien/5_Publikationen/Drogen_und_Sucht/Berichte/PINTA-Kurzbericht-Endfassung_140711_korr.pdf (accessed on 13 September 2017).
29. Montag, C.; Bey, K.; Sha, P.; Li, M.; Chen, Y.; Liu, W.; Zhu, Y.; Li, C.; Markett, S.; Keiper, J. Is it meaningful to distinguish between generalized and specific Internet addiction? Evidence from a cross-cultural study from Germany, Sweden, Taiwan and China. *Asia Pac. Psychiatry* **2015**, *7*, 20–26. [CrossRef] [PubMed]
30. Pontes, H.; Griffiths, M. Internet addiction disorder and internet gaming disorder are not the same. *J. Addict. Res. Ther.* **2014**, *5*, e124. [CrossRef]
31. Davis, R.A. A cognitive-behavioral model of pathological Internet use. *Comput. Hum. Behav.* **2001**, *17*, 187–195. [CrossRef]
32. Hosseinkhanzadeh, A.A.; Taher, M. The relationship between personality traits with life satisfaction. *Sociol. Mind* **2013**, *3*, 99–105. [CrossRef]
33. Specht, J.; Egloff, B.; Schmukle, S.C. Examining mechanisms of personality maturation the impact of life satisfaction on the development of the big five personality traits. *Soc. Psychol. Personal. Sci.* **2013**, *4*, 181–189. [CrossRef]
34. Steel, P.; Schmidt, J.; Shultz, J. Refining the relationship between personality and subjective well-being. *Psychol. Bull.* **2008**, *134*, 138–161. [CrossRef] [PubMed]
35. Gutiérrez, J.L.G.; Jiménez, B.M.; Hernández, E.G.; Pcn, C. Personality and subjective well-being: Big five correlates and demographic variables. *Personal. Individ. Differ.* **2005**, *38*, 1561–1569. [CrossRef]
36. Rammstedt, B.; John, O.P. Measuring personality in one minute or less: A 10-item short version of the Big Five Inventory in English and German. *J. Res. Personal.* **2007**, *41*, 203–212. [CrossRef]
37. John, O.P.; Donahue, E.M.; Kentle, R.L. *The Big Five Inventory: Versions 4a and 54, Institute of Personality and Social Research*; University of California: Berkeley, CA, USA, 1991.
38. Diener, E.; Emmons, R.A.; Larsen, R.J.; Griffin, S. The satisfaction with life scale. *J. Personal. Assess.* **1985**, *49*, 71–75. [CrossRef] [PubMed]
39. Siedler, T.; Schupp, J.; Spiess, C.K.; Wagner, G.G. The German socio-economic panel as reference data set. *Schmollers Jahrb.* **2009**, *129*, 367–374. [CrossRef]

40. Pawlikowski, M.; Altstötter-Gleich, C.; Brand, M. Validation and psychometric properties of a short version of Young's Internet Addiction Test. *Comput. Hum. Behav.* **2013**, *29*, 1212–1223. [CrossRef]
41. Field, A.; Miles, J. *Discovering Statistics Using SAS*; SAGE Publications Ltd.: Thousand Oaks, CA, USA, 2011.
42. Baretta, D.; Greco, A.; Steca, P. Understanding performance in risky sport: The role of self-efficacy beliefs and sensation seeking in competitive freediving. *Personal. Individ. Differ.* **2017**, *117*, 161–165. [CrossRef]
43. Ruedl, G.; Abart, M.; Ledochowski, L.; Burtscher, M.; Kopp, M. Self reported risk taking and risk compensation in skiers and snowboarders are associated with sensation seeking. *Accid. Anal. Prev.* **2012**, *48*, 292–296. [CrossRef] [PubMed]
44. Erikson, R. Descriptions of Inequality: The Swedish Approach to Welfare Research. In *The Quality of Life*; Clarendon Press: Oxford, UK, 1993; pp. 67–87.
45. Tomaszewski, W.; Perales, F. Who settles for less? Subjective dispositions, objective circumstances, and housing satisfaction. *Soc. Indic. Res.* **2014**, *118*, 181–203. [CrossRef]
46. Koslowsky, M.; Aizer, A.; Krausz, M. Stressor and personal variables in the commuting experience. *Int. J. Manpow.* **1996**, *17*, 4–14. [CrossRef]
47. Riedl, R. Mensch-Computer-Interaktion und Stress. *HMD Prax. Wirtsch.* **2013**, *50*, 97–106. [CrossRef]
48. Lachmann, B.; Sariyska, R.; Kannen, C.; Cooper, A.; Montag, C. Life satisfaction and problematic Internet use: Evidence for gender specific effects. *Psychiatry Res.* **2016**, *238*, 363–367. [CrossRef] [PubMed]
49. Sun, G.; Cui, T.; Yong, J.; Shen, J.; Chen, S. MLaaS: A cloud-based system for delivering adapative micro learning in mobile MOOC learning. *IEEE Trans. Serv. Comput.* **2015**. [CrossRef]
50. Montag, C.; Duke, É.; Markowetz, A. Toward Psychoinformatics: Computer science meets psychology. *Comput. Math. Methods Med.* **2016**, *2016*. [CrossRef] [PubMed]
51. Duke, É.; Montag, C. Smartphone Addiction. In *Internet Addiction*; Springer: Cham, Switzerland, 2017; pp. 359–372.
52. Montag, C.; Sindermann, C.; Becker, B.; Panksepp, J. An affective neuroscience framework for the molecular study of Internet addiction. *Front. Psychol.* **2016**, *7*, 1906. [CrossRef] [PubMed]
53. Duke, É.; Montag, C. Smartphone addiction, daily interruptions and self-reported productivity. *Addict. Behav. Rep.* **2017**, *6*, 90–95. [CrossRef]

International Journal of
Environmental Research and Public Health

MDPI

Article

Gaming Device Usage Patterns Predict Internet Gaming Disorder: Comparison across Different Gaming Device Usage Patterns

Soo-Hyun Paik, Hyun Cho, Ji-Won Chun, Jo-Eun Jeong and Dai-Jin Kim *

Department of Psychiatry, Seoul St. Mary's Hospital, The Catholic University of Korea, Seoul 06591, Korea; suehime@gmail.com (S.-H.P.); sonap1@hanmail.net (H.C.); cuneus@naver.com (J.-W.C.); goodi11@hanmail.net (J.-E.J.)

* Correspondence: kdj922@catholic.ac.kr; Tel.: +82-2-2258-7546

Received: 10 October 2017; Accepted: 2 December 2017; Published: 5 December 2017

Abstract: Gaming behaviors have been significantly influenced by smartphones. This study was designed to explore gaming behaviors and clinical characteristics across different gaming device usage patterns and the role of the patterns on Internet gaming disorder (IGD). Responders of an online survey regarding smartphone and online game usage were classified by different gaming device usage patterns: (1) individuals who played only computer games; (2) individuals who played computer games more than smartphone games; (3) individuals who played computer and smartphone games evenly; (4) individuals who played smartphone games more than computer games; (5) individuals who played only smartphone games. Data on demographics, gaming-related behaviors, and scales for Internet and smartphone addiction, depression, anxiety disorder, and substance use were collected. Combined users, especially those who played computer and smartphone games evenly, had higher prevalence of IGD, depression, anxiety disorder, and substance use disorder. These subjects were more prone to develop IGD than reference group (computer only gamers) (B = 0.457, odds ratio = 1.579). Smartphone only gamers had the lowest prevalence of IGD, spent the least time and money on gaming, and showed lowest scores of Internet and smartphone addiction. Our findings suggest that gaming device usage patterns may be associated with the occurrence, course, and prognosis of IGD.

Keywords: Internet gaming disorder; game device usage pattern; smartphone; comorbidity

1. Introduction

Playing online games is one of the most popular recreational activities. According to a Korean national survey conducted in 2016, 67.9% of the general population aged from 10 to 65 years old played online games [1]. Though gaming is a pleasurable and stimulatory activity, the dark side of excessive gaming is evident as well. Excessive use and loss of control over gaming has brought about various mental health and social concerns [2]. As numerous psychological and neurobiological correlates of excessive Internet gaming have been elucidated, such as impulsivity, reward sensitivity, and altered brain structure and function [3,4], the revised version of the Diagnostic and Statistical Manual of Mental Disorder, fifth edition (DSM-5) has listed the phenomenon of Internet gaming disorder (IGD) as a condition for further research [5].

To date, studies on IGD have mainly focused on personal computer (PC) games, especially massive multiplayer online role-playing games (MMORPGs) [3], and consistently reported that IGD was highly prevalent among male adolescents [6]. However, the spread of smartphones has explosively increased the number of gamers in female and in all age groups and subsequently changed the demographics and characteristics of online gamers. According to a Korean national survey conducted in 2015, among Korean smartphone users, 86.5% of the forties, 85.4% of the fifties, and 88.8% of females reported they had experiences of playing online games [7]. In addition, recent advances in smartphone game platforms have increased accessibility to various game genres including real-time strategy, MMORPG, or shooting games. Accordingly, the number of gamers who played with both devices has rapidly grown as well: 56.9% of the responders played smartphone games only while 20% played PC games only, and the remaining played both PC and smartphone games [7].

Given that each device has unique interface features and characteristics, gaming device usage patterns, such as single or combined use or time dedicated to each device, may play an important role on gaming behaviors and clinical characteristics and the occurrence of IGD. However, only a few studies have investigated the role of gaming device usage patterns on gaming-related attitudes or comorbid psychopathology. This study was designed to explore gaming behaviors and clinical characteristics across different gaming device usage patterns and their role on IGD.

2. Materials and Methods

2.1. Participants and Procedures

Data was collected from a large online survey conducted between April and September 2016 on online gaming and smartphone usage behaviors. Participants aged from 14 to 39 years were recruited from a pool of panelists registered for online panels at Panel Marketing Interactive (PMI), a research company that provides survey-related technology and data collection. The participants were given tokens that could be used as cybermoney as an incentive for their participation. In total, 9474 people were contacted and ultimately 7200 people (76% of those contacted) participated in our online survey. From the total 7200 responders, adults aged from 20 to 39 who both played online games and owned smartphones were included in this study (n = 3470). Since adolescents are considered to be more vulnerable to addictive disorders due to high novelty seeking and risk-taking temperament during adolescent period and immature cortical growth, which plays a critical role in cognitive control [8,9], we decided that the adult sample should be analyzed separately from adolescents. Though about 80% of smartphone users aged over 40 years had experiences of playing online games, the number of PC-based online gamers in this age are much lower than younger adults [1]. Thus, we set the upper age limit as 40 years old. We excluded responders who used game consoles (n = 412, 11.8% of total responders) for two reasons: (1) most of them also played PC and smartphone games (397, 96.3%), and (2) the aim of this study was to investigate the gaming characteristics among PC and smartphone gamers. Finally, 3058 subjects were selected (1548 males and 1510 females). The mean age was 26.95 years (standard deviation (SD) = 5.859 years).

All study procedures were performed in accordance with the guidelines of the Declaration of Helsinki. The Institutional Review Boards of Seoul St. Mary's Hospital approved the study protocol (KC15EISI0103). All subjects were informed about the study and all provided informed consent.

2.2. Measures

2.2.1. Demographic and Gaming Characteristics

Demographic information on education levels and occupation status was asked. Education levels were classified into two categories: (1) up to 12 years (up to high school graduates), and (2) more than 13 years (currently in university/college or higher education). Occupational status was classified into three categories: (1) current students; (2) individuals currently with full-time jobs, and (3) individuals currently without full-time jobs.

All participants were asked to answer dichotomously (1: Yes, 2: No) to nine diagnostic criteria questions for IGD according to the DSM-5. Participants who answered "yes" to five or more criteria for questions pertaining to the previous 12 months were defined as the IGD group in line with previous studies [10–12], and those with four or less affirmative answers were defined as the non-IGD group. Gaming device usage patterns were determined by reports from the responders on the time proportion dedicated to either PC or smartphone (SM) and were classified into five groups: (1) PC only group: individuals who played only PC games, 100% dedicated to PC games (n = 720); (2) PC > SM group: individuals who played more PC games than SM games, 60% to 99% dedicated to PC games and 1% to 40% to SM (n = 580); (3) PC = SM group: individuals who played PC and SM games evenly, 41% to 59% dedicated to both PC and SM games (n = 326); (4) PC < SM group: individuals who played SM games more than PC games, 60% to 99% dedicated to SM games and 1% to 40% to PC (n = 735), and (5) SM only group: individuals who played only SM games, 100% to SM games (n = 697). Additionally, time (minutes) and money (by Korean currency, KRW, per month) spent on gaming, whether they owned game community memberships, and whether they have any experience of attending offline meetings were asked. The most preferred game was asked and classified into five genres according to the White Paper on Korean Games [13]: (1) simulation and real-time strategy (i.e., League of Legends, StarCraft, and the Sims); (2) role-playing game (RPG) (i.e., World of Warcraft and Lineage); (3) sports and racing (i.e., FIFA, Winning Eleven, Need for Speed, and Tales Runner); (4) shooting and action (i.e., Sudden Attack, Counter Strike, and Virtual Fighter) and (5) puzzle, arcade, and board games (i.e., Candy Crush Saga, Monopoly, and rhythm games). The main motive for gaming was selected from the following categories: (1) for fun; (2) for killing time; (3) for relieving stress; (4) for need (i.e., to maintain interpersonal relationship), and (5) for sense of achievement.

2.2.2. Clinical Characteristics

Participants were asked to answer the following questionnaires in order to determine clinical characteristics: Young's Internet Addiction Test (YIAT), Smartphone Addiction Scale-Short Version (SAS-SV), Brief Self-Control Scale (BSCS), Patient Health Questionnaire-9 (PHQ-9), Generalized Anxiety Disorder-7 (GAD-7), Alcohol Use Disorder Identification Test (AUDIT), and Fagerstrom Test for Nicotine Dependence (FTND).

YIAT, a 20-item scale which is rated by a five-point Likert scale (1: Not at all, 5: Almost always) [14], was used to assess the severity of Internet addiction [15]. In South Korea, YIAT had acceptable internal consistency and reliability (Cronbach's alpha = 0.921) [16] and Cronbach's alpha for YIAT was 0.956 in this sample.

SAS-SV, a 10-item scale which is rated by a six-point Likert scale (1: Strongly disagree, 6: Strongly agree), was used to assess the severity of smartphone addiction [17]. SAS-SV has excellent concurrent validity and highly correlated with the original version (Cronbach's alpha = 0.958, $p < 0.001$). SAS-SV has been used in adult sample to assess the degree of smartphone addiction across various countries; a higher score indicated a higher degree of smartphone addiction [18,19]. Cronbach's alpha was 0.773 in this study.

BSCS, a 13-item questionnaire with a five-point Likert scale (1: Strongly disagree, 5: Strongly agree), was used to measure self-control ability. BSCS measures the ability to override or change one's inner response as well as to interrupt undesired behavioral tendencies and refrain from acting on them, with higher scores indicating lower self-control ability. BSCS had good internal consistency (Cronbach's alpha = 0.85) in the original study [20]. Cronbach's alpha was 0.757 in this study.

PHQ-9 is a nine-item depression rating scale which corresponds to the major depressive episode criteria of the DSM-IV with a four-point Likert scale (0: Not at all, 3: Almost every day) [19]. We defined individuals with scores of 10 or more as having depression according to Manea and colleagues [21]. The Korean version of PHQ-9 has proven to have excellent validity and reliability in primary care patients for detecting major depressive disorder (Cronbach's alpha = 0.852) [22]. Cronbach's alpha was 0.893 in this study.

GAD-7, a seven-item scale with a four-point Likert scale (0: Not at all, 3: Nearly every day), was used to screen generalized anxiety disorder (GAD) [23]. GAD-7 is particularly useful in assessing symptom severity. We defined individuals with scores of 10 or greater as having GAD according to Plummer and colleagues [24]. GAD-7 had excellent internal consistency (Cronbach's alpha = 0.92) [23]. Cronbach's alpha was 0.908 in this study.

AUDIT, developed by the World Health Organization, is a 10-item scale to identify alcohol-related problems [25]. Total scores range from 0 to 40 and the optimal cut-off point to identify at-risk alcohol users is 10 for males and 6 for females in South Korea [26]. In this study, alcohol use disorder (AUD) was defined for males as a score of 10 or more and for females as a score of 6 or more. AUDIT had good internal consistency (Cronbach's alpha = 0.80–0.93) and Cronbach's alpha was 0.859 in this study.

FTND is a six-item scale which is most widely used to measure nicotine dependence [27]. Total scores range from 0 to 10 and individuals with scores of four or more were determined to have nicotine dependence [28]. The Korean version of FTND was standardized and Cronbach's alpha was 0.6913, which was similar in this study, 0.636.

2.3. Statistical Analysis

The quantitative variables are presented as means ± SD and the qualitative data are presented as absolute numbers (N) and percentages (%). To compare differences between IGD and non-IGD groups, and across different gaming device usage pattern groups, non-parametric Mann-Whitney U tests and Kruskal-Wallis H tests were used, respectively. Additional post-hoc tests were done by the Mann-Whitney U tests. To examine the predictive values of each variable for IGD, binary logistic regression analysis was performed and the results are shown by odds ratios (ORs) and 95% confidential intervals (CIs). All statistical works were performed by using SPSS version 24.0 (SPSS Inc., Chicago, IL, USA).

3. Results

3.1. Comparison between the IGD and Control Groups

Table 1 shows differences in gaming behaviors and clinical characteristics between IGD and non-IGD subjects. From the total 3058 participants, 396 (12.9%) were classified as having IGD. IGD subjects had higher mean age, were more likely to be males, had lower prevalence of participants with higher education or who currently had full-time jobs, spent more time and money on gaming, and had more game community membership and experience of offline meeting attendance than non-IGD subjects. IGD was more prevalent in combined user groups (PC > SM, PC = SM, and PC < SM groups) than in the single user groups (PC and SM only groups) (66.4% vs. 51.7%, respectively). Preferred game genres were different; IGD subjects preferred simulation/strategy (28.5%) and RPG (27.3%) genres while non-IGD subjects did puzzle, arcade, and board games (31.0%). The main motives for gaming were different as well; the proportions of gamers who played for fun and for killing time were higher in non-IGD subjects (43.5% vs. 37.4%, 25.2% vs. 16.4%, respectively) while the proportion of gamers who played for relieving stress and for the sense of achievement was higher in IGD subjects (25.5% vs. 20.1%, and 15.2% vs. 8.2%, respectively). Scores of YIAT, SAS-SV, and BSCS and the prevalence of depression, GAD, AUD, and nicotine dependence were significantly higher among IGD subjects.

We performed two additional subgroup analyses to explore the gender role in IGD. First, we compared between IGD and non-IGD subjects for males and females separately (see Table S1 in the Supplementary Materials), and then between males and females in the IGD group (see Table S2 in the Supplementary Materials). The difference of education levels and occupational status between IGD and non-IGD groups was not observed in females, whereas that of preferred game genre was not found in males. Other findings were comparable to the findings of the whole sample. When compared between the male IGD group and female IGD group, the male IGD group had a higher proportion of individuals who currently had full-time jobs, game community membership, and nicotine dependence, and spent more money on gaming, while the female IGD group had higher scores for SAS-SV and BSCS, indicating lower self-control ability, and higher prevalence of AUD than the male IGD group. Gaming device usage patterns differed between genders. PC > SM (448, 28.9%) and PC only (431, 27.8%) patterns were more prevalent in males, and these patterns persisted in the male IGD group. Females had a higher proportion of PC < SM (417, 27.6%) and SM only (505, 33.4%) patterns, while the female IGD group showed higher PC < SM and PC = SM patterns.

Table 1. Comparison between the IGD and non-IGD groups.

Variables	IGD	Non-IGD	X^2/U	p
N (%)	396 (12.9%)	2662 (87.1%)		
Age	27.63 ± 5.797	26.85 ± 5.862	572,613.000	0.005 *
Male (%)	220 (55.6%)	1328 (49.9%)	4.431	0.035 *
Education levels				
Up to 12 years (high school)	63 (15.9%)	276 (10.4%)	10.731	0.001 **
More than 13 years	333 (84.1%)	2386 (89.6%)		
Occupational status				
Student	123 (31.1%)	1029 (38.7%)	8.472	0.014 *
Current full-time job	213 (53.8%)	1227 (48.0%)		
No currently full-time job	60 (15.2%)	356 (13.4%)		
Time spent on gaming				
Weekday (min)	167.79 ± 124.190	107.63 ± 96.227	722,949.000	0.000 **
Weekend (min)	253.76 ± 152.207	169.02 ± 131.868	729,658.000	0.000 **
Money spent on gaming (KRW)	32,270.45 ± 48,491.203	11,599.61 ± 27,190.750	752,279.500	0.000 **
Game community membership	253 (63.9%)	818 (30.7%)	166.566	0.000 **
Ever attended offline meeting	179 (45.2%)	352 (13.2%)	59.393	0.000 **
Gaming device usage pattern				
PC only	87 (22.0%)	633 (23.8%)	51.909	0.000 **
PC > SM	90 (22.7%)	490 (18.4%)		
PC = SM	71 (17.9%)	255 (9.6%)		
PC < SM	102 (25.8%)	633 (23.8%)		
SM only	46 (11.6%)	651 (24.5%)		
Preferred game genre				
Simulation/strategy	113 (28.5%)	708 (26.6%)	23.102	0.000 **
RPG	108 (27.3%)	521 (19.6%)		
Sports/racing	60 (15.2%)	413 (15.5%)		
Shooting/action	33 (8.3%)	196 (7.4%)		
Puzzle/arcade/board game	82 (20.7%)	824 (31.0%)		
Reason for gaming				
For fun	148 (37.4%)	1159 (43.5%)	44.352	0.000 **
For killing time	65 (16.4%)	671 (25.2%)		
For relieving stress	101 (25.5%)	535 (20.1%)		
For need	22 (5.6%)	80 (3.0%)		
For achievement	60 (15.2%)	217 (8.2%)		
YIAT	53.98 ± 26.267	36.80 ± 18.800	788,470.500	0.000 **
SAS-SV	39.97 ± 9.015	28.75 ± 10.114	838,582.000	0.000 **
BSCS	53.98 ± 26.267	36.80 ± 18.800	775,457.500	0.000 **
Depression (%)	252 (63.6%)	631 (23.7%)	267.652	0.000 **
Generalized anxiety disorder (%)	179 (45.2%)	383 (14.4%)	218.205	0.000 **
Alcohol use disorder (%)	139 (35.1%)	446 (16.8%)	75.003	0.000 **
Nicotine dependence (%)	55 (59.8%)	133 (31.7%)	25.675	0.000 **

Abbreviations: IGD: Internet Gaming Disorder; KRW: Korean Won; PC: Personal computer; SM: Smartphone; RPG: Role-playing game; YIAT: Young's Internet Addiction Test; SAS-SV: Smartphone Addiction Scale-Short Form; BSCS: Brief Self-Control Scale. * $p < 0.05$, ** $p < 0.005$.

3.2. Comparison across Different Gaming Device Usage Patterns

Table 2 shows the results of a comparison analysis across the five different gaming device usage groups. The mean age was highest in the SM only group and males were most prevalent in the PC only and PC > SM groups. PC only and PC = SM groups had higher prevalence of individuals with lower education. The PC only and PC > SM groups had higher proportions of students whereas the SM only and PC = SM groups had higher proportions of individuals who currently had full-time jobs. IGD was most prevalent in the PC = SM group (21.8%), followed by the PC > SM (15.5%) and PC < SM (13.9%) groups, and least prevalent in the SM only (6.6%) group. Time and money spent on gaming was highest in the PC > SM group and lowest in the SM only group. The PC > SM and PC = SM groups had the highest prevalence of game community membership and experience of offline meeting attendance.

Preferred game genres and motives for gaming were different across groups. The SM only group preferred puzzle, arcade, and board games while the other groups preferred simulation/strategy and RPG games. PC predominant groups (PC only, PC > SM, PC = SM groups) mainly played games "for fun" and "for relieving stress" while SM predominant groups (PC < SM and SM only groups) chose to play games more "for killing time". The SM only group had the lowest scores on YIAT, SAS-SV, and BSCS and proportions of depression, GAD, AUD, and nicotine dependence and PC = SM had the highest scores and proportions of comorbid psychopathology.

3.3. Predictive Value of the Gaming Device Usage Patterns

Table 3 shows the results of a binary logistic regression analysis predicting IGD using age, gender, time and money spent on gaming game community membership, device usage patterns, the scores of SAS-SV and BSCS, and the presence of depression, GAD, AUD, and nicotine dependence as covariates. Since the ranges of time and money spent on gaming were too broad to calculate odds ratios, these variables were divided dichotomously according to the median value: time spent on gaming during weekday (0: Minimum to 90 min, 1: 91 min to maximum) and weekend (0: Minimum to 150 min, 1: 151 min to maximum), and money (0: Minimum to KRW2000 (approximately 1.84 USD (1 USD = KRW1088.00, December 2017), 1: KRW2001 to maximum). When analyzed in a single equation, the input variables accounted for 39.5% of the total variance (Nagelkerke $R^2 = 0.395$) and the prediction success was 89.1%.

Individuals with higher age (B = 0.032, OR = 1.033), who spent 150 min or more on gaming during the weekend (B = 0.542, OR = 1.720), who spent more than KRW2001 on gaming per month (B = 0.729, OR = 2.072) and who owned a game community membership (B = 0.851, OR = 2.341) were likely to become IGD. As for the gaming device usage patterns, the PC = SM pattern increased the probability of IGD in comparison to the reference (PC only) group (B = 0.457, OR = 1.579). Individuals with higher SAS-SV and BSCS scores and depression were more prone to become IGD (B = 0.090, 0.043, and 0.667, OR = 1.094, 1.044, and 1.949 for SAS-SV, BSCS, and depression, respectively).

Table 2. Differences across device usage patterns.

Variables	PC Only	PC > SM	PC = SM	PC < SM	SM Only	X^2/H	p	Post-Hoc
N (%)	720 (23.5%)	580 (19.0%)	326 (10.7%)	735 (24.0%)	697 (22.8%)			
Age	25.70 ± 5.306	25.74 ± 5.458	27.16 ± 5.800	27.11 ± 5.890	28.95 ± 6.142	140.747	0.000 **	e > a,b,c,d, a,b < c,d
Male (%)	448 (62.2%)	431 (74.3%)	159 (48.8%)	318 (43.3%)	192 (27.5%)	333.801	0.000 **	
Education levels								
Up to 12 years	95 (13.2%)	66 (11.4%)	46 (14.1%)	65 (8.8%)	67 (9.6%)	11.608	0.021 *	
More than 13 years	625 (86.8%)	514 (88.6%)	280 (85.9%)	670 (91.2%)	630 (90.4%)			
Occupational status								
Student	330 (45.8%)	292 (50.3%)	107 (32.8%)	254 (34.6%)	169 (24.2%)			
Current full-time job	281 (39.0%)	217 (37.4%)	180 (55.2%)	383 (42.1%)	429 (61.5%)	134.837	0.000 **	
No current full-time job	109 (15.1%)	71 (12.2%)	39 (12.0%)	98 (13.3%)	99 (14.2%)			
IGD (%)	87 (12.1%)	90 (15.5%)	71 (21.8%)	102 (13.9%)	46 (6.6%)	51.909	0.000 **	
Time spent on gaming								
Weekday (min)	111.65 ± 103.703	139.56 ± 105.489	120.45 ± 112.919	110.11 ± 87.180	96.75 ± 103.262	122.254	0.000 **	e < a,b,c,d, b > a,c,d
Weekend (min)	194.60 ± 148.341	220.82 ± 134.656	180.53 ± 124.000	172.78 ± 123.798	137.61 ± 135.296	226.227	0.000 **	e < a,b,c,d, b > a,c,d
Money spent on gaming (KRW)	17,298.34 ± 37,995.779	24,993.36 ± 36,335.539	19,647.24 ± 34,119.854	11,973.13 ± 29,297.559	2082.50 ± 7421.816	577.364	0.000 **	e < a,b,c,d, b > a,c,d d < a,c
Game community membership	220 (30.6%)	304 (52.4%)	153 (46.9%)	273 (37.7%)	117 (16.8%)	207.871	0.000 **	
Ever attended offline meeting	101 (45.9%)	172 (56.6%)	88 (57.5%)	128 (46.2%)	42 (35.9%)	21.019	0.000 **	
Preferred game genre								
Simulation/strategy	255 (35.4%)	230 (39.7%)	77 (23.6%)	152 (20.7%)	107 (15.4%)			
RPG	183 (25.4%)	132 (22.8%)	79 (24.2%)	157 (21.4%)	78 (11.2%)			
Sports/racing	119 (16.5%)	89 (15.3%)	69 (21.2%)	123 (16.7%)	73 (10.5%)	591.480	0.000 **	
Shooting/action	88 (12.2%)	59 (10.2%)	21 (6.4%)	40 (5.4%)	12 (3.0%)			
Puzzle/arcade/board game	75 (10.4%)	70 (12.1%)	80 (24.5%)	263 (29.0%)	418 (60.0%)			
Reason for gaming								
For fun	237 (45.4%)	267 (46.0%)	147 (45.1%)	318 (43.3%)	248 (35.6%)			
For killing time	120 (16.7%)	88 (15.2%)	66 (20.2%)	181 (24.6%)	281 (38.2%)			
For relieving stress	178 (24.7%)	134 (23.1%)	69 (21.2%)	153 (20.8%)	102 (14.6%)	164.244	0.000 **	
For need	34 (4.7%)	24 (4.1%)	13 (4.0%)	21 (2.9%)	10 (1.4%)			
For achievement	61 (8.5%)	67 (11.6%)	31 (9.5%)	62 (8.4%)	56 (8.0%)			
YIAT	38.65 ± 20.899	43.56 ± 18.530	44.08 ± 21.511	40.73 ± 20.591	31.36 ± 19.930	145.760	0.000 **	e < a,b,c,d, a < b,c
SAS-SV	29.50 ± 10.646	30.19 ± 11.029	31.58 ± 11.149	31.60 ± 10.392	28.76 ± 10.657	31.614	0.000 **	e < c,d, a < c,d
BSCS	36.51 ± 6.663	36.17 ± 6.959	36.73 ± 6.920	36.49 ± 6.742	35.33 ± 6.795	17.479	0.002 **	e < a,c,d
Depression (%)	218 (30.3%)	156 (26.9%)	116 (35.6%)	233 (31.7%)	160 (23.0%)	23.687	0.000 **	
Generalized anxiety disorder (%)	141 (19.6%)	101 (17.4%)	74 (22.7%)	144 (19.6%)	102 (14.6%)	12.350	0.015 *	
Alcohol use disorder (%)	141 (31.6%)	84 (22.0%)	82 (36.4%)	160 (31.3%)	118 (26.3%)	19.325	0.001 **	
Nicotine dependence (%)	35 (31.0%)	50 (38.8%)	23 (44.2%)	60 (43.2%)	20 (25.3%)	10.006	0.040 *	

Abbreviations: PC: Personal computer; SM: Smartphone; IGD: Internet Gaming Disorder; KRW: Korean Won; RPG: Role-playing game; YIAT: Young's Internet Addiction Test; SAS-SV: Smartphone Addiction Scale Short Form; BSCS: Brief Self-Control Scale. * $p < 0.05$, ** $p < 0.005$.

Table 3. Logistic regression results predicting Internet gaming disorder.

Variables	B (s.e.)	OR	95% CI	p
Age	0.032 (0.011)	1.033	1.010–1.056	0.005 *
Gender (male)	0.260 (0.144)	1.296	0.977–1.719	0.072
Weekday gaming hour (>90 min)	0.262 (0.144)	1.300	0.931–1.816	0.124
Weekend gaming hour (>150 min)	0.542 (0.170)	1.720	1.232–2.399	0.001 **
Money spent on gaming (>KRW2000)	0.729 (0.144)	2.072	1.562–2.749	0.000 **
Game community membership	0.851 (0.137)	2.341	1.791–3.061	0.000 **
Gaming device usage pattern				
PC only		Reference		0.000 **
PC > SM	−0.085 (0.196)	0.919	0.626–1.349	0.666
PC = SM	0.457 (0.213)	1.579	1.040–2.397	0.032 *
PC < SM	0.019 (0.189)	1.019	0.703–1.476	0.921
SM only	−0.144 (0.229)	0.866	0.552–1.357	0.529
SAS-SV	0.090 (0.008)	1.094	1.076–1.112	0.000 **
BSCS	0.043 (0.013)	1.044	1.018–1.070	0.001 **
Depression	0.667 (0.168)	1.949	1.403–2.708	0.000 **
Generalized anxiety disorder	0.132 (0.173)	1.141	0.812–1.603	0.447
Alcohol use disorder	0.254 (0.150)	1.2899	0.961–1.728	0.090
Nicotine dependence	0.384 (0.223)	1.468	0.949–2.271	0.085

Abbreviations: s.e.: standard error; KRW: Korean Won; PC: Personal computer; SM: Smartphone; SAS-SV: Smartphone Addiction Scale Short Form; BSCS: Brief Self-Control Scale; OR: odds ratio; CI: confidence interval. * $p < 0.05$, ** $p < 0.005$.

4. Discussion

4.1. Characteristics of IGD in Smartphone Era

Overall, the prevalence of IGD was 12.9% in this study. The prevalence of IGD ranged from 0.6% to 46% depending on the sample and methods [29]. According to a Korean national survey conducted in 2015, the prevalence of Internet addiction was 5.8% in an adult population with ages ranging from 20 to 59 years old [7], which was lower than our results. Given that our sample was recruited from online survey responders who played online games and included those age ranged from 20 to 39, it seems reasonable that the prevalence of IGD is higher in our sample, though a direct comparison is impossible due to the differences in the sample collecting methods and applied diagnostic criteria. Rather, this result was in line with a previous study reporting that 38.7% of adult MMORPG players were IGD [30], since 20.5% of the responders played RPGs, most of which were MMORPGs. The prevalence of IGD is higher in Korea than in the other countries. The prevalence of IGD has been considered to be high in East Asian countries [31]. Cultural differences, such as Internet accessibility (speed of Internet access, availability of Wi-Fi, the cost paid for using Internet, or accessibility of Internet cafes), social norms for Internet gaming and device usage patterns, and government regulations for Internet gaming, as well as the sample recruitment and applied measurements may account for the different prevalence across different countries. Considering that IGD subjects in this study showed comparable clinical characteristics to previous studies, such that they owned game community memberships and attended offline meetings more frequently, spent more time and money on gaming, showed more preference for the simulation/strategy and RPG genres, were motivated to play games more to relieve stress, manifested higher degrees of smartphone addiction and lower levels of self-control, and had higher prevalence of depression, GAD, AUD, and nicotine dependence than non-IGD individuals [3,6,32–36], our sample may represent the general IGD population. In contrast to prior findings that IGD was frequently observed in male adolescents [37,38], the proportion of females in the IGD group was 44.4% and the mean age was higher among those with IGD. Given that our sample only included an adult population, this would be a unique feature of adult IGD, distinct from adolescent IGD. The spread of smartphone games, which provide easier accessibility and portability than PC games, may increase the number of female and middle-aged adult gamers.

Some gaming-related and clinical characteristics were significantly different between male and female participants. Preferred game genre and reason for gaming were different between genders, in line with previous findings [39–41]. Males displaying IGD spent more money on gaming and had more game community memberships, both classical features of IGD, while females displaying IGD had higher smartphone addiction severity and lower self-control ability, which may indicate the role of problematic smartphone use on IGD in females [42,43]. The proportion of nicotine dependence was higher among males with IGD, while that of AUD was higher among females with IGD. This may be due to an increased risk of AUD in the presence of comorbid psychiatric condition in females [44]. Gaming device usage patterns were significantly different between genders as well; males had a tendency to play PC games more than SM games while females seemed to play more SM games. The features of each device, which are described below, differences in time spent on gaming, preference of specific game genres that were more suitable for PC interface, and social expectation toward gaming behavior between genders may account for the results [41].

4.2. Characteristics across Different Gaming Device Usage Patterns

When comparing across different gaming device usage patterns, several intriguing findings were observed. First, IGD was more prevalent in combined user groups (PC > SM, PC = SM, and PC < SM groups), especially in the PC = SM group, than the single user groups (PC and SM only). IGD is a behavioral addiction that psychologically and neurobiologically resembles substance addiction [3]. As substance abusers frequently co-administer multiple substances simultaneously [45], IGD subjects may need multiple methods of playing games. In particular, the PC = SM group had the highest prevalence of IGD, possibly due to the differences in motivational background and gaming interface of each device. Though both PC and SM games provide a sense of reward and relatedness as well as escape from negative emotions [46,47], each device may satisfy different needs. Since PC games usually provide high quality sound and visual effects and necessitate substantial duration for playing, they may have the potential to provide a sense of immersion, achievement, and competitiveness [47]. Meanwhile, SM games may increase a sense of social relatedness and attenuate loneliness and negative emotion [48], since SM games are easily played in association with social networking service (SNS) applications. In this study, motives for gaming were significantly different across groups; PC predominant gamers played games "for fun", "for relieving stress", and "for achievement", while SM predominant gamers played games "for killing time", "for fun", and "for relieving stress". Considering that the PC = SM group manifested more comorbid psychopathology, which can be taken as being more defective in self-soothing ability, they may need both devices to satisfy different needs and become more indulged in gaming than the other groups.

Second, each combined user group (PC > SM, PC = SM, and PC < SM groups) had unique clinical characteristics that may modify the course and prognosis of IGD. Individuals in the PC > SM group showed typical behavioral manifestations of IGD [32]. They were younger, had a higher male proportion, spent more time and money on gaming, and had more game community memberships, but had less prevalence of comorbid psychopathology than other combined user groups. Meanwhile, the PC = SM group showed the highest prevalence of comorbid psychopathology, implying that they may be suffering from more difficulty in academic or occupational adaptation and impelled to continuously play games to soothe their negative emotions or to alleviate substance craving. The PC < SM group manifested similar gaming behaviors to the SM only group while comorbidity patterns were comparable to the PC = SM group. These findings suggest an important clinical implication that detailed investigations of the gaming device usage patterns may provide precise estimation of current status and prediction of prognosis.

Third, the PC only group spent less time and money on gaming and showed lower scores on YIAT and SAS-SV than the combined user groups. PC games usually need special environmental requisites, such as fast online connection speeds, high-resolution large screens, and charged memberships, which may paradoxically keep the PC gamers away from playing anytime and anywhere. In addition, since PC games offer excellent visual and sound effects and sophisticated game platforms, PC gamers may not be satisfied with SM games. Another interesting finding was that PC only gamers manifested as high prevalence of depression, GAD, and AUD as the PC < SM group. Considering that PC games had more addictive potential than SM games [49,50], PC only gamers may have difficulty in functioning as is the case for the combined users despite less behavioral disturbances.

Fourth, the SM only group had a higher mean age and proportion of females, spent less time and money on gaming, and had lower severity of Internet and smartphone addiction and prevalence of IGD, depression, GAD, AUD, and nicotine dependence and higher self-control ability than the other groups. Explanations can be drawn from the unique characteristics of SM game interfaces and smartphones themselves. First is the structural characteristic of smartphone games. Despite substantial technical advances that enable sophisticated games to be played on smartphone platforms, simpler games such as puzzle, arcade, or board games are still more suitable and preferred for the SM gamers, possibly due to unchangeable structures such as small screens and viewing angles, which interfere with the sense of immersion [51]. Secondly, the emergence of the social networking games (SNGs), a hybrid game genre that connects games with SNSs, should be considered. SNG players can play games and interact with online friends simultaneously within existing SNS applications. Many responders in the SM only group answered that they preferred to play SNGs, such as "Farmville" or "I Love Coffee", to mention but a few. Considering that getting something useful out of playing, such as the improvement of relationships, was the major motive to play SNGs [52], SM gamers, especially SNG players, may prefer to play games to get something that classical PC games cannot provide, such as enhancement of social relatedness and the alleviation of feelings of loneliness. Third, the unique properties of smartphones themselves can be considered. In addition to the portability and availability for easy and frequent access that leads to habitual checking behavior [53], the multi-tasking function enables SM users to play games while they are doing other things, such as searching on websites or sending messages. Though SM gamers may not spend as much time and money on gaming and feel as much immersion in games as PC gamers do, they may use games to "kill time" between tasks or during "empty" hours of waiting.

4.3. The Role of Gaming Device Usage Patterns on IGD

Logistic regression analysis revealed that the PC = SM pattern was a predictor of IGD, along with time and money spent on gaming, game community membership, the severity of smartphone addiction, self-control ability, and the presence of depression. As mentioned above, individuals in the PC = SM group had the highest prevalence of comorbidity and thus they may feel that it is difficult to quit games probably due to lack of tolerability for withdrawal symptoms or negative emotions. Another possibility lays in that the PC = SM pattern may have an intrinsic risk potential for IGD since no multicollinearity was observed among variables. Further studies are necessary to elucidate the neurobiology that underlies this distinct device usage pattern. An important finding that higher financial investment was associated with increased risk of IGD should be marked. This was in line with previous findings demonstrating that higher money spent on gaming was associated with IGD in adolescents and had a predictive value of IGD in adults [32,54]. Most online games are freemium services, which are free for download, but require payment for additional features or virtual goods [55]. Even though the amount is small, as much as $2, it could cause a big financial problem if accumulated, as is the case of gambling disorder, and could induce more commitment toward the games. Increased risk of IGD with individuals who spent more than 2000 won (approximately $1.84) for gaming per month would support the danger of accumulated micropayments.

There are some limitations that should be noted. First, the causality of IGD and the device usage patterns were not elucidated due to the cross-sectional nature of this study. Second, the distinction between the game genres may not be mutually exclusive because of the fast evolution of hybrid game genres. For example, Cookie Run is a hybrid form of running and action games. Likewise, the distinction between the games and other smartphone applications were not exclusive, such as with SNGs. Further research studies need to investigate the role of hybrid games or applications on IGD or other technological addiction. Third, IGD was determined by self-reports of endorsement to five or more DSM-5 IGD criteria, which was originally developed for professional usage. However, Lemmens and colleagues have demonstrated solid psychometric property and high practicality of The Short, Nine-Item IGD Scale, which assessed IGD using the self-reporting of nine items of DSM-5 IGD criteria with a dichotomous scale [12]. Despite deviation from original professional purposes, the self-rating IGD would be a valid tool for identifying IGD, especially in a large sample. Fourth, since comorbid psychopathology and gaming device usage patterns were assessed via self-report, over- or under-estimation of psychopathology and a recall bias may be present. Overestimation of IGD should be considered when determining IGD by the self-report as well. As mentioned above, the prevalence of IGD was higher in this study than in the national survey. Further research studies may need to validate the findings using diagnostic interviews and machinery collecting methods. Lastly, we did not include console users. In contrast to the United States of America and European countries, only a small number of gamers used video and portable consoles for gaming [1]. Console games had different features than PC and SM games, such as the need for buying consoles and software, and may manifest unique gaming-relate behaviors and characteristics. Further studies are necessary to investigate the characteristics of console gamers, especially in comparison with PC and SM gamers. Alternatively, was this meant to be "the United States of America"? If so, please specify to avoid potential ambiguity with the continents.

5. Conclusions

In this study, we investigated how gaming device usage patterns influenced gaming behaviors and clinical characteristics. To the best of our knowledge, this is the first study to elucidate the differences in gaming behavior and comorbid psychopathology across gaming device usage patterns and the role of specific usage patterns on IGD. Despite limitations, this study is notable for having been conducted using a large sample and having covered a diverse range of gaming behaviors and clinical characteristics, and thus contributes to deepening our understanding of IGD in the smartphone era. Our findings suggest that gaming device usage patterns may be associated with the occurrence, course, and prognosis of IGD. An important clinical implication can be drawn that the evaluation of gaming device usage patterns would help to determine the risk, predict outcome, and offer optimized treatment options for IGD.

Supplementary Materials: The following are available online at www.mdpi.com/1660-4601/14/12/1512/s1, Table S1. Comparison between the IGD and non-IGD groups in each gender group; Table S2. Comparison between males and females in the IGD group.

Acknowledgments: This work was supported by a grant from the Brain Research Program through the National Research Foundation of Korea (NRF) funded by the Ministry of Science, ICT & Future Planning (NRF-2014M3C7A1062893).

Author Contributions: All the authors contributed to the design and study planning and advised the course of study. S.-H.P. conducted the literature search, provided summaries of previous research studies, conducted the statistical analysis, and wrote the first draft of the manuscript. All authors contributed to the development of the manuscript, revised it critically, and approved the final manuscript.

Conflicts of Interest: The authors declare no conflict of interest.

References

1. Game Users Survey Report 2016. Available online: http://www.kocca.kr/cop/bbs/view/B0000147/1831102.do?menuNo=200904 (accessed on 18 July 2017).
2. Ferguson, C.J.; Coulson, M.; Barnett, J. A meta-analysis of pathological gaming prevalence and comorbidity with mental health, academic and social problems. *J. Psychiatr. Res.* **2011**, *45*, 1573–1578. [CrossRef] [PubMed]

3. Kuss, D.J.; Griffiths, M.D. Internet gaming addiction: A systematic review of empirical research. *Int. J. Ment. Health Addict.* **2012**, *10*, 278–296. [CrossRef]
4. Fauth-Bühler, M.; Mann, K. Neurobiological correlates of internet gaming disorder: Similarities to pathological gambling. *Addict. Behav.* **2017**, *64*, 349–356. [CrossRef] [PubMed]
5. American Psychiatric Association. *Diagnostic and Statistical Manual of Mental Disorders (DSM-5®)*; American Psychiatric Pub.: Arlington, VA, USA, 2013.
6. Ho, R.C.; Zhang, M.W.; Tsang, T.Y.; Toh, A.H.; Pan, F.; Lu, Y.; Cheng, C.; Yip, P.S.; Lam, L.T.; Lai, C.-M. The association between internet addiction and psychiatric co-morbidity: A meta-analysis. *BMC Psychiatr.* **2014**, *14*, 183. [CrossRef] [PubMed]
7. Survey on Internet Overdependence 2015. Available online: http://www.nia.or.kr/site/nia_kor/ex/bbs/View.do?cbIdx=65914&bcIdx=17132&parentSeq=17132 (accessed on 21 June 2016).
8. Kelley, A.E.; Schochet, T.; Landry, C.F. Risk taking and novelty seeking in adolescence: Introduction to part I. *Ann. N. Y. Acad. Sci.* **2004**, *1021*, 27–32. [CrossRef] [PubMed]
9. Crews, F.; He, J.; Hodge, C. Adolescent cortical development: A critical period of vulnerability for addiction. *Pharmacol. Biochem. Behav.* **2007**, *86*, 189–199. [CrossRef] [PubMed]
10. Kim, N.R.; Hwang, S.S.-H.; Choi, J.-S.; Kim, D.-J.; Demetrovics, Z.; Király, O.; Nagygyörgy, K.; Griffiths, M.; Hyun, S.Y.; Youn, H.C. Characteristics and psychiatric symptoms of internet gaming disorder among adults using self-reported DSM-5 criteria. *Psychiatr. Investig.* **2016**, *13*, 58–66. [CrossRef] [PubMed]
11. Wang, H.R.; Cho, H.; Dai-Jin, K. Prevalence and correlates of comorbid depression in a nonclinical online sample with DSM-5 internet gaming disorder. *J. Affect. Disord.* **2018**, *226*, 1–5. [CrossRef] [PubMed]
12. Lemmens, J.S.; Valkenburg, P.M.; Gentile, D.A. The Internet Gaming Disorder Scale. *Psychol. Assess.* **2015**, *27*, 567. [CrossRef] [PubMed]
13. Korean Creative Content Agency. *White Paper on Korean Games*; Korea Creative Content Agency: Seoul, Korea, 2013.
14. Young, K.S.; De Abreu, C.N. *Internet Addiction: A Handbook and Guide to Evaluation and Treatment*; John Wiley & Sons: Hoboken, NJ, USA, 2010.
15. Widyanto, L.; McMurran, M. The psychometric properties of the internet addiction test. *Cyberpsychol. Behav.* **2004**, *7*, 443–450. [CrossRef] [PubMed]
16. Gyeong, H.; Lee, H.-K.; Lee, K. Factor analysis of the Young's internet addiction test: In Korean College Students Group. *J. Korean Neuropsychiatr. Assoc.* **2012**, *51*, 45–51. [CrossRef]
17. Kwon, M.; Kim, D.-J.; Cho, H.; Yang, S. The smartphone addiction scale: Development and validation of a short version for adolescents. *PLoS ONE* **2013**, *8*, e83558. [CrossRef] [PubMed]
18. Hawi, N.S.; Samaha, M. To excel or not to excel: Strong evidence on the adverse effect of smartphone addiction on academic performance. *Comput. Educ.* **2016**, *98*, 81–89. [CrossRef]
19. Haug, S.; Castro, R.P.; Kwon, M.; Filler, A.; Kowatsch, T.; Schaub, M.P. Smartphone use and smartphone addiction among young people in Switzerland. *J. Behav. Addict.* **2015**, *4*, 299–307. [CrossRef] [PubMed]
20. Tangney, J.P.; Baumeister, R.F.; Boone, A.L. High self-control predicts good adjustment, less pathology, better grades, and interpersonal success. *J. Personal.* **2004**, *72*, 271–324. [CrossRef]
21. Manea, L.; Gilbody, S.; McMillan, D. Optimal cut-off score for diagnosing depression with the Patient Health Questionnaire (PHQ-9): A meta-analysis. *Can. Med. Assoc. J.* **2012**, *184*, E191–E196. [CrossRef] [PubMed]
22. Choi, H.S.; Choi, J.H.; Park, K.H.; Joo, K.J.; Ga, H.; Ko, H.J.; Kim, S.R. Standardization of the Korean version of Patient Health Questionnaire-9 as a screening instrument for major depressive disorder. *J. Korean Acad. Fam. Med.* **2007**, *28*, 114–119.
23. Spitzer, R.L.; Kroenke, K.; Williams, J.B.; Löwe, B. A brief measure for assessing generalized anxiety disorder: The GAD-7. *Arch. Intern. Med.* **2006**, *166*, 1092–1097. [CrossRef] [PubMed]
24. Plummer, F.; Manea, L.; Trepel, D.; McMillan, D. Screening for anxiety disorders with the GAD-7 and GAD-2: A systematic review and diagnostic metaanalysis. *Gen. Hosp. Psychiatry* **2016**, *39*, 24–31. [CrossRef] [PubMed]
25. Saunders, J.B.; Aasland, O.G.; Babor, T.F.; De la Fuente, J.R.; Grant, M. Development of the alcohol use disorders identification test (AUDIT): WHO collaborative project on early detection of persons with harmful alcohol consumption-II. *Addiction* **1993**, *88*, 791–804. [CrossRef] [PubMed]
26. Joe, K.H.; Chai, S.H.; Park, A.; Lee, H.K.; Shin, I.H.; Min, S.H. Optimum Cut-Off Score for Screening of Hazardous Drinking Using the Korean Version of Alcohol Use Disorder Identification Test(AUDIT-K). *J. Korean Addict. Psychiatry* **2009**, *13*, 34–40.

27. Heatherton, T.F.; Kozlowski, L.T.; Frecker, R.C.; Fagerstrom, K.O. The Fagerström test for nicotine dependence: A revision of the Fagerstrom Tolerance Questionnaire. *Addiction* **1991**, *86*, 1119–1127. [CrossRef]
28. Ahn, H.K.; Lee, H.J.; Jung, D.S.; Lee, S.Y.; Kim, S.W.; Kang, J.H. The reliability and validity of Korean version of questionnaire for nicotine dependence. *J. Korean Acad. Fam. Med.* **2002**, *23*, 999–1008.
29. Festl, R.; Scharkow, M.; Quandt, T. Problematic computer game use among adolescents, younger and older adults. *Addiction* **2013**, *108*, 592–599. [CrossRef] [PubMed]
30. Charlton, J.P.; Danforth, I.D. Distinguishing addiction and high engagement in the context of online game playing. *Comput. Hum. Behav.* **2007**, *23*, 1531–1548. [CrossRef]
31. Rho, M.J.; Jeong, J.-E.; Chun, J.-W.; Cho, H.; Jung, D.J.; Choi, I.Y.; Kim, D.-J. Predictors and patterns of problematic Internet game use using a decision tree model. *J. Behav. Addict.* **2016**, *5*, 500–509. [CrossRef] [PubMed]
32. Lemmens, J.S.; Hendriks, S.J. Addictive online games: Examining the relationship between game genres and internet gaming disorder. *Cyberpsychol. Behav. Soc. Netw.* **2016**, *19*, 270–276. [CrossRef] [PubMed]
33. Choi, S.-W.; Kim, D.-J.; Choi, J.-S.; Ahn, H.; Choi, E.-J.; Song, W.-Y.; Kim, S.; Youn, H. Comparison of risk and protective factors associated with smartphone addiction and Internet addiction. *J. Behav. Addict.* **2015**, *4*, 308–314. [CrossRef] [PubMed]
34. Yen, J.Y.; Ko, C.H.; Yen, C.F.; Chen, C.S.; Chen, C.C. The association between harmful alcohol use and Internet addiction among college students: Comparison of personality. *Psychiatry. Clin. Neurosci.* **2009**, *63*, 218–224. [CrossRef] [PubMed]
35. Kim, E.J.; Namkoong, K.; Ku, T.; Kim, S.J. The relationship between online game addiction and aggression, self-control and narcissistic personality traits. *Eur. Psychiatr.* **2008**, *23*, 212–218. [CrossRef] [PubMed]
36. Rehbein, F.; Kliem, S.; Baier, D.; Mößle, T.; Petry, N.M. Prevalence of internet gaming disorder in German adolescents: Diagnostic contribution of the nine DSM-5 criteria in a state-wide representative sample. *Addiction* **2015**, *110*, 842–851. [CrossRef] [PubMed]
37. Kuss, D.J. Internet gaming addiction: Current perspectives. *Psychol. Res. Behav. Manag.* **2013**, *6*, 125. [CrossRef] [PubMed]
38. Ko, C.-H.; Yen, J.-Y.; Chen, C.-C.; Chen, S.-H.; Yen, C.-F. Gender differences and related factors affecting online gaming addiction among Taiwanese adolescents. *J. Nerv. Ment. Dis.* **2005**, *193*, 273–277. [CrossRef] [PubMed]
39. Weiser, E.B. Gender differences in Internet use patterns and Internet application preferences: A two-sample comparison. *Cyberpsychol. Behav.* **2000**, *3*, 167–178. [CrossRef]
40. Winn, J.; Heeter, C. Gaming, gender, and time: Who makes time to play? *Sex Roles* **2009**, *61*, 1–13. [CrossRef]
41. Mok, J.-Y.; Choi, S.-W.; Kim, D.-J.; Choi, J.-S.; Lee, J.; Ahn, H.; Choi, E.-J.; Song, W.-Y. Latent class analysis on internet and smartphone addiction in college students. *Neuropsychiatr. Dis. Treat.* **2014**, *10*, 817. [PubMed]
42. Kim, Y.; Jeong, J.-E.; Cho, H.; Jung, D.-J.; Kwak, M.; Rho, M.J.; Yu, H.; Kim, D.-J.; Choi, I.Y. Personality factors predicting smartphone addiction predisposition: Behavioral inhibition and activation systems, impulsivity, and self-control. *PLoS ONE* **2016**, *11*, e0159788. [CrossRef] [PubMed]
43. Burns, L.; Teesson, M. Alcohol use disorders comorbid with anxiety, depression and drug use disorders: Findings from the Australian National Survey of Mental Health and Well Being. *Drug Alcohol Depend.* **2002**, *68*, 299–307. [CrossRef]
44. Barrett, S.P.; Darredeau, C.; Pihl, R.O. Patterns of simultaneous polysubstance use in drug using university students. *Hum. Psychopharmacol.* **2006**, *21*, 255–263. [CrossRef] [PubMed]
45. Demetrovics, Z.; Urbán, R.; Nagygyörgy, K.; Farkas, J.; Zilahy, D.; Mervó, B.; Reindl, A.; Ágoston, C.; Kertész, A.; Harmath, E. Why do you play? The development of the motives for online gaming questionnaire (MOGQ). *Behav. Res. Methods* **2011**, *43*, 814–825. [CrossRef] [PubMed]
46. Kuss, D.J.; Louws, J.; Wiers, R.W. Online gaming addiction? Motives predict addictive play behavior in massively multiplayer online role-playing games. *Cyberpsychol. Behav. Soc. Netw.* **2012**, *15*, 480–485. [CrossRef] [PubMed]
47. Park, N.; Lee, H. Social implications of smartphone use: Korean college students' smartphone use and psychological well-being. *Cyberpsychol. Behav. Soc. Netw.* **2012**, *15*, 491–497. [CrossRef] [PubMed]
48. Lee, C.; Kim, O. Predictors of online game addiction among Korean adolescents. *Addict. Res. Theory* **2017**, *25*, 58–66. [CrossRef]

49. Jeong, S.-H.; Kim, H.; Yum, J.-Y.; Hwang, Y. What type of content are smartphone users addicted to? SNS vs. games. *Comput. Hum. Behav.* **2016**, *54*, 10–17. [CrossRef]
50. Hou, J.; Nam, Y.; Peng, W.; Lee, K.M. Effects of screen size, viewing angle, and players' immersion tendencies on game experience. *Comput. Hum. Behav.* **2012**, *28*, 617–623. [CrossRef]
51. Park, E.; Baek, S.; Ohm, J.; Chang, H.J. Determinants of player acceptance of mobile social network games: An application of extended technology acceptance model. *Telemat. Inform.* **2014**, *31*, 3–15. [CrossRef]
52. Liu, C.-H.; Lin, S.-H.; Pan, Y.-C.; Lin, Y.-H. Smartphone gaming and frequent use pattern associated with smartphone addiction. *Medicine* **2016**, *95*, e4068. [CrossRef] [PubMed]
53. Dreier, M.; Wölfling, K.; Duven, E.; Giralt, S.; Beutel, M.; Müller, K. Free-to-play: About addicted Whales, at risk Dolphins and healthy Minnows. Monetarization design and internet gaming disorder. *Addict. Behav.* **2017**, *64*, 328–333. [CrossRef] [PubMed]
54. Evans, E. The economics of free: Freemium games, branding and the impatience economy. *Convergence* **2016**, *22*, 563–580. [CrossRef]
55. Kwon, M.; Nam, K.; Seo, B. *A Survey on Internet Overdependence, 2015*; Ministry of Science, ICT, and Future Planning: Seoul, Korea, 2016.

International Journal of
Environmental Research and Public Health

MDPI

Article

Emotional Regulation in Young Adults with Internet Gaming Disorder

Ju-Yu Yen [1,2], Yi-Chun Yeh [1,3,4], Peng-Wei Wang [1,4], Tai-Ling Liu [3,4], Yun-Yu Chen [1] and Chih-Hung Ko [1,3,4,5,*]

1 Department of Psychiatry, Faculty of Medicine, College of Medicine, Kaohsiung Medical University, Kaohsiung City 807, Taiwan; yenjuyu@cc.kmu.edu.tw (J.-Y.Y.); y7552156@gmail.com (Y.-C.Y.); wistar.huang@gmail.com (P.-W.W.); li52030tw@gmail.com (Y.-Y.C.)
2 Department of Psychiatry, Kaohsiung Municipal Ta-Tung Hospital, Kaohsiung Medical University, Kaohsiung City 801, Taiwan
3 Graduate Institute of Medicine, College of Medicine, Kaohsiung Medical University, Kaohsiung City 807, Taiwan; dai32155@gmail.com
4 Department of Psychiatry, Kaohsiung Medical University Hospital, Kaohsiung Medical University, Kaohsiung City 807, Taiwan
5 Department of Psychiatry, Kaohsiung Municipal Hsiao-Kang Hospital, Kaohsiung Medical University, Kaohsiung City 812, Taiwan
* Correspondence: chihhungko@gmail.com; Tel.: +886-7-803-6783 (ext. 3858)

Received: 5 November 2017; Accepted: 22 December 2017; Published: 25 December 2017

Abstract: People diagnosed with Internet gaming disorder (IGD) have been frequently reported to experience depression, anxiety, and hostility. Emotional regulation contributes to these mood symptoms. This study evaluated emotional regulation in subjects with IGD and examined relationships between emotional regulation, depression, anxiety, and hostility in young adults with IGD. We recruited 87 people with IGD and a control group of 87 people without a history of IGD. All participants underwent a diagnostic interview based on the IGD criteria of the Diagnostic and Statistical Manual of Mental Disorders, Fifth Edition, and they completed a questionnaire on emotional regulation, depression, anxiety, and hostility. We found that subjects with IGD were less likely to practice cognitive reappraisal and were more likely to suppress their emotions. Linear regression revealed the higher cognitive reappraisal and lower expressive suppression associated with depression, anxiety, and hostility among subjects with IGD. The emotional regulation strategies that characterize those with IGD could be contributing factors to the depression and hostility tendencies of these people. When treating patients with IGD, in addition to providing appropriate interventions to relieve depression and hostility, practitioners should effectively assess emotional regulation strategies and provide emotional regulation therapy to prevent a vicious cycle of negative emotions.

Keywords: Internet gaming disorder; IGD; emotional regulation; cognitive reappraisal; suppression; depression; hostility

1. Introduction

Diagnostic criteria for Internet gaming disorder (IGD), defined as an addiction to Internet games, are proposed as research criteria in section III of the Diagnostic and Statistical Manual of Mental Disorders, Fifth Edition (DSM-5) [1]. IGD is one type of internet addiction and has been associated with mood-related psychopathological symptoms, such as depression and irritability [2,3]. This comorbidity could contribute to treatment difficulties and a poor prognosis of addictive disorder [4], for example, the comorbidity of depression associated with the higher psychosocial burden among subjects with IGD [5]. Further, comorbidity could indicate a causal relationship between the disorders [6] or a common factors model [7], wherein a shared mechanism generally accounts for increased comorbidity. To intervene, the shared mechanism might benefit both disorders. Therefore, understanding the shared mechanism causing comorbidity between IGD and psychopathological symptoms could contribute to successfully developing treatments for them.

1.1. Association between IGD and Emotional Difficulties

The amount of time spent playing online games has been positively correlated with depressive symptoms [8,9]. The association between IGD, depression, and hostility were also demonstrated in recent studies [10,11]. Gentile et al. reported that IGD could be a cause of depression in adolescents [12]. Further, Ciarrochi et al. also reported that compulsive internet use predicted poor mental health among adolescents in longitudinal investigation [13]. These results might indicate that repeatedly excessive online gaming could contribute to emotional difficulties, possibly though impaired daily life functions or their negative consequences. On the other side, addictive behavior, such as online gaming [14], could be a way of coping with pre-existing emotional difficulties, such as depression [6]. Depression was reported to predict the incidence of internet addiction and support this claim [15]. This might suggest that emotional difficulties could possibly contribute to IGD; however, this has not been proved. The possible bidirectional effect between IGD and emotional difficulties deserves future prospective study. On the other hand, an underlying factor, such as emotional regulation, might be associated with both IGD and emotional difficulties, and could contribute to the comorbidity of IGD.

1.2. Emotional Regulation and Depression, Anxiety, Hostility, and IGD

Emotional regulation, also known as emotional self-regulation, was defined by [16] as the set of cognitive processes that influence emotional responses. Emotional regulation is a complex process that includes the initiation, inhibition, or modulation of aspects of emotion functioning. A previous review demonstrated that interventions that specifically target emotional regulation can not only promote positive emotional regulation but also attenuate associated psychopathological symptoms [17].

Two strategies are commonly used for downregulating emotion. The first, reappraisal, comes early in the emotion-generative process and entails changing how a situation is construed in order to reduce its emotional impact. The second, suppression, comes later in the emotion-generative process and entails the inhibition of outward signs of inner feelings [18]. The two types of emotional regulation are evaluated in The Emotional Regulation Questionnaire, which measures the habitual use of expressive suppression and cognitive re-evaluation. The scale includes items related to the regulation of positive and negative emotions [19]. According to this measurement, practicing reappraisal is associated with greater positive emotion, improved interpersonal functioning, and well-being. By contrast, practicing suppression is associated with negative emotions and poorer interpersonal functioning. These results suggest that strategies that act early in the emotion-generative process have a different profile of consequences than strategies that act later.

Emotional regulation was associated with depression [20] and anxiety [21]. The employment of adaptive emotional regulation strategies (e.g., reappraisal) causes a reduction in stress-induced emotions. Conversely, dysfunctional emotional regulation strategies, such as emotion suppression, appear to influence the pathogenesis of depression. For example, a structural equation modeling study found that expressive suppression mediated the relationship between intensity of negative affect and psychological distress [22]. In addition, emotional regulation therapy has been reported to be an effective treatment of emotional dysfunctions, such as anxiety or depression [17,23,24]. The literature demonstrates the role for emotional regulation in the development or maintenance of depression and anxiety [20,21].

Fewer studies have evaluated the relationship between emotional regulation and hostility than the relationship between emotional regulation and depression or anxiety. People with lower anger control can reasonably be assumed to display more aggressive behaviors [25]. A previous study demonstrated the relationship between emotional regulation and anger reactivity [26]. Hostile cognition is a major factor contributing to anger and aggressive behavior [27]. However, whether cognitive appraisal can attenuate the role of hostile cognition in depression has not been evaluated.

Depression and emotional regulation are considered risk factors for the development of addictive disorders [28]. Emotional regulation was reported to predict substance use disorder (specifically, alcohol use disorder [29]) and has been suggested to have a moderating role in addiction development [30]. IGD has been reported to be associated with depression, irritability, and anxiety [2,3,31]. Difficulties with emotional regulation are associated with these associated psychopathological symptoms [20,21]. Furthermore, poor emotional regulation might contribute to depression [20] that predicts IGD [15,32]. Moreover, excessive online gaming could have negative consequences that could result in stress for individuals with IGD. Appropriate emotional regulation mediates negative effects and psychological stress [22], whereas impaired emotional regulation might contribute to mood symptoms, such as depression and anxiety. Loton et al. revealed that the coping strategy had been reported to account for the association between video gaming addiction and depression [14]. It supported the claim that inappropriate emotional regulation might contribute to the association between psychopathological symptoms of IGD. However, the association between emotional regulation and these psychopathological symptoms has not been evaluated among subjects with IGD.

1.3. Study Hypothesis and Objectives

We hypothesized that emotional regulation, cognitive reappraisal, and suppression are associated with IGD, and that individuals with IGD practice less emotional regulation, use fewer reappraisal strategies, and tend to suppress emotions more than does the average person. Furthermore, the deficit in emotional regulation might be correlated with depression, hostility, and anxiety among subjects with IGD. Accordingly, this study evaluated the following: (1) cognitive reappraisal and expressive suppression among individuals with and without IGD, and (2) the associations among cognitive reappraisal, expressive suppression, depression, hostility, and anxiety among subjects with IGD.

2. Materials and Methods

2.1. Participants

Our participants, namely individuals with current IGD (the IGD group) and those with no history of IGD (the control group), were recruited through advertisements that demonstrated our recruitment critieria on campuses and bulletin board systems at universities in Taiwan between September 2012 and October 2013. Our recruitment criteria for the IGD group, which were based on an fMRI study for young adults with IGD, were as follows [32]: (1) aged 20–30 years with education of >9 years; (2) played Internet games for ≥4 h per day on weekdays and ≥8 h per day on weekends or for ≥40 h per week; and (3) had maintained an Internet gaming pattern for >2 years. The recruited participants spent most of their free time on Internet gaming. For participants fulfilling these criteria, a psychiatrist conducted an interview, during which the DSM-5 diagnostic criteria for IGD was used [1] in the interviewing room at laboratory. Participants that fulfilled the DSM-5 criteria of IGD were classified in the IGD group.

For every participant enrolled in the IGD group, a gender-, age- (within a range of 1 year), and education level-matched control participant was recruited according to the criteria that their nonessential Internet use was of <4 h per day in their daily life. The limitation on internet use was designed to prevent recruiting subjects with internet addiction in control group. Then, these participants also underwent a diagnostic interview with the psychiatrist based on DSM-5 criteria of IGD to confirm their recruitment in control group.

The diagnostic interview comprised two parts: (1) a diagnostic interview based on the Chinese version of the Mini-International Neuropsychiatric Interview (MINI) to reveal existing psychotic disorders, bipolar I disorder, and substance use disorders; and (2) a history-taking interview to determine psychotropic medication use, mental retardation, severe physical disorder, and brain injury. Individuals with psychotic disorders, bipolar I disorder, substance use disorders, psychotropic medication use, mental retardation, severe physical disorder, or brain injury were excluded. In total, 174 participants—87 in each group—were included after diagnostic interviewing and their informed consent was obtained. Then, study participants completed the assessment in this present study. This study was approved by the Institutional Review Board of Kaohsiung Medical University Hospital.

2.2. Measures

DSM-5 diagnostic criteria for IGD [1]. The DSM-5 IGD diagnostic criteria include comprises nine items: preoccupation, withdrawal, tolerance, unsuccessful attempts to control, loss or decrease of other interests, continued excessive use despite psychosocial problems, deceiving, escapism, and functional impairment [1]. We developed a semistructured interview for examining the DSM-5 criteria for IGD. Participants fulfilling $\geq$5 criteria were included in classified as the IGD group.

Chinese version of the MINI [33]. We conducted a diagnostic interview to rule out psychiatric disorders by using the modules of psychotic disorders, bipolar I disorder, and substance use disorders in the Chinese version of the MINI. Those with existing disorders were excluded from the study.

Emotional regulation questionnaire. The emotional regulation questionnaire (ERQ) is a 10-item scale designed to measure respondents' tendency to regulate their emotions in two ways: (1) cognitive reappraisal, assessed using a reappraisal scale (six items such as "When I want to feel less negative emotion (such as sadness or anger), I change what I'm thinking about"), and (2) expressive suppression, assessed using a suppression scale (four items such as "I control my emotions by not expressing them"). Respondents answer each item on a 7-point Likert-type scale from 1 (strongly disagree) to 7 (strongly agree). The alpha reliabilities were averaged 0.79 and 0.73 for the reappraisal and suppression scales, respectively. Test–retest reliability over 3 months was 0.69 for both scales in its original study [19]. There are several scales assessing emotional regulation. We utilized ERQ to assess the most important two strategy of emotional regulation because of its brief and convenient nature.

Depression, hostility, and anxiety were assessed by the Center for Epidemiological Studies' Depression Scale (CES-D) [34,35] Penn State Worry Questionnaire (PSWQ) [36] and the Buss–Durkee Hostility Inventory Chinese Version—Short Form (BDHIC-SF) [37]. Cronbach's alpha of CES-D, PSWQ, and BDHIC-SF in the present study were 0.92, 0.90, and 0.92, respectively. Higher score of CES-D, BDHIC-SF, and PSWQ indicates higher depression, hostility, and anxiety, respectively.

2.3. Statistical Analysis

We first evaluated the differences in cognitive reappraisal and expressive suppression between the IGD and control groups. Logistic regression was used to regress the diagnosis of IGD on the reappraisal and suppression while controlling for gender, age, and educational level. Then, linear regression was used to regress the depression on the cognitive reappraisal, and expressive suppression with control of gender, age, and educational level in both IGD and control group. The gender was set as female = 0 and male = 1 in the linear regression. The same method was used to evaluate the associations between reappraisal, suppression, and hostility or anxiety. $p < 0.05$ was considered significant in the analyses, all of which were performed using SPSS. The significant threshold of multiplicity was corrected using the Holm–Bonferroni methods. The Holm–Bonferroni method controls the familywise error rate (Type I errors) by adjusting the p value of the individual comparison [38].

3. Results

3.1. Gender, Age, and Education Levels

Eighty-seven people were recruited for each group. Their gender ($X^2 = 0$, $p = 1$), age ($t = 0.26$, $p = 0.80$), and education levels ($t = 1.15$, $p = 0.25$) did not differ significantly (Table 1).

Table 1. Age, educational level, emotional regulation, hostility, depression, and severity for the IGD and control groups.

Variables	IGD Diagnosis		X^2
	Yes (N = 87)	No (N = 87)	
	Mean ± SD	**Mean ± SD**	***t*-Test**
Gender			
Male	70 (80.5%)	70 (80.5%)	0.00
Female	17 (19.5%)	17 (19.5%)	
Age	23.29 ± 2.34	23.38 ± 2.40	0.26
Education level	15.93 ± 1.15	16.14 ± 1.22	1.15
Cognitive reappraisal [1]	31.09 ± 5.43	33.16 ± 4.87	−2.64 **
Expressive suppression [2]	19.22 ± 3.40	17.98 ± 3.74	2.292 *

* $p < 0.05$; ** $p < 0.01$; [1] Score of cognitive reappraisal subscale of *ERQ*; [2] Score of expressive suppression subscale of *ERQ*.

3.2. Emotional Regulation and IGD

The IGD group had significantly lower cognitive reappraisal strategies ($t = -2.64$, $p = 0.009$) and greater expressive suppression strategies ($t = 2.29$, $p = 0.02$) than did the control group (Table 1). Logistic regression (Table 2) revealed that cognitive reappraisal negatively predicts IGD (odds ratio; OR = 0.91; 95% CI = 0.85–0.97) and that expressive suppression positively predicts IGD (OR = 1.14; 95% CI = 1.04–1.25).

Table 2. Logistic regression to evaluate the predictive value of emotional regulation in IGD with control of gender, age, and educational level.

Variables	Wald	Exp(β)	95% CI
Among all subjects			
Gender	0.01	1.05	0.47–2.32
Age (year)	0.43	1.0550	0.91–1.22
Education level (year)	1.11	0.86	0.64–1.14
Cognitive reappraisal [1]	8.97 **	0.91	0.85–0.97
Expressive Suppression [2]	7.28 **	1.14	1.04–1.25

** $p < 0.01$; [1] Score of cognitive reappraisal subscale of *ERQ*; [2] Score of expressive suppression subscale of *ERQ*.

3.3. Within-Group Analysis for Emotional Regulation

Multiple linear regression analysis was used to test if the emotional regulation significantly predicted depression, anxiety, or hostility of subjects in IGD group (Table 3). The results indicated the model explained 19% of the variance in depression ($R^2 = 0.19$, $F_{(5,81)} = 3.74$). Cognitive reappraisal significant predicted depression ($B = -0.72$, $t = -3.66$, $p < 0.001$), as did expressive suppression ($B = 1.02$, $t = 3.24$, $p = 0.002$). Further, the model explained 18% of variance in anxiety ($R^2 = 0.18$, $F_{(5,81)} = 3.59$). Cognitive reappraisal significant predicted anxiety ($B = -0.69$, $t = -3.20$, $p = 0.002$), as did expressive suppression ($B = 0.91$, $t = 2.66$, $p = 0.01$). The model also explained 12% of variance in hostility ($R^2 = 0.12$, $F_{(5,81)} = 2.2$). Cognitive reappraisal significantly predicted hostility ($B = -0.75$, $t = -2.79$, $p = 0.007$), as did expressive suppression ($B = 1.09$, $t = 2.53$, $p = 0.01$). These results suggested that IGD subjects with lower cognitive reappraisal and higher expressive suppression had higher depression, anxiety, and hostility. We also provide the result in control group. It demonstrated the similar association between emotional regulation and depression, anxiety, and hostility in control group (Table 3).

Table 3. Multiple linear regression analysis for the predictive value of emotional regulation in depression, hostility, and CGI score among IGD group or control group.

Variables	IGD			Control		
	B	t	p	B	t	p
Depression [1]						
Gender	0.78	0.31	0.76	−1.76	−1.06	0.29
Age (year)	0.24	0.52	0.61	−0.24	−0.77	0.44
Education level (year)	−0.19	−0.20	0.84	0.21	0.34	0.73
Cognitive reappraisal [2]	−0.72	−3.66	<0.001	−0.73	−5.21	<0.001
Suppression [3]	1.02	3.24	0.002	1.06	6.01	<0.001
	$F_{(5,81)} = 3.74$	$R^2 = 0.19$		$F_{(5,81)} = 12.90$	$R^2 = 0.44$	
Anxiety [4]						
Gender	−1.86	−0.68	0.50	−2.75	−1.29	0.20
Age (year)	−0.79	−1.57	0.12	−0.72	−1.78	0.08
Education level (year)	1.39	1.34	0.19	1.58	1.98	0.05
Cognitive reappraisal	−0.69	−3.20	0.002	−0.96	−5.28	<0.001
Suppression	0.91	2.66	0.01	1.13	4.93	<0.001
	$F_{(5,81)} = 3.59$	$R^2 = 0.18$		$F_{(5,81)} = 11.71$	$R^2 = 0.42$	
Hostility [5]						
Gender	−1.19	−0.34	0.73	−3.61	−1.52	0.13
Age (year)	0.13	0.20	0.84	0.09	0.21	0.84
Education level (year)	−0.60	−0.46	0.65	0.15	0.17	0.87
Cognitive reappraisal	−0.75	−2.79	0.01	−0.72	−3.55	0.001
Suppression	1.09	2.53	0.01	1.45	5.69	<0.001
	$F_{(5,81)} = 2.20$	$R^2 = 0.12$		$F_{(5,81)} = 8.87$	$R^2 = 0.35$	

[1] Score of cognitive reappraisal subscale of *ERQ*; [2] Score of expressive suppression subscale of *ERQ*; [3] Score of Center for Epidemiological Studies' Depression Scale; [4] Score of PSWQ; [5] Score of the Buss–Durkee Hostility Inventory—Chinese Version—Short Form.

4. Discussion

People with poor emotional regulation often engage in maladaptive behavior to escape from their emotions, creating risks of a range of mood disorders and addictive disorders [39]. Thus, such people have been associated with various addictive disorders [29,30]. To our knowledge, no previous study has assessed emotional regulation among subjects with IGD. As expected, the present study demonstrated that subjects with IGD have lower cognitive reappraisal and higher expressive suppression. This result is similar to a previous report demonstrating lower cognitive reappraisal in gambling disorder [39]. Further, our study demonstrated that lower cognitive reappraisal and higher expressive suppression were associated with depression, anxiety, and hostility among subjects with IGD.

Our literature review suggested that those individuals experiencing depression or anxiety have ineffective emotional regulation and difficulties in processing negative emotions [20,21]. Cognitive reappraisal is a cognitively-oriented strategy for redefining emotional stimuli in unemotional terms or for reimagining depressive situations [40]. It comes early in the emotion-generative process and effectively decreases the experience of negative emotions [18]. By contrast, expressive suppression, coming later in the emotion-generative process, entails the inhibition of outward signs of inner feelings. Suppression is ineffective for down-regulating negative emotions, and people with a history of depression have been reported to spontaneously use this strategy [41]. Like these previous results, our results demonstrated that subjects with higher depression have lower cognitive reappraisal and higher expressive suppression among both subjects with IGD and controls.

People with IGD experience negative psychosocial consequences from excessive online gaming [42]. They also experience depression, anxiety, or irritation when they are prohibited from playing games online [1]. Thus, previous prospective study had suggested that internet gaming disorder or excessive online gaming [8,12] contributes to depression. They could reappraisal that this is a logical result of ceasing an excessive, self-gratifying behavior, and that the depression and restlessness could be avoided if they engaged in an alternative, appropriate activity such as exercise. However, without appropriate reappraisal, subjects with IGD could experience depression. Further, continuing to suppress negative emotions rather than reappraising them could leave these emotional difficulties unresolved. Thus, the lower cognitive reappraisal and higher suppression of subjects with IGD could partly account for their vulnerability to depression.

Although there is no report demonstrating the predictive effect of depression on internet gaming disorder, previous reports had suggested that depression predicted incidence of internet addiction [32]. Subjects with lower cognitive reappraisal that were habituated to use suppression could experience depression under stress [20,22]. Online gaming could provide a virtual world for people to escape from their negative emotions [43] and could buffer stress [44]. However, if the gaming time could not be well controlled, the repeatedly excessive gaming could result in further negative consequences among vulnerable subjects. It could create a vicious cycle and lead to repeated engagement in online gaming, resulting in increased addiction risk. Anyway, this claim should be further evaluated in prospective study.

Subjects with higher anxiety were more likely to pay attention to threat-related stimuli rather than neutral stimuli [45]. The continued attention to threat increases their cognitive and emotional response, contributing to anxiety symptoms. The manner in which information was processed in emotional regulation could determine anxiety severity [24]. The use of suppression as a regulatory mechanism and limited access to emotional regulation strategies, such as cognitive reappraisal, were associated with anxiety [46]. Thus, dysfunctional emotional regulation contributes to the development of anxiety disorder [24]. In this study, the anxiety of subjects with IGD is negatively associated with cognitive reappraisal and positively associated with expressive suppression.

In addition, reappraisal facilitates the adaptive processing of anger-inducing situations and contributes to anger regulation [47]. However, the suppression of anger could increase hostility under stress [48]. As expected, subjects with IGD habitually suppress emotions, or those unlikely to reappraise their negative cognition exhibited higher levels of hostility in this study. Moreover, suppression of hostility could increase sympathetic activity [49], as well as the risk of cardiovascular disorder [50]. Thus, emotional suppression and hostility of subjects with IGD might result in not only emotional difficulty but also cardiovascular risk.

Cognitive control ability is essential and contributes to emotional regulation, such as reappraisal [40]. Subjects with IGD had impaired cognitive control [51], similar to people with gambling disorders [52] and addictive disorder, such as cocaine use disorder [53]. The impaired cognitive control ability could associate with their impaired cognitive reappraisal in subjects with IGD. Further study is necessary to understand the neurocognitive mechanism of the impaired emotional regulation, such as cognitive control, among subjects with IGD.

4.1. Clinical Implication

The dysfunctional emotional regulation of subjects with IGD was associated with depression, anxiety, and hostility [32]. Emotional regulation should be well assessed and intervened in among young adults with IGD. Three key steps—emotional awareness, emotional regulation, and exchanging one emotion for another—help people to modify the state, the belief, and the behavior in response to emotion-eliciting events. These interventions for emotional regulation [23] have been recommended for the treatment of depression [20]. Evidence-based emotion management strategies, such as emotion-focused therapy [54], could be provided to young adults with IGD to promote cognitive reappraisal and attenuate expressive suppression strategies and responses. They must become aware that their negative emotions result from the negative consequences of gaming or from the conflicts in their lives. Alternative activities, physical exercise, and further psychological support should be offered to help relieve negative emotions. Furthermore, information and guidance on reappraisal should be provided so that positive thinking can replace negative thinking. This intervention to promote reappraisal and prevent suppression may attenuate their depression, anxiety, and hostility, and prevent the vicious cycle of IGD. However, these claims for the effects of emotional regulation therapy should be evaluated with future clinical research.

4.2. Limitations

This study has three limitations. First, emotional regulation was assessed using a questionnaire only and not through the investigation of real situations. Second, IGD was diagnosed only through diagnostic interviews with the participants, and supplementary information from family members or partners, which could have contributed to verifying the validity of the diagnoses, was not collected. Third, our cross-sectional research design could not confirm causal relationships between emotional regulation and IGD. Besides, the structure equation model had not been utilized to test hypothesized model because of unconfirmed causal relationship.

5. Conclusions

People with IGD practice less cognitive reappraisal and more suppression. In this study, people practicing less cognitive reappraisal and more suppression had more symptoms of depression, anxiety, and hostility, suggesting that impaired emotional regulation might exacerbate negative mood symptoms in people with IGD. Thus, emotional regulation should be effectively assessed when treating people with IGD. Furthermore, this group should be given interventions to promote cognitive reappraisal and attenuate expressive suppression in order to avoid a vicious cycle of negative emotions.

Acknowledgments: This study was supported by grants from the National Science Council (MOST105-2314-B-037-027-MY2), Kaohsiung Municipal Ta-Tung Hospital (kmtth-102-016; kmtth-103-018), and the Kaohsiung Medical University Hospital (KMUH103-3R62). These institutions had no role in the design, process, analysis, and production of the present study.

Author Contributions: Chih-Hung Ko conceived and designed the experiments; Tai-Ling Liu and Yun-Yu Chen performed the experiments; Yi-Chun Yeh and Peng-Wei Wang analyzed the data; Ju-Yu Yen wrote the paper.

Conflicts of Interest: The authors declare no conflict of interest.

References

1. American Psychiatric Association. *Diagnostic and Statistical Manual of Mental Disorders*, 5th ed.; American Psychiatric Association: Arlington, TX, USA, 2013.
2. Ko, C.H.; Liu, T.L.; Wang, P.W.; Chen, C.S.; Yen, C.F.; Yen, J.Y. The exacerbation of depression, hostility, and social anxiety in the course of Internet addiction among adolescents: A prospective study. *Compr. Psychiatry* **2014**, *55*, 1377–1384. [CrossRef] [PubMed]
3. Ko, C.H.; Yen, J.Y.; Yen, C.F.; Chen, C.S.; Chen, C.C. The association between Internet addiction and psychiatric disorder: A review of the literature. *Eur. Psychiatry* **2012**, *27*, 1–8. [CrossRef] [PubMed]
4. Benaiges, I.; Prat, G.; Adan, A. Neuropsychological aspects of dual diagnosis. *Curr. Drug Abuse Rev.* **2010**, *3*, 175–188. [CrossRef] [PubMed]
5. Wang, H.R.; Cho, H.; Kim, D.J. Prevalence and correlates of comorbid depression in a nonclinical online sample with DSM-5 internet gaming disorder. *J. Affect. Disord.* **2017**, *226*, 1–5. [CrossRef] [PubMed]
6. Kessler, R.C. The epidemiology of dual diagnosis. *Biol. Psychiatry* **2004**, *56*, 730–737. [CrossRef] [PubMed]
7. Mueser, K.T.; Drake, R.E.; Wallach, M.A. Dual diagnosis: A review of etiological theories. *Addict. Behav.* **1998**, *23*, 717–734. [CrossRef]
8. Hellstrom, C.; Nilsson, K.W.; Leppert, J.; Aslund, C. Effects of adolescent online gaming time and motives on depressive, musculoskeletal, and psychosomatic symptoms. *Upsala J. Med. Sci.* **2015**, *120*, 263–275. [CrossRef] [PubMed]
9. Wei, H.T.; Chen, M.H.; Huang, P.C.; Bai, Y.M. The association between online gaming, social phobia, and depression: An internet survey. *BMC Psychiatry* **2012**, *12*, 92. [CrossRef] [PubMed]

10. Yeh, Y.C.; Wang, P.W.; Huang, M.F.; Lin, P.C.; Chen, C.S.; Ko, C.H. The procrastination of Internet gaming disorder in young adults: The clinical severity. *Psychiatry Res.* **2017**, *254*, 258–262. [CrossRef] [PubMed]
11. Yen, J.Y.; Liu, T.L.; Wang, P.W.; Chen, C.S.; Yen, C.F.; Ko, C.H. Association between Internet gaming disorder and adult attention deficit and hyperactivity disorder and their correlates: Impulsivity and hostility. *Addict. Behav.* **2017**, *64*, 308–313. [CrossRef] [PubMed]
12. Gentile, D.A.; Choo, H.; Liau, A.; Sim, T.; Li, D.; Fung, D.; Khoo, A. Pathological video game use among youths: A two-year longitudinal study. *Pediatrics* **2011**, *127*, e319–e329. [CrossRef] [PubMed]
13. Ciarrochi, J.; Parker, P.; Sahdra, B.; Marshall, S.; Jackson, C.; Gloster, A.T.; Heaven, P. The development of compulsive internet use and mental health: A four-year study of adolescence. *Dev. Psychol.* **2016**, *52*, 272–283. [CrossRef] [PubMed]
14. Loton, D.; Borkoles, E.; Lubman, D.; Polman, R. Video game addiction, engagement and symptoms of stress, depression and anxiety: The mediating role of coping. *Int. J. Mental Health Addict.* **2016**, *14*, 14. [CrossRef]
15. Ko, C.H.; Yen, J.Y.; Chen, C.S.; Yeh, Y.C.; Yen, C.F. Predictive values of psychiatric symptoms for internet addiction in adolescents: A 2-year prospective study. *Arch. Pediatr. Adolesc. Med.* **2009**, *163*, 937–943. [CrossRef] [PubMed]
16. Gross, J.J. Antecedent- and response-focused emotion regulation: Divergent consequences for experience, expression, and physiology. *J. Personal. Soc. Psychol.* **1998**, *74*, 224–237. [CrossRef]
17. Sloan, E.; Hall, K.; Moulding, R.; Bryce, S.; Mildred, H.; Staiger, P.K. Emotion regulation as a transdiagnostic treatment construct across anxiety, depression, substance, eating and borderline personality disorders: A systematic review. *Clin. Psychol. Rev.* **2017**, *57*, 141–163. [CrossRef] [PubMed]
18. Gross, J.J. Emotion regulation: Affective, cognitive, and social consequences. *Psychophysiology* **2002**, *39*, 281–291. [CrossRef] [PubMed]
19. Gross, J.J.; John, O.P. Individual differences in two emotion regulation processes: Implications for affect, relationships, and well-being. *J. Personal. Soc. Psychol.* **2003**, *85*, 348–362. [CrossRef]
20. Compare, A.; Zarbo, C.; Shonin, E.; Van Gordon, W.; Marconi, C. Emotional Regulation and Depression: A Potential Mediator between Heart and Mind. *Cardiovasc. Psychiatry Neurol.* **2014**, *2014*, 324374. [CrossRef] [PubMed]
21. Amstadter, A. Emotion regulation and anxiety disorders. *J. Anxiety Disord.* **2008**, *22*, 211–221. [CrossRef] [PubMed]
22. Lynch, T.R.; Robins, C.J.; Morse, J.Q.; Krause, E.D. A mediational model relating affect intensity, emotion inhibition, and psychological distress. *Behav. Ther.* **2001**, *32*, 519–536. [CrossRef]
23. Mennin, D.S.; Fresco, D.M.; Ritter, M.; Heimberg, R.G. An Open Trial of Emotion Regulation Therapy for Generalized Anxiety Disorder and Cooccurring Depression. *Depress. Anxiety* **2015**, *32*, 614–623. [CrossRef] [PubMed]
24. Esbjorn, B.H.; Bender, P.K.; Reinholdt-Dunne, M.L.; Munck, L.A.; Ollendick, T.H. The development of anxiety disorders: Considering the contributions of attachment and emotion regulation. *Clin. Child Fam. Psychol. Rev.* **2012**, *15*, 129–143. [CrossRef] [PubMed]
25. Sullivan, T.N.; Helms, S.W.; Kliewer, W.; Goodman, K.L. Associations between Sadness and Anger Regulation Coping, Emotional Expression, and Physical and Relational Aggression among Urban Adolescents. *Soc. Dev.* **2010**, *19*, 30–51. [CrossRef] [PubMed]
26. Harrist, A.W.; Hubbs-Tait, L.; Topham, G.L.; Shriver, L.H.; Page, M.C. Emotion regulation is related to children's emotional and external eating. *J. Dev. Behav. Pediatr.* **2013**, *34*, 557–565. [CrossRef] [PubMed]
27. DeWall, C.N.; Twenge, J.M.; Gitter, S.A.; Baumeister, R.F. It's the thought that counts: The role of hostile cognition in shaping aggressive responses to social exclusion. *J. Personal. Soc. Psychol.* **2009**, *96*, 45–59. [CrossRef] [PubMed]
28. Nikmanesh, Z.; Kazemi, Y.; Khosravy, M. Study role of different dimensions of emotional self-regulation on addiction potential. *J. Fam. Reprod. Health* **2014**, *8*, 69–72.
29. Wilens, T.E.; Martelon, M.; Anderson, J.P.; Shelley-Abrahamson, R.; Biederman, J. Difficulties in emotional regulation and substance use disorders: A controlled family study of bipolar adolescents. *Drug Alcohol Depend.* **2013**, *132*, 114–121. [CrossRef] [PubMed]
30. Wills, T.A.; Pokhrel, P.; Morehouse, E.; Fenster, B. Behavioral and emotional regulation and adolescent substance use problems: A test of moderation effects in a dual-process model. *Psychol. Addict. Behav.* **2011**, *25*, 279–292. [CrossRef] [PubMed]

31. Yu, H.; Cho, J. Prevalence of Internet Gaming Disorder among Korean Adolescents and Associations with Non-psychotic Psychological Symptoms, and Physical Aggression. *Am. J. Health Behav.* **2016**, *40*, 705–716. [CrossRef] [PubMed]
32. Ko, C.H.; Hsieh, T.J.; Wang, P.W.; Lin, W.C.; Yen, C.F.; Chen, C.S.; Yen, J.Y. Altered gray matter density and disrupted functional connectivity of the amygdala in adults with Internet gaming disorder. *Prog. Neuropsychopharmacol. Biol. Psychiatry* **2015**, *57*, 185–192. [CrossRef] [PubMed]
33. Sheehan, D.V.; Lecrubier, Y.; Sheehan, K.H.; Amorim, P.; Janavs, J.; Weiller, E.; Herqueta, T.; Baker, R.; Dunbar, G.C. The Mini-International Neuropsychiatric Interview (M.I.N.I.): The development and validation of a structured diagnostic psychiatric interview for DSM-IV and ICD-10. *J. Clin.* **1998**, *59*, 22–33.
34. Chien, C.P.; Cheng, T.A. Depression in Taiwan: Epidemiological survey utilizing CES-D. *Seishin Shinkeigaku Zasshi* **1985**, *45*, 335–338.
35. Radloff, L.S. The CES-D Scale: A self-report depression scale for research in the general population. *Appl. Psychol. Meas.* **1977**, *1*, 16. [CrossRef]
36. Meyer, T.J.; Miller, M.L.; Metzger, R.L.; Borkovec, T.D. Development and validation of the Penn State Worry Questionnaire. *Behav. Res. Ther.* **1990**, *28*, 487–495. [CrossRef]
37. Lin, T.K.; Weng, C.Y.; Wang, W.C.; Chen, C.C.; Lin, I.M.; Lin, C.L. Hostility trait and vascular dilatory functions in healthy Taiwanese. *J. Behav. Med.* **2008**, *31*, 517–524. [CrossRef] [PubMed]
38. Aickin, M.; Gensler, H. Adjusting for multiple testing when reporting research results: The Bonferroni vs. Holm methods. *Am. J. Public Health* **1996**, *86*, 726–728. [CrossRef] [PubMed]
39. Williams, A.D.; Grisham, J.R.; Erskine, A.; Cassedy, E. Deficits in emotion regulation associated with pathological gambling. *Br. J. Clin. Psychol.* **2012**, *51*, 223–238. [CrossRef] [PubMed]
40. Joormann, J.; Gotlib, I.H. Emotion Regulation in Depression: Relation to Cognitive Inhibition. *Cogn. Emot.* **2010**, *24*, 281–298. [CrossRef] [PubMed]
41. Ehring, T.; Tuschen-Caffier, B.; Schnulle, J.; Fischer, S.; Gross, J.J. Emotion regulation and vulnerability to depression: Spontaneous versus instructed use of emotion suppression and reappraisal. *Emotion* **2010**, *10*, 563–572. [CrossRef] [PubMed]
42. Ko, C.H.; Yen, J.Y.; Chen, S.H.; Wang, P.W.; Chen, C.S.; Yen, C.F. Evaluation of the diagnostic criteria of Internet gaming disorder in the DSM-5 among young adults in Taiwan. *J. Psychiatr. Res.* **2014**, *3*, 103–110. [CrossRef] [PubMed]
43. Kazakova, S.; Cauberghe, V.; Pandelaere, M.; De Pelsmacker, P. Players' expertise and competition with others shape the satisfaction of competence needs, gaming gratifications, and contingent self-esteem in a gaming context. *Cyberpsychol. Behav. Soc. Netw.* **2014**, *17*, 26–32. [CrossRef] [PubMed]
44. Reinecke, L. Games and recovery: The use of video and computer games to recuperate from stress and strain. *J. Media Psychol. Theor. Methods Appl.* **2009**, *21*, 126–142. [CrossRef]
45. Bar-Haim, Y.; Lamy, D.; Pergamin, L.; Bakermans-Kranenburg, M.J.; van Ijzendoorn, M.H. Threat-related attentional bias in anxious and nonanxious individuals: A meta-analytic study. *Psychol. Bull.* **2007**, *133*, 1–24. [CrossRef] [PubMed]
46. Campbell-Sills, L.; Barlow, D.H.; Brown, T.A.; Hofmann, S.G. Acceptability and suppression of negative emotion in anxiety and mood disorders. *Emotion* **2006**, *6*, 587–595. [CrossRef] [PubMed]
47. Denson, T.F.; Moulds, M.L.; Grisham, J.R. The effects of analytical rumination, reappraisal, and distraction on anger experience. *Behav. Ther.* **2012**, *43*, 355–364. [CrossRef] [PubMed]
48. Quartana, P.J.; Burns, J.W. Painful consequences of anger suppression. *Emotion* **2007**, *7*, 400–414. [CrossRef] [PubMed]
49. Giese-Davis, J.; Conrad, A.; Nouriani, B.; Spiegel, D. Exploring Emotion-Regulation and Autonomic Physiology in Metastatic Breast Cancer Patients: Repression, Suppression, and Restraint of Hostility. *Personal. Individ. Differ.* **2008**, *44*, 226–237. [CrossRef] [PubMed]
50. Vogele, C.; Jarvis, A.; Cheeseman, K. Anger suppression, reactivity, and hypertension risk: Gender makes a difference. *Ann. Behav. Med.* **1997**, *19*, 61–69. [CrossRef] [PubMed]
51. Cai, C.; Yuan, K.; Yin, J.; Feng, D.; Bi, Y.; Li, Y.; Yu, D.; Jin, C.; Qin, W.; Tian, J. Striatum morphometry is associated with cognitive control deficits and symptom severity in internet gaming disorder. *Brain Imaging Behav.* **2016**, *10*, 12–20. [CrossRef] [PubMed]

52. Moccia, L.; Pettorruso, M.; De Crescenzo, F.; De Risio, L.; di Nuzzo, L.; Martinotti, G.; Bifone, A.; Janiri, L.; Di Nicola, M. Neural correlates of cognitive control in gambling disorder: A systematic review of fMRI studies. *Neurosci. Biobehav. Rev.* **2017**, *78*, 104–116. [CrossRef] [PubMed]
53. Ide, J.S.; Hu, S.; Zhang, S.; Yu, A.J.; Li, C.S. Impaired Bayesian learning for cognitive control in cocaine dependence. *Drug Alcohol. Depend.* **2015**, *151*, 220–227. [CrossRef] [PubMed]
54. Greenberg, L. *Emotion-Focused Therapy, Coaching Client to Work through Their Feelings*; American Psychiatric Association: Washington, DC, USA, 2002.

International Journal of
Environmental Research and Public Health

MDPI

Article

Risk Factors for Internet Gaming Disorder: Psychological Factors and Internet Gaming Characteristics

Mi Jung Rho [1,2,†], Hyeseon Lee [3,†], Taek-Ho Lee [3], Hyun Cho [4,5], DongJin Jung [5,6], Dai-Jin Kim [5,6,*,‡] and In Young Choi [1,2,*,‡]

1. Department of Medical Informatics, College of Medicine, The Catholic University of Korea, Seoul 06591, Korea; rhomijung@gmail.com
2. Catholic Institute for Healthcare Management and Graduate School of Healthcare Management and Policy, The Catholic University of Korea, Seoul 06591, Korea
3. Department of Industrial & Management Engineering, Pohang University of Science and Technology, Pohang 37673, Korea; hyelee@postech.ac.kr (H.L.); dlxorgh2@postech.ac.kr (T.-H.L.)
4. Department of Psychology, Korea University, Seoul 02841, Korea; sonap1@hanmail.net
5. Addiction Research Institute, Department of Psychiatry, Seoul St. Mary's Hospital, College of Medicine, The Catholic University of Korea, Seoul 06591, Korea; forever0851@naver.com
6. Department of Psychiatry, Seoul St. Mary's Hospital, College of Medicine, The Catholic University of Korea, Seoul 06591, Korea

* Correspondence: kdj922@catholic.ac.kr (D.-J.K.); iychoi@catholic.ac.kr (I.Y.C.); Tel.: +82-2-2258-6086 (D.-J.K.); +82-2-2258-7870 (I.Y.C.)

† Both authors contributed equally to this work.

‡ Both corresponding authors contributed equally to this work.

Received: 18 October 2017; Accepted: 19 December 2017; Published: 27 December 2017

Abstract: *Background*: Understanding the risk factors associated with Internet gaming disorder (IGD) is important to predict and diagnose the condition. The purpose of this study is to identify risk factors that predict IGD based on psychological factors and Internet gaming characteristics; *Methods*: Online surveys were conducted between 26 November and 26 December 2014. There were 3568 Korean Internet game users among a total of 5003 respondents. We identified 481 IGD gamers and 3087 normal Internet gamers, based on Diagnostic and Statistical Manual for Mental Disorders (DSM-5) criteria. Logistic regression analysis was applied to identify significant risk factors for IGD; *Results*: The following eight risk factors were found to be significantly associated with IGD: functional and dysfunctional impulsivity (odds ratio: 1.138), belief self-control (1.034), anxiety (1.086), pursuit of desired appetitive goals (1.105), money spent on gaming (1.005), weekday game time (1.081), offline community meeting attendance (2.060), and game community membership (1.393; $p < 0.05$ for all eight risk factors); *Conclusions*: These risk factors allow for the prediction and diagnosis of IGD. In the future, these risk factors could also be used to inform clinical services for IGD diagnosis and treatment.

Keywords: internet gaming disorder; Dickman Impulsivity Inventory-Short Version (DII); Brief Self-Control Scale (BSCS); Symptom Checklist-90-Revised (SCL-90-R); Behavioral Inhibition System/Behavioral Activation System (BIS/BAS); Diagnostic and Statistical Manual for Mental Disorders (DSM-5)

1. Introduction

Since Internet games became widespread in the 2000s [1], Internet game usage has experienced rapid growth among both youth and adults. According to a report by the Entertainment Software

Association (ESA) [2], 155 million Americans play video games, of which 42% play video games regularly. In 2015 alone, American game consumers spent more than US$22.41 billion on game content, hardware, and accessories [2]. Worldwide Internet game usage and gaming money has been rapidly increasing. As a result, Internet Gaming Disorder (IGD) has become a major social problem and important research topic. The World Health Organization (WHO) has proposed a new category named "Gaming Disorder" for the 11th Revision of the International Classification of Diseases (ICD-11) [3]. The ability to predict, diagnose, and manage IGD in advance is critical to the prevention of IGD. To do that, the risk factors associated with IGD need to be better understood.

Firstly, the psychological factors associated with IGD need to be understood. IGD can be considered a behavioral addiction [4–8] and has been found to be related to a number of psychological and health problems, including depression, social anxiety, fatigue, loneliness, negative self-esteem, and impulsivity [9–12]. IGD co-occurs with various psychiatric conditions and can lead to a range of negative outcomes. For example, IGD can cause social problems such as lower academic achievement [10,11,13–17]. In addition, IGD shares many similarities with other addictions, such as substance use disorder [18].

Secondly, the Internet gaming characteristics associated with IGD need to be better understood. Research in this area has increased in both quantity and quality. In order to predict, diagnose, and manage IGD, researchers have attempted to identify the causes and negative consequences of excessive gaming as well as risk factors of IGD. Some research, however, has only focused on psychological factors [16,19] or Internet gaming characteristics, such as the level of Internet usage, money spent on gaming, and type of game device [20]. A comprehensive approach based on both psychological factors and Internet gaming characteristics is needed to better understand IGD. Accordingly, the purpose of the present study was to identify risk factors that predict IGD, based on psychological factors and Internet gaming characteristics.

2. Materials and Methods

2.1. Participants

Online surveys were conducted using an existing survey company online panel (Hankook Research, Inc., Seoul, South Korea between 26 November and 26 December 2014. Online informed consent was obtained from all participants, prior to their participation. The online panel consisted of native Koreans aged 20–49 years, from metropolitan areas in South Korea. Among a total of 5003 respondents, 3881 Internet game users were identified. The final sample size comprised 3568 Internet game users, which did not include missing values.

Using the DSM-5 criteria to diagnose IGD is controversial [3,21]. Some researchers have attempted to overcome this confusion [21,22]. Because there are very few criteria for IGD in the DSM-5, it was used to evaluate IGD in the present study. In addition, DSM-5 criteria were validated from discussions among an expert group. Based on DSM-5 criteria, Internet game users with scores above 5 were evaluated as the IGD group [20,23]. Thus, in the final sample, there were 481 IGD gamers (13.48% of the sample) and 3087 normal Internet gamers (86.52%).

2.2. Measures and Procedure

Twenty independent variables were measured as potential risk factors for IGD. Independent variables consisted of participants' demographic characteristics, Internet gaming characteristics, and psychological variables.

In the case of Internet gaming characteristics, there were very few related studies, so related variables could not be chosen from the existing literature. Internet gaming characteristics were therefore derived from the Internet Addiction Survey 2013 conducted by the Korea National Information Society Agency [24]. The specific items were identified from discussions among an expert group. The expert group consisted of psychiatrists, psychologists, and data scientists of medical informatics who had more than 3 years' experience in addiction. Psychological variables were derived from previous research and were again collected from discussions among an expert group. The reliability of all variables was determined by the expert group.

Participants' demographic characteristics consisted of five factors: gender, age, job, score on the Alcohol Use Disorder Identification Test (AUDIT-K) [25], and score on the Fagerström Test for Nicotine Dependence (FTND) [26]. Participant data were divided into three groups based on AUDIT-K and FTND scores, as summarized in Appendix A (Table A1). The AUDIT-K is a ten-item questionnaire developed for male and female drinkers at a high risk of alcohol abuse. It is composed of three scores that are dependent on gender: male (0–9: normal drinker, 10–19: mild-to-moderate drinker, and $\geq$20: heavy drinker) and female (0–5: normal drinker, 6–9: mild-to-moderate drinker, and $\geq$10: heavy drinker). The FTND test is a six-item questionnaire designed to measure nicotine dependence. It is composed of three scores (0–3: low, 4–6: intermediate, and $\geq$7: high).

Seven Internet gaming characteristics were also measured: money spent on gaming, weekday game time, weekend game time, game device, game venue, offline game club attendance, and game club membership status.

Finally, eight psychological variables were measured, including the Dickman Impulsivity Inventory-Short Version (DII), Brief Self-Control Scale (BSCS) [27], Symptom Checklist-90-Revised (SCL-90-R) [28] and Behavioral Inhibition System/Behavioral Activation System (BIS/BAS) [29,30], as summarized in Table 1. The DII measures the personality trait of impulsivity [31]. The response options for each item are true (1) or false (0). The BSCS assesses dispositional self-control [27]. Each BSCS item is rated on a five-point scale, from 1 (strongly disagree) to 5 (strongly agree). The SCL-90-R consists of 90 items and assesses psychological distress [32,33]. Each of the items is rated on a five-point scale of distress, from 0 (no distress) to 4 (extreme distress). In the present study, 23 items from the SCL-90-R were adapted to evaluate depression (13 items) and anxiety (10 items).

The behavioral inhibition system (BIS) and a behavioral activation system (BAS) underlie behavior and affect [30]. The BIS scale estimates reactions to anticipated punishment and the BAS scale assesses positive responses to rewards. The BAS Drive scale estimates the pursuit of desired goals. The BAS Fun Seeking scale examines the tendency to seek and impulsively engage in potentially rewarding activities [30,34]. The BIS/BAS consists of a four-point scale, from 1 (not at all) to 4 (strongly agree). The total scores of the BIS/BAS scales range from zero to 80.

Questions related to cost, gaming time, and age were self-reported questions and free text which yielded a continuous value. The rest were multiple choice questions based on predefined categories.

Table 1. Description of Internet gaming characteristics and psychological factors.

	Variables	# of Items
Demographic characteristics	Gender, age, job	3
	AUDIT-K	10
	FTND	6
Internet gaming characteristics	Money spent on gaming (/month), Weekday game time (/day), Weekend game time (/day), Game device, Game venue, Offline game club attendance, Game club membership status	7
Psychological factors	DII	12
	BSCS	13
	SCL depression	13
	SCL anxiety	10
	BIS	7
	BAS reward responsiveness	5
	BAS drive	4
	BAS fun seeking	4

AUDIT-K: Alcohol Use Disorder Identification Test; FTND: Fagerström Test for Nicotine Dependence; DII: Dickman Impulsivity Inventory-Short Version; BSCS: Brief Self-Control Scale; SCL: Symptom Checklist; BIS: behavioral inhibition system; BAS: behavioral activation system.

2.3. Statistical Analysis

Out of 3881 respondents who identified as Internet game users, cases with missing responses were excluded, and all analyses were performed for 3568 respondents. We conducted *t*-tests and Chi-square tests to compare the IGD group to the control group in terms of demographic and Internet gaming characteristics. Multiple regression analysis was used to identify risk factors for the IGD group. The data were analyzed using SAS 9.4 (SAS Institute, Inc., Cary, NC, USA).

2.4. Ethics

The study procedures were carried out in accordance with the Declaration of Helsinki and were approved by the Institutional Review Board of Catholic University (IRB number: KC15EISI0103). Participants' data were de-identified.

3. Results

Out of 3568 participants, 481 (13.5%) were included in the IGD group and 3087 (86.5%) were included in the control group. The respondents' age ranged from 20 to 49, and 1559 (43.7%) were between the ages of 30 and 39. There were 2036 (57.1%) males and 1532 (42.9%) females (Table 2). Office workers and professional technicians comprised 67.8% of the sample, and college students comprised 15%. There were similar proportions of individuals in each group with a marital status of either single or married. There were no significant differences in demographic characteristics between the two groups; however, males were more likely to be in the IGD group than females. For income level, there were more people from the control group in the middle class, while low and high income classes showed slightly higher dependence.

Table 2. Participants' characteristics.

Variables		Total	IGD Group	Control Group	Chi-Square (p-Value)
		n (%)	n (%)	n (%)	
Gender	Mal	2036 (57.1)	290 (60.3)	1746 (56.6)	2.36 (0.124)
	Female	1532 (42.9)	191 (39.7)	1341 (43.4)	
Age	20–29 years	1259 (35.3)	170 (35.3)	1089 (35.3)	0.43 (0.808)
	30–39 years	1559 (43.7)	215 (44.7)	1344 (43.5)	
	40–49 years	750 (21.0)	96 (20.0)	654 (21.2)	
Education	High school graduate or less	1053 (29.5)	134 (27.9)	919 (29.8)	0.76 (0.683)
	College graduate	2130 (59.7)	295 (61.3)	1835 (59.4)	
	Graduate school	385 (10.8)	52 (10.8)	333 (10.8)	
Job	Office worker, et al. [1]	2418 (67.8)	334 (69.4)	2084 (67.5)	0.86 (0.835)
	Student	535 (15.0)	67 (13.9)	468 (15.2)	
	etc.	217 (6.1)	27 (5.6)	190 (6.2)	
	Unemployed/housewife	398 (11.2)	53 (11.0)	345 (11.2)	
Marital status	Couple [2]	1867 (52.3)	241 (50.1)	1626 (52.7)	1.10 (0.294)
	Single [2]	1701 (47.7)	240 (49.9)	1461 (47.3)	
Income level	Low	1567 (43.9)	219 (45.5)	1348 (43.7)	3.52 (0.172)
	Middle	1557 (43.6)	193 (40.1)	1364 (44.2)	
	High	444 (12.4)	69 (14.3)	375 (12.2)	
Total		3568 (100)	481 (13.5)	3087 (86.5)	

[1] Office worker et al.: office worker, administrative position, service industry, professional technician and production employee; [2] Single: never married, divorced, separated or widowed, Couple: married or living with a partner; IGD: Internet gaming disorder.

Differences in Internet gaming characteristics for all variables except game playing were significant between the IGD group and the control group (Table 3). Among all participants, 57.8% of the IGD group had a game club membership, while 35.4% of the control group had a game club membership. The respondents having a game club membership showed higher IGD than the control group (57.8% vs. 35.4%). Most of the Internet game users played at home, and there was no difference between the IGD group and control group (76.1% vs. 77.2%). In the case of playing in a gaming Internet cafe, the IGD group was much higher than the control group (17.5% vs. 10.2%). For game devices, the IGD group used a personal computer (PC) more than the control group (53.0% vs. 37.9%). For game partners, those playing with friends or online partners showed higher dependence than the control group (29.1% vs. 21.5%). Both the IGD and the control group had perceptions of addictiveness. For offline club game attendance, the IGD group's attendance was much higher than that of the control group (57.3% vs. 26.6%). For the onset of Internet games, 48.3% of respondents began in middle or high school. The IGD group spent more time gaming than the control group (2.85 vs. 1.97 h on weekdays and 4.12 vs. 2.92 h on weekends, respectively) and spent more money on gaming than the control group ($31.4 vs. $11.0, respectively).

Table 3. Internet gaming characteristics.

Variables		Total	IGD Group	Normal Group	Test Statistics (p-Value)
		n (%)	n (%)	n (%)	
Game club membership	No	2198 (61.6)	203 (42.2)	1995 (64.6)	88.45 (<0.001)
	Yes	1370 (38.4)	278 (57.8)	1092 (35.4)	
Game playing	Playing one game intensively	2098 (58.8)	302 (62.8)	1796 (58.2)	3.65 (0.056)
	Playing various games	1470 (41.2)	179 (37.2)	1291 (41.8)	
Game venue	Home	2748 (77.0)	366 (76.1)	2382 (77.2)	32.85 (<0.001)
	Gaming Internet cafe	400 (11.2)	84 (17.5)	316 (10.2)	
	Others [1]	420 (11.8)	31 (6.4)	389 (12.6)	
Game device	PC	1424 (39.9)	255 (53.0)	1169 (37.9)	42.39 (<0.001)
	Console	63 (1.8)	11 (2.3)	52 (1.7)	
	Mobile device [2]	2080 (58.3)	215 (44.7)	1865 (60.4)	
Game partner	Alone	2593 (72.7)	321 (66.7)	2272 (73.6)	14.07 (0.003)
	Family	169 (4.7)	20 (4.2)	149 (4.8)	
	Friends	280 (7.9)	52 (10.8)	228 (7.4)	
	Online partner	526 (14.7)	88 (18.3)	438 (14.2)	
Self-perceptions of addictiveness	Not at all	203 (5.7)	19 (4.0)	184 (6.0)	85.69 (<0.001)
	A little	1077 (30.2)	90 (18.7)	987 (32.0)	
	Much	1979 (55.5)	285 (59.3)	1694 (54.9)	
	Very much	309 (8.7)	87 (18.1)	222 (7.2)	
Offline game club attendance	Not attend	2469 (69.2)	205 (42.6)	2264 (73.3)	185.63 (<0.001)
	Sometimes	1032 (28.9)	256 (53.2)	776 (25.1)	
	Very often	67 (1.9)	20 (4.2)	47 (1.5)	
Onset of Internet game	Under middle school	842 (23.6)	122 (25.4)	720 (23.3)	11.42 (0.009)
	Middle or high school	882 (24.72)	142 (29.5)	740 (24.0)	
	After graduating high school	1056 (29.6)	131 (27.2)	925 (30.0)	
	30s or 40s	788 (22.09)	86 (17.9)	702 (22.7)	
Gaming time/day	Weekdays	2.09	2.85	1.97	7.21 (<0.001)
	Weekends and holidays	3.08	4.12	2.92	7.19 (<0.001)
	Maximum	4.07	5.93	3.78	6.30 (<0.001)
Money spent on gaming/month		$13.76	$31.36	$11.02	8.23 (<0.001)

Time unit: hours, the exchange rate for Korean won to the U.S. dollar is 1100.00 won (September 2016), t-statistics for continuous variable, and chi-square value for categorical variables. [1] Others: School, play station room, the outside including bus, substation; [2] Mobile device: Smartphone and Tablet.

Risk Factors Predicting IGD

The results of the multivariate logistic regression analysis are shown in Table 4. Firstly, demographic characteristics were shown not to be risk factors. All variables included in the logistic regression model do not show muliticollinearity. Secondly, with regard to Internet gaming characteristics, money spent on gaming (OR = 1.005), weekday game time (OR = 1.081), offline game club attendance (OR = 2.060), and game club membership status (OR = 1.393) were significant behavioral factors predicting IGD. Thirdly, DII (OR = 1.138), BSCS (OR = 1.034), anxiety (OR = 1.086), and BAS-Drive (OR = 1.105) were significant psychological predictors of IGD. Those who had one unit score higher for DII were 1.138 times more likely to be dependent. Additionally, with one unit score higher for the BSCS, Anxiety, and BAS-Drive factors, the probability of dependence increased by 1.034, 1.086, and 1.105 times, respectively. One of measures for model performance in a general linear model, Nagelkerke's R^2 is 0.3012 which showed it was a better model than others [35].

Table 4. Risk factors predicting IGD.

Variables		Estimate (SE)		*p*-Value	OR 95% CI
Intercept		−5.452 (0.602)			-
Gender		0.023 (0.139)		0.869	1.023 (0.779–1.344)
Age		0.138	0.090	0.125	1.148 (0.962–1.37)
Job	Office worker, et al. [1]	−0.167	0.193	0.387	0.846 (0.579–1.236)
	Student	−0.017	0.248	0.944	0.983 (0.604–1.599)
	etc.	−0.260	0.291	0.373	0.771 (0.436–1.365)
AUDIT	Normal drinker	−0.136	0.163	0.404	0.873 (0.634–1.201)
	Mild-to-moderate drinker	−0.313	0.166	0.059	0.731 (0.528–1.012)
	Heavy drinker	0.171	0.161	0.289	1.186 (0.865–1.626)
FTND	Low	−0.201	0.171	0.240	0.818 (0.585–1.144)
	Intermediate	0.177	0.195	0.362	1.194 (0.815–1.748)
	High	0.358	0.307	0.243	1.431 (0.784–2.611)
Money spent on gaming ***		0.005	0.002	<0.001 ***	1.005 (1.002–1.008)
Weekday game time ***		0.078	0.027	0.003 ***	1.081 (1.026–1.139)
Weekend game time		0.004	0.019	0.843	1.004 (0.968–1.041)
Game device	PC	0.160	0.132	0.224	1.174 (0.907–1.519)
	Console	0.239	0.413	0.563	1.270 (0.565–2.853)
Game venue	Home	0.324	0.217	0.135	1.383 (0.905–2.114)
	Gaming Internet cafe	0.282	0.270	0.296	1.326 (0.781–2.25)
Offline game club attendance ***		0.723	0.130	<0.001 ***	2.060 (1.597–2.658)
Game club membership status **		0.332	0.125	0.008 **	1.393 (1.09–1.78)
DII ***		0.129	0.022	<0.001 ***	1.138 (1.09–1.188)
BSCS **		0.034	0.012	0.006 **	1.034 (1.01–1.059)
SCL Depression		−0.008	0.012	0.496	0.992 (0.968–1.016)
SCL Anxiety ***		0.082	0.015	<0.001 ***	1.086 (1.054–1.118)
BIS		−0.031	0.025	0.215	0.969 (0.923–1.018)
BAS reward responsiveness		0.005	0.039	0.908	1.005 (0.93–1.085)
BAS drive *		0.100	0.041	0.015 *	1.105 (1.02–1.198)
BAS fun seeking		−0.063	0.042	0.133	0.939 (0.865–1.019)

SE: standard error; * $p < 0.1$, ** $p < 0.05$, *** $p < 0.01$; [1] Office worker, et al.: office worker, administrative position, service industry, professional technician and Production employee.

4. Discussion

We identified risk factors predicting IGD, specifically examining psychological and Internet gaming characteristics as potential risk factors. Based on the results of the present study, we draw the following conclusions.

Firstly, examination of psychological factors yielded meaningful results. Users with IGD perceived themselves as being obsessed with Internet gaming (Table 3) and that they had difficulty quitting the game. Thus, social support may be needed to prevent IGD and support treatment efforts. Psychological risk factors related to IGD included impulsivity, low self-control, anxiety, and pursuit of desired appetitive goals. Past research has shown that IGD has similarities to other addictions, such as gambling and substance use disorder [18,36,37]. In particular, impulsivity and self-control are important psychological factors affecting addiction [38,39]. Impulsivity has been reported as a risk factor in addiction to social networking sites or smartphones [29,40] and lack of self-control is related to addictions such as substance use disorder [27] and Internet use [41–43]. Anxiety may be relevant psychopathological symptom to detect Internet, smartphone, and video game addiction [44–46]. Lastly, BAS Drive was a risk factor associated with IGD. The level of BAS Drive represents the tendency to pursue desired goals actively [34] and has been shown to be one of the personality factors associated with smartphone addiction [29]. This shows that to predict and diagnose IGD, research on the associated psychological risk factors is needed.

Secondly, a number of Internet gaming characteristics were significant in predicting IGD. Users with IGD mainly played games at home. In the case of playing games in a gaming Internet cafe, the proportion of individuals with dependence was higher than normal (17.5% vs. 10.2%). Game users mainly played using a PC compared to a mobile device (53.0% vs. 37.9%) since high specification desktops were needed. However, the control group played games more frequently using mobile devices compared to PCs. With regard to the onset of Internet gaming, 48.3% of respondents began in middle or high school. Users with IGD tended to start playing Internet games at a relatively early age. This finding suggests that early initiation of game playing may be a risk factor for IGD. Accordingly, diverse approaches are needed early on to prevent adolescent and adult IGD. Offline game club attendance and game club membership status were also risk factors for IGD. Users with IGD were more likely to be game club members than those in the control group (57.3% vs. 26.6%) and were more likely to attend offline clubs, with 73.3% of the control group having never attended offline game meetings. On average, users with IGD were thought to have no social relationships and to be more isolated. However, they did attend offline game clubs and have game club memberships. There were some social users with IGD.

Additional risk factors of IGD were money spent on gaming and weekday game time. In the case of game time, Internet game users spent an average of 2.09 h on weekends playing games. Users with IGD spent more time than normal gamers playing Internet games (2.85 vs. 1.97 h on weekdays and 4.12 vs. 2.92 h on weekends). According to the Ministry of Science ICT and Future Planning (MSIP) report, Korean gamers spent an average of 1.1 h on weekends playing games. Users with Internet over-dependence spent 0.3 more hours playing on weekends than normal users (1.4 vs. 1.1 h) [47]. The results from our study show that game time was higher in our sample. The MSIP report focused on individuals ranging in age from early childhood (3 years) to 59 years whereas our results came from a sample of adults between the ages of 20 and 49. This higher game time suggests that IGD is more serious in adults. Users with IGD spent more money on gaming than the control group ($31.4 vs. $11.0). Previous research has reported that spending extreme amounts of time and money is a predictor of IGD [20,48–52]. Lo et al., (2005) found that the amount of time spent playing online games is directly correlated with levels of social anxiety [50]. Rau et al., (2006) proposed that many game players have difficulty in controlling game time [49]. Accordingly, approaches are needed for IGD among adults and controlling time and money is important to preventing and managing IGD.

5. Conclusions

This study had several limitations. Data on Internet gaming characteristics were self-reported, including money spent on gaming, weekday game time, and weekend game time. If technology could be developed, such as the Smartphone Overdependence Management System (SOMS), to collect time or money data automatically [53], future research may provide more accurate and realistic results. We collected data using an online survey. This was based on an existing online panel from a survey company. Online panel respondents were native Koreans aged 20–49 years, from metropolitan areas in South Korea. Using an online survey based on an existing panel was a useful way to collect a large amount of data; however, this may have resulted in some recruitment bias. Future research should involve data collected from the entire Korea area. The present study was designed to be cross-sectional because it is difficult to collect time-series data from Internet gamers. As a result, our findings are limited in their ability to reflect fast-changing Internet gaming trends. Future research could incorporate time-series data from longitudinal studies. Future research could also involve a more accurate diagnosis of IGD based on a clinical interview. The results showed that depression has no significant relationship with IGD. This is contrary to other published studies that have found video or internet game addiction to be related to depression [10,45,54]. We used the SCL-90-L to evaluate depression; however, there are many other scales to measure depression, such as the 21-item Depression Anxiety Stress Scale (DASS-21) [55] and the Hopkins Symptom Checklist (HSCL) [56]. Future studies should evaluate depression using other measures. This study targeted respondents aged between 20 and 49, of which 1559 (43.7%) were between the ages of 30 and 39. Therefore, the reported results could be influenced by demographic characteristics.

Despite these limitations, the present study yielded a valuable contribution to our understanding of risk factors for IGD by using a comprehensive approach based on psychological factors and Internet gaming characteristics. These findings can be used to develop clinical services for the diagnosis and treatment of IGD.

Acknowledgments: This research was supported by the Brain Research Program through the National Research Foundation of Korea (NRF) funded by the Ministry of Science, ICT & Future Planning (NRF-2014M3C7A1062893). In addition, this research by Hyeseon Lee partially was supported by the Basic Science Research Program through the National Research Foundation of Korea (NRF-2017R1A2B4002944).

Author Contributions: All authors participated in the study concept and design. Hyeseon Lee and Taek-Ho Lee performed the statistical analysis. Mi Jung Rho performed interpretation of the data and drafted the manuscript. Hyun Cho and DongJin Jung participated in collecting the data. In Young Choi and Dai-Jin Kim participated in the study supervision.

Conflicts of Interest: The authors declare no conflict of interest.

Appendix A

Table A1. Criteria and score in the AUDIT and the FTND tests.

AUDIT Test			FTND Test	
Category	Male	Female		
	Score		Category	Score
Normal drinker	≤9	≤5	Low risk	≤3
Mild-to-moderate drinker	10~19	6~9	Intermediate risk	4~6
Heavy drinker	≥20	≥10	High risk	≥7

References

1. Kuss, D.J. Internet gaming addiction: Current perspectives. *Psychol. Res. Behav. Manag.* **2013**, *6*, 125–137. [CrossRef] [PubMed]
2. Entertainment software association (ESA). *2015 Essential Facts about the Computer and Video Game Industry*; ESA: Washington, DC, USA, 2015.
3. Aarseth, E.; Bean, A.M.; Boonen, H.; Colder Carras, M.; Coulson, M.; Das, D.; Deleuze, J.; Dunkels, E.; Edman, J.; Ferguson, C.J. Scholars' open debate paper on the world health organization ICD-11 gaming disorder proposal. *J. Behav. Addict.* **2017**, *6*, 267–270. [CrossRef] [PubMed]
4. Choi, S.-W.; Kim, H.; Kim, G.-Y.; Jeon, Y.; Park, S.; Lee, J.-Y.; Jung, H.; Sohn, B.; Choi, J.-S.; Kim, D.-J. Similarities and differences among internet gaming disorder, gambling disorder and alcohol use disorder: A focus on impulsivity and compulsivity. *J. Behav. Addict.* **2014**, *3*, 246–253. [CrossRef] [PubMed]
5. Na, E.; Lee, H.; Choi, I.; Kim, D.J. Comorbidity of internet gaming disorder and alcohol use disorder: A focus on clinical characteristics and gaming patterns. *Am. J. Addict.* **2017**, *26*, 326–334. [CrossRef] [PubMed]
6. Cho, H.; Kwon, M.; Choi, J.-H.; Lee, S.-K.; Choi, J.S.; Choi, S.-W.; Kim, D.-J. Development of the internet addiction scale based on the internet gaming disorder criteria suggested in DSM-5. *Addict. Behav.* **2014**, *39*, 1361–1366. [CrossRef] [PubMed]

7. Demetrovics, Z.; Urbán, R.; Nagygyörgy, K.; Farkas, J.; Griffiths, M.D.; Pápay, O.; Kökönyei, G.; Felvinczi, K.; Oláh, A. The development of the problematic online gaming questionnaire (POGQ). *PLoS ONE* **2012**, *7*, e36417. [CrossRef] [PubMed]
8. Petry, N.M.; O'brien, C.P. Internet gaming disorder and the DSM-5. *Addiction* **2013**, *108*, 1186–1187. [CrossRef] [PubMed]
9. Männikkö, N.; Billieux, J.; Kääriäinen, M. Problematic digital gaming behavior and its relation to the psychological, social and physical health of finnish adolescents and young adults. *J. Behav. Addict.* **2015**, *4*, 281–288. [CrossRef] [PubMed]
10. Brunborg, G.S.; Mentzoni, R.A.; Frøyland, L.R. Is video gaming, or video game addiction, associated with depression, academic achievement, heavy episodic drinking, or conduct problems? *J. Behav. Addict.* **2014**, *3*, 27–32. [CrossRef] [PubMed]
11. Van Rooij, A.J.; Kuss, D.J.; Griffiths, M.D.; Shorter, G.W.; Schoenmakers, T.M.; Van de Mheen, D. The (co-) occurrence of problematic video gaming, substance use, and psychosocial problems in adolescents. *J. Behav. Addict.* **2014**, *3*, 157–165. [CrossRef] [PubMed]
12. Park, J.H.; Han, D.H.; Kim, B.-N.; Cheong, J.H.; Lee, Y.-S. Correlations among social anxiety, self-esteem, impulsivity, and game genre in patients with problematic online game playing. *Psychiatry Investig.* **2016**, *13*, 297–304. [CrossRef] [PubMed]
13. Kardefelt-Winther, D. A conceptual and methodological critique of internet addiction research: Towards a model of compensatory internet use. *Comput. Hum. Behav.* **2014**, *31*, 351–354. [CrossRef]
14. Lemmens, J.S.; Valkenburg, P.M.; Peter, J. Psychosocial causes and consequences of pathological gaming. *Comput. Hum. Behav.* **2011**, *27*, 144–152. [CrossRef]
15. Liu, M.; Peng, W. Cognitive and psychological predictors of the negative outcomes associated with playing mmogs (massively multiplayer online games). *Comput. Hum. Behav.* **2009**, *25*, 1306–1311. [CrossRef]
16. Caplan, S.E. Relations among loneliness, social anxiety, and problematic internet use. *Cyberpsychol. Behav.* **2006**, *10*, 234–242. [CrossRef] [PubMed]
17. Kuss, D.J.; Griffiths, M.D. Internet gaming addiction: A systematic review of empirical research. *Int. J. Ment. Health Addict.* **2012**, *10*, 278–296. [CrossRef]
18. Kardefelt-Winther, D. A critical account of DSM-5 criteria for internet gaming disorder. *Addict. Res. Theory* **2015**, *23*, 93–98. [CrossRef]
19. Hyun, G.J.; Han, D.H.; Lee, Y.S.; Kang, K.D.; Yoo, S.K.; Chung, U.-S.; Renshaw, P.F. Risk factors associated with online game addiction: A hierarchical model. *Comput. Hum. Behav.* **2015**, *48*, 706–713. [CrossRef]
20. Rho, M.J.; Jeong, J.-E.; Chun, J.-W.; Cho, H.; Jung, D.J.; Choi, I.Y.; Kim, D.-J. Predictors and patterns of problematic internet game use using a decision tree model. *J. Behav. Addict.* **2016**, *5*, 500–509. [CrossRef] [PubMed]
21. Kuss, D.J.; Griffiths, M.D.; Pontes, H.M. Chaos and confusion in DSM-5 diagnosis of internet gaming disorder: Issues, concerns, and recommendations for clarity in the field. *J. Behav. Addict.* **2017**, *6*, 103–109. [CrossRef] [PubMed]
22. Kuss, D.J.; Griffiths, M.D.; Pontes, H.M. DSM-5 diagnosis of internet gaming disorder: Some ways forward in overcoming issues and concerns in the gaming studies field: Response to the commentaries. *J. Behav. Addict.* **2017**, *6*, 133–141. [CrossRef] [PubMed]
23. Petry, N.M.; Rehbein, F.; Gentile, D.A.; Lemmens, J.S.; Rumpf, H.J.; Mößle, T.; Bischof, G.; Tao, R.; Fung, D.S.; Borges, G. An international consensus for assessing internet gaming disorder using the new DSM-5 approach. *Addiction* **2014**, *109*, 1399–1406. [CrossRef] [PubMed]
24. Agency, N.I.A. *Internet Addiction Survey*; NIA: Daegu Metropolitan City, South Korea, 2013.
25. Lee, B.; Lee, C.; Lee, P.; Choi, M.; Namkoong, K. Development of korean version of alcohol use disorders identification test (AUDIT-K): Its reliability and validity. *J. Korean Acad. Addict. Psychiatry* **2000**, *4*, 83–92.
26. Fagerstrom, K.-O.; Schneider, N.G. Measuring nicotine dependence: A review of the fagerstrom tolerance questionnaire. *J. Behav. Med.* **1989**, *12*, 159–182. [CrossRef] [PubMed]
27. Tangney, J.P.; Baumeister, R.F.; Boone, A.L. High self-control predicts good adjustment, less pathology, better grades, and interpersonal success. *J. Personal.* **2004**, *72*, 271–324. [CrossRef]
28. Derogatis, L. *Manual for the Symptom Checklist 90 Revised (SCL-90-R)*; The Johns Hopkins University School of Medicine: Baltimore, MD, USA, 1986.

29. Kim, Y.; Jeong, J.-E.; Cho, H.; Jung, D.-J.; Kwak, M.; Rho, M.J.; Yu, H.; Kim, D.-J.; Choi, I.Y. Personality factors predicting smartphone addiction predisposition: Behavioral inhibition and activation systems, impulsivity, and self-control. *PLoS ONE* **2016**, *11*, e0159788. [CrossRef] [PubMed]
30. Carver, C.S.; White, T.L. Behavioral inhibition, behavioral activation, and affective responses to impending reward and punishment: The BIS/BAS scales. *J. Personal. Soc. Psychol.* **1994**, *67*, 319–333. [CrossRef]
31. Dickman, S.J. Functional and dysfunctional impulsivity: Personality and cognitive correlates. *J. Personal. Soc. Psychol.* **1990**, *58*, 95–102. [CrossRef]
32. Franke, G. *SCL-90-R: Die Symptom-Check-Liste von Derogatis-Deutsche Version*; Beltz Test Gesellschaft Google Scholar: Göttingen, Germany, 1995.
33. Derogatis, L.R.; Cleary, P.A. Factorial invariance across gender for the primary symptom dimensions of the SCL-90. *Br. J. Soc. Clin. Psychol.* **1977**, *16*, 347–356. [CrossRef] [PubMed]
34. Gray, J.A.; McNaughton, N. *The Neuropsychology of Anxiety: An Enquiry into the Function of the Septo-Hippocampal System*; Oxford University Press: Oxford, UK, 2003.
35. Nagelkerke, N.J.D. A note on a general definition of the coefficient of determination. *Biometrika* **1991**, *78*, 691–692. [CrossRef]
36. Kaptsis, D.; King, D.L.; Delfabbro, P.H.; Gradisar, M. Withdrawal symptoms in internet gaming disorder: A systematic review. *Clin. Psychol. Rev.* **2016**, *43*, 58–66. [CrossRef] [PubMed]
37. Yen, J.Y.; Ko, C.H.; Yen, C.F.; Chen, S.H.; Chung, W.L.; Chen, C.C. Psychiatric symptoms in adolescents with internet addiction: Comparison with substance use. *Psychiatry Clin. Neurosci.* **2008**, *62*, 9–16. [CrossRef] [PubMed]
38. Reynolds, B.; Ortengren, A.; Richards, J.B.; de Wit, H. Dimensions of impulsive behavior: Personality and behavioral measures. *Personal. Individ. Differ.* **2006**, *40*, 305–315. [CrossRef]
39. Baumeister, R.F. Ego depletion and self-regulation failure: A resource model of self-control. *Alcohol. Clin. Exp. Res.* **2003**, *27*, 281–284. [CrossRef] [PubMed]
40. Wu, A.M.; Cheung, V.I.; Ku, L.; Hung, E.P. Psychological risk factors of addiction to social networking sites among chinese smartphone users. *J. Behav. Addict.* **2013**, *2*, 160–166. [CrossRef] [PubMed]
41. Mei, S.; Yau, Y.H.; Chai, J.; Guo, J.; Potenza, M.N. Problematic internet use, well-being, self-esteem and self-control: Data from a high-school survey in china. *Addict. Behav.* **2016**, *61*, 74–79. [CrossRef] [PubMed]
42. LaRose, R.; Lin, C.A.; Eastin, M.S. Unregulated internet usage: Addiction, habit, or deficient self-regulation? *Media Psychol.* **2003**, *5*, 225–253. [CrossRef]
43. Park, J.-A.; Park, M.-H.; Shin, J.-H.; Li, B.; Rolfe, D.T.; Yoo, J.-Y.; Dittmore, S.W. Effect of sports participation on internet addiction mediated by self-control: A case of korean adolescents. *Kasetsart J. Soc. Sci.* **2016**, *37*, 164–169. [CrossRef]
44. Tonioni, F.; D'Alessandris, L.; Lai, C.; Martinelli, D.; Corvino, S.; Vasale, M.; Fanella, F.; Aceto, P.; Bria, P. Internet addiction: Hours spent online, behaviors and psychological symptoms. *Gen. Hosp. Psychiatry* **2012**, *34*, 80–87. [CrossRef] [PubMed]
45. Loton, D.; Borkoles, E.; Lubman, D.; Polman, R. Video game addiction, engagement and symptoms of stress, depression and anxiety: The mediating role of coping. *Int. J. Mental Health Addict.* **2016**, *14*, 565–578. [CrossRef]
46. Elhai, J.D.; Dvorak, R.D.; Levine, J.C.; Hall, B.J. Problematic smartphone use: A conceptual overview and systematic review of relations with anxiety and depression psychopathology. *J. Affect. Disord.* **2017**, *207*, 251–259. [CrossRef] [PubMed]
47. The Ministry of Science and ICT. *(The) 2015 Survey on Internet Overdependence*; The Ministry of Science and ICT: Gwacheon-si, Korea, 2016.
48. Allison, S.E.; von Wahlde, L.; Shockley, T.; Gabbard, G.O. The development of the self in the era of the internet and role-playing fantasy games. *Am. J. Psychiatry* **2006**, *163*, 381–385. [CrossRef] [PubMed]
49. Rau, P.-L.P.; Peng, S.-Y.; Yang, C.-C. Time distortion for expert and novice online game players. *Cyberpsychol. Behav.* **2006**, *9*, 396–403. [CrossRef] [PubMed]
50. Lo, S.-K.; Wang, C.-C.; Fang, W. Physical interpersonal relationships and social anxiety among online game players. *Cyberpsychol. Behav.* **2005**, *8*, 15–20. [CrossRef] [PubMed]
51. Wood, R.T.; Griffiths, M.D.; Parke, A. Experiences of time loss among videogame players: An empirical study. *Cyberpsychol. Behav.* **2007**, *10*, 38–44. [CrossRef] [PubMed]

52. Wood, R.T.; Griffiths, M.D. Time loss whilst playing video games: Is there a relationship to addictive behaviours? *Int. J. Ment. Health Addict.* **2007**, *5*, 141–149. [CrossRef]
53. Lee, S.-J.; Rho, M.J.; Yook, I.H.; Park, S.-H.; Jang, K.-S.; Park, B.-J.; Lee, O.; Lee, D.K.; Kim, D.-J.; Choi, I.Y. Design, development and implementation of a smartphone overdependence management system for the self-control of smart devices. *Appl. Sci.* **2016**, *6*, 440. [CrossRef]
54. Kim, D.J.; Kim, K.; Lee, H.-W.; Hong, J.-P.; Cho, M.J.; Fava, M.; Mischoulon, D.; Heo, J.-Y.; Jeon, H.J. Internet game addiction, depression, and escape from negative emotions in adulthood: A nationwide community sample of korea. *J. Nerv. Ment. Dis.* **2017**, *205*, 568–573. [CrossRef] [PubMed]
55. Lovibond, P.F.; Lovibond, S.H. The structure of negative emotional states: Comparison of the depression anxiety stress scales (DASS) with the beck depression and anxiety inventories. *Behav. Res. Ther.* **1995**, *33*, 335–343. [CrossRef]
56. Derogatis, L.R.; Lipman, R.S.; Rickels, K.; Uhlenhuth, E.H.; Covi, L. The hopkins symptom checklist (HSCL): A self-report symptom inventory. *Syst. Res. Behav. Sci.* **1974**, *19*, 1–15. [CrossRef]

International Journal of
Environmental Research and Public Health

MDPI

Article

Analysis of Gambling in the Media Related to Screens: Immersion as a Predictor of Excessive Use?

Jean-Jacques Rémond * and Lucia Romo *

Psychology Department, Laboratory EA CLIPSYD 4430, Paris Nanterre University, 92000 Nanterre, France
* Correspondance: jack-storm@orange.fr (J.-J.R.); romodesprez@gmail.com (L.R.)

Received: 12 November 2017; Accepted: 27 December 2017; Published: 2 January 2018

Abstract: This study investigates the intricacies between the player interface proposed by the screens, (in particular on smartphone applications or in video games) and gambling. Recent research indicates connections between "immersion" and excessive screen practice. We want to understand the causal-effects between online gambling and the "immersion" variable and understand their relationship and its contingencies. This article empirically investigates whether and how it is possible to observe immersion with its sub-dimensions in gambling on different screens. The objective of this study was to analyze: (1) the costs and benefits associated with gambling practice on screens (2) the link between gambling practice and screen practice (video game, Internet, mobile screen); (3) to observe the propensity to immersion for individuals practicing gambling on screens; and (4) to examine the comorbidities and cognitive factors associated with the practice of gambling on screen. A total of 432 adults (212 men, 220 women), recruited from Ile-de-France (France), responded to a battery of questionnaires. Our study suggests that immersion variables make it possible to understand the cognitive participation of individuals towards screens in general, the practice of gambling on screens and the excessive practice of screens.

Keywords: gambling; video-game addiction; screen addiction; immersion; problematic Internet use; comorbidity; cognitive distortion

1. Introduction

This innovative study is an extension of previous studies based on gamblers commitment to screens on slot machine operated electronically [1–4]. The notion of "gambler" for its part, is understood very broadly and includes any person who would bet on chance [5].

In this study, we consider the practice of the lottery, bingo, sports betting, card games, dice games and electronic slot machines. These practices may include the virtual format.

1.1. Gambling

Through the literature, it is easily recognized that these practices have a high participation rate in Western cultures [6–8]. The excessive practice of gambling can generate problems that affect all spheres of a person's life. The negative consequences can be significant and include financial debts, bankruptcy, family disputes and dissolution, criminal behavior and suicidal acts [9–13]. Gambling addiction has recently been reclassified in the DSM-5 [14] in the category of "substance-related addictions and behavioral addictions". As such, excessive gambling and gambling practice are recognized [15]. Gambling practice exists on a continuum ranging from casual, recreational, problematic and excessive gambling.

Epidemiological research on problematic gambling practice has revealed predominant risk factors, including age, whereby young people are more prone to these risks; living in urban areas than rural areas; socially and economically disadvantaged and with easy access to gambling [16].

Some authors [17–20] have argued that well-established risk factors are age, gender, cognitive distortions (misperceptions, illusion of control), sensory characteristics, schedules, comorbidities (personality disorder), AD/HD, substance use, depression, anxiety and illegal acts. The difficulty in regulating emotions is also a risk factor. Authors have found that low self-control, emotional dysregulation, a need to change mood, fill an existential or emotional void (lack of social interaction, relieve boredom), need for approval and avoid difficulties, increased the severity of gambling problems [21–23].

The distinction in gender composition for the practice of gambling has been particularly studied in the literature [24–28]. The practice of gambling has traditionally been regarded as an essentially masculine activity, but studies have shown a significant prevalence of women [25,26,29]. Men are more likely to play for excitement or sensation seeking, while for women, gambling can be related to emotional regulation. Women also begin gambling practice at a later age than men and generally have a faster rate of gambling than men [30]. The distinction between the reasons for gambling practice between men and women may reflect differential comorbidities between the gender.

Concerning comorbidities, a number of studies have revealed that pathological gamblers are more likely to experience problems of psychiatric comorbidity [9,31,32], among adults and teenagers [33–38]. The most common comorbidities found are depression, anxiety, alcoholism and obsessive-compulsive behavior. Studies also report that compulsive gamblers are more likely to report mood disorders, attention deficit disorders with or without hyperactivity [39–44]. Clarke found that pathological gamblers were more depressed than gamblers with no gambling problems [45]. This is reflected in recent studies [36]. Pathological gamblers may also experience cognitive distortions, such as denial, superstition, overconfidence, sense of control, etc. [46,47]. Irrational beliefs would be reinforced when the player reaches a high level of immersion while playing on the Internet or on electronic machines [48–53].

Even before practicing gambling on modern screens through the Internet, the first interactive mediation with gambling involved electronic slot machines. The first studies on immersion and gambling were based on the interactive format of these machines. They show lower rates of engagement of the general population, in contrast to traditional gambling, but studies show that they are more closely associated with problematic use [48]. Studies describe a gambling trajectory, going as far as problematic practice, more accelerated for players playing on an electronic slot machine (on average one year), in contrast to traditional gambling players (on average three and a half years) [54]. These gambling practices on the screens, within casinos, are also more strongly related to psychological distress than other forms of gambling [55]. This acceleration of short-term commitment to gambling by these media can be explained by the immersive and continuous audiovisual experience that can stimulate the pursuit of the game phases [48,56]. These devices also use a faster play rate compared to other traditional forms of gambling. These variables can be found in gambling formats on the Internet via modern screens.

Individual differences in the tendency to be immersed in gambling via electronic machines can contribute to the development of behaviors towards gambling. Individuals in a state of excessive immersion may no longer be able to perceive external stimuli outside the game [57], future appointments or physiological evidence (such as the need for food or to urinate). Just like the traditional gambling, players report relieving chronic stress or negative emotions [44,57]. If positive reinforcement (earning money) is the motivation, the most common among players, negative reinforcement (avoidance) is a significant predictor of gambling and its severity [58]. An avoidance copying style revolves around leaving or avoiding the stressful situation. Typically, this includes distraction via specific tasks or social engagement. Based on the pathways model, individuals who are emotionally vulnerable endorse an escape pattern of gambling where the primary motivation for gambling is avoidance of negative emotions and situations. Although few empirical studies have observed the "immersion" variable on gambling's electronic machines, Schull argues that for some players, this mental state of avoidance may be the consequence of gaming behavior on electronic machines [57].

1.2. Gambling on Screens

When the mobile game is discussed in research, it was often included under the aegis of "Internet games" [6,59–63], regardless of the different platforms and user behavior. It is essential to distinguish the behavior of a gamer of Internet with elements of gambling according to whether a gamer practices this activity on a fixed screen, or on a mobile screen, or on software dedicated to gambling, or on platforms Video games integrating these gambling forms. Studies have already shown that mobile gambling is associated with a high risk of problem gambling [64].

Gambling software or applications on mobile monitors and screens such as smartphones offer greater interactivity [65,66]. The advent of gambling practice on a smartphone increased the practice of betting and lottery playing [67]. Players now have instant access to their favorite game type. To understand the immersion variable of the gambling player on the mobile screen, it is necessary to understand the playful modality of the media proposed for the player on this screen, as well as the playful mode of the related technology. Numerous studies have examined these two modalities [68–71].

Gambling on mobile phones includes several ways to access gambling. This can be through an application, an optimized site, a game on the phone, or written messages. The status of these activities as gambling remains uncertain in a regulatory and legislative framework. Gainsbury et al. [59] give an online gambling taxonomy that separates different activities depending on whether the payment is mandatory or optional, depending on luck or skill, the platform on which the game is played and the proposed theme in connection with gambling. In Gainsbury's taxonomy, the term "online gambling" refers to "gambling on the Internet" where the notions of virtual gambling and gambling are integrated into a gaming format of video games, so the gambler on the Internet will spend the money via the game for a reward and the appearance will be likened to a gambling, but also a wider range of activities such as multiplayer casino video games, practice games, competitions or tournaments based on gambling (for example, poker).

The study also investigate if any model of research on video games on mobile screens can't get closer to models on gambling when it has concealed a financial incentive and chance. The visual, sound and hedonic appeal would thus could thus to influence the player. Immersion theories largely contribute to explain an individual's commitment to the screen [72–75].

1.3. Immersion and Gambling

If we consider studying gambling-related immersion, then we have to consider the notion of the player's commitment and desire to play the game on the screen. Immersion can be defined as "the physical experience of being immersed (...) the sensation of being surrounded by a completely different reality, (...), the whole of our perceptive apparatus" [76]. Immersion also comes from the Latin term "immergo" meaning to be damaged in the sense of a complete absorption that can harm the individual.

The immersion uses the flow principles based on the Csikszentmihalyi theory [77,78]. "Flow is the term given to a constellation of subjective experiences reported by people engaged in fulfilling activities" [72]. The flow experience is multifaceted. First, it involves the attenuation of the different attentional processes; Csikszentmihalyi argues for a definition that would be the fusion of action and sensations, a decrease in the perception of time and a distortion of consciousness [77]. Csikszentmihalyi refers, on several occasions, to flow as a state representing "the highest level of well-being" [77]. However, this assumption is based on the assumption that the activity from which the immersion is derived would be positive or without consequences for excessive use. If gambling activities on screens with their fast pace of play can provide an immersive experience, gamers may unconsciously pursue gambling activities until this results in significant financial losses. If the individual is directed toward a specific goal, evaluative thinking is mitigated during the immersion phase, individuals may be less able to discern whether the accumulated financial losses exceed the expected limits. This conception of immersion related to gambling on screens or on electronic gambling machines has been the subject of few studies. Dixon et al. [79] have shown that the introduction of "concealed losses into victories" in modern media such as electronic machines or screens favors a model of empowering the experience [57]. As it was, the losses still permitted to acquire points of experience or even multiple losses made it possible to obtain all the same credits to replay. Indeed, this function creates a continuous stream of small victories, as opposed to large occasional gains separated by long phases of losses. In the study by Dixon et al. [79], network participants reported an increase in sensation related to immersion. There is a limit to the assessment and interpretation of immersion degree on electronic slot machines. Indeed, few studies assess immersion in relation to excessive practice and propose a holistic model of excessive behaviors related to the cognitive processes in interaction during the phases of viewing and practice of gambling.

For this study, we distinguish the processes of suggestibility to immersion, cognitive processes related to the quality of immersion, as demonstrated by Psotka and Davison [80]. Thus, self-awareness, concentration, attention, self-control are variables belonging to the domain of suggestibility processes of immersion, while sensory skills, the search for new experimentation, persistence of the object, distractibility are processes related to the quality of immersion. A dissociative state could help players cope with stressors and provide relief from aversive states like anxiety, depression or boredom. A number of studies [81–86] have shown a link between pathological gambling and the dissociation experiment during gambling phases. Cartmill et al. studied the relationship between anxiety state, dissociation states and gambling problems depending on the severity of gambling problems [81].

Recent studies support the idea that basic attention is disrupted in problem gamblers. Similarly, studies have shown that gambling-related stimuli create interference in the processing of information. In these studies, pathological gamblers had to perform the Stroop test [87,88]. In the Diskin and Hodgins studies, the pathological gambler group responded significantly more slowly to the objectives presented than the casual gamblers. This difference in latency was interpreted as an attentional narrowing resulting from the fact that pathological gamblers overvalued the visual stimuli of the game to the detriment of peripheral events.

In addition to attentional bias and visual processing deficits, studies have shown that players have distractibility [89]. Distractibility can interfere with the processing of affective information, which reduces the empathic analysis of pain stimuli (Kam, Xu, and Handy [90]). It is possible to consider that players with mood disorders pre-existing in the practice of gambling may seek, in the gambling on the screens, a form of escape. These perceptual deficits can cause functional impairment. The elderly present a high risk of idiopathy and report a higher frequency of distractibility. This suggests that distractibility can lead to physiological consequences via a reduction in the processing of environmental information [91].

This study focuses on the immersion process during the phases of cast projection of the screen and gambling stimuli on the screens can be modulated in order to adapt prevention messages during the phases of game on the screens to the gambling players [92], in particular for the new media related to screens that encourage a first gambling experience [93].

2. Objectives and Hypothesis

Our study focuses on the link between gambling and the new medias related to screens, with "immersion" as a moderator variable. By examining the costs and degree of gambling practice on the screens, we may presume, on the one hand, the importance of evaluating the practices on the screens as a priority, and on the other hand, to consider evaluating the immersion variable and its sub-dimensions. In order to show the interest of studying the immersion variable, we have to consider that the immersion degree is higher for people at risk of excessive use of screens and even higher for people practicing excessively screens and having a problematic practice of gambling.

Hypothesis

Following the examination of previous researches in the field, five hypotheses have been formulated:

Hypothesis 1 (H1). *The economic costs of purchasing applications or gambling games on the screens are more important in gamblers with problem gambling than non-problem gamblers.*

Hypothesis 2 (H2). *Individuals with problem gambling practice have higher scores on scales assessing excessive gambling practices.*

Hypothesis 3 (H3). *The propensity to immersion scores are higher: for people with gambling addiction [1], than for those with no problem gambling practice; And for people with gambling addiction and high scores on scales evaluating practice on screens [2].*

Hypothesis 4 (H4). *Individuals with problem gambling and high scores on scales assessing screen practices, unlike those with only a problem gambling practice, have higher scores on scales assessing anxiety and depression; On the scale assessing impulsivity; To scale assessing emotional regulation; On the scale assessing cognitive distortions.*

Hypothesis 5 (H5). *Individuals with very high scores on the immersion scale (QPI) and low scores at the ICJP scale had significantly lower scores at scales assessing screen practices.*

3. Methods

3.1. Data Collection and Sampling

According to the criteria of the Helsinki Declaration on Consent to Research and Clinical Practice, a paper and online protocol was proposed to a French adult population on a voluntary and anonymous basis. The paper version of the protocol has been distributed mainly to students from universities in Île-de-France region of France (bachelor degree) (n = 222; 51.27%) and to those coming from various socio-professional categories (n = 211; 48.73%). The online questionnaire was distributed to various social media. No financial incentives were offered. The online questionnaire was filled out on different types of screens (computer, tablet, and smartphone). Four hundred thirty-three questionnaires were collected. The final sample (n = 432) comprised 212 men (49%) and 220 women (51%), with an average age of 21.94 years (SD = 5.51).

3.2. Scales Related to Games of Gambling

The Canadian Problem Gambling Index (ICJE) is a measure of the severity of gambling problems over the past 6 months. This instrument, developed in English and French, has nine questions rated 0–3 (never, sometimes, most of the time, always). A comparison between the CPGI and another questionnaire commonly used in prevalence studies of pathological gambling, the South Oaks Gambling Screen (SOGS, Lesieur and Blume, [94]) reveals that the two instruments generate identical rates (Ladouceur, Jacques, Chevalier, Sévigny, and Hamel, [95]).

The Gambling Related Cognition Scale (GRCS) is a translated and validated tool of the Gambling Related Cognitions Scale (GRCS). The ECJ has a good psychometric quality and has 23 items on a seven-point Lickert scale that identify a variety of game-related cognitions. It identifies the gambler's expectations of the game, his control illusions, the predictive power he believes to have on the game, his perception of his ability to resist a desire to play, and the interpretations gained through motivating its continuation of the game.

3.3. Immersion

The Questionnaire on the Propensity to Immersion (QPI) (Bouchard et al. [96,97]). The French QPI scale, initially named ITQ, consists of 18 items, on a seven-point Lickert scale, and is a self-questionnaire measuring the propensity for immersion. With a Cronbach α of 0.88, the scale proposes to evaluate the total propensity to immersion and contain four variables (focus, implication, play, and emotion).

3.4. Scales Assessing Screen Practices

The Problematic Internet Use Questionnaire (PIUQ) scale has been chosen to evaluate the excessive use of the Internet, validated in French by Kern and Acier [98]). The version used consists of twelve items, on a six-point Lickert scale, and four dimensions (self-control, negative consequences, psychological weaning and Internet concerns). The Cronbach alpha of the overall score is between 0.87 and 0.91 according to the studies of Demrovics, Szeredi and Razsa, [99].

The Smartphone Addiction Scale (SAS) of Kwon et al. [100] was chosen to evaluate smartphone practices. The internal consistency and concurrent validity were analyzed with a Cronbach alpha of 0.97. This scale consists of 34 items on a six-point Lickert scale and was constructed from the Y-scale (scale assessing Internet addiction), a visual analog scale and the diagnostic criteria for abuse and of the psychoactive substance dependence of DSM-IV-TR. A more recent Belgian-French version (Smartphone Addiction Scale Short version or SAS-SV, Psychological Sciences Research Institute, Louvain-la-Neuve, Belgium) (Lopez-Fernandez [101])) exists for adolescents with ten items with a Cronbach alpha of 0.90.

The Video Game Addiction Test was selected to assess the excessive use of video games (Van Rooij et al. [102]). The scale consists of 14 items, on a five-point Lickert scale, where five variables (loss of control, preoccupation, withdrawal symptoms, adjustment/mood modification, conflict) can be visualized. It has good internal consistency with a 0.93 Cronbach alpha.

Additional items were added to the initial protocol to glimpse the different modes of screen-related practices, as well as the financial costs associated with screen practices.

3.5. Impulsiveness

"Impulsivity" variable was measured using the UPPS Impulsive Behavior Scale (UPPS, Whiteside, Lynam, Miller and Reynolds, [103]). This evaluation questionnaire was validated in the French version by Van der Linden et al. [104]. The French scale includes 20 items on a four-point Lickert scale and has good psychometric qualities with a Cronbach α of 0.70 to 0.84 for the internal consistency of the different subscales.

3.6. Symptomatology

Data on anxiety and depression were collected using Hospital Anxiety and Depression (HAD), a self-rated scale for anxiety-depressive disorders in non-psychiatric populations. The HAD (Zigmond and Snaith, [105]) contains 14 items, divided into two factors (depression: seven items, and anxiety: seven items). The internal consistency varies from 0.68 to 0.93 (mean 0.83) for anxiety, and from 0.67 to 0.90 (mean 0.82) for depression. This scale shows good sensitivity, and specificity for identifying anxiety-depressive disorders.

3.7. Emotion

The Emotional Regulation Questionnaire (ERQ) scale was developed by Gross and John [106]). It consists of ten items on a seven-point Lickert scale with two variables (cognitive re-evaluation and cognitive repression). Internal validity has a Cronbach alpha of 0.79 for re-evaluation and 0.73 for cognitive repression.

4. Results

Correlations analyses were made between the combinations of variables considered in this study. In order to test the aforementioned hypotheses, statistical analyzes were carried out using the Statistica software (v.13.2, StatSoft society, Tulsa, OK, USA). Descriptive statistics helped to identify the characteristics of the population in our sample. An analysis of variance and correlations (Chi2 and Student's t-tests) were performed. Finally, a multiple regression models and a principal component analysis (PCA) were proposed.

4.1. Descriptive Statistics

The average individual in our sample is a French male ($M = 1.5$), aged 22 ($M = 21.9$), with no children ($M = 0.2$), who has at least a bachelor level, and is predominantly student. On average, they spend between

2 and 4 h on the computer, 1–2 h on the tablets, 1–2 h on game consoles and 2–4 h on smartphones. In our sample, 54.04% reported feeling dependent on the computer, 6.7% on tablets, 4.6% on game consoles and 58.66% on smartphones. The costs and wagers according to the gambling practice on the screens were analyzed (Table 1). We also estimated excessive media practices related to screens (Table 2).

Table 1. Costs and prevalence of gambling.

Cost Buying by Material	Total (n = 432)		ICJP NPb (n = 397) (A)		ICJP Ar (n = 19) (B)		CPGI Path (n = 8) (C)		T p A vs. B	T p A vs. C	T p B vs. C
	M (€)	SD (€)									
(Total sample, Mean)	n = 56, 12.93%		n = 37, 9.32%		n = 5, 26.32%		n = 1, 12.5%		0.27	0.33	0.93
Mobile Application	12.23	38.04	12.95	40.92	9	8.94	2	/	0.78	0.73	0.37
Cost buying	n = 35, 8.08%		n = 31, 7.81%		n = 8, 42.11%		/		3.45	0.27	0.51
Bets	0.91	3.74	1.03	3.96	15.5	35.36	/	/	<0.001	0.78	0.62
(Total sample, Mean)	n = 58, 13.39%		n = 40, 10.08%		n = 6, 31.58%		n = 2, 25%		3.21	1.75	0.23
Computer	55.31	88.07	45.12	60.74	113.33	192.94	85	91.92	0.001	0.08	0.82
Cost buying	n = 34, 7.85%		n = 29, 7.3%		n = 4, 21.05		n = 1, 12.5%		3.88	6.6	0.21
Bets	21.03	86.1	3.1	9.58	143.75	239.21	50	/	<0.001	<0.001	0.84
(Total sample, Mean)	n = 34, 7.85%		n = 22, 5.54%		n = 3, 15.79%		n = 2, 0.46%		1.66	0.43	1.21
Game console	118.33	141.79	112.5	142.59	200	180.28	60	56.57	0.1	0.66	0.24
Cost buying	n = 19, 4.39%		n = 16, 4.03%		/		/				
Bets	1.05	3.15	1.25	3.42	/	/	/	/			

CPGI: Canadian Pathological Game Index, (NPb): Non-problematic; (Ar): At risk; (Path): Pathological.

Table 2. Scores of the intensity of the practices on the screens (PIUQ, SAS, VAT) for gambling players (ICJP).

Scales	ICJP NPb (n = 397)		ICJP Ar (n = 19)		ICJP Path (n = 8)		Anova 1		Anova 2	
PIUQ practices	total sample (n)	Mean (%)	n	M (%)	n	M (%)				
PIUQ NP	131	33	/		/					
PIUQ a little	239	62.2	10	52.63	3	37.5				
PIUQ Prob	27	6.8	9	47.37	5	62.5				
PIUQ PrSign	9	2.27	/		/					
	M.	ET.	M.	ET.	M.	ET.	F	p	F	p
PIUQ Tot	28.93	9.99	41.16	7.68	44.63	5.53	21.1	0.001	11.9	0.001
AC	2.89	1.17	3.44	1.01	4	0.56	5.41	0.004	4.28	0.02
CN	2.59	0.99	3.58	0.77	3.63	0.98	13.05	0.001	7.59	0.001
SP	2.09	1.1	3.47	1.25	3.5	0.76	20.34	0.001	5.8	0.005
P	2.11	0.89	2.89	0.99	3.75	0.64	19.81	0.001	6.92	0.002
DT	2.41	0.83	3.35	0.64	3.72	0.46	21.04	0.001	11.99	0.001
SAS practices	Total Sample (n)	Mean (%)	Total Sample (n)	Mean (%)	Total Sample (n)	Mean (%)				
SAS NP	392	98.74	16	84.21	6	75				
SAS Dp	11	2.77	2	10.53	2	25				
SAS Abus	3	0.76	1	5.26	/					
	M.	ET.	M.	ET.	M.	ET.	F	p	F	p
SAS Tot	69.43	28.69	90.26	32.55	103.88	25.52	10.04	0.001	5.45	0.007
PVQ	9.46	4.59	12.42	5.29	16.13	4.52	11.57	0.001	8.49	0.001
AP	15.98	7.82	20.26	9.1	23.38	7.19	5.97	0.002	2.8	0.07
RS	12.15	6.15	15.21	6.36	19.38	3.89	7.49	0.001	2.98	0.05
ROC	13.63	6.48	17.68	8.23	21.25	5.7	8.52	0.001	3.12	0.05
SU	10.75	4.97	14.11	5.04	12.75	3.58	4.7	0.009	4.13	0.02
T	7.45	4.24	10.58	4.27	11	4.78	7.41	0.001	3.32	0.04
VAT practices	Total Sample (n)	Mean (%)	Total Sample (n)	Mean (%)	Total Sample (n)	Mean (%)				
VAT NP	353	88.92	8	42.11	3	37.5				
VAT Path	53	13.35	11	57.89	5	62.5				
	M.	ET.	M.	ET.	M.	ET.	F	p	F	p
VAT Tot	23.11	11.43	36	10.37	35.38	15.95	15.49	0.001	3.87	0.03
PC	7.28	3.81	10.89	3.41	9.88	5.25	9.65	0.001	5.2	0.009
Pré	4.81	2.67	7.47	2.74	7.75	3.81	13.13	0.001	2.3	0.11
SS	1.69	1.11	2.47	1.43	2.63	1.6	6.85	0.001	0.65	0.65
AMH	3.14	1.8	5.11	2.08	5.75	2.66	17.89	0.001	1.09	0.34
C	6.19	3.41	10.05	3.24	9.38	4.72	14.51	0.001	5.44	0.007

Table 2: CPGI: Canadian Pathological Game Index, (NPb): Non-problematic; (Ar): At risk; (Path): Pathological; PIUQ: Problematic Internet Use Questionnaire, (NP): Non pathological; (Little): little risk; Prob: Problems; PrSign: significant problem; (SC): Self-checking; (NC): Negative consequences; (PW): Psychological weaning; (C): concern; (TD): Total dimension. SAS: Smartphone Addiction Scale, (NP): Non-pathological; (PD): Probable dependence;

(Abuse): excessive abuse with repercussions on daily life; (DDL): Disruption of daily life; (PA): Positive Anticipation; (WS): Withdrawal Social; (COR): Cyberspace-oriented relationship; (OU): Over-use; (T): Tolerance. VAT: Video Game Addiction Test, (NP): Non-pathological; (Path): Pathological; (LoC): Loss of Control; (C): Concern; (WS): Weaning syndrome; (AMM): Adaptation/Modification of Mood; (C): Conflicts. Anova 2 with variable control "gambling players playing on at least one of the three screen types" (smartphone/touch pad, computer, console). The significance threshold is the threshold $p < 0.05$.

Analysis of the correlations on the total sample allowed us to see that the intensity of the gambling practice was positively correlated with the expenses related to the purchase of applications, computer games, video games, and consoles, as well as bets offered on applications, on the computer, and in video games; To the set of cognitive distortions described in the GRCS scale; The intensity of Internet practice (PIUQ); The positive and negative UPPS urgency; The intensity of impulsivity (UPPS); To all the variables making up the intensity of the practice of mobile screens (SAS); To the set of variables of the intensity of the practice of video games; Psychological distress including depression (HAD-D); and to the "game" variable of the immersion scale.

4.2. Costs and Prevalence of Gambling/Gambling on Screens

The maximum bet amount differs significantly depending on the intensity of the gambler's practice. The difference is significant for computerization between non-pathological gamblers and pathological gamblers ($t = 6.6$, $p < 0.001$). The use of mobile and computer screens between non-pathological gamblers and problem gamblers is also significant ($t = 3.45$, $p = 0.001$ for mobile screens and $t = 3.88$, $p = 0.001$). The purchase of computer games is also significant between non-pathological gamblers and problem gamblers ($t = 3.22$; $p = 0.001$). For gambling players on the computer, bets between non-pathological and pathological gambler's are significantly different ($t = 2.97$, $p = 0.01$) (Table 1). Linked to the screens, players' bets are also significantly different. Individuals with a problematic practice have higher gambling bets than those with moderate Internet practice ($t = 2.95$; $p = 0.003$). Similarly, those with problem gambling practice are more likely than non-pathologists ($t = 2.14$, $p = 0.03$).

Table 2 illustrates the different practices related to gambling on screens and the intensity of this practice. Anova analysis revealed a significant difference between the different groups representing the intensity of Internet practice (PIUQ) and the score obtained at the ICJP scale ($F = 23.91$, $p < 0.001$). Similarly, we observed a significant difference between the groups representing the intensity of the practice of the mobile screens (SAS) and the scores obtained at the ICJP scale ($F = 12.27$, $p < 0.001$). This significant difference is found in the weighting variables "gambling gamblers on mobile screen applications" ($F = 11.02$, $p < 0.001$), "gambling players on computers" ($F = 36.24$, $p < 0.001$) and "gambling players on mobile, console and computer applications" ($F = 83.05$, $p < 0.001$). Finally, there was a significant difference between the groups representing the intensity of the video games practice and the score obtained at the ICJP scale ($F = 39.45$, $p < 0.001$). The difference is significant only with the weighting variable "players with a gambling practice on mobile screen applications" ($F = 7.26$, $p = 0.01$).

4.3. Gambling on Screens and Comorbidities

For gambling on screens, the first elaborated multiple linear regression models consists of 34 variables ($F = 2.42$, $p < 0.01$). These included comorbidities (impulsivity, depression, emotional regulation), cognitive distortions, and scaling variables that evaluated excessive screen-related practices. The variable to be predicted is the practical variable of gambling on the screens with the ICJP scale and the positive variable to play the games of gambling on a screen. This model accounts for 78% of the total variance.

The variable perturbation of daily life (of the SAS scale) ($p = 0.004$) is the best predictor of pathological gambling on screens and especially on mobile screens. Other variables have a predictive value, although lower such as the "lack of premeditation" variable (from the UPPS scale) ($p = 0.04$);

The variable "game expectations" ($p = 0.01$) and the "inability to abstain from playing" variable ($p = 0.01$) (from the GRCS scale). The second linear regression model, with 13 closest variables, accounts for 64% of the total variance. The variable "lack of premeditation" ($p = 0.004$), "sensation seeking" ($p = 0.02$), "expectation of the game" ($p = 0.08$), and the total dimensions of the PIUQ ($p = 0.03$) are the predictors of pathological gambling on the screens (Table 3).

Table 3. Comorbidities and gambling on screens.

Scales and Sub-Dimensions	CPGI NPb (n = 397)		CPGI Ar (n = 19)		CPGI Path (n = 8)		Anova 1 Traditional Gambling		Anova 2 Gambling on Console and Mobile Screen		Anova 3 Gambling on Computer		Anova 4 Gambling on All Media Together	
	M.	SD.	M.	SD.	M.	SD.	F	p	F	p	F	p	F	p
							2.07	0.001	2.37	0.33	2.04	0.006	53.21	0.02
GRCS	43.61	26.08	50.75	33.69	93.5	12.5	4.07	0.02	3.98	0.14	6.51	0.003	1.11	0.37
- ALJ	8.67	5.23	10.75	7.44	14.75	3.77	1.41	0.25	7.1	0.07	2.54	0.09	1.47	0.28
- IC	6.13	4.05	18.5	3.87	5.38	4.69	8.32	0.001	0.85	0.5	17.12	0.001	1.35	0.31
- PP	12.43	8.5	13.38	8.21	26	3.74	4.3	0.01	15.32	0.02	4.94	0.01	1.1	0.37
- IA	9.02	6.77	12.13	9.4	16.75	5.38	2.46	0.09	3.05	0.19	2.61	0.08	1.51	0.27
- IF	7.35	4.73	9.13	7.4	17.5	1.91	5.64	0.003	3.22	0.18	7.56	0.001	1.43	0.29
UPPS	46.76	8.9	51.75	11.34	/		0.02	0.89	2.34	0.24	1.04	0.36	2.94	0.1
- UN	9.46	3.42	10.63	2.88	/		0.22	0.64	1.52	0.35	0.48	0.62	0.45	12.82
- UP	11.26	3.01	11	2.45	/		1.5	0.22	0.52	0.64	0.22	0.8	1.31	0.32
- RS	11.37	2.82	12	2.73	/		0.64	0.43	5.75	0.09	0.79	0.46	1.51	0.27
- MPer	7.22	2.8	9.38	2.67	/		5.06	0.03	1.97	0.28	2.21	0.13	6.17	0.02
- MPré	7.46	2.36	8.75	2.71	/		4.08	0.04	1.96	0.29	1.48	0.24	7.88	1.17
ERQ	4.28	1.11	4.13	1.1	4.93	0.74	1.39	0.25	5.2	0.11	0.76	0.47	0.3	0.75
- RC	4.34	1.42	4.29	1.5	4.96	0.94	1.96	0.14	3.41	0.17	0.37	0.69	0.1	0.91
- RE	4.21	1.24	3.88	1.19	4.88	1.09	0.35	0.7	8.78	0.06	0.89	0.42	0.88	0.17
HAD	11.23	5.4	10.2	1.08	16.5	8	4.09	0.02	32.03	0.009	3.82	0.03	3.67	0.07
Anxiety	6.54	3.38	5	3.21	9	3.37	0.29	0.75	5.55	0.09	1.91	0.16	1.83	0.21
Depression	4.02	2.92	3	1.6	7.75	6.24	0.82	0.44	8.07	0.06	3.36	0.04	5.08	0.03

CPGI: Canadian Pathological Game Index, (NPb): Non-problematic; (Ar): At risk; (Path): Pathological; HAD: Hospital Anxiety and Depression; GRCS: Gambling Related Cognition Scale; IF: Interpretation Favorable; IC: Illusion of Control; PP: Predictive power; ERG: Expectations Related to the Game; AI: Inability to Stop Playing; ERQ: Emotion Relationship Questionnaire; CR: Cognitive reassessment; ER: Expressive repression; UPPS: Impulsive Behavior Scale; NU: Negative urgency; PE: Positive Emergency; SS: Sensation Seeking; LPe: Lack of perseverance; LPr: Lack of premedication. The Anova for the UPPS concerned only problematic and non-pathological gamblers. Anova 2 with control of the variable "practice of gambling on console and mobile screens". Anova 3 with control of the variable "practice of gambling on a computer". Anova 4 with control of the variable "practice of gambling all confused screens". The significance threshold is the threshold $p < 0.05$.

In the case of comorbidities (anxiety, depression, emotional regulation, impulsivity, cognitive distortions), the only predictive variable for pathological gambling on the screens was "control illusion" ($p = 0.03$). The 14 variables accounted for 78% of the total variance.

Conversely, the linear regression model suggests that the predictors of excessive use of the Internet shows that 30 variables explain 54% of the total variance, six variables were significant. The variables of gambling practice ($p = 0.001$), tolerance ($p < 0.001$), preoccupation ($p = 0.003$), focus ($p = 0.01$), perturbation of daily life, ($p = 0.04$) and implication ($p = 0.04$) were predictive of excessive use of the Internet.

It is also noted that the predictive variables of excessive video gaming practice when we gamble regularly are different from the excessive practice of pure video gaming practice. For example, the variables "lack of perseverance" (UPPS), "lack of premeditation" (UPPS), the "game" variable of the immersion scale (QPI), the "inability to abstain from playing" GRCS) and "tolerance" (SAS) are predictive of the excessive gambling practice associated with regular gambling (82% of the total variance with 31 variables) (Table 3).

4.4. Immersion, Gambling on Screens and Comorbidities

I was observed that immersion scores were significantly higher for people with gambling addiction (Table 4). A Student test was carried out to glimpse the link between immersion and the use of screens. The scores for the VAT, PIUQ and SAS scales are significantly different according to the three immersion thresholds (low, moderate, high), respectively for the PIUQ scale ($t = 2.51$, $p = 0.01$), for the scale SAS ($t = 3.26$, $p = 0.001$) for the VAT scale ($t = 4.03$, $p = 0.001$). The surface analysis of the responses shows that, starting from a high immersion threshold, the individual is less likely to have a high score on scales assessing the intensity of Internet practice and on the scale evaluating the intensity of gambling practice.

Table 4. Scores related to propensity to immersion according to the intensity of gambling alone/ gambling on screens.

Immersion Variables	CPGI NPb (n = 397)		CPGI Ar (n = 19)		CPGI Path (n = 8)		Anova 1 Traditional Gambling		Anova 2 Gambling on Console and Mobile Screen		Anova 3 Gambling on Computer		Anova 4 Gambling on All Media Together	
	M.	SD.	M.	SD.	M.	SD.	F	p	F	p	F	p	F	p
Immersion (QPI)	53.46	18.22	55.16	19.17	63.5	11.98	2.26	0.01	34.84	0.001	24.85	0.001	80.10	0.001
Focus	17.88	6.53	16.37	6.61	18.38	3.54	0.52	0.59	21.47	0.001	45.65	0.001	26.86	0.001
Implication	14.66	6.58	15.37	6.19	17	3.63	0.6	0.55	12.5	0.001	10.85	0.001	31.53	0.001
Emotion	11.96	5.84	11.47	4.34	14.13	3.36	0.63	0.53	23.69	0.001	25.98	0.001	26.45	0.001
Game	5.61	4.23	8.47	4.25	10.25	2.87	8.68	0.001	286.94	0.001	142.33	0.001	760.52	0.001

CPGI: Canadian Pathological Game Index, (NPb): Non-problematic; (Ar): At risk; (Path): Pathological; QPI: Immersion propensity questionnaire. Anova 2 with control of the variable "Intensity of practices on the Internet". Anova 3 with control of the variable "Intensity of the practices of video games". Anova 4 with control of the variable "Intensity of practices on mobile screens". The significance threshold is the threshold $p < 0.05$.

Multiple linear regression analysis was used to analize the predictive power of immersion for the practice of gambling on screens in relation with comorbidities. In our model, 14 variables were integrated which explained 48% of the variance ($F = 2.84$, $p = 0.004$) with impulsivity, emotional regulation, anxious and depressive comorbidities and cognitive distortions. Predictors of immersion in the practice of games of chance on the screens are positive urgency (UPPS), anxiety (HAD), predictive power (GRCS), inability to Refraining from Playing (GRCS) and Interpretation for Game Continuation (GRCS).

A principal component analysis (PCA) was carried out to study the relationship between immersion according to the intensity of the gambling practice on the screens and the other variables of our study. Three axes were selected, representing 52.92% of the total variance. The first axis ($\lambda = 26.51\%$) is characterized and particularly correlated, on the negative side, by all cognitive distortions (GRCS); By the variables implication, emotion, game and total dimensions of immersion; The practice of mobile displays (SAS); The Negative Emergency and the Positive Urgency of Impulsivity (UPPS), and Internet Practice (PIUQ). This axis is a "size" factor, ordering individuals according to their scores on all scales. The two axis ($\lambda = 14.98\%$) is characterized by anxiety (HAD), all variables of immersion on the positive side and the inability to abstain from playing (GRCS) Impulsivity (UPPS), including sensation seeking (UPPS) and negative urgency (UPPS). The three-axis ($\lambda = 11.43\%$) is characterized by negative urgency (UPPS), positive urgency (UPPS) and sensation seeking (UPPS), total impulsivity (UPPS) and the illusion of control and interpretation favorable to the pursuit of the game on the negative side.

5. Discussion

The objective of this study was to assess the practice of gambling on screens, and to understand the impact of the perceived intensity of immersion by an individual on the practice of screens and gambling on the screens, as well as determining the associated comorbidities. The results obtained in this study highlight certain psychological and psychopathological specificities of gamblers on screens related to immersion.

First, the prevalence of problematic practices is consistent with the literature, particularly for Internet practice (2.27% versus 1–3.2% in the literature for European countries including France and Germany) [107–109], particularly for the practice of gambling (i.e., 1.85% in our study against 0.2–3% according to the estimate of the prevalence studies in France) [107–110], for the practice of video games (13.35% for our study against 14% for the most recent French study) and for the practice of mobile screens (2.77% lower for our study unlike the study Validation of the SAS scale which was at 9.6%) (Tablets, smartphones) [111].

The results of our study confirmed that pathological and problematic gamblers are more prone to spend more money than non-pathological gamblers and also significantly different depending on the interface chosen, whether it is a fixed screen or mobile, via a console or a computer. These increased costs can be explained by the ease of access, the possibility of betting during uninterrupted periods and the immersive interactivity of gambling software on the Internet [3,112–114]. Our first hypothesis is confirmed.

We also showed that individuals who were tending to excessive gambling had significantly higher scores on scales assessing fixed-screen and mobile screens (whose behaviors had been assessed by a recent scale that focused solely on self-action directed to a mobile display, including the smartphone or touchpad) (Table 2). The practice of gambling could thus participate in the propensity of excessive practices related to screens. It is important, however, to see the same cognitive distortions found in these different types of excessive practices [115,116]. The intensity of immersion would therefore be a predominant factor in the detection of the excessive practice related to the screen. This confirms the high level of commitment generated during excessive practice phases, as well as the role of the dissociation contained in the "focus" variable [117,118].

Our fourth hypothesis is only partially confirmed. Concerning, the practice of gambling on mobile screens, the only cognitive distortion variable, "predictive power" was significantly different according to the three groups. Regarding only the practice of computer gambling, the set of cognitive distortion scores, with the exception of gambling expectations, were significantly different between the groups. Concerning the practice of gambling on all the screens combined, the variable cognitive distortion was not significant. However, depressive variables (HAD) and lack of perseverance (UPPS) were significantly different across groups. Although the literature demonstrates that cognitive distortions can predict future commitment to excessive gambling, gaming expectations differ according to the interface used, and cognitive distortions are much less pronounced in practice on mobile screens [47,119,120].

The analysis of the area of responses between three variables (PIUQ, ICJP and QPI) showed that, starting from a high immersion threshold, the individual is less likely to have a high score at the scales evaluating the "Intensity of Internet and gambling practice". Immersion could thus be perceived as a protection factor from a certain threshold.

Finally, variables explaining pathological gambling on the screens allow us to adjust the prevention modalities for the gambling on the screens. Indeed, taking into account the format of the screen (for mobile or fixed purposes), daily life disruption (SAS) variables, lack of premeditation (UPPS), sensation seeking (GRCS), expectations related to the (GRCS) and impulsivity (total PIUQ sums) were predictive of excessive gambling.

There are limitations to this study. The research was constructed from self-assessment, dealing with general behavioral behavior with respect to screens and the Internet and not only on immersion experienced during a main task (a task that would be delimited by a specific action on a continuum within a virtual universe of gambling). This may be a bias in assessing the prevalence of excessive screen-related practices in this population. The average age of our population is also low, our sample could have been enlarged by older people. Also, we could have chosen scales assessing anxiety-depression disorders regularly used in clinical practice such as "Generalized Anxiety Disorder scale (GAD-7) or Patient Health Questionnaire (PHQ). Finally, the distribution of practices according to their intensities is heterogeneous. Better distribution of groups would have been preferable.

Neurophysiological measurements and longitudinal analysis would allow a more precise glimpse of the cognitive processes related to immersion variables.

6. Conclusions

This study explored the involvement of gambling gamblers in the screens. It also showed the implication of the immersion variable in relation to the intensity of the practice of gambling on the screens. Immersion thus plays a mediating role between cognitive resources and the impact on screen practices.

The practice of gambling begins in adolescence. Studies also indicate that those who gamble during childhood will be more likely to become compulsive gamblers later in life [85,121,122]. It would therefore be essential to reinforce our understanding of the negative consequences of gambling-related behaviors on the screens and also to increase the methods of prevention and intervention [123].

The interest is also to be able to modulate the variables of immersion in the therapeutic management of excessive practice or, upstream, to improve prevention by proposing prevention messages at precise level thresholds Immersion according to the variables composing it [124]. It is also a possibility for game developers to modulate cognitive factors related to immersion leading to excessive practice and thus avoid future problems of excessive and impulsive behavior.

Acknowledgments: No funding was associated with this research. The laboratory of Paris-Nanterre University (Paris, FRANCE) financially supported this research both in the costs related to research and those related to publication.

Author Contributions: Lucia Romo and Jean-Jacques Rémond have conceived and designed the experimentation and more specifically participated in the recruitment of study participants. Jean-Jacques Rémond analyzed the data; Jean-Jacques Rémond and Lucia Romo wrote the paper.

Conflicts of Interest: The authors declare no conflict of interest.

References

1. Griffiths, M. Behavioural addiction: An issue for everybody? *Empl. Counc. Today* **1996**, *8*, 19–25. [CrossRef]
2. Maclin, O.H.; Dixon, M.R.; Hayes, L.J. A computerized slot machine simulation to investigate the variables involved in gambling behavior. *Behav. Res. Methods Instrum. Comput.* **1999**, *31*, 731–734. [CrossRef] [PubMed]
3. Gainsbury, S.M.; King, D.L.; Russell, A.M.; Delfabbro, P.; Derevensky, J.; Hing, N. Exposure to and engagement with gambling marketing in social media: Reported impacts on moderate-risk and problem gamblers. *Psychol. Addict. Behav.* **2016**, *30*, 270. [CrossRef] [PubMed]
4. Adams, P.J.; Wiles, J. Gambling machine annexes as enabling spaces for addictive engagement. *Health Place* **2017**, *43*, 1–7. [CrossRef] [PubMed]
5. Bolen, D.W.; Boyd, W.H. Gambling and the gambler: A review and preliminary findings. *Arch. Gen. Psychiatry* **1968**, *18*, 617–630. [CrossRef] [PubMed]
6. Kairouz, S.; Paradis, C.; Nadeau, L.; Tovar, M.-L.; Pousset, M. A cross-cultural comparison of population gambling patterns and regulatory frameworks: France and Québec. *J. Public Health Policy* **2016**, *37*, 467–482. [CrossRef] [PubMed]
7. Kessler, R.C.; Hwang, I.; LaBrie, R.; Petukhova, M.; Sampson, N.A.; Winters, K.C.; Shaffer, H.J. DSM-IV pathological gambling in the National Comorbidity Survey Replication. *Psychol. Med.* **2008**, *38*, 1351–1360. [CrossRef] [PubMed]
8. Volberg, R.A. Fifteen years of problem gambling prevalence research: What do we know? Where do we go? *J. Gambl. Issues* **2004**. [CrossRef]
9. Shaffer, H.J.; Martin, R. Disordered gambling: Etiology, trajectory, and clinical considerations. *Annu. Rev. Clin. Psychol.* **2011**, *7*, 483–510. [CrossRef] [PubMed]
10. Bouju, G.; Grall-Bronnec, M.; Landreat-Guillou, M.; Venisse, J.-L. Jeu pathologique: Facteurs impliqués. *L'Encéphale* **2011**, *37*, 322–331. (In French) [CrossRef] [PubMed]

11. Schmit, S.; Chauchard, E.; Chabrol, H.; Sejourne, N. Évaluation des caractéristiques sociales, des stratégies de coping, de l'estime de soi et de la symptomatologie dépressive en relation avec la dépendance aux jeux vidéo en ligne chez les adolescents et les jeunes adultes. *L'Encéphale* **2011**, *37*, 217–223. (In French) [CrossRef] [PubMed]
12. Chan, C.C.; Li, W.W.L.; Leung, E.C.I. The Etiology of Problem Gambling. In *Problem Gambling in Hong Kong and Macao*; Springer: New York, NY, USA, 2016; pp. 75–94.
13. Husky, M.M.; Michel, G.; Richard, J.-B.; Guignard, R.; Beck, F. Gender differences in the associations of gambling activities and suicidal behaviors with problem gambling in a nationally representative French sample. *Addict. Behav.* **2015**, *45*, 45–50. [CrossRef] [PubMed]
14. American Psychiatric Association. *Diagnostic and Statistical Manual of Mental Disorders (DSM-5®)*; American Psychiatric Pub.: Washington, DC, USA, 2013.
15. Clark, L. Disordered gambling: The evolving concept of behavioral addiction. *Ann. N. Y. Acad. Sci.* **2014**, *1327*, 46–61. [CrossRef] [PubMed]
16. Dowling, N.A.; Merkouris, S.S.; Greenwood, C.J.; Oldenhof, E.; Toumbourou, J.W.; Youssef, G.J. Early risk and protective factors for problem gambling: A systematic review and meta-analysis of longitudinal studies. *Clin. Psychol. Rev.* **2016**, *51*, 109–124. [CrossRef] [PubMed]
17. Griffiths, M.; Wood, R.T. Risk factors in adolescence: The case of gambling, videogame playing, and the Internet. *J. Gambl. Stud.* **2000**, *16*, 99–225.
18. Hing, N.; Russell, A.; Tolchard, B.; Nower, L. Risk factors for gambling problems: An analysis by gender. *J. Gambl. Stud.* **2016**, *32*, 511–534. [CrossRef] [PubMed]
19. Welte, J.W.; Barnes, G.M.; Wieczorek, W.F.; Tidwell, M.-C.O.; Parker, J.C. Risk factors for pathological gambling. *Addict. Behav.* **2004**, *29*, 323–335. [CrossRef] [PubMed]
20. Johansson, A.; Grant, J.E.; Kim, S.W.; Odlaug, B.L.; Götestam, K.G. Risk factors for problematic gambling: A critical literature review. *J. Gambl. Stud.* **2009**, *25*, 67–92. [CrossRef] [PubMed]
21. Maniaci, G.; Picone, F.; van Holst, R.J.; Bolloni, C.; Scardina, S.; Cannizzaro, C. Alterations in the emotional regulation process in gambling addiction: The role of anger and alexithymia. *J. Gambl. Stud.* **2017**, *33*, 633–647. [CrossRef] [PubMed]
22. Navas, J.F.; Perandres-Gomez, A.; Perales, J.C. OR-78: Associations between adaptive emotion regulation strategies and gambling-related cognitive biases in gambling disorder patients. *J. Behav. Addict.* **2016**, *5*, 32–34. [CrossRef] [PubMed]
23. Jauregui, P.; Estevez, A.; Urbiola, I. Pathological gambling and associated drug and alcohol abuse, emotion regulation, and anxious-depressive symptomatology. *J. Behav. Addict.* **2016**, *5*, 251–260. [CrossRef] [PubMed]
24. Potenza, M.N.; Steinberg, M.A.; McLaughlin, S.D.; Wu, R.; Rounsaville, B.J.; O'Malley, S.S. Gender-related differences in the characteristics of problem gamblers using a gambling helpline. *Am. J. Psychiatry* **2001**, *158*, 1500–1505. [CrossRef] [PubMed]
25. González-Ortega, I.; Echeburúa, E.; de Corral, P.; Polo-López, R. Pathological gambling: Clinical gender differences. In *Psychopathology in Women*; Springer: New York, NY, USA, 2015; pp. 713–726.
26. McCormack, A.; Shorter, G.W.; Griffiths, M.D. An empirical study of gender differences in online gambling. *J. Gambl. Stud.* **2014**, *30*, 71–88. [CrossRef] [PubMed]
27. Grant, J.E.; Kim, S.W. Gender differences in pathological gamblers seeking medication treatment. *Compr. Psychiatry* **2002**, *43*, 56–62. [CrossRef] [PubMed]
28. Merkouris, S.S.; Thomas, A.C.; Shandley, K.A.; Rodda, S.N.; Oldenhof, E.; Dowling, N.A. An update on gender differences in the characteristics associated with problem gambling: A systematic review. *Curr. Addict. Rep.* **2016**, *3*, 254–267. [CrossRef]
29. Casey, E. *Women, Pleasure and the Gambling Experience*; Routledge: Abingdon, UK, 2016.
30. Wenzel, H.G.; Dahl, A.A. Female pathological gamblers—A critical review of the clinical findings. *Int. J. Ment. Health Addict.* **2009**, *7*, 190–202. [CrossRef]
31. Crockford, D.N.; El-Guebaly, N. Psychiatric comorbidity in pathological gambling: A critical review. *Can. J. Psychiatry* **1998**, *43*, 43–50. [CrossRef] [PubMed]
32. Hartmann, M.; Blaszczynski, A. The longitudinal relationships between psychiatric disorders and gambling disorders. *Int. J. Ment. Health Addict.* **2016**, 1–29. [CrossRef]
33. Griffiths, M.; Sutherland, I. Adolescent gambling and drug use. *J. Community Appl. Soc. Psychol.* **1998**, *8*, 423–427. [CrossRef]

34. Van Rooij, A.J.; Kuss, D.J.; Griffiths, M.D.; Shorter, G.W.; Schoenmakers, T.M.; Van de Mheen, D. The (co-) occurrence of problematic video gaming, substance use, and psychosocial problems in adolescents. *J. Behav. Addict.* **2014**, *3*, 157–165. [CrossRef] [PubMed]
35. Carbonneau, R.; Vitaro, F.; Brendgen, M.; Tremblay, R.E. Trajectories of gambling problems from mid-adolescence to age 30 in a general population cohort. *Psychol. Addict. Behav.* **2015**, *29*, 1012–1021. [CrossRef] [PubMed]
36. Dussault, F.; Brendgen, M.; Vitaro, F.; Carbonneau, R.; Boivin, M.; Tremblay, R.E. Co-morbidity between gambling problems and depressive symptoms: A longitudinal perspective of risk and protective factors. *J. Gambl. Stud.* **2016**, *32*, 547–565. [CrossRef] [PubMed]
37. Cook, S.; Turner, N.E.; Ballon, B.; Paglia-Boak, A.; Murray, R.; Adlaf, E.M.; Ilie, G.; den Dunnen, W.; Mann, R.E. Problem gambling among Ontario students: Associations with substance abuse, mental health problems, suicide attempts, and delinquent behaviours. *J. Gambl. Stud.* **2015**, *31*, 1121–1134. [CrossRef] [PubMed]
38. Palaiologou, A.; Kotzia, D.; Anagnostopoulos, D.C.; Lazaratou, H. Risk factors and comorbidity of pathological gambling in adolescence. *J. Adolesc. Psychol. Psychiatry* **2015**, *2015*, 3. [CrossRef]
39. Fatseas, M.; Alexandre, J.-M.; Vénisse, J.-L.; Romo, L.; Valleur, M.; Magalon, D.; Chéreau-Boudet, I.; Luquiens, A.; Guilleux, A.; Groupe, J.E.U.; et al. Gambling behaviors and psychopathology related to Attention-Deficit/Hyperactivity Disorder (ADHD) in problem and non-problem adult gamblers. *Psychiatry Res.* **2016**, *239*, 232–238. [CrossRef] [PubMed]
40. Porteret, R.; Bouchez, J.; Baylé, F.J.; Varescon, I. ADH/D and impulsiveness: Prevalence of impulse control disorders and other comorbidities, in 81 adults with attention deficit/hyperactivity disorder (ADH/D). *L'Encephale* **2016**, *42*, 130–137. [CrossRef] [PubMed]
41. Romo, L.; Rémond, J.J.; Coeffec, A.; Kotbagi, G.; Plantey, S.; Boz, F.; Kern, L. Gambling and attention deficit hyperactivity disorders (ADHD) in a population of french students. *J. Gambl. Stud.* **2015**, *31*, 1261–1272. [CrossRef] [PubMed]
42. Romo, L.; Legauffre, C.; Guilleux, A.; Valleur, M.; Magalon, D.; Fatséas, M.; Chéreau-Boudet, I.; Luquiens, A.; Vénisse, J.L.; JEU Group; et al. Cognitive distortions and ADHD in pathological gambling: A national longitudinal case-control cohort study. *J. Behav. Addict.* **2016**, *5*, 649–657. [CrossRef] [PubMed]
43. American Psychiatric Association. *DSM-IV-TR: Diagnostic and Statistical Manual of Mental Disorders, Text Revision*; American Psychiatric Association: Washington, DC, USA, 2000; p. 75.
44. Lister, J.J.; Milosevic, A.; Ledgerwood, D.M. Psychological characteristics of problem gamblers with and without mood disorder. *Can. J. Psychiatry* **2015**, *60*, 369–376. [CrossRef] [PubMed]
45. Clarke, D. Impulsivity as a mediator in the relationship between depression and problem gambling. *Personal. Individ. Differ.* **2006**, *40*, 5–15. [CrossRef]
46. Taylor, R.N.; Parker, J.D.; Keefer, K.V.; Kloosterman, P.H.; Summerfeldt, L.J. Gambling related cognitive distortions in adolescence: Relationships with gambling problems in typically developing and special needs students. *J. Gambl. Stud.* **2015**, *31*, 1417–1429. [CrossRef] [PubMed]
47. Griffiths, M.D. The role of cognitive bias and skill in fruit machine gambling. *Br. J. Psychol.* **1994**, *85*, 351–369. [CrossRef]
48. MacLaren, V.; Ellery, M.; Knoll, T. Personality, gambling motives and cognitive distortions in electronic gambling machine players. *Personal. Individ. Differ.* **2015**, *73*, 24–28. [CrossRef]
49. Harrigan, K.; MacLaren, V.; Brown, D.; Dixon, M.J.; Livingstone, C. Games of chance or masters of illusion: Multiline slots design may promote cognitive distortions. *Int. Gambl. Stud.* **2014**, *14*, 301–317. [CrossRef]
50. Ladouceur, R.; Mayrand, M.; Dussault, R.; Letarte, A.; Tremblay, J. Illusion of control: Effects of participation and involvement. *J. Psychol.* **1984**, *117*, 47–52. [CrossRef]
51. Ladouceur, R.; Mayrand, M. Evaluation of the "illusion of control": Type of feedback, outcome sequence, and number of trials among regular and occasional gamblers. *J. Psychol.* **1984**, *117*, 37–46. [CrossRef]
52. Ladouceur, R.; Walker, M. A cognitive perspective on gambling. *Trends Cogn. Behav. Ther.* **1996**, *5*, 89–120.
53. Ladouceur, R.; Sévigny, S. Structural characteristics of video lotteries: Effects of a stopping device on illusion of control and gambling persistence. *J. Gambl. Stud.* **2005**, *21*, 117–131. [CrossRef] [PubMed]
54. Breen, R.B.; Zimmerman, M. Rapid onset of pathological gambling in machine gamblers. *J. Gambl. Stud.* **2002**, *18*, 31–43. [CrossRef] [PubMed]

55. Khanbhai, Y.; Smith, D.; Battersby, M. Gender by preferred gambling activity in treatment seeking problem gamblers: A comparison of subgroup characteristics and treatment outcomes. *J. Gambl. Stud.* **2017**, *33*, 99–113. [CrossRef] [PubMed]
56. Murch, W.S.; Clark, L. Games in the brain neural substrates of gambling addiction. *Neuroscientist* **2016**, *22*, 534–545. [CrossRef] [PubMed]
57. Schüll, N.D. *Addiction by Design: Machine Gambling in Las Vegas*; Princeton University Press: Princeton, NJ, USA, 2012.
58. Cookman, M.L.; Weatherly, J.N. Investigating possible effects of ethnicity and age on gambling as an escape. *J. Gambl. Stud.* **2016**, *32*, 499–509. [CrossRef] [PubMed]
59. Gainsbury, S.M.; Hing, N.; Delfabbro, P.H.; King, D.L. A taxonomy of gambling and casino games via social media and online technologies. *Int. Gambl. Stud.* **2014**, *14*, 196–213. [CrossRef]
60. Gainsbury, S.; Wood, R.; Russell, A.; Hing, N.; Blaszczynski, A. A digital revolution: Comparison of demographic profiles, attitudes and gambling behavior of Internet and non-Internet gamblers. *Comput. Hum. Behav.* **2012**, *28*, 1388–1398. [CrossRef]
61. Kairouz, S.; Nadeau, L.; Paradis, C. *Portrait of Gambling in Quebec: Prevalence, Incidence and Trajectories over Four Years*; Concordia University: Montréal, QC, Canada, 2011.
62. Williams, R.J.; Wood, R.T.; Parke, J. *Routledge International Handbook of Internet Gambling*; Routledge: Abingdon, UK, 2012.
63. Yani-de-Soriano, M.; Javed, U.; Yousafzai, S. Can an industry be socially responsible if its products harm consumers? The case of online gambling. *J. Bus. Ethics* **2012**, *110*, 481–497. [CrossRef]
64. Gainsbury, S.M.; Delfabbro, P.; King, D.L.; Hing, N. An exploratory study of gambling operators' use of social media and the latent messages conveyed. *J. Gambl. Stud.* **2016**, *32*, 125–141. [CrossRef] [PubMed]
65. Allen, J.L.; Massing, S.A.; Moyles, B.P.; Pacey, L.J.; Ward, M.J. Mobile Device Applications for Casinos. U.S. Patent 20140045586 A1, 13 February 2014.
66. Hing, N.; Gainsbury, S.; Blaszczynski, A.; Wood, R.; Lubman, D.; Russel, A. *Interactive Gambling*; Australian Gambling Research Centre, Australian Institute of Family Studies: Melbourne, Australia, 2014.
67. Griffiths, M. Mobile phone gambling. In *Encyclopedia of Mobile Computing and Commerce*; IGI Global: Hershey, PA, USA, 2007; pp. 553–556.
68. Korhonen, H.; Koivisto, E.M. Playability heuristics for mobile games. In Proceedings of the 8th Conference on Human-Computer Interaction with Mobile Devices and Services, Helsinki, Finland, 12–25 September 2006; ACM: New York, NY, USA, 2006; pp. 9–16.
69. Wakefield, R.L.; Whitten, D. Mobile computing: A user study on hedonic/utilitarian mobile device usage. *Eur. J. Inf. Syst.* **2006**, *15*, 292–300. [CrossRef]
70. Turel, O.; Serenko, A.; Bontis, N. User acceptance of hedonic digital artifacts: A theory of consumption values perspective. *Inf. Manag.* **2010**, *47*, 53–59. [CrossRef]
71. Wunnava, S. Mobile commerce usage: Application of theory of reasoned action (TRA) and technology acceptance model (TAM). *World J. Soc. Sci.* **2015**, *5*, 41–50.
72. Csikszentmihalyi, M. *Flow: The Psychology of Optimal Experience*; Harper & Row: New York, NY, USA, 1990; 303p.
73. Csikszentmihalyi, M.; Csikszentmihalyi, I.S. *Optimal Experience: Psychological Studies of Flow in Consciousness*; Cambridge University Press: Cambridge, UK, 1992.
74. Csikszentmihalyi, M.; Asakawa, K. Universal and cultural dimensions of optimal experiences. *Jpn. Psychol. Res.* **2016**, *58*, 4–13. [CrossRef]
75. Csikszentmihalyi, M.; Khosla, S.; Nakamura, J. Flow at Work. In *The Wiley Blackwell Handbook of the Psychology of Positivity and Strengths-Based Approaches at Work*; Oades, L.G., Steger, M.F., Fave, A.D., Passmore, J., Eds.; John Wiley & Sons, Ltd.: Hoboken, NJ, USA, 2016; pp. 99–109.
76. Randolph, L.; Murray, J.; Lanham, R. Hamlet on the holodeck: The future of narrative in cyberspace. *JSTOR* **1998**, *72*, 187–190.
77. Csikszentmihalyi, M. Toward a psychology of optimal experience. In *Flow and the Foundations of Positive Psychology*; Springer: New York, NY, USA, 2014; pp. 209–226.
78. Nakamura, J.; Csikszentmihalyi, M. The concept of flow. In *Flow and the Foundations of Positive Psychology*; Springer: New York, NY, USA, 2014; pp. 239–263.
79. Dixon, M.J.; Graydon, C.; Harrigan, K.A.; Wojtowicz, L.; Siu, V.; Fugelsang, J.A. The allure of multi-line games in modern slot machines. *Addiction* **2014**, *109*, 1920–1928. [CrossRef] [PubMed]

80. Psotka, J.; Davison, S. *Cognitive Factors Associated with Immersion in Virtual Environments*; Army Research Institute: Alexandria, VA, USA, 1993.
81. Cartmill, T.; Slatter, T.; Wilkie, B. The role of anxiety and dissociation in young Australian gamblers. *J. Gambl. Stud.* **2015**, *31*, 1215–1226. [CrossRef] [PubMed]
82. Diskin, K.M.; Hodgins, D.C. Narrowing of attention and dissociation in pathological video lottery gamblers. *J. Gambl. Stud.* **1999**, *15*, 17–28. [CrossRef] [PubMed]
83. Diskin, K.M.; Hodgins, D.C. Narrowed focus and dissociative experiences in a community sample of experienced video lottery gamblers. *Can. J. Behav. Sci. Can. Sci. Comport.* **2001**, *33*, 58–64. [CrossRef]
84. Diskin, K.M.; Hodgins, D.C. Psychophysiological and subjective arousal during gambling in pathological and non-pathological video lottery gamblers. *Int. Gambl. Stud.* **2003**, *3*, 37–51. [CrossRef]
85. Gupta, R.; Derevensky, J.L. Adolescent gambling behavior: A prevalence study and examination of the correlates associated with problem gambling. *J. Gambl. Stud.* **1998**, *14*, 319–345. [CrossRef] [PubMed]
86. Wood, R.T.; Griffiths, M.D. The acquisition, development and maintenance of lottery and scratchcard gambling in adolescence. *J. Adolesc.* **1998**, *21*, 265–273. [CrossRef] [PubMed]
87. Boyer, M.; Dickerson, M. Attentional bias and addictive behaviour: Automaticity in a gambling-specific modified Stroop task. *Addiction* **2003**, *98*, 61–70. [CrossRef] [PubMed]
88. McCusker, C.G.; Gettings, B. Automaticity of cognitive biases in addictive behaviours: Further evidence with gamblers. *Br. J. Clin. Psychol.* **1997**, *36*, 543–554. [CrossRef] [PubMed]
89. Handy, T.C.; Kam, J.W. Mind wandering and selective attention to the external world. *Can. J. Exp. Psychol.* **2015**, *69*, 183. [CrossRef] [PubMed]
90. Kam, J.W.; Xu, J.; Handy, T.C. I don't feel your pain (as much): The desensitizing effect of mind wandering on the perception of others' discomfort. *Cogn. Affect. Behav. Neurosci.* **2014**, *14*, 286–296. [CrossRef] [PubMed]
91. Nagamatsu, L.S.; Kam, J.W.; Liu-Ambrose, T.; Chan, A.; Handy, T.C. Mind-wandering and falls risk in older adults. *Psychol. Aging* **2013**, *28*, 685–691. [CrossRef] [PubMed]
92. Harris, A.; Parke, A.; Griffiths, M.D. The case for using personally relevant and emotionally stimulating gambling messages as a gambling harm-minimisation strategy. *Int. J. Ment. Health Addict.* **2016**, 1–10. [CrossRef]
93. Savard, A.-C.; Turcotte, D.; Tremblay, J. <<La première fois que j'ai joué...>>: L'expérience d'adolescents considérés comme ayant des difficultés avec les jeux de hasard et d'argent. *J. Gambl. Issues* **2016**, 44–67. (In French) [CrossRef]
94. Lesieur, H.R.; Blume, S.B. The South Oaks Gambling Screen (SOGS): A new instrument for the identification of pathological gamblers. *Am. J. Psychiatry* **1987**, *144*, 9.
95. Ladouceur, R.; Jacques, C.; Chevalier, S.; Sévigny, S.; Hamel, D. Prevalence of pathological gambling in Quebec in 2002. *Can. J. Psychiatry* **2005**, *50*, 451–456. [CrossRef] [PubMed]
96. Bouchard, S.; Robillard, G.; Renaud, P. *Questionnaire sur la Propension à L'immersion. Lab Cyberpsychologie L'UQO*; Laboratoire de Cyberpsychologie de l'UQO: Outaouais, QC, Canada, 2002.
97. Robillard, G.; Bouchard, S.; Renaud, P.; Cournoyer, L.G. Validation canadienne-française de deux mesures importantes en réalité virtuelle: l'Immersive Tendencies Questionnaire et le Presence Questionnaire. Presented at the 25e Congrès Annu Société Québécoise Pour Rech En Psychol SQRP, Trois-Rivières, QC, Canada, 1–3 November 2002.
98. Kern, L.; Acier, D. Adaptation française de l'échelle problematic internet use questionnaire. *Lévolution psychiatr.* **2013**, *78*, 357–371. [CrossRef]
99. Demetrovics, Z.; Szeredi, B.; Rózsa, S. The three-factor model of Internet addiction: The development of the Problematic Internet Use Questionnaire. *Behav. Res. Methods* **2008**, *40*, 563–574. [CrossRef] [PubMed]
100. Kwon, M.; Lee, J.-Y.; Won, W.-Y.; Park, J.-W.; Min, J.-A.; Hahn, C.; Gu, X.; Choi, J.-H.; Kim, D.-J. Development and validation of a smartphone addiction scale (SAS). *PLoS ONE* **2013**, *82*, e56936. [CrossRef] [PubMed]
101. Lopez-Fernandez, O. Short version of the smartphone addiction scale adapted to Spanish and French: Towards a cross-cultural research in problematic mobile phone use. *Addict. Behav.* **2017**, *64*, 275–280. [CrossRef] [PubMed]
102. Van Rooij, A.J.; Schoenmakers, T.M.; Van den Eijnden, R.J.; Vermulst, A.A.; van de Mheen, D. Video game addiction test: Validity and psychometric characteristics. *Cyberpsychol. Behav. Soc. Netw.* **2012**, *15*, 507–511. [CrossRef] [PubMed]
103. Whiteside, S.P.; Lynam, D.R.; Miller, J.D.; Reynolds, S.K. Validation of the UPPS impulsive behaviour scale: A four-factor model of impulsivity. *Eur. J. Personal.* **2005**, *19*, 559–574. [CrossRef]

104. Van der Linden, M.; d'Acremont, M.; Zermatten, A.; Jermann, F.; Larøi, F.; Willems, S.; Juillerat, A.-C.; Bechara, A. A French adaptation of the UPPS impulsive behavior scale. *Eur. J. Psychol. Assess.* **2006**, *22*, 38–42. [CrossRef]
105. Zigmond, A.S.; Snaith, R.P. The hospital anxiety and depression scale. *Acta Psychiatr. Scand.* **1983**, *67*, 361–370. [CrossRef] [PubMed]
106. Gross, J.J.; John, O.P. Individual differences in two emotion regulation processes: Implications for affect, relationships, and well-being. *J. Pers. Soc. Psychol.* **2003**, *85*, 348. [CrossRef] [PubMed]
107. Planzer, S.; Gray, H.M.; Shaffer, H.J. Associations between national gambling policies and disordered gambling prevalence rates within Europe. *Int. J. Law Psychiatry* **2014**, *37*, 217–229. [CrossRef] [PubMed]
108. Costes, J.-M.; Pousset, M.; Eroukmanoff, V.; Le Nezet, O.; Richard, J.-B.; Guignard, R.; Beck, F.; Arwidson, P. Les niveaux et pratiques des jeux de hasard et d'argent en 2010. *Tendances* **2011**, *77*, 8.
109. Costes, J.-M.; Kairouz, S.; Eroukmanoff, V.; Monson, E. Gambling patterns and problems of gamblers on licensed and unlicensed sites in France. *J. Gambl. Stud.* **2016**, *32*, 79–91. [CrossRef] [PubMed]
110. Grall-Bronnec, M.; Bouju, G.; Landréat-Guillou, M.; Vénisse, J.-L. Évaluation sociodémographique, clinique et du parcours de jeu d'un échantillon de joueurs pathologiques français. *L'Encéphale* **2010**, *36*, 452–460. (In French) [CrossRef] [PubMed]
111. Patesson, R. Enquête sur L'addiction des Jeunes aux Smartphones. In *XIe Colloque du Réseau International et Interdisciplinaire Pour les Enjeux et Usages des Technologies de L'information et de la Communication (EUTIC) «Les Ecosystèmes Numériques et la Démocratisation Informationnelle: Intelligence Collective, Développement Durable, Interculturalité, Transfert de Connaissances»*. 2015. Available online: https://hal-uag.archives-ouvertes.fr/hal-01375819/ (accessed on 14 February 2017).
112. Monaghan, S. Responsible gambling strategies for internet gambling: The theoretical and empirical base of using pop-up messages to encourage self-awareness. *Comput. Hum. Behav.* **2009**, *25*, 202–207. [CrossRef]
113. Gainsbury, S.M.; Russell, A.; Wood, R.; Hing, N.; Blaszczynski, A. How risky is Internet gambling? A comparison of subgroups of Internet gamblers based on problem gambling status. *New Media Soc.* **2015**, *17*, 861–879. [CrossRef]
114. Schull, N.D. Digital Gambling: The coincidence of desire and design. *Ann. Am. Acad. Political Soc. Sci.* **2005**, *597*, 65–81. [CrossRef]
115. McBride, J.; Derevensky, J. Gambling and video game playing among youth. *J. Gambl. Issues* **2016**, 156–178. [CrossRef]
116. Forrest, C.J.; King, D.L.; Delfabbro, P.H. The measurement of maladaptive cognitions underlying problematic video-game playing among adults. *Comput. Hum. Behav.* **2016**, *55*, 399–405. [CrossRef]
117. Mazzoni, E.; Cannata, D.; Baiocco, L. Focused, not lost: The mediating role of temporal dissociation and focused immersion on problematic internet use. *Behav. Inf. Technol.* **2017**, *36*, 11–20. [CrossRef]
118. Lehenbauer-Baum, M.; Klaps, A.; Kovacovsky, Z.; Witzmann, K.; Zahlbruckner, R.; Stetina, B.U. Addiction and engagement: An explorative study toward classification criteria for internet gaming disorder. *Cyberpsychol. Behav. Soc. Netw.* **2015**, *18*, 343–349. [CrossRef] [PubMed]
119. King, D.L.; Delfabbro, P.H. Is preoccupation an oversimplification? A call to examine cognitive factors underlying internet gaming disorder. *Addiction* **2014**, *109*, 1566–1567. [CrossRef] [PubMed]
120. Yakovenko, I.; Hodgins, D.C.; El-Guebaly, N.; Casey, D.M.; Currie, S.R.; Smith, G.J.; Williams, R.J.; Schopflocher, D.P. Cognitive distortions predict future gambling involvement. *Int. Gambl. Stud.* **2016**, *16*, 175–192. [CrossRef]
121. Griffiths, M. *Adolescent Gambling*; Psychology Press: Hove, UK, 1995.
122. King, D.L.; Delfabbro, P.H. Early exposure to digital simulated gambling: A review and conceptual model. *Comput. Hum. Behav.* **2016**, *55*, 198–206. [CrossRef]
123. St-Pierre, R.; Derevensky, J.L. Youth gambling behavior: Novel approaches to prevention and intervention. *Curr. Addict. Rep.* **2016**, *3*, 157–165. [CrossRef]
124. Davies, B.; Blake, E. Evaluating existing strategies to limit video game playing time. *IEEE Comput. Graph. Appl.* **2016**, *36*, 47–57. [CrossRef] [PubMed]

International Journal of
Environmental Research and Public Health

MDPI

Article

Problematic Smartphone Use, Deep and Surface Approaches to Learning, and Social Media Use in Lectures †

Dmitri Rozgonjuk [1,2,*], Kristiina Saal [1] and Karin Täht [1]

[1] Institute of Psychology, University of Tartu, Tartu 50409, Estonia; kristiinasaal@gmail.com (K.S.); karin.taht@ut.ee (K.T.)
[2] Department of Psychology, University of Toledo, Toledo, OH 43606, USA
* Correspondence: dmroz@ut.ee; Tel.: +372-737-5912
† The preliminary results of this study were presented on the ICBA 2017 Conference on Proneness to Smartphone Addiction, Internet Addiction, and Approaches to Learning, Haifa, Israel, 20–22 February 2017.

Received: 22 November 2017; Accepted: 4 January 2018; Published: 8 January 2018

Abstract: Several studies have shown that problematic smartphone use (PSU) is related to detrimental outcomes, such as worse psychological well-being, higher cognitive distraction, and poorer academic outcomes. In addition, many studies have shown that PSU is strongly related to social media use. Despite this, the relationships between PSU, as well as the frequency of social media use in lectures, and different approaches to learning have not been previously studied. In our study, we hypothesized that both PSU and the frequency of social media use in lectures are negatively correlated with a deep approach to learning (defined as learning for understanding) and positively correlated with a surface approach to learning (defined as superficial learning). The study participants were 415 Estonian university students aged 19–46 years (78.8% females; age M = 23.37, SD = 4.19); the effective sample comprised 405 participants aged 19–46 years (79.0% females; age M = 23.33, SD = 4.21). In addition to basic socio-demographics, participants were asked about the frequency of their social media use in lectures, and they filled out the Estonian Smartphone Addiction Proneness Scale and the Estonian Revised Study Process Questionnaire. Bivariate correlation analysis showed that PSU and the frequency of social media use in lectures were negatively correlated with a deep approach to learning and positively correlated with a surface approach to learning. Mediation analysis showed that social media use in lectures completely mediates the relationship between PSU and approaches to learning. These results indicate that the frequency of social media use in lectures might explain the relationships between poorer academic outcomes and PSU.

Keywords: problematic smartphone use; smartphone addiction; social media; approaches to learning; deep approach to learning; surface approach to learning

---

1. Introduction

Since their introduction in 2009, smartphones have had a significant impact on daily life across the world. The worldwide ownership of smartphones is around 43% [1], and more than 50% of Estonians own a smartphone [2]. Smartphones have several advantages in educational settings, allowing one to take notes, browse for information, communicate with others, and use specific applications for learning skills [3]. Several studies have shown, however, that there are instances where the excessive use of smartphones leads to the development of "problematic smartphone use" (PSU), a phenomenon characterized by the occurrence of addictive-like symptoms [4]. PSU has been associated with psychopathological symptoms [5] as well as poor academic outcomes [6,7]. Similar findings have been reported with excessive social media use [8]. The aim of this paper is to investigate how

PSU is related to approaches to learning. Specifically, this study addresses the questions of how PSU is related to learning for understanding (known as the deep approach to learning) and learning motivated by external incentives (for example, grades; known as a surface approach to learning) [9]. In addition, the mediating effect of social media use in lectures is analyzed, as it may be the case that the relationship between PSU and approaches to learning is dependent on the frequency of social media use.

Previous works regarding smartphone addiction is often seen as controversial. It has been debated whether or not smartphone "addiction" is a behavioral addiction, and whether or not it may be considered as an actual addiction from the perspective of contemporary addiction theories [10–12]. Nevertheless, one cannot neglect the growing body of evidence that suggests that problematic smartphone usage is related to several psychopathological symptoms, such as depression and anxiety [5,13], poorer health and sleep quality [14], and lower academic achievements [6]. In addition to smartphone addiction, the following terms have been used to describe conceptually the same phenomenon: proneness to smartphone addiction [15,16], smartphone overuse [17,18], excessive smartphone use [19], problematic mobile phone use [20], mobile phone dependence [21], and mobile phone addiction [22,23]. In essence, these terms characterize the excessive use of smartphones accompanied by symptoms resembling those found in contemporary addiction models: dependence, withdrawal, tolerance, and functional impairment [4]; therefore, we conceptualize the phenomenon as problematic smartphone use.

PSU as a maladaptive coping method [24] has received relatively little attention in academic contexts in scientific literature. Although studies have found that the higher levels of PSU are related to poor academic outcomes [6,7], different approaches to learning as potential causal factors for academic outcomes have not yet been researched in the light of problematic technology use.

Empirical findings have led Marton and Säljö [9] to propose a distinction between approaches to learning; these were categorized as surface and deep approaches to learning. Accordingly, one uses a deep approach to learning to fully understand the content studied. In contrast, the surface approach to learning is characterized by instrumental learning which aims to minimally fulfill the requirements of study; with the latter, only the basics of the study material are learned [25]. Biggs [26] defined both of these approaches as motives and strategies. While a deep motive involves intrinsic interest in the content that is learned, and the aim is to develop competence, a surface motive is to meet requirements minimally. In addition, deep strategy is to discover meaning by reading widely, relating one's previous knowledge with learned material, whereas surface strategy is to limit the amount of work one needs to do (through rote learning).

It has been shown that a deep approach to learning facilitates critical thinking, finding causal relationships, creativity, and autonomous thinking. Therefore, this type of approach is an essential ingredient in successful learning, as it leads to a more systematic and thorough comprehension of information [27]. Authors [28] have found that those with a deep approach to learning have better developed organization skills, more spontaneously expressed ideas, and clearer, longer, and more detailed explanations. Those with a deep approach to learning may also make more associations with their work and prior knowledge and experiences [28]. Other works have linked deep learning with a higher need to understand material and a drive to satisfy one's curiosity [29,30]. On the other hand, a surface approach to learning is characterized by incentive-oriented strategy, according to which the student aims to do only as much as is necessary to study, often compulsory, material [25,29]. Those who are prone to this approach tend to focus on learning facts, isolated details, examples and illustrations without connecting the pieces of information to their knowledge to provide a broader perspective [30]. This leads to a focus on understanding simpler principles [27]. Surface learners are rather passive and their attitude towards learning is to regard it as something which inevitably happens. In comparison, a deep approach to learning leads to an increased sense of control over the learning process [9].

As the information suggests, relationships between academic success and learning strategies are well researched. Several studies have shown that surface learners have lower academic outcomes and are less successful in school, whereas those with a deep approach to learning are more successful in school [31–35]. Additionally, the probability of graduating from higher education institutions is lower for students who study superficially [36].

In addition, it has been found that more social media use (on the example of Facebook) is related to worse academic outcomes [37]; also, engaging with social distractors (social media with its functionalities could be considered as one) have shown to be related to poorer concentration and procrastination [38]. Unsurprisingly, one of the main functions of smartphones is social media use [39–42], allowing for almost constant access to continually changing content and engaging with social distractors.

Though some studies have shown that higher levels of PSU and other excessive technology use (e.g., the Internet) are associated with poorer academic outcomes [6,7,43,44], it has not been studied if different levels of PSU are related to deep and surface approach to learning. Furthermore, the role of social media use in lectures is not investigated in the potential relationships between the levels of PSU, and deep and surface approaches to learning.

The aims of this study are (a) to investigate how deep and surface approaches to learning are related to the levels of PSU and social media use in lectures, and (b) to clarify if social media use in lectures mediates the relationship between the levels of PSU and approaches to learning. Based on the empirical findings discussed earlier, we have posed the following hypotheses:

Hypothesis 1 (H1). *The levels of PSU are negatively related to a deep approach to learning and positively related with a surface approach to learning. PSU is a maladaptive coping method [24] that has been found to be associated with several other dysfunctional behaviors or detrimental outcomes, including in academic settings [6,7]. Therefore, we expect the levels of PSU to be correlated negatively to the deep approach to learning (adaptive learning strategy) and positively to the surface approach to learning (maladaptive learning strategy).*

Hypothesis 2 (H2). *The frequency of social media use in lectures is negatively correlated to a deep approach to learning and positively related to a surface approach to learning. Studies have shown that more social media use is related to poorer academic outcomes [37]. Similarly to the previous hypothesis, we expect social media use in lectures to be related to approaches to learning.*

Hypothesis 3 (H3). *The levels of PSU are positively correlated with the frequency of social media use in lectures. The levels of PSU have been shown to be related to the social use of smartphones [39–42]. We expect that is also the case with social media use in lectures, specifically.*

Hypothesis 4 (H4). *Social media use in lectures mediates the relationship between the levels of PSU and approaches to learning. It has been found that levels of PSU are related to social media use and social use in general [15,42,45]. It has also been shown that more social media use is associated with lower academic outcomes [37]. We hypothesize that higher levels of PSU is related to more social media use in lectures, and this is related with a less deep and more surface approach to learning.*

2. Materials and Methods

2.1. Sample and Procedure

Four hundred fifteen Estonian university students (ages ranging from 19 to 46 years, M_{age} = 23.37 ± 4.19 years; 78.8% were female) participated in an online survey in Estonian. Ten participants had missing data in their responses and were, therefore, removed from further analyses. The characteristics of the effective sample are the following: ages ranging from 19 to 46 years, M_{age} = 23.33 ± 4.21 years; 79.0% were female. Therefore, the efficient sample comprised 405 participants (age range from 19 to 46 years, M_{age} = 23.33 ± 4.21 years; 79.0% were female). Three hundred ninety-three (97.04%) participants marked their mother tongue as Estonian, 11 (2.7%) as Russian, and 1 as "other."

The participants were recruited through university mailing lists and via social media. Prior to taking the survey, participants filled out an informed consent form in accordance with the Declaration of Helsinki. The participants were asked to complete a merged questionnaire that consisted of the following parts: (a) general socio-demographics, (b) the Estonian Revised Study Process Questionnaire, (c) questions about social media use in lectures, (d) the Estonian Smartphone Addiction Proneness Scale.

2.2. Explanations of Measures

Further details of the measures used are provided.

2.2.1. General Socio-Demographics

These included items about the participants' age (in full years), gender (male/female/other), and mother tongue (Estonian/Russian/other).

2.2.2. The Revised Study Process Questionnaire (Estonian Adaptation)

(R-SPQ-2F; [46]): this is a modified two-dimensional 16-item questionnaire with Likert-type scale (from 1 = "do not agree at all" to 5 = "totally agree") that is inspired by the Revised Study Process Questionnaire developed by Biggs, Kember, and Leung [25]; see [46] for further details about the adaptation process and Table S1 for the items of the scale. The scale measures the extent to which students use a surface (e.g., "I generally restrict my study to what is specifically set as I think it is unnecessary to do anything extra") or deep (e.g., "I find that at times studying gives me a feeling of deep personal satisfaction") approach to learning as their learning strategy. The internal reliability was acceptable for the surface approach to learning subscale (Cronbach's $\alpha = 0.72$), and for the deep approach to learning subscale (Cronbach's $\alpha = 0.75$).

2.2.3. Social Media Use in Lectures

Participants were asked two questions about social media use in lectures, derived from the items and scales used in [47]: firstly, "How frequently do you keep track of social media in lectures?" (1 = "never" ... 5 = "all the time") and, secondly, "How often do you communicate with friends using social networking sites (Facebook, Twitter, etc.) in lectures?" (1 = "never" ... 5 = "all the time"). These items were highly correlated ($r = 0.74$, $p < 0.0001$) and were aggregated into one social media use index.

2.2.4. The Estonian Smartphone Addiction Proneness Scale

(E-SAPS18; [15]: The E-SAPS18 is a five-dimensional questionnaire consisting of 18 items measuring the frequency of smartphone addiction symptoms on a 6-point Likert scale (from 1 = "strongly disagree" to 6 = "strongly agree"). E-SAPS18 consists of five factors (tolerance, positive anticipation, cyberspace-oriented relationships, withdrawal, and physical symptoms), with a higher-order factor of smartphone addiction proneness. The internal reliability of E-SAPS18 is very good (Cronbach's $\alpha = 0.87$), and E-SAPS18 has been validated against other instruments measuring Internet and smartphone addiction proneness.

2.3. Data Analysis

RStudio version 3.2.3 (R Core Team, Vienna, Austria) was used for descriptive statistics and correlation analysis. The skewness and kurtosis of distributions of variables were among the acceptable ranges (from −2 to +2) [48,49]. One-tailed Pearson product–moment correlation was used to analyze bivariate relationships between variables, as we expected directional relationships in each correlation. Mplus version 8 was used for mediation analysis.

We used the weighted least squares estimation with a mean- and variance-adjusted chi-square (WLSMV), because it has been shown to be less biased and more accurate than similar estimation methods (e.g., robust maximum likelihood, or MLR) for the data type [50]. Using the WLSMV estimation method, we treated the items of social media use in lectures and E-SAPS18 as categorical/ordinal data, thus involving a polychoric covariance matrix and probit regression coefficients [51]. The summed scale scores for deep and surface approach to learning were treated as observed variables in the models. The cross-products of direct effects to compute mediation effect were used, applying the Delta method for computing indirect effect standard errors, with non-parametric bootstrapping across 1000 samples [52]. Goodness of fit was judged by standard parameters: (a) Comparative Fit Index (CFI) ≥ 0.90, (b) Tucker–Lewis Index (TLI) ≥ 0.90, and (c) root mean square error of approximation (RMSEA) ≤ 0.08.

We used SPSS version 24 to compute Mahalanobis distances to detect the outliers in the multivariate dataset. Altogether, there were 19 cases (4.7%) where $p < 0.05$ (4 cases, or 1% of the sample, with $p < 0.01$). The analyses described above were re-ran in order to detect if these statistical outliers influenced the results cardinally. However, as the results were virtually the same (signs and magnitudes of coefficients, statistical significance of the predictors, and the mediating variable), we retained these multivariate outliers in the sample.

3. Results

3.1. Correlation Analysis

Summary and scale reliability statistics for the study sample are highlighted in Table 1. In order to analyze the relationships between PSU, approaches to learning, and social media use in lectures, Pearson product–moment correlation analyses were carried out. The results are shown in Table 1.

Table 1. Pearson product–moment correlations, means, standard deviations, and internal consistencies of scales and subscales.

Variable		1	2	3	Min	Max	M	SD	α
1	PSU				18	76	35.52	10.73	0.86
2	SMUL	0.326 ***			2	10	5.92	2.10	0.85
3	DA	−0.109 *	−0.304 ***		11	39	26.80	4.62	0.78
4	SA	0.177 ***	0.297 ***	−0.425 ***	8	34	20.12	4.93	0.75

Notes: $N = 415$. PSU = problematic smartphone use measured by the E-SAPS18; SMUL = Social media use in lectures; DA = deep approach to learning; SA = surface approach to learning. * $p \leq 0.05$, *** $p \leq 0.001$.

It could be observed that all variables are statistically significantly inter-related. Specifically, PSU (as an E-SAPS18 score) is in positive correlation with social media use in lectures and a surface approach to learning; the latter two variables are, in turn, also positively correlated with each other. Additionally, both PSU and social media use in lectures are negatively correlated with a deep approach to learning.

3.2. Mediation Analysis

Confirmatory factor analysis showed that the five-factor E-SAPS18 with one higher-order factor had an acceptable fit: CFI = 0.93; TLI = 0.92; RMSEA = 0.09. The fit indices for two-factor R-SPQ-2F (Estonian adaptation) were CFI = 0.76, TLI = 0.72, and RMSEA = 0.13. Because the latter scale did not show a great fit, we used observed variables (summed scores for deep and surface approach to learning) as outcome variables.

Three mediation models were constructed to investigate if social media use in lectures mediates the relationship between PSU and approaches to learning. In the first model (Model 1), we used only the deep approach to learning as the outcome variable. In the second model (Model 2), we used the surface approach to learning as the outcome model. In both of these models, the mediating variable was controlled for age and gender, as it has been shown that age and gender might impact social media engagement [53]. The third model (Model 3) included both the deep and surface approaches to learning as outcome variables predicted by the levels of PSU and mediated by social media use in lectures. Again, the mediator was controlled for gender; furthermore, deep and surface approaches to learning were specified to covary in that model. As mentioned earlier, measures of PSU and social media use in lectures were modeled as latent (categorical/ordinal), and the summed scores for deep and surface approaches to learning and the covariates (age, gender) were treated as observed variables.

The fit indices for Model 1 were CFI = 0.94, TLI = 0.93, and RMSEA = 0.07; for Model 2, they were CFI = 0.94, TLI = 0.93, and RMSEA = 0.07; and for Model 3: CFI = 0.94, TLI = 0.93, and RMSEA = 0.07. The coefficients of the models are presented in Table 2.

Table 2. The results of mediation models.

Model 1 (Outcome DA)				
Covariates	**Bivariate B (SE)**	t	**Multivariate B (SE)**	t
PSU	0.001 (0.079)	0.016	0.001 (0.061)	0.016
SMUL	−0.345 (0.060)	−5.751 ***	−0.417 (0.063)	−6.630 ***
PSU -> SMUL	0.547 (0.098)	5.573 ***	0.348 (0.059)	5.874 ***
Age -> SMUL	−0.095 (0.012)	−7.799 ***	−0.440 (0.053)	−8.349 ***
Gender -> SMUL	0.103 (0.139)	0.744	0.046 (0.061)	0.754
PSU -> SMUL -> DA	−0.189 (0.050)	−3.757 ***	−0.145 (0.035)	−4.115 ***
Model 2 (Outcome SA)				
Covariates	**Bivariate B (SE)**	t	**Multivariate B (SE)**	t
PSU	0.088 (0.061)	1.457	0.093 (0.061)	1.513
SMUL	0.205 (0.050)	4.136 ***	0.336 (0.076)	4.410 ***
PSU -> SMUL	0.548 (0.094)	5.820 ***	0.350 (0.056)	6.247 ***
Age -> SMUL	−0.082 (0.011)	−7.369 ***	−0.382 (0.047)	−8.055 ***
Gender -> SMUL	0.103 (0.124)	0.829	0.046 (0.055)	0.839
PSU -> SMUL -> SA	0.112 (0.035)	3.193 ***	0.118 (0.035)	3.340 ***
Model 3 (Outcomes DA and SA)				
Covariates	**Bivariate B (SE)**	t	**Multivariate B (SE)**	t
PSU -> DA	0.002 (0.084)	0.024	0.001 (0.060)	0.024
SMUL -> DA	−0.361 (0.060)	−6.008 ***	−0.419 (0.064)	−6.604 ***
PSU -> SA	0.112 (0.071)	1.572	0.097 (0.061)	1.594
SMUL -> SA	0.243 (0.057)	4.261 ***	0.336 (0.074)	4.569 ***
PSU -> SMUL	0.562 (0.102)	5.526 ***	0.353 (0.060)	5.848 ***
Age -> SMUL	−0.097 (0.013)	−7.609 ***	−0.447 (0.053)	−8.364 ***
Gender -> SMUL	0.053 (0.146)	0.365	0.024 (0.064)	0.370
PSU -> SMUL -> DA	−0.203 (0.055)	−3.668 ***	−0.148 (0.037)	−3.969 ***
PSU -> SMUL -> SA	0.137 (0.044)	3.095 **	0.119 (0.037)	3.222 ***
DA <-> SA	−0.228 (0.037)	−6.094 ***	−0.526 (0.051)	−10.365 ***

Notes $N = 415$. PSU = problematic smartphone use measured by the E-SAPS18; SMUL = Social media use in lectures; DA = deep approach to learning; SA = surface approach to learning. ** $p \leq 0.01$, *** $p \leq 0.001$.

It could be observed from all three models that social media use in lectures mediates the relationship between problematic smartphone use and approaches to learning. In Model 1, PSU does not have a direct effect on deep approach to learning; however, social media use in lectures completely mediates the relationship between PSU and deep approach to learning. Furthermore, the indirect effect is negative.

Similarly, social media use in lectures completely mediates the relationship between PSU and surface approach to learning (see Model 2). However, in this case, the indirect effect is positive. Higher PSU leads to higher social media use in lectures which, in turn, is associated with more surface approach to learning.

These effects are present even if both surface and deep approach to learning are included in the same model as outcome variables. Interestingly, gender is not associated with social media use in lectures, whereas in all models, younger people tend to use more social media in lectures.

4. Discussion

According to previous research, PSU and excessive social media use relate to poorer academic outcomes [6]. However, these relationships are often examined in relation to grades or grade point average (GPA), not to deep and surface approaches to learning. As such, in our study, we hypothesized that PSU would be negatively related to a deep approach to learning and positively related to a surface approach to learning. Indeed, the results showed that a higher PSU is associated with less commitment to a deep approach to learning and with greater commitment to a surface approach to learning. We expected, too, that more frequent social media use in lectures would be negatively correlated with a deep approach to learning and positively associated with a surface approach to learning. Our results also confirmed this hypothesis. The relationship between frequent, engaged smartphone use and multitasking [54,55], which may be detrimental to academic achievement [56], may potentially explain these findings. Those with additional access to technology may be more prone to multitasking [57], which may in turn lead to less in-depth or more superficial learning. Future studies should try to take into account multitasking as a potential factor in influencing the relationships reported in this study.

According to our third hypothesis, we expected PSU to be positively associated with the frequency of social media use in lectures. This hypothesis was confirmed by the results. This result is not surprising, as one of the most popular activities on smartphones is social media usage. Therefore, it is natural that those with higher PSU are also more prone to use social media in lectures.

As a second aim of the study, we presented a conceptual mediation model where the relationship between PSU and approaches to learning are mediated by the frequency of social media use in lectures. Our previous results provided the statistical basis for the models [58]. According to the models, social media use in lectures completely mediated the relationship between PSU and approaches to learning. In other words, though PSU and approaches to learning were related in bivariate correlations, these relationships could be explained by the effect of social media use frequency. Indeed, it has been indicated that the most frequent activity carried out on smartphones is social media use [15,45]. Smartphones allow for mobility and availability in terms of online interactions and content consumption from social media platforms.

Several limitations are worth mentioning. Firstly, self-report measures were used, and this might affect the reliability and ecological validity of the findings. It would be interesting to contrast our results with behavioral social media usage data (e.g., the actual measured time of social media usage) and objective academic data (e.g., GPA, test scores, and SAT scores). This kind of approach would allow for more accurate and reliable data, and better generalization. Secondly, the study design was cross-sectional; therefore, one has to be cautious in making causal interpretations. This is especially important to keep in mind while interpreting the results of mediation models. Our hypothesized models are based on the logic that PSU, as a more generic construct, encompasses social media use and that it therefore could be a precondition for more social media consumption. However, it could also be the case that more social media use causes more PSU. Additionally, students' approaches to learning could be influenced by both PSU and social media use; however, it could also be the case that approaches to learning impact how much social media a student consumes in lectures. Finally, we did not ask the participants how much time they spent online (per day or per week) and why the participants engaged in activities involving social media use in lectures. Though studies have shown that self-reported estimated time of technology use is typically inaccurate in comparison to objective use [59–61], knowing learning about the objective usage time and could greatly improve the understanding of these results. Additionally, distinguishing between the activities in which students engage on social media (e.g., learning vs. entertainment that is not related to class content) could specify the potential effects. We encourage the reader to be mindful of these limitations, and we consider experimental and/or longitudinal studies to be helpful in clarifying this issue; we hope the current paper inspires studies of that nature.

This study has both theoretical and practical implications. From a theoretical perspective, this is the first study to investigate the relationships between approaches to learning and PSU. Though it has been shown that a deep approach to learning is positively related to academic outcomes, and a surface approach to learning is found to have negative effects on academic outcomes [31–34], this study provides potential evidence to the findings where PSU has been associated with poor academic outcomes [6]. Specifically, poorer outcomes might be due to a less deep and/or more surface approach to learning. In addition, the findings between PSU and academic outcomes could be explained by students engaging in more social media use.

Among the potential practical implications, this paper suggests that social media use in lectures might be the driver in the relationship between approaches to learning and PSU. Lecturers might consider restricting access to social media use in lectures or to advise their students that engaging in excessive smartphone use and especially social media use in lectures could have detrimental effects on their academic outcomes. However, this idea needs further testing—preferably by implementing experimental and, if possible, longitudinal study design to better understand if this practical suggestion derived from the results of this study have beneficial effects.

5. Conclusions

Recent advances in technology offer many promising advantages in the classroom. However, problematic technology use has been shown to be negatively related to academic outcomes. In the current study, we found that both PSU and social media use in lectures are related to less commitment to a deep approach to learning (learning to fully understand the content studied) in place of a more surface approach to learning (instrumental learning to meet the requirements of one's learning outcomes). Additionally, we proposed hypothetical models where the frequency of social media use in lectures mediates the relationship between PSU and approaches to learning. We found that social media use in lectures completely mediates the aforementioned relationship. Further studies should address the limitations of this study, such as using self-reported measures and cross-sectional study design.

Supplementary Materials: The following are available online at www.mdpi.com/1660-4601/15/1/92/s1, Table S1: The 16-Item Revised Study Process Questionnaire (Estonian Adaptation).

Acknowledgments: This study was not funded. The authors did not receive funds for covering the costs to publish in open access.

Author Contributions: Dmitri Rozgonjuk, Kristiina Saal, and Karin Täht conceived and designed the study; Dmitri Rozgonjuk and Kristiina Saal performed the data collection; Dmitri Rozgonjuk and Karin Täht analyzed the data; Dmitri Rozgonjuk, Kristiina Saal, and Karin Täht wrote the paper.

Conflicts of Interest: The authors declare no conflict of interest.

References

1. Poushter, J. Smartphone Ownership and Internet Usage Continues to Climb in Emerging Economies. 2016. Available online: http://www.pewglobal.org/2016/02/22/smartphone-ownership-and-internet-usage-continues-to-climb-in-emerging-economies/ (accessed on 10 October 2017).
2. Statista. Forecast of the Smartphone User Penetration Rate in Estonia from 2015 to 2022. 2017. Available online: https://www.statista.com/statistics/568089/predicted-smartphone-user-penetration-rate-in-estonia/ (accessed on 11 February 2017).
3. Godwin-Jones, R. Mobile apps for language learning. *Lang. Learn. Technol.* **2011**, 15, 2–11.
4. Billieux, J.; Maurage, P.; Lopez-Fernandez, O.; Kuss, D.J.; Griffiths, M.D. Can disordered mobile phone use be considered a behavioral addiction? An update on current evidence and a comprehensive model for future research. *Curr. Addict. Rep.* **2015**, 2, 156–162. [CrossRef]

5. Elhai, J.D.; Dvorak, R.D.; Levine, J.C.; Hall, B.J. Problematic smartphone use: A conceptual overview and systematic review of relations with anxiety and depression psychopathology. *J. Affect. Disord.* **2017**, *207*, 251–259. [CrossRef] [PubMed]
6. Samaha, M.; Hawi, N.S. Relationships among smartphone addiction, stress, academic performance, and satisfaction with life. *Comput. Hum. Behav.* **2016**, *57*, 321–325. [CrossRef]
7. Hawi, N.S.; Samaha, M. To excel or not to excel: Strong evidence on the adverse effect of smartphone addiction on academic performance. *Comput. Educ.* **2016**, *98*, 81–89. [CrossRef]
8. Al-Menayes, J.J. Social Media Use, Engagement and Addiction as Predictors of Academic Performance. *Int. J. Psychol. Stud.* **2015**, *7*, 86–94. [CrossRef]
9. Marton, F.; Säljö, R. On qualitative differences in learning: I—Outcome and process. *Br. J. Educ. Psychol.* **1976**, *46*, 4–11. [CrossRef]
10. Kardefelt-Winther, D.; Heeren, A.; Schimmenti, A.; van Rooij, A.; Maurage, P.; Carras, M.; Edman, J.; Blaszczynski, A.; Khazaal, Y.; Billieux, J. How can we conceptualize behavioural addiction without pathologizing common behaviours? *Addiction* **2017**, *112*, 1709–1715. [CrossRef] [PubMed]
11. Sussman, S.; Rozgonjuk, D.; van den Eijnden, R. Substance and behavioral addictions may share a similar underlying process of dysregulation. *Addiction* **2017**, *112*, 1717–1718. [CrossRef] [PubMed]
12. Griffiths, M.D. Behavioural addiction and substance addiction should be defined by their similarities not their dissimilarities. *Addiction* **2017**, *112*, 1718–1720. [CrossRef] [PubMed]
13. Elhai, J.D.; Levine, J.C.; Dvorak, R.D.; Hallde, B.J. Fear of missing out, need for touch, anxiety and depression are related to problematic smartphone use. *Comput. Hum. Behav.* **2016**, *63*, 509–516. [CrossRef]
14. Demirci, K.; Akgonul, M.; Akpinar, A. Relationship of smartphone use severity with sleep quality, depression, and anxiety in university students. *J. Behav. Addict.* **2015**, *4*, 85–92. [CrossRef] [PubMed]
15. Rozgonjuk, D.; Rosenvald, B.; Janno, S.; Täht, K. Developing a shorter version of the Estonian Smartphone Addiction Proneness Scale (E-SAPS18). *Cyberpsychol. J. Psychosoc. Res. Cyberspace* **2016**, *10*, 4. [CrossRef]
16. Kim, D.; Lee, Y.; Lee, J.; Nam, J.K.; Chung, Y. Development of Korean Smartphone addiction proneness scale for youth. *PLoS ONE* **2014**, *9*, e97920. [CrossRef] [PubMed]
17. Inal, E.E.; Demİrcİ, K.; Çetİntürk, A.; Akgönül, M.; Savaş, S. Effects of smartphone overuse on hand function, pinch strength, and the median nerve. *Muscle Nerve* **2015**, *52*, 183–188. [CrossRef] [PubMed]
18. Lee, H.K.; Kim, J.H.; Fava, M.; Mischoulon, D.; Park, J.H.; Shim, E.J.; Lee, E.H.; Lee, J.H.; Jeon, H.J. Development and validation study of the Smartphone Overuse Screening Questionnaire. *Psychiatry Res.* **2017**, *257*, 352–357. [CrossRef] [PubMed]
19. Chen, J.; Liang, Y.; Mai, C.; Zhong, X.; Qu, C. General Deficit in Inhibitory Control of Excessive Smartphone Users: Evidence from an Event-Related Potential Study. *Front. Psychol.* **2016**, *7*, 511. [CrossRef] [PubMed]
20. Billieux, J.; Van der Linden, M.; d'Acremont, M.; Ceschi, G.; Zermatten, A. Does impulsivity relate to perceived dependence on and actual use of the mobile phone? *Appl. Cogn. Psychol.* **2007**, *21*, 527–537. [CrossRef]
21. Wang, C.; Wang, S.Y.; Li, W.H.; Dong, X.W.; Chi, G.B. Study on the mobile phone dependence syndrome and its distribution among 2213 college students in Guangzhou. *Zhonghua Liu Xing Bing Xue Za Zhi* **2013**, *34*, 949–952. [PubMed]
22. Choliz, M. Mobile phone addiction: A point of issue. *Addiction* **2010**, *105*, 373–374. [PubMed]
23. Pedrero Perez, E.J.; Rodriguez Monje, M.T.; Ruiz Sanchez De Leon, J.M. Mobile phone abuse or addiction. A review of the literature. *Adicciones* **2012**, *24*, 139–152. [PubMed]
24. Elhai, J.D.; Tiamiyu, M.F.; Weeks, J.W. Depression and Social Anxiety in Relation to Problematic Smartphone Use: The Prominent Role of Rumination. *Internet Res.* **2017**, in press.
25. Biggs, J.; Kember, D.; Leung, D.Y. The revised two-factor Study Process Questionnaire: R-SPQ-2F. *Br. J. Educ. Psychol.* **2001**, *71*, 133–149. [CrossRef] [PubMed]
26. Biggs, J. *Study Process Questionnaire Manual. Student Approaches to Learning and Studying*; Australian Council for Educational Research: Hawthorn, Australia, 1987.
27. Warburton, K. Deep learning and education for sustainability. *Int. J. Sustain. High. Educ.* **2003**, *4*, 44–56. [CrossRef]
28. Chin, C.; Brown, D.E. Learning in science: A comparison of deep and surface approaches. *J. Res. Sci. Teach.* **2000**, *37*, 109–138. [CrossRef]

29. Rogaten, J.; Moneta, G.B.; Spada, M.M. Academic performance as a function of approaches to studying and affect in studying. *J. Happiness Stud.* **2013**, *14*, 1751–1763. [CrossRef]
30. Hoeksema, L.H. *Learning Strategy as a Guide to Career Success in Organizations*; DSWO Press, Leiden University: Leiden, The Netherlands, 1995.
31. Arquero, J.L.; Fernández-Polvillo, C.; Hassall, T.; Joyce, J. Vocation, motivation and approaches to learning: A comparative study. *Educ. Train.* **2015**, *57*, 13–30. [CrossRef]
32. Gynnild, V.; Myrhaug, D. Revisiting approaches to learning in science and engineering: A case study. *Eur. J. Eng. Educ.* **2012**, *37*, 458–470. [CrossRef]
33. Heikkilä, A.; Lonka, K. Studying in higher education: Students' approaches to learning, self-regulation, and cognitive strategies. *Stud. High. Educ.* **2006**, *31*, 99–117. [CrossRef]
34. Malie, S.; Akir, O. Bridging the gaps between learning and teaching through recognition of students' learning approaches: A case study. *Res. Educ.* **2012**, *87*, 75–94. [CrossRef]
35. Salamonson, Y.; Weaver, R.; Chang, S.; Koch, J.; Bhathal, R.; Khoo, C.; Wilson, L. Learning approaches as predictors of academic performance in first year health and science students. *Nurse Educ. Today* **2013**, *33*, 729–733. [CrossRef] [PubMed]
36. Rocconi, L.M.; Ribera, A.K.; Laird, T.F.N. College Seniors' Plans for Graduate School: Do Deep Approaches Learning and Holland Academic Environments Matter? *Res. High. Educ.* **2015**, *56*, 178–201. [CrossRef]
37. Kirschner, P.A.; Karpinski, A.C. Facebook® and academic performance. *Comput. Hum. Behav.* **2010**, *26*, 1237–1245. [CrossRef]
38. Dewitte, S.; Schouwenburg, H.C. Procrastination, temptations, and incentives: The struggle between the present and the future in procrastinators and the punctual. *Eur. J. Personal.* **2002**, *16*, 469–489. [CrossRef]
39. Lopez-Fernandez, O.; Kuss, D.J.; Romo, L.; Morvan, Y.; Kern, L.; Graziani, P.; Rousseau, A.; Rumpf, H.J.; Bischof, A.; Gässler, A.K.; et al. Self-reported dependence on mobile phones in young adults: A European cross-cultural empirical survey. *J. Behav. Addict.* **2017**, *6*, 168–177. [CrossRef] [PubMed]
40. Oulasvirta, A.; Rattenbury, T.; Ma, L.; Raita, E. Habits make smartphone use more pervasive. *Pers. Ubiquitous Comput.* **2011**, *16*, 105–114. [CrossRef]
41. Elhai, J.D.; Levine, J.C.; Dvorak, R.D.; Hall, B.J. Non-social features of smartphone use are most related to depression, anxiety and problematic smartphone use. *Comput. Hum. Behav.* **2017**, *69*, 75–82. [CrossRef]
42. Elhai, J.D.; Hall, B.J.; Levine, J.C.; Dvorak, R.D. Types of smartphone usage and relations with problematic smartphone behaviors: The role of content consumption vs. social smartphone use. *Cyberpsychol. J. Psychosoc. Res. Cyberspace* **2017**, *11*, 3. [CrossRef]
43. Rozgonjuk, D.; Täht, K. To what extent does Internet use affect academic performance? Using Evidence from the large-scale PISA study. *Annu. Rev. Cyberther. Telemed.* **2017**, *15*, 39–44.
44. Jiang, Q. Internet addiction among young people in China. *Internet Res.* **2014**, *24*, 2–20. [CrossRef]
45. Kwon, M.; Lee, J.Y.; Won, W.Y.; Park, J.W.; Min, J.A.; Hahn, C.; Gu, X.; Choi, J.H.; Kim, D.J. Development and validation of a smartphone addiction scale (SAS). *PLoS ONE* **2013**, *8*, e56936. [CrossRef] [PubMed]
46. Valk, A.; Marandi, T. How to support deep learning at a university. In Proceedings of the International Conference on Education 2005, Beijing, China, 9–10 April 2015.
47. Junco, R. The relationship between frequency of Facebook use, participation in Facebook activities, and student engagement. *Comput. Educ.* **2012**, *58*, 162–171. [CrossRef]
48. Trochim, W.M.K.; Donnelly, J.P. *The Research Methods Knowledge Base*, 3rd ed.; Atomic Dog: Cincinnati, OH, USA, 2006.
49. Gravetter, F.; Wallnau, L. *Essentials of Statistics for the Behavioral Sciences*, 8th ed.; Wadsworth: Belmont, CA, USA, 2014.
50. Li, C.H. Confirmatory factor analysis with ordinal data: Comparing robust maximum likelihood and diagonally weighted least squares. *Behav. Res. Methods* **2016**, *48*, 936–949. [CrossRef] [PubMed]
51. DiStefano, C.; Morgan, G.B. A Comparison of Diagonal Weighted Least Squares Robust Estimation Techniques for Ordinal Data. *Struct. Equ. Model. Multidiscip. J.* **2014**, *21*, 425–438. [CrossRef]
52. MacKinnon, D.P. *Introduction to Statistical Mediation Analysis*; Routledge Academic: New York, NY, USA, 2008.
53. Kuss, D.J.; Griffiths, M.D. Online social networking and addiction—A review of the psychological literature. *Int. J. Environ. Res. Public Health* **2011**, *8*, 3528–3552. [CrossRef] [PubMed]
54. Grinols, A.B.; Rajesh, R. Multitasking with smartphones in the college classroom. *Bus. Prof. Commun. Q.* **2014**, *77*, 89–95. [CrossRef]

55. Lim, S.; Shim, H. Who Multitasks on Smartphones? Smartphone Multitaskers' Motivations and Personality Traits. *Cyberpsychol. Behav. Soc. Netw.* **2016**, *19*, 223–227. [CrossRef] [PubMed]
56. Walsh, J.L.; Fielder, R.L.; Carey, K.B.; Carey, M.P. Female college students' media use and academic outcomes: Results from a longitudinal cohort study. *Emerg. Adulthood* **2013**, *1*, 219–232. [CrossRef] [PubMed]
57. Kononova, A.; Chiang, Y.H. Why do we multitask with media? Predictors of media multitasking among Internet users in the United States and Taiwan. *Comput. Hum. Behav.* **2015**, *50*, 31–41. [CrossRef]
58. Baron, R.M.; Kenny, D.A. The moderator-mediator variable distinction in social psychological research: Conceptual, strategic, and statistical considerations. *J. Pers. Soc. Psychol.* **1986**, *51*, 1173–1182. [CrossRef] [PubMed]
59. Boase, J.; Ling, R. Measuring Mobile Phone Use: Self-Report versus Log Data. *J. Comput.-Mediat. Commun.* **2013**, *18*, 508–519. [CrossRef]
60. Kobayashi, T.; Boase, J. No Such Effect? The Implications of Measurement Error in Self-Report Measures of Mobile Communication Use. *Commun. Methods Meas.* **2012**, *6*, 126–143. [CrossRef]
61. Elhai, J.D.; Tiamiyu, M.F.; Weeks, J.W.; Levine, J.C.; Picard, K.J.; Hall, B.J. Depression and emotion regulation predict objective smartphone use measured over one week. *Personal. Individ. Differ.* **2017**. [CrossRef]

International Journal of
Environmental Research and Public Health

MDPI

Article

Problematic Smartphone Use: Investigating Contemporary Experiences Using a Convergent Design

Daria J. Kuss [1,*], Lydia Harkin [1], Eiman Kanjo [2] and Joel Billieux [3,4]

1 International Gaming Research Unit, Psychology Department, Nottingham Trent University, Nottingham NG1 4FQ, UK; lydia.harkin02@ntu.ac.uk
2 Computing and Technology Department, Nottingham Trent University, Nottingham NG1 4FQ, UK; eiman.kanjo@ntu.ac.uk
3 Addictive and Compulsive Behaviour Lab., Institute for Health and Behaviour, Integrative Research Unit on Social and Individual Development (INSIDE), University of Luxembourg, Esch-sur-Alzette, L-4365 Luxembourg, Luxembourg; joel.billieux@uni.lu
4 Addiction Division, Department of Mental Health and Psychiatry, University Hospitals of Geneva, 44041 Geneva, Switzerland
* Correspondence: daria.kuss@ntu.ac.uk; Tel.: +44-1158-484-153

Received: 19 December 2017; Accepted: 11 January 2018; Published: 16 January 2018

Abstract: Internet-enabled smartphones are increasingly ubiquitous in the Western world. Research suggests a number of problems can result from mobile phone overuse, including dependence, dangerous and prohibited use. For over a decade, this has been measured by the Problematic Mobile Phone Use Questionnaire (PMPU-Q). Given the rapid developments in mobile technologies, changes of use patterns and possible problematic and addictive use, the aim of the present study was to investigate and validate an updated contemporary version of the PMPU-Q (PMPU-Q-R). A mixed methods convergent design was employed, including a psychometric survey ($N = 512$) alongside qualitative focus groups ($N = 21$), to elicit experiences and perceptions of problematic smartphone use. The results suggest the PMPU-Q-R factor structure can be updated to include smartphone dependence, dangerous driving, and antisocial smartphone use factors. Theories of problematic mobile phone use require consideration of the ubiquity and indispensability of smartphones in the present day and age, particularly regarding use whilst driving and in social interactions.

Keywords: smartphone; problematic mobile phone use; convergent design; focus group; survey

1. Introduction

The Western world has seen a significant increase in mobile technology use in the last decade. In 2016, the communications regulator Ofcom [1] referred to the UK as a "smartphone society"; 93% of the population own a smartphone, and users spend more time accessing the Internet via a phone than through other devices, such as laptops and desktop-computers. These recent trends suggest mobiles and the Internet have become intimately intertwined to enable "on-the-go" access to a range of facilities, including web-browsing, communication, shopping, banking, and gaming [1].

Recent research suggests a number of problems can result from smartphone overuse, including addiction-like symptoms and feelings of dependence [2,3], dangerous use, particularly whilst driving [4,5], and forbidden or prohibited use in areas such as libraries, classrooms, or public transport [6]. Accumulating evidence also connects excessive mobile phone use with increasing psychopathological symptoms, such as those related to depression and anxiety [7]. In other words, research suggests excessive mobile phone use can result from psychopathology and constitute a dysfunctional strategy to cope with adverse emotions. Similarly, King et al. [8] suggested that mobile phone checking can constitute a safety behaviour in anxious individuals. Internet-enabled devices may encourage checking behaviours by hosting a range of applications (or apps) with regular updates and notifications. Thus, mobile Internet use may increase habitual checking behaviours, which may contribute to developing and maintaining symptoms of psychopathology, such as addictive use [9]. Consequently, a growing number of studies are conducted to determine whether smartphone overuse constitutes a genuine addictive disorder (e.g., [10]), which is in line with the inclusion of a behavioural addiction category in the latest edition of the Diagnostic and Statistical Manual of Mental Disorders (DSM-5; [11]). Yet, to date, the evidence supporting problematic smartphone use as an addictive disorder is scarce, and the studies emphasizing behavioural and neurobiological similarities between problematic smartphone use and other types of recognised addictive disorders are limited [12].

1.1. Gaining a Contemporary View of Smartphone Behaviours

Past research on problematic mobile phone and addictive smartphone behaviours employed quantitative methodologies to examine negative consequences associated with smartphone use. Various ways of measuring problematic smartphone use have been proposed considering different criteria and sources, including empirical evidence [13,14], substance abuse criteria [4,15–20], pathological gambling criteria [19,21], reviews of the relevant literature [2,17,18,22–24], or Internet addiction criteria [23]. When it comes to determining when smartphone use becomes problematic, it is important to be aware that time spent using these devices is not a sufficient indicator. For instance, it has been found that time spent socialising on mobile apps left users with positive mood [25]. Thus, the types of smartphone interactions appear to have varying impacts on user wellbeing. However, merely reading, removing, and scrolling through messages leaves users with negative emotions [25]. In addition to utilising a quantitative research approach, an experiential perspective based on users' own perceptions and understanding of their smartphone use may offer significant insights into what constitutes problematic smartphone use and how it is experienced on an individual level. User perceptions of smartphones can help to define what aspects of this technology are beneficial or problematic.

However, experiential evidence of mobile devices is outdated. Surveys capturing smartphone perspectives have failed to keep up with the speed of technological advancement and often do not reflect the full range of behaviours possible on modern smartphones [25]. Relatively recent smartphone interactions, particularly those which are supported by 'on-the -go' Internet technology, have not been accounted for and may influence problematic smartphone experiences. An experiential perspective based on users' own perceptions and understanding of their smartphone use may offer significant insights into what constitutes problematic use and how it is experienced on an individual level. The present research aims to fill this gap in knowledge by using a mixed methods convergent design incorporating a qualitative exploration of perspectives on contemporary smartphone use.

1.2. Existing Measures of Problematic Smartphone Use

A theory of problematic mobile phone use [12] suggests that there are three pathways which may result in negative and pathological smartphone behaviours, namely (i) the excessive reassurance pathway, (ii) the impulsive-antisocial pathway, and (iii) an extraversion pathway. These pathways suggest that personality, psychopathological symptoms, and frequency of smartphone use can have particular problematic consequences. The Problematic Mobile Phone Use Questionnaire (PMPU-Q) [2] was developed to assess various facets of problematic mobile phone use. The original questionnaire included four subscales: (1) prohibited use; (2) dangerous use; (3) dependent use, and (4) financial problems resulting from use.

Contemporary publications and theoretical reflections on problematic smartphone use take different perspectives relative to Billieux et al.'s proposed model [12]. For instance, financial implications may no longer be considered a contemporary problematic use of smartphones. Recent evidence links the evolution of mobile phones to smart technology with many benefits; social applications such as WhatsApp and Skype can now facilitate communication with little cost to the user, and apps are available, which support financial and banking activities [26,27]. In addition, the US Department of Transportation reported smartphone technology as a key distractor which can deflect the attention of pedestrians and drivers, leading to potential collisions [28]. Considering this, previous survey measures of problematic behaviours excluding such contemporary activities may only partially record problematic experiences. Given the rapid developments in mobile technologies, changes of use patterns and possible problematic and addictive use, the aim of the present study was to test and validate an updated contemporary version of the original PMPU-Q using a rigorous and innovative convergent parallel design. In order to investigate the efficacy of the existing measure of these phenomena, a psychometric survey was included in this study which featured the PMPU-Q and validated measures of smartphone affect.

2. Methods

2.1. Design

This study used a mixed methods methodological approach with a convergent parallel design. A mixed methods approach allowed for bridging two research traditions regarding problematic smartphone use by means of integrating large-scale psychometric inquiry with a qualitative analysis on personalized experiences, allowing for a better understanding of the validity of the PMPU-Q-R. Fetters et al. [29] identified a convergent parallel design as a suitable mixed method for investigating the validity of quantitative measures. Design convergence in this case refers to decreasing PMPU-Q measurement uncertainty by using different methods [29]. The updated PMPU-Q (PMPU-Q-R) was administered to a sample of smartphone users, together with a number of relevant other validated psychometric measures, to determine the construct validity and internal consistency of the measure. In the second phase, perceptions and experiences of smartphone use, including the respective usages (i.e., dependent, dangerous and prohibited), were explored using focus groups. This concurrent procedure was time efficient, and meant that interpretive analysis of each individual dataset informed the other [29]. This was important in the present study as interpretation required innovative, evidence-based reconceptualising of an evolving technology. Figure 1 demonstrates the convergent design employed in this study.

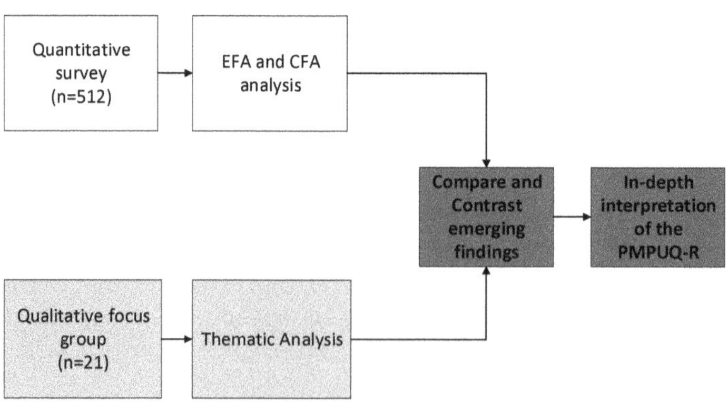

Figure 1. Convergent study design.

2.2. Study Recruitment

Smartphone users were recruited to the quantitative survey during December 2016 to March 2017 using opportunity and snowball sampling. Study advertisements encouraged smartphone users to follow a weblink to the survey hosted through Qualtrics in the UK. Offline advertisements were posted throughout university networks, and online advertisements were shared within student portals and social media networks, which focused on smartphone use. This social media dissemination included forums, Twitter, Facebook, and Reddit networks. Participants were considered eligible if they were smartphone users. Focus group participants were obtained from an opportunity sample of survey participants.

2.3. Study Procedures

2.3.1. The PMPU-Q

The PMPU-Q was originally developed by Billieux and colleagues [2] and investigated four dimensions of problematic smartphone use: (1) prohibited use, (2) dangerous use, (3) dependent use, and (4) financial problems resulting from use. In this study, the financial problem scale was excluded as financial implications are no longer considered a contemporary problematic smartphone use [26,27]. In addition, items concerning pedestrian safety were included as dangerous items in the adapted PMPU-Q, as seen in Table 1. These adaptations produced the PMPU-Q-R, which was administered as a 17-item questionnaire. Responses were measured on a four-point Likert scale (ranging from 1 = strongly disagree to 4 = strongly agree).

Table 1. Items added to the PMPUQ.

Item Question
I use my mobile phone whilst crossing the road.
I have found myself in risky situations because I have used my mobile phone whilst walking.

2.3.2. Validated Measures for Comparative Analysis

1. Smartphone Addiction and Social Media Disorder

This survey included the Smartphone Addiction Scale (SAS) [16] and the Social Media Disorder Scale (SMD) [30]. These two scales measure excessive smartphone use as an addictive behaviour, and thus included items adapted from the substance abuse literature. The SAS consisted of ten items assessing symptoms of smartphone addiction. Statements relating to smartphone addiction were rated on a seven-point Likert scale ranging from 1 = strongly disagree to 7 = strongly agree. The SAS has previously demonstrated good internal consistency and concurrent validity [16]. Cronbach's alpha for this scale in the present study indicated good reliability ($\alpha = 0.88$). Recent research indicates a strong association between social media and smartphone use [31], supporting the inclusion of a psychometric tool assessing social media addiction. The SMD scale consisted of nine items representing eight aspects of social media disorder: preoccupation, tolerance, withdrawal, displacement, escape, problems, deception, displacement, and conflict. Participants were asked to rate, on a five-point Likert scale, how often they had experienced a symptom of social media disorder. The SMD scale has previously demonstrated appropriate internal consistency, good convergent and criterion validity, and sufficient test-retest reliability [30]. Cronbach's alpha in the present study indicated good reliability ($\alpha = 0.88$).

2. Psychopathology

The survey also included measures of psychopathological symptoms (depression, anxiety, stress, and ADHD), impulsivity, and the big five personality traits (i.e., neuroticism, extraversion, conscientiousness, agreeableness and openness) to assure content validity as the PMPU-Q has previously been linked to wellbeing and personality [2,4,32]. The Depression-Anxiety-Stress Scale (DASS-21; [33]) consisted of 21 measures of depression, anxiety, and stress symptoms experienced in the previous two weeks rated on a scale ranging from 0 (symptom did not apply to me at all) to 3 (applied to me very much, or most of the time). Impulsivity was rated using a short form of the Barratt Impulsiveness Scale (BIS-15) [34,35]. The DASS-21 and BIS-15 are both well established and highly cited scales, consistently demonstrating strong validity, reliability, and excellent internal validity [33,35,36]. Measures of Attention Deficit Hyperactivity Disorder (ADHD) symptoms approved by the World Health Organisation were included in the survey [37] as ADHD has been associated with smartphone addiction [38].

3. Focus Group Procedure

Three independent focus groups were used including two trained facilitators (LH and DK), and were scheduled for approximately 90 min, and held in a quiet university research room. Focus group questions were designed to ask general questions about smartphone experiences (e.g., "Can anyone tell us your favourite/least favourite smartphone uses, and why?") and to probe both beneficial and problematic aspects of smartphone experiences (e.g., "Can you tell us more about [experience]"). Eight prompt images were used on a PowerPoint slideshow to encourage discussion amongst participants, including images of smartphones used in different settings (e.g., on a train, in a library), and artistic depictions of smartphone use (e.g., cartoons and street art of smartphone use).

2.4. Analyses

2.4.1. Quantitative Survey Analysis

The underlying structure of the PMPU-Q-R was assessed through exploratory factor analysis (EFA). This analysis was used to identify the latent constructs of smartphone experience underlying the variance in scores on measurements originally designed to measure mobile phone use. EFA analysis was conducted in IBM-SPSS with principle component analysis (PCA) and Direct Oblimin rotation, as recommended when establishing preliminary solutions [39,40].

CFA analysis was conducted to verify the factor structure of the variables. CFA was conducted in R package Lavaan using maximum likelihood estimators for CFA as they can effectively handle interactions between latent variables with multiple indicators [41,42]. To understand the fit of the model to the data, models were compared to threshold fit indices recommended by Hu and Bentler [43]. A model showing good fit to the data was expected to report CFI > 0.93, TLI > 0.93, RMSEA < 0.05 for very good fit, <0.084 for acceptable fit, and SRMR < 0.09.

The data set was randomly partitioned by a ratio 60:40 to run the factor analysis tests. Exploratory tests were performed on 60% of the data, and the proposed factor structure of the PMPUQ-R was examined on the remaining 40% of the data to test the proposed model on a second sample. This cross-validation approach is recommended to investigate the proposed model structure on independent datasets [44].

2.4.2. Qualitative Thematic Analysis

Focus groups were audio recorded, and data were input into QSR-NVIVO, and analysed using thematic analysis [45] to explore smartphone experience. This involved familiarisation with the data by listening to and reading transcripts, generating initial codes for key features in the transcripts, searching for connections between codes (or themes), and reviewing and defining the identified themes based on the evidence presented across focus groups. Analysis was primarily conducted by Lydia Harkin. Themes were discussed amongst the wider team to assess the developing logic, in addition to discussing the apparent similarities and divergences within the qualitative and qualitative data.

2.5. Ethical Considerations

Participants provided informed consent to take part in this study. Only participants who agreed to be contacted for focus groups were invited for group interview. All contact information was stored securely, and after the study was completed, all identifying information was removed from the transcripts and destroyed. This study was given ethical approval by the Business, Law, and Social

Sciences Ethics Committee at Nottingham Trent University and abided by the ethical codes of the British Psychological Society.

3. Results

Participants in the quantitative survey were 512 smartphone users. The sample had a mean age of 25.5 (range = 13–68 years), were primarily female (78.5%), university students (67%), and from the United Kingdom (91.8%). Table 2 demonstrates the breakdown of demographics represented in this sample.

Table 2. Survey and Focus Group Participant Demographics.

Participant Demographics	Survey Participants *N* (%) *	FG Participants *N* (%)
Gender		
Male	401 (20.9)	10 (47.6)
Female	107 (78.3)	11 (52.4)
Prefer not to say	3 (0.6)	0
Country of Origin		
United Kingdom	470 (91.8)	
USA	10 (2)	21 (100)
Ireland	4 (0.8)	
Other	49 (6.4)	
Level of Education		
No formal qualifications	4 (0.8)	0
GCSEs	13 (2.5)	0
Further education	286 (55.9)	0
Vocational qualification	8 (1.6)	10 (47.6)
Higher education	122 (23.8)	2 (9.5)
Postgraduate degree	79 (15.4)	9 (42.9)
Self-reported calls per day		
0–1 year	262 (51.2)	
2–5 years	214 (41.8)	
5–10 years	27 (5.3)	
>10 years	8 (1.6)	
Self-reported texts per day		
0–5	50 (9.8)	1 (4.8)
5–10	64 (12.5)	6 (28.6)
10–20	82 (16)	4 (19)
20–30	79 (15.4)	2 (9.5)
30–40	39 (7.6)	3 (14.3)
>40	198 (38.7)	4 (19)
Self-reported time spent on phone p/day		
<30 min	10 (2)	0
30 min–1 h	34 (6.6)	4 (19)
1–2 h	134 (26.2)	8 (38)
3–5 h	219 (42.8)	7 (33.3)
5–10 h	92 (18)	2 (9.5)
>10 h	22 (4.3)	0

* Note. Rounding may have led to percentages that do not equal 100.

3.1. Exploratory Factor Analysis

Firstly, 16 of the 17 items on the PMPU-Q-R correlated over 0.3 with at least one other item, indicating reasonable factorability [46,47]. Secondly, a Kaiser-Meyer-Olkin measure of sampling adequacy was 0.859, above the commonly recommended value of 0.6. Thirdly, Bartlett's test of sphericity was significant (χ^2 (120) = 1682.441, $p < 0.001$), indicating the present sample had a suitable size for factor analysis. Finally, the communalities of 16 of the 17 items were above 0.3, confirming that these 16 items shared common variance with other items. One item from the original 'prohibited use' scale did not share variance with the body of items, and was excluded from further analyses ("When using my mobile phone on public transport, I try not to talk too loud"). The EFA revealed a three-factor solution which explained 54% of the variance in scores. Cronbach's alpha indicated the items consistently measured a closely related set of concepts (α = 0.86). All latent variables positively correlated with one another. For factors one and two, this correlation was moderate to strong (R^2 = 0.436), whilst factor two correlated weakly with factor three (R^2 = 0.172). However, factor three correlated very weakly with factor one (R^2 = 0.043). Thus, the factors appeared to measure related, but distinct, concepts.

As Table 3 indicates, the pattern structure produced in the data did not correspond to the theoretical structure of the PMPU-Q-R. In line with the predefined PMPU-Q-R structure, all dependence items loaded highly together on one factor, with no cross-loadings onto factors two or three. The dependence factor explained 35% of variance in overall scores, and demonstrated high reliability (α = 0.89). A combination of seven items from the original prohibited and dangerous mobile phone use subscales loaded highly onto one factor, explaining 12% of the variance, suggesting the factor labels of 'prohibited' and 'dangerous' smartphone use could not be applied to the items within the scale for this population. On face value, the items contributing to this factor did not demonstrate an immediately apparent underlying theoretical property as demonstrated in Table 3. However, alpha scores indicated a high level of shared variance in scores (α = 0.77). Finally, two items from the dangerous subscale loaded highly onto factor three, explaining 8% of the score variance. A Cronbach's alpha calculation is not meaningful for a two item factor, and therefore a Pearson correlation coefficient was calculated, showing a significant low to moderate correlation (R^2 = 0.33, $p < 0.001$), indicating two distinct measures of a related concept. A review of these items revealed that they were the only PMPU-Q-R items to refer to driving behaviours: "I use my mobile phone while driving" and "I try to avoid using my mobile phone when driving on the motorway" (R^2 = 0.43, $p < 0.001$).

Table 3. Factor loadings for the PMPUQ-R items.

Original Item Subscale		Factor Loading		
		1	2	3
Dependence	I can easily live without my mobile phone *	0.869		
Dependence	I feel lost without my mobile phone	0.844		
Dependence	It is hard for me to turn my mobile phone off	0.769		
Dependence	It is easy for me to spend all day not using my mobile phone	0.747		
Dependence	I get irritated when I am forced to turn my mobile phone off	0.699		
Dependence	I don't attach a lot of importance to my mobile phone *	0.694		
Dependence	Is it hard for me not to use my mobile phone when I feel like it	0.495		
Prohibited use	I don't use my mobile phone when it is completely forbidden to use it *		0.701	
Prohibited use	I don't use my mobile phone in a library, cinema, or hospital *		0.700	
Prohibited use	I use my mobile phone where it is forbidden to do so		0.673	
Danger	I have found myself in risky situations because I have used my mobile phone whilst walking		0.583	
Danger	I use my mobile phone whilst crossing the road		0.574	
Prohibited use	I try to avoid using my mobile phone where people need silence *		0.568	
Danger	I use my mobile phone in situations that would qualify as dangerous		0.563	
Danger	I use my mobile phone while driving			0.758
Danger	I try to avoid using my mobile phone when driving on the motorway			0.751

* Reversed item.

This factor structure was tested using a CFA. As expected, all items showed significant positive factor loadings with standardised coefficients ranging from 0.482 to 0.805 (see Table 4). Additionally, modification indices were low (<50), which indicated that items corresponded to the proposed structure of the PMPUQ-R and did not covary too strongly with other items. The model showed an adequate fit to the data, with most of the indices of model fit falling within the acceptable values for the CFA model to fit the variance in scores (χ^2 (101) = 190.424, $p < 0.000$, CFI = 0.927, TLI = 0.906, RMSEA = 0.062, SRMR = 0.054).

Table 4. Beta coefficients and significance values of PMPUQ-R factor loadings.

Latent Factor	Item	B	Standard Error	Beta (Standardised)	Sig
Dependence	x1	1.000		0.504	**
Dependence	x2	0.941	0.278	0.485	***
Dependence	x3	1.000		0.661	***
Dependence	x4	0.847	0.148	0.474	***
Dependence	x5	1.053	0.136	0.627	***
Dependence	x6	0.763	0.13	0.483	***
Dependence	x7	0.978	0.141	0.599	***
Factor 2	x8	0.895	0.126	0.602	***
Factor 2	x9	0.745	0.124	0.482	***
Factor 2	x10	1.000		0.79	***
Factor 2	x11	0.570	0.07	0.568	***
Factor 2	x12	0.797	0.075	0.72	***
Factor 2	x13	0.774	0.073	0.714	***
Factor 2	x14	0.954	0.075	0.805	***
Danger	x15	0.839	0.08	0.714	***
Danger	x16	0.849	0.075	0.752	***

*** $p < 0.001$; ** $p = 0.001$.

3.2. Construct Analyses

To shed light on the theoretical constructs underpinning the PMPUQ factor structure, Spearman's correlations were calculated with the new factor structure and validated measures connected to problematic technology and smartphone behaviours. These were summed measures of smartphone and addictive social media use, psychopathology (i.e., depression, stress, anxiety; and ADHD), and impulsivity. These scales were correlated with the three-factor structure emerging from our survey. Only correlations which were significant and within the low to high range (0.2–1) are reported here. Notably, no significant correlations were found for the dangerous driving-related item scores.

An increase in dependence-related items strongly correlated with increased smartphone addiction symptoms (R^2 = 0.67, $p < 0.001$), self-reported smartphone addiction (R^2 = 0.66, $p < 0.001$), and moderately correlated with social media addiction symptoms (R^2 = 0.44, $p < 0.001$). Additionally, dependence items weakly correlated with stress symptoms (R^2 = 0.22, $p < 0.001$), and attention impulsivity (R^2 = 0.26, $p < 0.001$).

Higher scores on the second, undefined factor strongly correlated with smartphone addiction symptom scores (R^2 = 0.52, $p < 0.001$), and with social media addiction symptoms (R^2 = 0.44, $p < 0.001$), self-reported smartphone addiction (R^2 = 0.42, $p < 0.001$), attention impulsivity symptoms (R^2 = 0.33, $p < 0.001$), motor impulsivity symptoms (R^2 = 0.25, $p < 0.001$), and ADHD symptoms (R^2 = 0.24, $p < 0.001$).

3.3. Thematic Analysis

Three focus groups were attended by a total of 21 individuals (11 females). Focus groups lasted an average of 96 min. Participant self-reported smartphone use ranged considerably, as highlighted in Table 2. The focus groups produced lively discussions about the integration of smartphones in participants' lives. Thematic analysis of the focus group data revealed three distinct themes across perceptions of problematic smartphone use. These themes were smartphone dependence, dangerous driving, and antisocial smartphone use.

3.3.1. Smartphone Dependence

Across all three focus groups, participants discussed the dependence their lives had on smartphones and applications. This theme incorporated both problematic and beneficial use. Many participants awoke with their smartphone alarms, read news or played games as they travelled to work or university, used email apps to support out-of-hours' work, and to contact friends through social media. Thus, participants associated smartphones with many aspects of their day, much beyond the capabilities of traditional texting or calling. This was not perceived as necessarily problematic, and many participants emphasised that they benefitted from the integration of smartphones in day-to-day life.

> *P8: "My Smartphone is literally my life (laugh). Self-professed addict. I know the whole discussion has been generally negative, but erm, I still feel the positives outweigh any negatives that we get ... (holds up phone) this is my calendar, this is how I communicate with everybody, how I organise my life, how I get up in the morning. It is my adult pacifier." Focus group 1.*

Smartphone dependence was discussed as an act of balancing benefits with potentially problematic aspects. Daily smartphone use could lead to habitual checking behaviours and wasted time. Therefore, several participants made attempts to delete particular apps or groups of apps. For instance, across two focus groups, discussions around the utility of *Facebook* and other social media arose. These apps were perceived as 'addictive' because participants compulsively checked their social media, despite limited beneficial interactions or information. Most participants were unsuccessful in attempts to reduce time on smartphones due to the convenience and functionality of the devices. One participant had successfully stopped using all social media for 29 days prior to the focus group, and yet he found himself searching for other apps to use on his phone in replacement of his prior social media use. Therefore, the dependence on smartphones in daily life seemed to have a profound impact on the thoughts and actions of participants.

> *P5: "I have moulded my phone around me. My music, my pictures, everything. And I would probably feel lost without it." Focus group 2.*

> *P1: (After quitting social media for 29 days) "So yeah my immediate reaction was like finding a replacement, not necessarily thinking oh I don't have social media I need a replacement for it, but it was kind of what can I go on now? So it didn't occur to me that I could just not go on my phone." Focus group 2.*

3.3.2. Dangerous Driving Behaviours

Of the problematic behaviours described by participants in the focus groups, driving whilst actively engaging with a smartphone was described as unequivocally problematic and dangerous. These types of actions were differentiated from some other potentially harmful smartphone uses. Participants were open to accepting smartphone use behaviours, such as crossing the road and walking, as an inevitable outcome of the proliferation of smartphone use. Discussions around emotionally dangerous behaviours, such as trolling online, were also perceived as a necessary evil of having constant exposure to the Internet. Thus, dangerous driving stood out distinctly as a problematic smartphone behaviour.

> *P2: "It (seeing people driving whilst on smartphones) infuriates you even when you see it and you know they are driving ridiculously." Focus group 1.*

> *P5: "My best friend always does that (uses smartphone) in the car. I am like ... you are on Instagram. (Name) what are you doing? You are going to kill me. I have to take her phone, I am holding your phone." Focus group 3.*

> *P3: "Maybe you should learn how to cross the road safely while using your phone because that is what adult people do. It is like crossing the road with the red light on. We all do it. I do it at least, and I think I do it safely ... I wouldn't say don't use it. I would say just be careful. Know how to use it." Focus group 1.*

3.3.3. Antisocial Smartphone Use

The antisocial properties of smartphone use emerged as a strong indicator of problematic behaviours, according to participants in this study. Participants were concerned that smartphone use often replaced face-to-face interactions. Groups of friends or couples were observed together in the 'real world', whilst seemingly disengaged from one another because they were using smartphones. This was perceived as indicative of poor social functioning. Similarly, several participants described feeling rejected or upset when their friends became distracted from their face-to-face conversations by their smartphones.

> *P2: "I feel personally ... even if it's in a group and someone starts taking their phone out I feel like a sting of rejection, so I'll like sting them back. So I'm looking on my phone even if I've got no reason to. I'll just try and find a reason to look on my phone and then hope that they see that I don't need them either." Focus Group 2.*

> *P6: "My partner absolutely drives me crazy on the phone. We go out for dinner ... and he just sits there like that (mime sitting with phone in front). And that is all he does. And it is a nightmare going for dinner. And he never used to do it. All I can see is him like that (mime holding phone in front of face) constantly. It drives me mad." Focus group 1.*

The antisocial properties of regular smartphone use seemed to govern and moderate participants' behaviours. In particular, participants were concerned about how they would be perceived by others. Using smartphones too frequently or in dangerous or prohibited situations was perceived as embarrassing and reflective of their character. Many participants did not want to be viewed as someone who 'needs' their smartphone. For this reason, a few participants had enforced rules on how they, their partner, or friends were allowed to use their smartphones. For instance, several participants 'banned' smartphone use in restaurants during romantic meals, and others endeavoured not to use their smartphones while walking in the street.

> *P5: "I don't want to be seen as someone who, like in a restaurant me and my boyfriend never have ours out ... because I don't want people to look at me and be like 'oh god, she needs her phone out" Focus group 1.*

> *P9: "I hate the idea of being thought of as somebody who can't put my phone down." Focus group 3.*

4. Discussion

The present study aimed to investigate and validate an updated contemporary version of the original PMPU-Q, the PMPU-Q-R, using a rigorous and innovative convergent design. The PMPU-Q-R was tested to determine how many factors emerged from the scale, and how this corresponded with the theoretical underpinnings of the original PMPU-Q subscales [2]. Construct validity of the PMPU-Q-R items was investigated, alongside existing contemporary measures of problematic smartphone behaviours, and psychopathology. The quantitative data inquiry using an EFA revealed the pattern structure did not correspond to the expected and predefined structure of the PMPU-Q. Whilst the dependence factor was explained well by the data, a second factor was made up by a combination of items of the prohibited and dangerous subscales, suggesting the factor labels of 'prohibited' and 'dangerous' smartphone use could not be applied to the items within the scale for this sample. An explanation for this may be the rapid expansion and development of mobile technology since the development of the original PMPU-Q [2], which may have contributed to the results. Smartphone functionality has significantly increased in this time period, including the availability of high-quality satellite navigation and location-based augmented reality games (e.g., *Pokémon-GO*), changing the possible risks related to engaging in smartphone activities.

An increase in dependence-related items strongly correlated with increased self-reported smartphone addiction symptoms (measured via items considering excessive smartphone use as addictive disorder), and moderately correlated with social media addiction symptoms. This provides support for the construct validity of the dependence subscale. Previous research [31] has suggested that addictive smartphone use may be part of social media addiction. According to the pathway model of problematic mobile phone use [12], an addictive pattern of smartphone use is characterized by the use of specific applications, including calls and instant messaging. This definition could be extended by evidence from the focus group discussions; participants found that features, such as social media, emails, and games, contributed to increased feelings of dependence. This suggests that rather than being an addictive medium per se, mobile technologies including smartphones and tablets are media that enable the engagement in potentially addictive activities, including social media use. Similarly, it has been argued that individuals do not become addicted to the medium of the Internet per se, but to the activities they engage in on the Internet [48], such as gaming [49] or social media use [50]. With the advent and ubiquity of mobile technologies, this supposition appears particularly pertinent. Using social networking sites is a particularly popular activity on smartphones, with around 80% of social media used via mobile technologies [51], and around 75% of *Facebook* users access *Facebook* via their smartphones [52]. Consequently, social media use and smartphone use appear inherently intertwined [31], suggesting future research should pay additional attention to the forms and functions of specific smartphone use.

Additionally, dependence items weakly correlated with stress symptoms, and attention impulsivity. With regards to stress, it has been shown that the increased use of smartphones was related to general distress, anxiety and depression [13,53,54]. Further research [9] also highlighted that stress predicts addictive smartphone use. Individuals may use their smartphones to cope with everyday stressors (e.g., social situations, relationship problems), and using smartphones as coping mechanism can be considered dysfunctional, similar to using the Internet to cope with life problems [55], resembling symptoms traditionally associated with substance-related addictions [56].

Considering attention impulsivity, previous research [2] has shown that impulsivity was a strong predictor of problematic smartphone use, specifically with regards to the subscales urgency, lack of premeditation and lack of perseverance, which appear related to attention impulsivity [57]. Similarly, the present research found that higher scores on the dependence factor strongly correlated with ADHD symptoms, which is in line with previous research [58] in children. Alternatively, it has been suggested that particular types of activities engaged in on smartphones, e.g., gaming, may lead to the development of ADHD symptoms, suggesting future research may be necessary to disentangle the differential impact of specific smartphone application use on possible dependence.

In addition to this, the thematic analysis applied in the present research revealed that there appears to be a strong awareness that using smartphones whilst driving can be dangerous both for the self and for others, which corroborates the quantitative data regarding the PMPU-Q-R dangerous driving factor. This suggests the driving factor is a valid and reliable factor that contributes to explaining contemporary problematic smartphone use, and should therefore be retained in future analyses of problematic smartphone use.

Findings from the qualitative and quantitative analysis suggested that dangerous driving stands out as a distinct form of problematic smartphone behaviour. This corroborates the US Department of Transportation's report showing smartphone use can distract pedestrians and drivers, leading to potential collisions [28]. In 2014, over 3000 individuals died in the US as a consequence of being distracted while driving, leading the US National Highway Traffic Safety Administration (NHTSA) to issue voluntary guidelines for smartphone developers, which aim to restrict the functions of smartphones being used by a driver. A recent report by the American Automobile Association

Foundation found that using smartphones, including Apple's voice control system Siri, is very dangerous in the context of driving as it leads to cognitive distraction [59].

The thematic analysis of the focus group data revealed three distinct themes across perceptions of problematic smartphone use, namely smartphone dependence, dangerous driving, and antisocial smartphone use. With regards to the first theme, smartphone dependence, the thematic analysis indicated smartphones are essential elements of individuals' lives as they are being used for their many functions, going beyond phone calls and texting, including other entertainment functions (e.g., music, pictures), as well as organisational functions (i.e., calendar, alarm). Participants perceived particular smartphone applications as being potentially addictive, including social media, which they were checking compulsively, although there were limited advantages of doing so. Participants found it difficult to reduce the time they spent on their smartphones as these were perceived to be very convenient and functional.

Antisocial smartphone use emerged as a key problematic behaviour, as evidenced through the thematic analysis. Importantly, when looking at the prohibited use items of the PMPU-Q-R, clear links between the qualitative and quantitative analyses emerged. Indeed, most of the prohibited items refer to situations where using smartphones is banned, implying that scoring highly on these items depicts engaging in antisocial behaviours. Beside prohibited use per se, this was a key observation as some individuals use their smartphones in social contexts, which may similarly appear as antisocial, and can consequently impact negatively on their overall social functioning, both in terms of the quality of interaction with others, and with regards to perceptions of rudeness and rejection in interpersonal contexts. Along the same lines, recent research focused on the "phubbing" phenomenon, defined as the act of snubbing someone in a social setting by using one's phone instead of interacting, and research has shown that such types of antisocial smartphone use are linked to lack of self-control [60] and lower relationship satisfaction among romantic partners [61].

In the current study, participants discussed how they disapproved of friends or couples disengaging from one another whilst engaging with their smartphones, leading to feeling devalued. Similar situations and behaviours have been observed in the context of young people disconnecting from their offline contacts for the sake of connecting online, which has been linked to a preference for online social interaction [62]. This was tied to an awareness of public perceptions on the individuals' smartphone use, often leading to behavioural change in terms of limiting use in particular situations and contexts, as found in the present research.

With regards to the integration of both methods using the adopted convergent design, the findings confirm that a combination of prohibited and dangerous items from the PMPU-Q-R may be explained by antisocial smartphone use. Further research is necessary to inquire about motivations for smartphone use, as well as the norms of smartphone use in public, given that stigma and public perceptions appear to significantly contribute to how smartphone use is perceived by the users regarding being prohibited or dangerous. The focus group data analysis furthermore revealed that public perceptions may lead to behavioural change in terms of how individuals engage with their smartphones, emphasising the need to assess problematic smartphone use within its sociocultural context, bearing in mind the cultural and behavioural norms associated with smartphone use. Using anthropological and cultural studies may aid our understanding and study of the impacts of technology use as it has been shown to be particularly insightful in the study of specific technology use, such as gaming, given it allows for an assessment of the behavioural norms and practices surrounding a concrete behaviour [63,64]. The individual's context is a significant factor that can mark the dividing line between problematic smartphone use and potential smartphone "addiction", and the smartphone use context can gain particular importance for users, depending on their life situation (i.e., the meaning they attach to their smartphone) and smartphone use preferences (i.e., particular types of apps used and activities engaged in). Moreover, the cultural context is significant because it embeds the smartphone user in a community with shared beliefs and practices, endowing their use with particular meaning as well as possible stigma. The context of the individual, the specific smartphone use and the smartphone use environment, as well as the broader framework of the respective culture the user is situated in are relevant in the study of problematic smartphone use and are therefore recommended to be used in the context of future smartphone use research [10].

Regarding the methodology that has been utilised in the present research, mixed methods have been employed, integrating quantitative with qualitative techniques. There was an existing body of theory to draw on [12], but it needed updating, so mixed methods allowed us to combine previous theory with present experiential understanding. The mixing of quantitative with qualitative methods is a challenging endeavour, particularly as these methods can be understood as separate scientific paradigms. One could claim these methods are incommensurate as their unit of analysis (i.e., words versus numbers), their epistemological position (i.e., knowledge derived from meaning versus behaviours), and their source of scientific knowledge (i.e., induction versus deduction) are inherently incompatible. Accordingly, the adherence to a single method could be seen as the epitome of normal science, indicating a scientific revolution is necessary to integrate the seemingly incongruous positions of quantitative and qualitative research. This integration overcomes the limitations of a single methodology, i.e., its inevitable incompleteness. The usage of mixed methods, on the other hand, allows for the corroboration, elaboration, and complementation of findings [65]. The present study corroborated the PMPU-Q-R structure in terms of the dependence factor, and suggested that dangerous driving is a distinct factor that needs considering when studying problematic smartphone use. Moreover, the qualitative element of this research complemented the quantitative findings with regards to the combination of the dangerous and prohibited factor by elaborating on how antisocial use and public perceptions may contribute to individual perceptions of norms surrounding smartphone use in different contexts.

5. Conclusions

The updated version of the PMPU-Q, the PMPU-Q-R, is a valid and reliable tool for measuring contemporary smartphone use and problems associated with this use, concerning dependence, antisocial use and dangerous driving. Future research is encouraged to discern user motivations and perceptions of usage norms and meanings applied to smartphone use to delineate the impact social stigma may have on smartphone use. The convergent design used in the present study appeared to offer a corroborative and complementary perspective on contemporary knowledge of problematic smartphone use, and has expanded the knowledge base in the field of behavioural technological addictions. Taken together, contemporary smartphone use may become problematic if engaged in excessively. Nonetheless, contemporary smartphone users often actively seek to modify their behaviours when they are being perceived as problematic, suggesting users appear aware and conscious of their usage patterns, which may limit the extent of problematic use.

Acknowledgments: We would like to acknowledge Nottingham Trent University (grant number Kickstarter_DK1), and the British Academy and the Leverhulme Trust (grant number SG162119) for funding this research.

Author Contributions: D.J.K. designed the study, collected the data and wrote the manuscript, including subsequent iterations. L.H. collected the data, completed the data analyses, and wrote the results section. E.K. co-designed the study. All authors contributed to the final write-up.

Conflicts of Interest: The authors declare no conflict of interest.

References

1. The UK Is Now a Smartphone Society. Available online: http://media.ofcom.org.uk/news/2015/cmr-uk-2015/ (accessed on 28 August 2016).
2. Billieux, J.; Van der Linden, M.; Rochat, L. The role of impulsivity in actual and problematic use of the mobile phone. *Appl. Cogn. Psychol.* **2008**, *22*, 1195–1210. [CrossRef]
3. Chóliz, M. Mobile phone addiction: A point of issue. *Addiction* **2010**, *105*, 373–374. [PubMed]
4. Bianchi, A.; Phillips, J.G. Psychological predictors of problem mobile phone use. *Cyberpsychol. Behav.* **2005**, *8*, 39–51. [CrossRef] [PubMed]

5. White, M.P.; Eiser, J.R.; Harris, P.R. Risk perceptions of mobile phone use while driving. *Risk Anal.* **2004**, *24*, 323–334. [CrossRef] [PubMed]
6. Nickerson, R.C.; Isaac, H.; Mak, B. A multi-national study of attitudes about mobile phone use in social settings. *Int. J. Mob. Commun.* **2008**, *6*, 541–563. [CrossRef]
7. Elhai, J.D.; Levine, J.C.; Dvorak, R.D.; Hall, B.J. Non-social features of smartphone use are most related to depression, anxiety and problematic smartphone use. *Comput. Hum. Behav.* **2017**, *69*, 75–82. [CrossRef]
8. King, A.L.S.; Valença, A.M.; Silva, A.C.O.; Baczynski, T.; Carvalho, M.R.; Nardi, A.E. Nomophobia: Dependency on virtual environments or social phobia? *Comput. Hum. Behav.* **2013**, *29*, 140–144. [CrossRef]
9. Jeong, S.-H.; Kim, H.; Yum, J.-Y.; Hwang, Y. What type of content are smartphone users addicted to? SNS vs. games. *Comput. Hum. Behav.* **2016**, *54*, 10–17. [CrossRef]
10. Lopez-Fernandez, O.; Kuss, D.J.; Romo, L.; Morvan, Y.; Kern, L.; Graziani, P.; Rousseau, A.; Rumpf, H.J.; Bischof, A.; Gässler, A.K.; et al. Self-reported dependence on mobile phones in young adults: A European cross-cultural empirical survey. *J. Behav. Addict.* **2017**, *6*, 168–177. [CrossRef] [PubMed]
11. American Psychiatric Association. *Diagnostic and Statistical Manual of Mental Disorders (DSM-5)*; American Psychiatric Association: Arlington, VA, USA, 2013.
12. Billieux, J.; Maurage, P.; Lopez-Fernandez, O.; Kuss, D.J.; Griffiths, M.D. Can disordered mobile phone use be considered a behavioral addiction? An update on current evidence and a comprehensive model for future research. *Curr. Addict. Rep.* **2015**, *2*, 156–162. [CrossRef]
13. Ha, J.H.; Chin, B.; Park, D.H.; Ryu, S.H.; Yu, J. Characteristics of excessive cellular phone use in Korean adolescents. *Cyberpsychol. Behav.* **2008**, *11*, 783–784. [CrossRef] [PubMed]
14. Toda, M.; Monden, K.; Kubo, K.; Morimoto, K. Cellular phone dependence tendency of female university students. *Nihon Eiseigaku Zasshi* **2004**, *59*, 383–386. [CrossRef] [PubMed]
15. Chóliz, M. Mobile-phone addiction in adolescence: The Test of Mobile Phone Dependence (TMD). *Prog. Health Sci.* **2012**, *21*, 33–44.
16. Kwon, M.; Lee, J.; Won, W.; Park, J.; Min, J.; Hahn, C.; Kim, D. Development and validation of a Smartphone Addiction Scale. *PLoS ONE* **2013**, *8*, e56936. [CrossRef] [PubMed]
17. Lopez-Fernandez, O.; Honrubia-Serrano, L.; Freixa-Blanxart, M.; Gibson, W. Prevalence of problematic mobile phone use in British adolescents. *Cyberpsychol. Behav. Soc. Netw.* **2014**, *17*, 91–98. [CrossRef] [PubMed]
18. Walsh, S.P.; White, K.M.; Young, R.M. Needing to connect: The effect of self and others on young people's involvement with their mobile phones. *Aust. J. Psychol.* **2010**, *62*, 194–203. [CrossRef]
19. Xu, H.; Wu, X.; Lan, Y.; Chen, Y. Development of mobile phone dependence inventory for college students. *Chin. J. Clin. Psychol. Health* **2008**, *16*, 26–27.
20. Yen, C.F.; Tang, T.C.; Yen, J.Y.; Lin, H.C.; Huang, C.F.; Liu, S.C.; Ko, C.H. Symptoms of problematic cellular phone use, functional impairment and its association with depression among adolescents in Southern Taiwan. *J. Adolesc.* **2009**, *32*, 863–873. [CrossRef] [PubMed]
21. Jenaro, C.; Flores, N.; Gomez-Vela, M.; Gonzalez-Gil, F.; Caballo, C. Problematic internet and cell-phone use: Psychological, behavioral, and health correlates. *Addict. Res. Theory* **2007**, *15*, 309–320. [CrossRef]
22. Güzeller, C.O.; Cosguner, T. Development of a problematic mobile phone use scale for Turkish adolescents. *Cyberpsychol. Behav. Soc. Netw.* **2012**, *15*, 205–211. [CrossRef] [PubMed]
23. Kim, D.; Lee, Y.; Lee, J.; Nam, J.; Chung, Y. Development of Korean smartphone addiction proneness scale for youth. *PLoS ONE* **2014**, *9*, e97920. [CrossRef] [PubMed]
24. Martinotti, G.; Villella, C.; Di Thiene, D.; Di Nicola, M.; Bria, P.; Conte, G.; Cassano, M.; Petruccelli, F.; Corvasce, N.; Janiri, L.; et al. Problematic mobile phone use in adolescence: A cross-sectional study. *J. Public Health* **2011**, *19*, 545–551. [CrossRef]
25. Kanjo, E.; Kuss, D.J.; Ang, C.S. NotiMind: Responses to smartphone notifications as affective sensors. *IEEE Access* **2017**. [CrossRef]
26. Hanafizadeh, P.; Keating, B.; Khedmatgozar, H.R. A systematic review of Internet banking adoption. *Telemat. Inform.* **2014**, *31*, 492–510. [CrossRef]
27. Lyman, T.; Pickens, M.; Porteous, D. *Regulating Transformational Branchless Banking: Mobile Phones and Other Technology to Increase Access to Finance*; CGAP & Deparment for International Development: London, UK, 2008.
28. Scopatz, R.A.; Zhou, Y. *Effect of Electronic Device Use on Pedestrian Safety: A Literature Review*; National Highway Traffic Safety Administration: Washington, DC, USA, 2016.

29. Fetters, M.D.; Curry, L.A.; Creswell, J.W. Achieving integration in mixed methods designs—Principles and practices. *Health Serv. Res.* **2013**, *48*, 2134–2156. [CrossRef] [PubMed]
30. Van den Eijnden, R.J.J.M.; Lemmens, J.S.; Valkenburg, P.M. The Social Media Disorder Scale: Validity and psychometric properties. *Comput. Hum. Behav.* **2016**, *61*, 478–487. [CrossRef]
31. Kuss, D.J.; Griffiths, M.D. Social Networking Sites and Addiction: Ten lessons learned. *Int. J. Environ. Res. Public Health* **2017**, *14*, 311. [CrossRef] [PubMed]
32. Billieux, J. Problematic Mobile Phone Use: A literature review and a pathways model. *Curr. Psychiatry Rev.* **2012**, *8*, 299–307. [CrossRef]
33. Lovibond, S.H.; Lovibond, P.F. *Manual for the Depression Anxiety Stress Scales*, 2nd ed.; Psychological Foundation: Sydney, Australia, 1995.
34. Patton, J.H.; Stanford, M.S.; Barratt, E.S. Factor structure of the Barratt Impulsiveness Scale. *J. Clin. Psychol.* **1995**, *51*, 768–774. [CrossRef]
35. Spinella, M. Normative data and a short form of the Barratt Impulsiveness Scale. *Int. J. Neurosci.* **2007**, *117*, 359–368. [CrossRef] [PubMed]
36. Henry, J.D.; Crawford, J.R. The short-form version of the Depression Anxiety Stress Scales (DASS-21): Construct validity and normative data in a large non-clinical sample. *Br. J. Clin. Psychol.* **2005**, *44*, 227–239. [CrossRef] [PubMed]
37. Kessler, R.C.; Adler, L.; Ames, M.; Demler, O.; Faraone, S.; Hiripi, E.; Howes, M.J.; Jin, R.; Secnik, K.; Spencer, T.; et al. The World Health Organization Adult ADHD Self-Report Scale (ASRS): A short screening scale for use in the general population. *Psychol. Med.* **2005**, *35*, 245–256. [CrossRef] [PubMed]
38. Khang, H.; Kim, J.K.; Kim, Y. Self-traits and motivations as antecedents of digital media flow and addiction: The Internet, mobile phones, and video games. *Comput. Hum. Behav.* **2013**, *29*, 2416–2424. [CrossRef]
39. Beavers, G.A.; Iwata, B.A.; Lerman, D.C. Thirty years of research on the functional analysis of problem behavior. *J. Appl. Behav. Anal.* **2013**, *46*, 1–21. [CrossRef] [PubMed]
40. Pett, M.A.; Lackey, N.R.; Sullivan, J. *Making Sense of Factor Analysis: The Use of Factor Analysis for Instrument Development in Health Care Research*; Sage: London, UK, 2003.
41. Muthén, L.K.; Muthén, B.O. *Mplus User's Guide*, 6th ed.; Muthén & Muthén: Los Angeles, CA, USA, 2011.
42. Rosseel, Y. lavaan: An R Package for Structural Equation Modeling. *J. Stat. Softw.* **2012**, *48*, 1–36. [CrossRef]
43. Hu, L.T.; Bentler, P.M. Cutoff criteria for fit indexes in covariance structure analysis: Conventional criteria versus new alternatives. *Struct. Equ. Model. Multidiscip. J.* **1999**, *6*, 1–55. [CrossRef]
44. Bandalos, D.L. Factors influencing cross-validation of confirmatory factor analysis models. *Multivar. Behav. Res.* **1993**, *28*, 351–374. [CrossRef] [PubMed]
45. Braun, V.; Clarke, V. Using thematic analysis in psychology. *Qual. Res. Psychol.* **2006**, *3*, 77–101. [CrossRef]
46. Tabachnick, B.G.; Fidell, L.S. *Using Multivariate Statistics*, 4th ed.; Allyn and Bacon: Boston, MA, USA, 2001.
47. Williams, B.; Onsman, A.; Brown, T. Exploratory factor analysis: A five-step guide for novices. *J. Emerg. Prim. Health Care* **2010**, *8*, 1–13.
48. Starcevic, V.; Billieux, J. Does the construct of Internet addiction reflect a single entity or a spectrum of disorders? *Clin. Neuropsychiatry* **2017**, *14*, 5–10.
49. Kuss, D.J.; Griffiths, M.D. Internet gaming addiction: A systematic review of empirical research. *Int. J. Ment. Health Addict.* **2012**, *10*, 278–296. [CrossRef]
50. Kuss, D.J.; Griffiths, M.D. Online social networking and addiction—A review of the psychological literature. *Int. J. Environ. Res. Public Health* **2011**, *8*, 3528–3552. [CrossRef] [PubMed]
51. Marketing Land Nearly 80 Percent of Social Media Time Now Spent on Mobile Devices. Available online: http://marketingland.com/facebook-usage-accounts-1-5-minutes-spent-mobile-171561 (accessed on 29 January 2017).
52. Statista Share of Facebook Users Worldwide Who Accessed Facebook via Mobile from 2013 to 2018. Available online: https://www.statista.com/statistics/380550/share-of-global-mobile-facebook-users/ (accessed on 29 January 2017).
53. Hong, F.-Y.; Chiu, S.-I.; Huang, D.-H. A model of the relationship between psychological characteristics, mobile phone addiction and use of mobile phones by Taiwanese university female students. *Comput. Hum. Behav.* **2012**, *28*, 2152–2159. [CrossRef]

54. Panova, T.; Lleras, A. Avoidance or boredom: Negative mental health outcomes associated with use of Information and Communication Technologies depend on users' motivations. *Comput. Hum. Behav.* **2016**, *58*, 249–258. [CrossRef]
55. Kuss, D.J.; Dunn, T.J.; Wölfling, K.; Müller, K.W.; Hędzelek, M.; Marcinkowski, J. Excessive Internet use and psychopathology: The role of coping. *Clin. Neuropsychiatry* **2017**, *14*, 73–81.
56. Kuss, D.J.; Shorter, G.W.; van Rooij, A.J.; Griffiths, M.D.; Schoenmakers, T. Assessing Internet addiction using the parsimonious Internet addiction components model—A preliminary study. *Int. J. Ment. Health Addict.* **2014**, *12*, 351–366. [CrossRef]
57. Billieux, J.; Van der Linden, M.; D'Acremont, M.; Ceschi, G.; Zermatten, A. Does impulsivity relate to perceived dependence on and actual use of the mobile phone? *Appl. Cogn. Psychol.* **2007**, *21*. [CrossRef]
58. Byun, Y.H.; Ha, M.; Kwon, H.J.; Hong, Y.C.; Leem, J.H.; Sakong, J.; Kim, S.Y.; Lee, C.G.; Kang, D.; Choi, H.D.; et al. Mobile phone use, blood lead levels, and attention deficit hyperactivity symptoms in children: A longitudinal study. *PLoS ONE* **2013**, *8*, e59742. [CrossRef] [PubMed]
59. Strayer, D.L.; Turrill, J.; Coleman, J.R.; Ortiz, E.V.; Cooper, J.M. *Measuring Cognitive Distraction in the Automobile II: Assessing in-Vehicle Voice-Based Interactive Technologies*; American Automobile Association Foundation: Salt Lake City, UT, USA, 2014.
60. Chotpitayasunondh, V.; Douglas, K.M. How "phubbing" becomes the norm: The antecedents and consequences of snubbing via smartphone. *Comput. Hum. Behav.* **2016**, *63*, 9–18. [CrossRef]
61. Roberts, J.A.; David, M.E. My life has become a major distraction from my cell phone: Partner phubbing and relationship satisfaction among romantic partners. *Comput. Hum. Behav.* **2016**, *54*, 134–141. [CrossRef]
62. Caplan, S.E. Theory and measurement of generalized problematic Internet use: A two-step approach. *Comput. Hum. Behav.* **2010**, *26*, 1089–1097. [CrossRef]
63. Karlsen, F. *A World of Excesses; Online Games and Excessive Playing*; Ashgate: Farnham, UK, 2013.
64. Kuss, D.J. *For the Horde! How Playing World of Warcraft Reflects Our Participation in Popular Media Culture*; LAP LAMBERT Academic Publishing: Saarbrücken, Germany, 2013.
65. Brannen, J. Mixing methods: The entry of qualitative and quantitative approaches into the research process. *Int. J. Soc. Res. Methodol.* **2005**, *8*, 173–184. [CrossRef]

International Journal of
Environmental Research and Public Health

MDPI

Article

The Relationship between Impulsivity and Internet Gaming Disorder in Young Adults: Mediating Effects of Interpersonal Relationships and Depression

Hyera Ryu [1], Ji-Yoon Lee [1], Aruem Choi [1], Sunyoung Park [1], Dai-Jin Kim [2] and Jung-Seok Choi [1,3,*]

1 Department of Psychiatry, SMG-SMU Boramae Medical Center, Seoul 07061, Korea; hyera.ryu12@gmail.com (H.R.); idiyuni91@gmail.com (J.-Y.L.); choiar90@gmail.com (A.C.); spark.37eme@gmail.com (S.P.)

2 Department of Psychiatry, Seoul St. Mary's Hospital, College of Medicine, The Catholic University of Korea, Seoul 06591, Korea; kdj922@catholic.ac.kr

3 Department of Psychiatry and Behavioral Science, Seoul National University College of Medicine, Seoul 03080, Korea

* Correspondence: choijs73@gmail.com; Tel.: +82-2-870-3461

Received: 25 January 2018; Accepted: 2 March 2018; Published: 6 March 2018

Abstract: *Background:* This study aimed to explore relationships between impulsivity, interpersonal relationships, depression, and Internet Gaming Disorder (IGD) symptoms. *Methods:* A total of 118 young adults participated in this study: 67 IGD patients who met five or more of the DSM-5 diagnostic criteria for IGD and 56 healthy controls. We administered questionnaires to assess IGD symptoms (Young's Internet Addiction Test; Y-IAT), impulsivity (Barratt Impulsiveness Scale; BIS-11), interpersonal relationship (Relationship Change Scale; RCS), and depression (Beck Depression Inventory; BDI). We used PROCESS macro in SPSS to perform mediation analysis. *Results:* IGD symptom was positively related to depression and impulsivity, and negatively related to the quality of interpersonal relationships. Mediation analysis revealed full mediation effects of interpersonal relationships and depression on the association between impulsivity and IGD symptoms in the IGD group. Specifically, even after adjusting for gender as a covariate, high impulsivity was associated with greater difficulty with interpersonal relationships; which further affected depression and increased the risk of IGD. *Conclusions:* These results demonstrate the importance of early intervention in IGD patients, particularly in young adults with high impulsivity. When intervening in adults' IGD, we should consider not only individual factors (e.g., depression) but also socioenvironmental factors (e.g., interpersonal relationships).

Keywords: internet gaming disorder; impulsivity; depression; interpersonal relationships; serial mediation

1. Introduction

Internet Gaming Disorder (IGD) is a kind of behavioral addiction that has been defined as a loss of control, and persistent and recurrent use of internet games leading to significant impairment in psychosocial functioning [1]. In particular, Diagnostic and Statistical Manual of Mental Disorders-Fifth Edition (DSM-5) has added an IGD as one of the conditions for further study in Section 3 [1], due to increased social interest worldwide. In previous studies, the prevalence of IGD, diagnosed according to DSM-5 criteria, was 1.16% in Germany [2], 2.5% in Slovenia [3], and 2.9% in Hungary [4]. However, higher estimates of prevalence (5.9% [5] and 10.8% [6]), have been reported in Korea. Because the risk of IGD is high in Korea, there is an urgent need to explore the characteristics of IGD in Korea, with further research in particular with respect to personal and environmental factors.

Choi [7] and Choi et al. [8], who proposed an integrated pathway model of IGD based on the model of gambling problems developed by Blaszczynski and Nower [9], suggested that IGD is a complex disorder caused by interactions among different bio-psycho-social factors. One of the biological factors is trait impulsiveness. The impulsivity is a tendency to behave voluntarily with little or no prior consideration of consequences [10], and could be measured by the Barratt Impulsiveness Scale (BIS-11), which is one of the oldest and most widely used a measurement of impulsive personality traits. In the BIS-11, impulsivity could be classified cognitive, motor, and nonplanning impulsiveness. Cognitive impulsiveness is the propensity to respond or make decisions during problem solving without thinking; motor impulsiveness is the tendency of impulsive behaviors, such as manifesting difficulty with impulse control or acting without thinking. Finally, nonplanning impulsiveness is a lack of foresight and a tendency not to plan or consider consequences before starting something [11]. In many previous studies, impulsivity has been regarded as a marker for vulnerability to IGD [12–16]. Dalbudak et al. found that IGD was correlated with the severity of impulsivity among Turkish university students, and Lee et al. showed that trait impulsivity is a vulnerable factor of IGD in young adults. In particular, impulsivity assessed by neuropsychological tests, such as the Stop Signal Test and Go/No-Go Task, was related to IGD symptoms [13,17,18]. In addition, several studies showed that impulsivity is related to depression [19,20] and that it indirectly predicts loneliness and poor interpersonal relationships [21]. Thus, impulsivity is not only a core feature of IGD symptoms but also an important factor affecting individuals' emotional and social functioning.

The quality of interpersonal relationships such as deficient social support and loneliness is among the social factors reported to be risk factors for IGD [22–24]. Previous studies have shown that loneliness increases the difficulty of maintaining healthy social interactions and may increase a preference for online social interaction, which can lead to IGD [22,25–28]. Furthermore, social contacts with family and friends help to reduce IGD symptoms [24,29]. In previous studies, loneliness, lack of social support, and lack of a sense of belonging have been shown to predict depression [30,31]. Thus, IGD could be assumed to be related to poor interpersonal relationships, which is a risk factor for depression and IGD symptoms.

Depression is one of the psychological factors associated with IGD. Prior research has shown that depression is a psychiatric disorder that is often comorbid with IGD [29,32,33]. In one study, 7% of adult IGD patients had a comorbid dysthymic disorder [34], and Ko et al. [35] identified a relationship between IGD and major depressive disorder or dysthymic disorder in college students. However, some studies have reported inconsistent results regarding the relationship between depression and IGD symptoms [29,36–40]. Ha et al. [41] suggested that participants who reported depression tended to seek cyberspace to avoid negative emotions and difficulties in daily life, and they had a high likelihood of being addicted to internet games because of the emotional support they found in cyberspace. Furthermore, a 2-year longitudinal study found that participants who overused internet games tended to be more depressed than those who did not [42]. Given these inconsistent results, it is necessary to clarify the relationship between depression and IGD. In addition, no previous studies have examined the relationship between bio-psycho-social factors and IGD symptoms by distinguishing between IGD and HC groups. Thus, it is important to examine the pathways connecting impulsivity, interpersonal relationships, depression, and IGD symptoms.

This study aimed to clarify relationships between impulsivity, interpersonal relationships, depression, and IGD symptoms and to identify the mediating effects of interpersonal relationships and depression on the relationship between impulsivity and IGD symptoms by distinguishing between IGD and HC group. We hypothesized that participants in the IGD group who also showed higher impulsivity would have more difficulty in interpersonal relationships, which would increase depression and increase the risk of IGD symptoms.

2. Materials and Methods

2.1. Participants and Procedure

A total of 123 young adults participated in this study, including 67 patients diagnosed with IGD and 56 healthy controls. The patients with IGD were seeking treatment at the outpatient clinics of SMG-SNU Boramae Medical Center in Seoul due to excessive internet gaming. They were diagnosed with IGD by a clinically experienced psychiatrist according to the DSM-5 criteria (more than five items). Also, the Structured Clinical interview was administered by a psychiatrist to identify past and current psychiatric disorders and only individuals with a history of intellectual disability or psychotic disorder were excluded. Of 67 patients with IGD, 10 had major depressive disorders, and two and one displayed social anxiety disorder and bipolar I disorder, respectively. Healthy controls (HC), who were recruited through advertisements, had no history of any psychiatric disorder.

To screen for the participants' intelligence quotient (IQ), the Korean-Wechsler Adult Intelligence Scale-IV (K-WAIS-IV) was administered, and five subjects with an IQ < 80 were excluded. Thus, 118 participants were included in the final analyses, including 62 in the IGD (male = 55; age = 25.54 ± 5.29 years) and 56 in the HC group (male = 38; age = 24.23 ± 3.92 years). All subjects completed informed consent forms before participating in the study. This study was conducted in accordance with the Declaration of Helsinki, and the protocol was approved by the Institutional Review Board of the SMG-SNU Boramae Medical Center (16-2014-139).

2.2. Measures

2.2.1. Demographic Variables

All participants answered a questionnaire to provide basic information such as age, gender, education year, and Internet use time (weekday and weekend).

2.2.2. Young's Internet Addiction Test (Y-IAT)

The Y-IAT was developed by Young (1998) [28], and has been validated in Korea [43]. It is 20-item self-report questionnaire and each item was answered using a 5-point scale ranging from 1 (very rarely) to 5 (very frequently). The total score ranged from 20 to 100, with higher scores reflecting a greater tendency of IGD symptoms. Cronbach's alpha was 0.96 in this study.

2.2.3. Barratt Impulsiveness Scale-11 (BIS-11)

The BIS-11 is an 11-item, revised version of the original Barratt Impulsiveness Scale, which is used to assess the degree of impulsivity [11]. This scale includes three subscales: cognitive, motor, and nonplanning impulsiveness. Cronbach's alpha was 0.76 in this study.

2.2.4. Relationship Change Scale (RCS)

The RCS consists of 25-item and 5-point Likert scale, which was originally developed by Schlein et al. [44], and was later translated into Korean by Mun (1980) and revised according to the Korean culture by Chun [45]. The RCS measures interpersonal relationships, and higher scores indicate better interpersonal relationships. The total score ranges from 25 to 125. Cronbach's alpha was 0.93 in this study.

2.2.5. Beck Depression Inventory-II (BDI-II)

The BDI, developed by Beck et al. [46], is a 21-item self-report questionnaire that measures the severity of particular symptoms experienced over the past week. Total scores range from 0 to 63, and higher scores reflect more severe depression. The BDI-II has previously been validated in Korean [47]. Cronbach's alpha was 0.92 in this study.

2.3. Statistical Analysis

Chi-square and *t*-tests were performed to compare the demographic and clinical characteristics of the HC and IGD groups. Pearson's correlation analysis was conducted to examine relationships between IGD symptoms (Y-IAT), impulsivity (BIS-11), depression (BDI), and interpersonal relationships (RCS) in the HC and IGD groups, respectively. To examine whether the quality of interpersonal relationships and depression mediated the relationship between impulsivity and IGD symptoms, we performed serial mediation analysis using the SPSS PROCESS macro, version 2.16 (model 6), developed by Hayes [48]. Serial mediation assumes a causal chain linking the mediators, with a specified direction of causal flow [49]. In particular, we analyzed using bootstrapping method, because there were limitations of Sobel's test (e.g., need a large sample). 5000 bootstrapping was used to identify indirect effects in the mediation models and analyzed with 95% confidence interval. SPSS software version 21.0 (SPSS, Inc., Chicago, IL, USA) was used for all data analyses.

3. Results

3.1. Demographic and Clinical Characteristics

In the total group, the mean age was 24.92 ± 4.71 years, and 78.8% (n = 93) of the sample were male. A comparison of the demographic and clinical characteristics of the IGD and HC groups showed that the percentage of males [$x^2(1) = 7.662$, $p < 0.05$], the internet gaming use time on weekdays [$t(69.23) = 9.088$, $p < 0.001$], internet gaming use time on weekends [$t(71.52) = 10.979$, $p < 0.001$], Y-IAT scores [$t(99.305) = 9.855$, $p < 0.001$], BIS-11 scores [$t(116) = 4.673$, $p < 0.001$], and BDI scores [$t(89.97) = 6.261$, $p < 0.001$] were significantly higher in the IGD than in the HC group, and the RCS score [$t(108.75) = -5.033$, $p < 0.001$] was significantly lower in the IGD group. Results are shown in Table 1.

Table 1. Demographics and Clinical characteristics (n = 118).

Variables	IGD (n = 62) M ± SD	HC (n = 56) M ± SD	$Ø^2/t$	p Value
Gender (male, %)	55 (88.7%)	38 (67.9%)	7.662 *	0.007
Age (years)	25.54 ± 5.29	24.23 ± 3.92	1.545	0.125
Education (years)	13.88 ± 1.70	13.92 ± 1.51	−0.139	0.890
Time for weekday (h/day)	3.89 ± 3.02	0.28 ± 0.73	9.088 **	<0.001
Time for weekend (h/day)	5.72 ± 3.53	0.44 ± 1.09	10.979 **	<0.001
Y-IAT	53.71 ± 15.93	30.34 ± 9.15	9.855 **	<0.001
RCS	83.43 ± 14.48	94.87 ± 9.99	−5.033 **	<0.001
BIS-11	64.87 ± 9.65	57.33 ± 7.59	4.673 **	<0.001
BIS-11_Cognitive	19.03 ± 2.98	17.60 ± 2.39	2.840 *	0.005
BIS-11_Motor	18.09 ± 4.12	14.17 ± 2.84	6.053 **	<0.001
BIS-11_Nonplanning	27.74 ± 4.75	25.55 ± 3.82	2.735 *	0.007
BDI	12.04 ± 8.58	4.28 ± 4.30	6.261 **	<0.001

* $p < 0.05$, ** $p < 0.001$. Time for weekday/weekend = Average internet gaming use time per day on weekday and weekend; IGD = Internet Gaming Disorders; HC = Healthy Control; Y-IAT = Young's Internet Addiction Test; RCS = Relationship Change scale; BIS-11 = Barratt Impulsiveness Scale; BDI = Beck Depression Inventory.

3.2. Association between IGD Symptoms and Clinical Variables in the IGD and HC Groups

In both IGD and HC group, IGD symptoms as measured by Y-IAT were significantly correlated with depression (IGD: $r = 0.472$, $p < 0.001$; HC: $r = 0.363$, $p < 0.001$), and interpersonal problems (IGD: $r = -0.285$, $p < 0.05$; HC: $r = -0.268$, $p < 0.05$). However, there was a significant relationship between IGD symptoms and impulsivity only in the IGD group ($r = 0.306$, $p < 0.05$), specifically, IGD symptoms were related to cognitive impulsiveness ($r = 0.375$, $p < 0.001$) and nonplanning impulsiveness ($r = 0.275$, $p < 0.05$), but not to motor impulsiveness ($r = 0.129$, $p = 0.318$). In the HC group, impulsivity was not related to the IGD symptoms (Table 2).

Table 2. Correlation Analysis among the variables in IGD and HC group.

IGD (n = 62)	Y-IAT	BIS-11	BIS-11_Cognitive	BIS-11_Motor	BIS-11_Nonplanning	BDI	RCS
Y-IAT	1						
BIS-11	0.306 *	1					
BIS_cognitive	0.375 **	0.783 **	1				
BIS_Motor	0.129	0.765 **	0.388 **	1			
BIS_Nonplanning	0.275 *	0.875 **	0.626 **	0.441 **	1		
BDI	0.472 **	0.334 **	0.407 **	0.136	0.308 *	1	
RCS	−0.285 *	−0.407 **	−0.487 **	−0.103	−0.432 **	−0.641 **	1
HC (n = 56)	**Y-IAT**	**BIS-11**	**BIS-11_Cognitive**	**BIS-11_Motor**	**BIS-11_Nonplanning**	**BDI**	**RCS**
Y-IAT	1						
BIS-11	−0.023	1					
BIS-11_cognitive	0.070	0.744 **	1				
BIS-11_Motor	0.062	0.845 **	0.489 **	1			
BIS-11_Nonplanning	−0.136	0.893 **	0.489 **	0.630 **	1		
BDI	0.393 **	0.340 *	0.414 **	0.369 **	0.143	1	
RCS	−0.323 *	−0.317 *	−0.453 **	−0.289 *	−0.133	−0.586 **	1

* $p < 0.05$, ** $p < 0.001$. IGD = Internet Gaming Disorders; HC = Healthy Control; Y-IAT = Young's Internet Addiction Test; RCS = Relationship Change scale; BIS-11 = Barratt Impulsiveness Scale; BDI = Beck Depression Inventory.

3.3. Relationships between Impulsivity, Interpersonal Relationships, Depression, and IGD Symptoms

The model depicting serial mediation of the relationship between impulsivity and IGD symptoms by interpersonal relationships and depression in the IGD group is shown in Figure 1. We added sex as a covariate in the model. The serial mediation model was significant [$F(2,58) = 5.9481$, $p < 0.05$] and explained about 17% of the variance in IGD symptoms in the IGD group. Specifically, both the total effect of impulsivity on IGD ($c = 0.47$, SE = 0.19, $t = 2.39$, $p < 0.05$) and the direct effect of impulsivity on interpersonal relationships ($a_1 = -0.60$, SE = 0.17, $t = -3.39$, $p < 0.05$) as mediating variables were significant. However, there was no significant direct effect of impulsivity on depression (path a_2 in Figure 1). The direct effect of interpersonal relationships, the first mediating variable, on depression ($d_{21} = -0.35$, SE = 0.06, $t = -5.53$, $p < 0.001$), the second mediating variable, was also significant. Furthermore, the direct effect of depression on IGD symptoms ($b_2 = 0.76$, SE = 0.28, $t = 2.65$, $p < 0.05$) was also statistically significant, whereas that of interpersonal relationships (path b_1 in Figure 1) was not. Finally, no significant direct effect of impulsivity on IGD was found ($c' = 0.29$, SE = 0.20, $t = 1.42$, $p = 0.16$) when impulsivity and both mediating variables were simultaneously entered into the equation. These results showed that interpersonal relationships and depression fully mediated the relationship between impulsivity and IGD symptoms in IGD group. Furthermore, the findings suggest that, in the IGD group only, high impulsivity leads to poor interpersonal relationships, and increasing depression, which in turn increases IGD symptoms. Additionally, the results of the bootstrapping to verify the indirect effects were significant only for the impact of impulsivity on IGD symptoms through interpersonal relationships and depression (B = 0.10, BCa 95% CI [0.0282, 0.2573]). In contrast, in the HC group, relationships between impulsivity, interpersonal relationships, depression, and IGD symptoms were not significant. A summary of the serial mediation results in the IGD group is shown in Table 3.

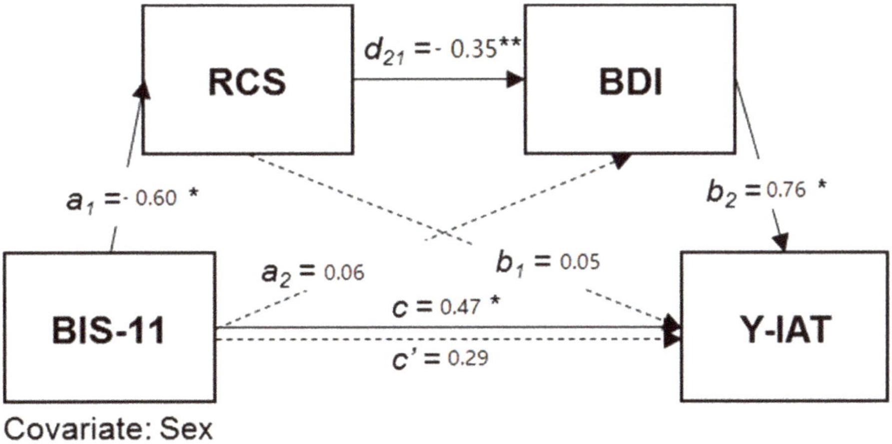

Figure 1. Serial mediation effects on problematic internet use in IGD group (n = 61). BIS-11 = Barratt Impulsiveness Scale; RCS = Relationship Change scale; BDI = Beck Depression Inventory; Y-IAT = Young's Internet Addiction Test; Solid lines showed significant paths with standardized path coefficients and dashed line represents a nonsignificant path. c' means direct effect of impulsivity (X) on internet addiction (Y) and c means indirect effect of impulsivity (X) on internet addiction (Y) through interpersonal relationship (M1) and depression (M2) in serial. Results showed that high impulsivity affected more difficulty in interpersonal relationships, which further affected depression and increased the risk of internet gaming disorder symptoms in IGD group. * $p < 0.05$, ** $p < 0.001$.

Table 3. Summary of serial mediation analysis of interpersonal relationship and depression between impulsivity and IGD symptoms in IGD group. (n = 61, bootstrap = 5000).

Effect	Paths	B	SE	t	BCa 95% CI	
					Lower	Upper
Direct effects	Impulsivity → Relationship	−0.60	0.17	−3.39 *		
	Impulsivity → Depression	0.06	0.09	0.72		
	Relationship → Depression	−0.35	0.06	−5.53 **		
	Relationship → Internet Addiction	0.05	0.16	0.30		
	Depression → Internet Addiction	0.76	0.28	2.65 *		
	Impulsivity → Internet Addiction	0.29	0.20	1.42		
Indirect effect	Impulsivity → Relationship → Internet Addiction	−0.01	0.07		−0.1759	0.1243
	Impulsivity → Relationship → Depression → Internet Addiction	0.10	0.05		0.0282	0.2573
	Impulsivity → Depression → Internet Addiction	0.03	0.04		−0.0268	0.1514

* $p < 0.05$, ** $p < 0.001$. BCa = Biased-Corrected and Accelerated 5000 bootstrapping; Covariate = sex.

4. Discussion

In this study, we investigated the well-known association between impulsivity and IGD symptoms by examining a mechanism linking interpersonal relationships and depression by distinguishing between the IGD and HC groups. As no previous studies have examined the pathways between impulsivity, interpersonal relationship, depression, and IGD symptoms by distinguishing between IGD and HC groups, the main purpose of this study was to examine the mechanism of association among these variables. The findings revealed serial mediation effects of depression and difficulty with interpersonal relationships on the relationship between impulsivity and IGD symptoms in the IGD group, but not in HC group. Furthermore, there were only full mediation effects. These results suggest a pathway in the IGD group whereby high impulsivity was related to difficulty with interpersonal relationships, which increased depression and, thereby, the risk of IGD symptoms. Thus, impulsivity was identified as one of the reasons for problems with interpersonal relationships among IGD patients, and the effect on IGD symptoms is mediated by difficulties with interpersonal relationships and depression, rather than having direct effects on IGD symptoms. This result suggests that IGD symptoms can be alleviated by reducing interpersonal problems and depression. This conclusion should be considered in developing treatment programs for IGD patients who report high impulsivity. In particular, Cognitive-Behavioral Therapy (CBT), Interpersonal Psychotherapy of Depression, Acceptance and commitment (ACT), and group therapy might be included as an element of treatment programs to alleviate interpersonal problems and depression.

In addition, one of the main results of our study was those that differentiated between the IGD and HC groups. Participants with IGD were significantly more likely to be male. They reported significantly higher rates of IGD symptoms including internet gaming use time on weekdays and weekends, poor interpersonal relationships, high impulsivity, and depressive symptoms compared with the healthy control group. Bakken et al. [50] and Tsai et al. [51] found that male gender was a predictor of IGD symptoms, with a higher proportion of males in the IGD than in the normal group.

Choi (2012) [7] argued that IGD could be caused by bio-psycho-social factors. In keeping with that study, IGD symptoms were related to impulsivity as a biological factor, interpersonal relationships as a social factor, and depression as a psychological factor in the present study. Specifically, consistent with previous studies, our results showed significant relationships among IGD symptoms measured by the Y-IAT, depression, and difficulty in interpersonal relationships in both the IGD and HC group. Dalbudak et al. [14] found that depression, as measured by the Symptom Checklist-revised (SCL-90-R) was related to the risk for IGD. Similarly, IGD has been found to be associated with many psychiatric disorders and symptoms, including depression, anxiety, ADHD, hostility, interpersonal sensitivity, and paranoid ideation [32,33,41,52,53]. In addition, interpersonal relationships were also associated with IGD symptoms in our study, suggesting that poor interpersonal relationships are associated with IGD. Several previous studies have also found that interpersonal problems were known to cause IGD symptoms [24,31], and Caplan [54] found that people who had psychological problems preferred online interaction and used the internet to cope with loneliness.

However, the relationship between impulsivity, including all subscales, and IGD symptoms was significant in the IGD group, but not in the HC group. In many previous studies, impulsivity was a significant risk factor for IGD [13,14,16,55]. In particular, high impulsivity has been shown to increase the severity of IGD symptoms; this relationship was also found in a study of Korean adults by Lee et al. [15]. However, our results showed that there was no significant correlation between impulsivity and IGD symptoms in HC group. In the previous studies, impulsivity differed significantly by gender, being more common among males [56,57]. The ratio of males to females in the HC group was significantly lower than that in the IGD group in this study. Therefore, the non-significant correlation between impulsivity and IGD symptoms in the HC group may reflect differences between groups in the gender ratio.

There are several limitations to this study. First, it is difficult to generalize the results to other populations because this study was conducted with a small sample of young adults. Thus, it is necessary to include larger and more diverse samples in future studies. Second, this was a cross-sectional study, and all variables were measured at one point in time. MacKinnon et al. [58] suggested that longitudinal research would provide richer information about mediation and would be useful to identify causal relationships when analyzing mediation effects. Therefore, future studies should be longitudinal study so that clear pathways and causal relationships could be more readily identified.

5. Conclusions

The findings of this study clarify the main factors to be considered when intervening in IGD, particularly in the case of IGD patients with high impulsivity. Additionally, the present findings highlight the importance of the role of bio-psycho-social factors in relation to IGD symptoms. Thus, we should focus on reducing interpersonal problems and depression when performing interventions to improve IGD symptoms in individuals with high impulsivity.

Acknowledgments: This study was funded by the National Research Foundation of Korea (NRF-2014M3C7A1062894), Republic of Korea.

Author Contributions: All authors participated in the study and read the final manuscript; Hyera Ryu analyzed and interpreted the data and drafted the manuscript; Ji-Yoon Lee, Aruem Choi, and Sunyoung Park participated in collecting the data; Dai Jin Kim and Jung-Seok Choi conceived and designed the experiments, and the result interpretation and discussion; Jung-Seok Choi reviewed and edited the manuscript.

Conflicts of Interest: The authors declare no conflict of interest.

References

1. American Psychiatric Association. *Diagnostic and Statistical Manual of Mental Disorders (DSM-5®)*, 5th ed.; American Psychiatric Association: Washington, DC, USA, 2013, ISBN 978-0-89042-555-8.
2. Rehbein, F.; Kliem, S.; Baier, D.; Mossle, T.; Petry, N.M. Prevalence of internet gaming disorder in German adolescents: Diagnostic contribution of the nine DSM-5 criteria in a state-wide representative sample. *Addiction* **2015**, *110*, 842–851. [CrossRef] [PubMed]
3. Pontes, H.M.; Macur, M.; Griffiths, M.D. Internet gaming disorder among Slovenian primary schoolchildren: Findings from a nationally representative sample of adolescents. *J. Behav. Addict.* **2016**, *5*, 304–310. [CrossRef] [PubMed]
4. Király, O.; Sleczka, P.; Pontes, H.M.; Urbán, R.; Griffiths, M.D.; Demetrovics, Z. Validation of the ten-item Internet Gaming Disorder gest (IGDT-10) and evaluation of the nine DSM-5 Internet Gaming Disorder criteria. *Addict. Behav.* **2017**, *64*, 253–260. [CrossRef] [PubMed]
5. Yu, H.; Cho, J. Prevalence of internet gaming disorder among Korean adolescents and associations with non-psychotic psychological symptoms, and physical aggression. *Am. J. Health Behav.* **2016**, *40*, 705–716. [CrossRef] [PubMed]

6. Wang, H.R.; Cho, H.; Kim, D.J. Prevalence and correlates of comorbid depression in a nonclinical online sample with DSM-5 Internet Gaming Disorder. *J. Affect. Disord.* **2018**, *226*, 1–5. [CrossRef] [PubMed]
7. Choi, S.W. Internet addiction: Why we become addicted to the internet? *Asia Pac. J. Clin. Oncol.* **2012**, *4*, 12.
8. Choi, S.W.; Lee, H.K.; Kim, H.S.; Lee, K.S. In the modified integrated pathways model of internet addiction. *J. Behav. Addict.* **2013**, *2*, 10.
9. Blaszczynski, A.; Nower, L. A pathways model of problem and pathological gambling. *Addiction* **2002**, *97*, 487–499. [CrossRef] [PubMed]
10. VandenBos, G.R. *APA Dictionary of Psychology*; APA: Washington, DC, USA, 2007.
11. Patton, J.H.; Stanford, M.S.; Barratt, E.S. Factor structure of the Barratt Impulsiveness Scale. *J. Clin. Psychol.* **1995**, *51*, 768–774. [CrossRef]
12. Cao, F.; Su, L.; Liu, T.; Gao, X. The relationship between impulsivity and internet addiction in a sample of Chinese adolescents. *Eur. Psychiatry* **2007**, *22*, 466–471. [CrossRef] [PubMed]
13. Choi, J.S.; Park, S.M.; Roh, M.S.; Lee, J.Y.; Park, C.B.; Hwang, J.Y.; Gwak, A.R.; Jung, H.Y. Dysfunctional inhibitory control and impulsivity in internet addiction. *Psychiatry Res.* **2014**, *215*, 424–428. [CrossRef] [PubMed]
14. Dalbudak, E.; Evren, C.; Topcu, M.; Aldemir, S.; Coskun, K.S.; Bozkurt, M.; Evren, B.; Canbal, M. Relationship of internet addiction with impulsivity and severity of psychopathology among Turkish university students. *Psychiatry Res.* **2013**, *210*, 1086–1091. [CrossRef] [PubMed]
15. Lee, H.W.; Choi, J.S.; Shin, Y.C.; Lee, J.Y.; Jung, H.Y.; Kwon, J.S. Impulsivity in internet addiction: A comparison with pathological gambling. *Cyberpsychol. Behav. Soc. Netw.* **2012**, *15*, 373–377. [CrossRef] [PubMed]
16. Lin, M.P.; Ko, H.C.; Wu, J.Y. Prevalence and psychosocial risk factors associated with internet addiction in a nationally representative sample of college students in Taiwan. *Cyberpsychol. Behav. Soc. Netw.* **2011**, *14*, 741–746. [CrossRef] [PubMed]
17. Ding, W.N.; Sun, J.H.; Sun, Y.W.; Chen, X.; Zhou, Y.; Zhuang, Z.G.; Li, L.; Zhang, Y.; Xu, J.R.; Du, Y.S. Trait impulsivity and impaired prefrontal impulse inhibition function in adolescents with internet gaming addiction revealed by a go/no-go fMRI study. *Behav. Brain Funct.* **2014**, *10*, 20. [CrossRef] [PubMed]
18. Dong, G.; Zhou, H.; Zhao, X. Impulse inhibition in people with internet addiction disorder: Electrophysiological evidence from a go/nogo study. *Neurosci. Lett.* **2010**, *485*, 138–142. [CrossRef] [PubMed]
19. Granö, N.; Keltikangas-Jarvinen, L.; Kouvonen, A.; Virtanen, M.; Elovainio, M.; Vahtera, J.; Kivimäki, M. Impulsivity as a predictor of newly diagnosed depression. *Scand. J. Psychol.* **2007**, *48*, 173–179. [CrossRef] [PubMed]
20. Swann, A.C.; Steinberg, J.L.; Lijffijt, M.; Moeller, F.G. Impulsivity: Differential relationship to depression and mania in bipolar disorder. *J. Affect. Disord.* **2008**, *106*, 241–248. [CrossRef] [PubMed]
21. Savci, M.; Aysan, F. Relationship between impulsivity, social media usage and loneliness. *Educ. Process Int. J.* **2016**, *5*, 106–115. [CrossRef]
22. Kim, J.; LaRose, R.; Peng, W. Loneliness as the cause and the effect of problematic internet use: The relationship between internet use and psychological well-being. *Cyberpsychol. Behav.* **2009**, *12*, 451–455. [CrossRef] [PubMed]
23. Nalwa, K.; Anand, A.P. Internet addiction in students: A cause of concern. *Cyberpsychol. Behav.* **2003**, *6*, 653–656. [CrossRef] [PubMed]
24. Yao, M.Z.; Zhong, Z.-J. Loneliness, social contacts and internet addiction: A cross-lagged panel study. *Comput. Hum. Behav.* **2014**, *30*, 164–170. [CrossRef]
25. Ceyhan, A.A.; Ceyhan, E. Loneliness, depression, and computer self-efficacy as predictors of problematic internet use. *Cyberpsychol. Behav.* **2008**, *11*, 699–701. [CrossRef] [PubMed]
26. Meerkerk, G.J.; Van Den Eijnden, R.J.; Vermulst, A.A.; Garretsen, H.F. The compulsive internet use scale (CIUS): Some psychometric properties. *Cyberpsychol. Behav.* **2009**, *12*, 1–6. [CrossRef] [PubMed]
27. Thatcher, A.; Goolam, S. Defining the South African internet 'addict': Prevalence and biographical profiling of problematic internet users in South Africa. *S. Afr. J. Psychol.* **2005**, *35*, 766–792. [CrossRef]
28. Young, K.S. Internet addiction: The emergence of a new clinical disorder. *Cyberpsychol. Behav.* **1998**, *1*, 237–244. [CrossRef]
29. Young, K.S.; Rogers, R.C. The relationship between depression and internet addiction. *Cyberpsychol. Behav.* **1998**, *1*, 25–28. [CrossRef]

30. Adams, K.B.; Sanders, S.; Auth, E.A. Loneliness and depression in independent living retirement communities: Risk and resilience factors. *Aging Ment. Health* **2004**, *8*, 475–485. [CrossRef] [PubMed]
31. Hagerty, B.M.; Williams, R.A. The effects of sense of belonging, social support, conflict, and loneliness on depression. *Nurs. Res.* **1999**, *48*, 215–219. [CrossRef] [PubMed]
32. Ko, C.H.; Yen, J.Y.; Yen, C.F.; Chen, C.S.; Chen, C.C. The association between internet addiction and psychiatric disorder: A review of the literature. *Eur. Psychiatry* **2012**, *27*, 1–8. [CrossRef] [PubMed]
33. Yen, J.Y.; Ko, C.H.; Yen, C.F.; Wu, H.Y.; Yang, M.J. The comorbid psychiatric symptoms of internet addiction: Attention deficit and hyperactivity disorder (ADHD), depression, social phobia, and hostility. *J. Adolesc. Health* **2007**, *41*, 93–98. [CrossRef] [PubMed]
34. Bernardi, S.; Pallanti, S. Internet addiction: A descriptive clinical study focusing on comorbidities and dissociative symptoms. *Compr. Psychiatry* **2009**, *50*, 510–516. [CrossRef] [PubMed]
35. Ko, C.H.; Yen, J.Y.; Chen, C.S.; Chen, C.C.; Yen, C.F. Psychiatric comorbidity of internet addiction in college students: An interview study. *CNS Spectr.* **2008**, *13*, 147–153. [CrossRef] [PubMed]
36. Bahrainian, S.A.; Alizadeh, K.H.; Raeisoon, M.R.; Gorji, O.H.; Khazaee, A. Relationship of internet addiction with self-esteem and depression in university students. *J. Prev. Med. Hyg.* **2014**, *55*, 86–89. [PubMed]
37. Dieris-Hirche, J.; Bottel, L.; Bielefeld, M.; Steinbüchel, T.; Kehyayan, A.; Dieris, B.; Te Wildt, B. Media use and internet addiction in adult depression: A case-control study. *Comput. Hum. Behav.* **2017**, *68*, 96–103. [CrossRef]
38. Jang, K.S.; Hwang, S.Y.; Choi, J.Y. Internet addiction and psychiatric symptoms among Korean adolescents. *J. Sch. Health* **2008**, *78*, 165–171. [CrossRef] [PubMed]
39. Seifi, A.; Ayati, M.; Fadaei, M. The study of the relationship between internet addiction and depression, anxiety and stress among students of Islamic Azad University of Birjand. *Int. J. Econ. Manag. Soc. Sci.* **2014**, *3*, 28–32.
40. Shapira, N.A.; Goldsmith, T.D.; Keck, P.E.; Khosla, U.M.; McElroy, S.L. Psychiatric features of individuals with problematic internet use. *J. Affect. Disord.* **2000**, *57*, 267–272. [CrossRef]
41. Ha, J.H.; Yoo, H.J.; Cho, I.H.; Chin, B.; Shin, D.; Kim, J.H. Psychiatric comorbidity assessed in Korean children and adolescents who screen positive for internet addiction. *J. Clin. Psychiatry* **2006**, *67*, 821–826. [CrossRef] [PubMed]
42. Ko, C.H.; Yen, J.Y.; Chen, C.S.; Yeh, Y.C.; Yen, C.F. Predictive values of psychiatric symptoms for internet addiction in adolescents: A 2-year prospective study. *Arch. Pediatr. Adolesc. Med.* **2009**, *163*, 937–943. [CrossRef] [PubMed]
43. Kim, E.; Lee, S.; Oh, S. The validation of korean adolescent internet addiction scale (K-AIAS). *Korean J. Clin. Psychol.* **2003**, *22*, 125–139.
44. Schlein, S.; Guerney, B.; Stover, L. The Interpersonal Relationship Scale. Ph.D. Thesis, Pennsylvania State University, State College, PA, USA, 1971. Unpublished work.
45. Chun, S. The social skills training for social adjustment of the schizophrenic patients. *Ment. Health Soc. Work* **1995**, *2*, 33–50.
46. Beck, A.T.; Steer, R.A.; Brown, G.K. *Manual for the Beck Depression Inventory-II*; Psychological Corporation: San Antonio, TX, USA, 1996.
47. Sung, H.; Kim, J.; Park, Y.; Bai, D.; Lee, S.; Ahn, H. A study on the reliability and the validity of Korean version of the beck depression inventory-II (BDI-II). *J. Korean Soc. Biol. Ther. Psychiatry* **2008**, *14*, 201–212.
48. Hayes, A. *Process for SPSS (Version 2.16) [Macros]*; The Guilford Press: New York, NY, USA, 2016.
49. Hayes, A. PROCESS: A Versatile Computational Tool for Observed Variable Mediation, Moderation, and Conditional Process Modeling. 2012. Available online: http://www.afhayes.com/public/process2012.pdf (accessed on 16 May 2016).
50. Bakken, I.J.; Wenzel, H.G.; Götestam, K.G.; Johansson, A.; ØREN, A. Internet addiction among Norwegian adults: A stratified probability sample study. *Scand. J. Psychol.* **2009**, *50*, 121–127. [CrossRef] [PubMed]
51. Tsai, H.F.; Cheng, S.H.; Yeh, T.L.; Shih, C.-C.; Chen, K.C.; Yang, Y.C.; Yang, Y.K. The risk factors of internet addiction—A survey of university freshmen. *Psychiatry Res.* **2009**, *167*, 294–299. [CrossRef] [PubMed]
52. Carli, V.; Durkee, T.; Wasserman, D.; Hadlaczky, G.; Despalins, R.; Kramarz, E.; Wasserman, C.; Sarchiapone, M.; Hoven, C.W.; Brunner, R. The association between pathological internet use and comorbid psychopathology: A systematic review. *Psychopathology* **2013**, *46*, 1–13. [CrossRef] [PubMed]
53. Koc, M. Internet addiction and psychopathology. *Turk. Online J. Educ. Technol.* **2011**, *10*, 143–148.

54. Caplan, S.E. Preference for online social interaction: A theory of problematic internet use and psychosocial well-being. *Commun. Res.* **2003**, *30*, 625–648. [CrossRef]
55. Bargeron, A.H.; Hormes, J.M. Psychosocial correlates of internet gaming disorder: Psychopathology, life satisfaction, and impulsivity. *Comput. Hum. Behav.* **2017**, *68*, 388–394. [CrossRef]
56. Chapple, C.L.; Johnson, K.A. Gender differences in impulsivity. *Youth Violence Juv. Justice* **2007**, *5*, 221–234. [CrossRef]
57. Cross, C.P.; Copping, L.T.; Campbell, A. Sex differences in impulsivity: A meta-analysis. *Psychol. Bull.* **2011**, *137*, 97–130. [CrossRef] [PubMed]
58. MacKinnon, D.P.; Fairchild, A.J.; Fritz, M.S. Mediation analysis. *Annu. Rev. Psychol.* **2007**, *58*, 593–614. [CrossRef] [PubMed]

International Journal of
Environmental Research and Public Health

MDPI

Article

Problematic Use of the Internet and Smartphones in University Students: 2006–2017

Xavier Carbonell [1,*], Andrés Chamarro [2,3], Ursula Oberst [1], Beatriz Rodrigo [1] and Mariona Prades [1]

1 FPCEE Blanquerna, Universitat Ramon Llull, 08022 Barcelona, Spain; ursulao@blanquerna.url.edu (U.O.); beatrizrc2@blanquerna.url.edu (B.R.); marionapo@blanquerna.url.edu (M.P.)
2 Departamento de Psicología, Universitat Autónoma de Barcelona, Cerdanyola del Vallès, 08193 Barcelona, Spain; andres.chamarro@uab.es
3 Serra Hunter Program, Generalitat de Catalunya, 08029 Barcelona, Spain
* Correspondence: xaviercs@blanquerna.url.edu; Tel.: +34-932-533-000

Received: 24 January 2018; Accepted: 7 March 2018; Published: 8 March 2018

Abstract: It has been more than a decade since a concern about the addictive use of the Internet and mobile phones was first expressed, and its possible inclusion into the lists of mental disorders has recently become a popular topic of scientific discussion. Thus, it seems to be a fitting moment to investigate the prevalence of this issue over time. The aim of the present study was to analyze the prevalence of the perception of problematic Internet and smartphone use in young people over the period 2006–2017. To this end, a questionnaire on Internet use habits and two questionnaires on the negative consequences of Internet and smartphone use were administered to a sample of 792 university students. The scores were then compared with the results of former studies that had used these questionnaires. The perception of problematic Internet and mobile phone use has increased over the last decade, social networks are considered responsible for this increase, and females are perceived to be more affected than males. The current study shows how strong smartphone and Internet addiction and social media overlap. Participants from 2017 report higher negative consequences of both Internet and mobile phone use than those from 2006, but long-term observations show a decrease in problematic use after a sharp increase in 2013. We conclude that the diagnosis of technological addictions is influenced by both time and social and culture changes.

Keywords: Internet addiction; mobile phone addiction; online social network; university students; technological addictions; behavioral addictions; CERI; CERM

1. Introduction

Ever since Young [1] presented *Internet Addiction: the Emergence of a New Disorder* at the Congress of the American Psychological Association in Toronto, Internet addiction has been a widely discussed disorder in the media and in scientific literature [2]. The interest in the possible addiction to the Internet, video games, online role-playing games, television, and mobile phones has given rise to a new field of study: technological addictions [3]. In fact, the DSM-5 [4] included addiction to video games in the list of disorders that should receive further research. The negative consequences of this problem include the possible increase of stress, anxiety, and/or the "paradox" of a lack of communication, despite being more connected, especially among young people and adolescents [5–7].

Internet use and Internet access habits have recently evolved. For example, in 2015, in Spain, there was a clear preference for smartphones (88.2%) over computers (78.2%) for accessing the Internet, especially in 14–19 year olds. This evolution of preference for the phone over the computer was also observed for the accessing of leisure activities, which decreases the dominance of the computer in the professional and educational spheres [8]. Because of its popularity and because it is a relatively new device (compared to the "classical" mobile phone), the smartphone has raised concerns about its potential to be addictive [9–12] as happens with other possible behavioral addictions, such as those to the Internet or to social networking sites [13].

The prevalence of smartphone addiction has been established using different self-report measures. Among the most frequently used, and actually one of the first self-report questionnaires, there was the Mobile Phone Problem Usage Scale (MPPUS) [9,14], later translated and adapted to the Spanish population [15]. The Problematic Mobile Phone Use Questionnaire (PMPUQ) [16] has also been employed in several studies [17]. Another questionnaire used in the Spanish speaking context is the *Cuestionario de Experiencias Relacionadas con el Móvil* [Questionnaire for mobile phone-related experiences] (CERM) [18], used to collect data from young people [19] and university students [20,21]. Recently, several other questionnaires were adapted to asses smartphone addiction, like the Smartphone Addiction Scale (SAS) [22], the Smartphone Addiction Inventory (SPAI) [23], or the one used in Saudi Arabia [24]. All of them have shown their usefulness and good validity and reliability, but it is difficult to assess if problematic mobile phone use has increased over the years since researchers use different measurements and instruments.

The term 'Internet addiction' or 'smartphone addiction' is not used consistently in the literature. A review shows the use of different expressions; there is, for instance, 'digital addiction' [25], 'problematic Internet use' [26], or 'Internet-use disorder' [27]. Even the same authors use different expressions in different papers [27,28]. It is not the purpose of the present article to discuss the convenience of one term or another, nor to discuss if the negative consequences can be interpreted as an addiction or not; however, we will use the term 'addiction' for the studied phenomenon, because it was one of the first terms used [1] and because in the studies that we will compare, this term has also been used from the beginning.

Added to the scientific relevance of this issue, the media tend to echo and spread negative information about the use of mobile phones. Results of these alerts are concepts such as the so-called technostress [29], smombie (a combination of "smartphone" and "zombie") [30], fear of missing out ("FoMO") [31], and nomophobia ("no-mobile-phone phobia") [32]. However, studies on the addictive consequences of both the 'old' mobile phone [9,33–36], the current smartphone [22,24,37–40], and the Internet are cross-sectional, and therefore the temporal evolution of their addictive impact on the population is still unknown. On the basis of the aforementioned research, the objectives of the current study were, first, to explore the perception of problematic Internet and smartphone use in young people in 2017 and, second, to compare these results with former those of studies that used the same measurement instruments, in order to analyze the evolution of the negative effects of Internet and mobile phone use over a period of ten years. As Internet and smartphone use has increased considerably during the past decade, we also expect negative effects to increase over the years (H1). As shown in the previous literature, also in our study women reported a greater use of social networking sites, whereas men used more videogames and adult pages (H2). We also expect women to experiment stronger negative consequences than males (H3).

2. Materials and Methods

2.1. Participants

In the present study, 792 students from Universitat Ramon Llull of Barcelona (Spain) participated in the study in May 2017. They were studying Psychology (30.7%), Physical Education and Sports Sciences (17.2%), Education (47.6%), and Speech Therapy (4.5%). The mean age was 21.6 years (SD = 3.3), and 76.5% were women.

The data on problematic Internet and mobile phone use obtained in the present study (called Cohort 6 hereinafter) was compared with data obtained from other cohorts of university students who answered the CERI and CERM questionnaires in studies conducted by our team between 2006 and 2017:

- Cohort 1: 322 students from the fields of Psychology, Physical Education, Nursing, Physiotherapy, or Communication in Universitat Ramon Llull of Barcelona. The mean age was 19.71 years (SD = 1.73), and 72.7% were women. The data were collected during the 2005–2006 academic year [18].
- Cohort 2: 318 Psychology students from the University of Illinois at Urbana-Champaign in the United States, 51% of which were women, aged between 17 and 21 years. The data were collected during the 2013–2014 academic year [20].
- Cohort 3: 425 Psychology students from the University of Illinois at Urbana-Champaign with an average age of 19.5 (SD = 1.5); 65.4% were women. The data were collected during the 2015–2016 academic year [21].
- Cohort 4: 308 Psychology students from the Universitat Ramon Llull. The mean age was 22.2 years (SD = 4.1), and 77.9% were women. The data were collected during the 2015–2016 academic year [21].
- Cohort 5: 308 psychology students from Ibagué University in Ibagué, Colombia. The mean age was 19.8 years (SD = 3.03), and 65.6% were women. The data were collected during the 2015–2016 academic year [21].

2.2. Instruments

The following instruments were used:

- Sociodemographic data and Internet use habits were assessed with an ad hoc questionnaire. This questionnaire collected sociodemographic data (age, sex, and university degree) and frequency and type of Internet use (e.g., gambling, social networks, etc.) in a five-point Likert scale. The questionnaire also included a Likert-type question about the user's degree of agreement with the statement: "I am addicted to the Internet" and one question about gender and addiction: "Do you think that girls are more Internet-addicted than boys?"
- Addictive behaviors related to the Internet were assessed with the Cuestionario de Experiencias Relacionadas con Internet (CERI) [Questionnaire on Internet-related experiences] [18]. This questionnaire consisted of 10 items about Internet use that were answered on a four-point Likert scale. Item example: "Piensas que la vida sin Internet es aburrida, vacía y triste?" (Do you think that life without the Internet is boring, empty, and sad?). The reliability (Cronbach's alpha) in the present study was 0.76; in the original study it was 0.77.
- Addictive behaviors related to the mobile phone were assessed with the Cuestionario de Experiencias Relacionadas al Móvil (CERM) (Questionnaire on experiences related to the mobile phone) (CERM) [18]. This questionnaire consisted of 10 items about mobile phone use that were answered on a four-point Likert scale. Item example: "Hasta qué punto te sientes inquieto cuando no recibes mensajes o llamadas?" (To what extent do you feel anxious when you do not receive messages or calls?). In the present study, Cronbach's alpha was 0.73. Cronbach's alpha in the original study was 0.80. Other studies have also reported reliability indexes of 0.80 [41].

For both CERI and CERM, the scores were calculated by adding up the answers to all the items, to a maximum of 40 points; cut-off points were established in a former study [42]. CERI and CERM have been used in several studies on adolescents' excessive Internet and mobile phone use [19,43].

2.3. Procedure

Eligible participants were invited to participate in the present study by means of an email containing a link to a Google Docs form. No personal information was requested, and it was not possible to connect any of the data from the questionnaires to academic records. The participants had to click on a box to give their informed consent and continue with the study. The students did not receive any monetary or academic reward for their participation. The study was approved by the Committee of Ethics and Research of the FPCEE Blanquerna, Universitat Ramon Llull.

2.4. Data Analysis

Normality checks were run on the data. Student's t-tests were run to assess gender differences in relation to: (a) the type of use they engage in on the Internet; (b) their scores on the CERI and CERM; (c) the degree of agreement that users expressed regarding the question "I'm addicted to the Internet." To check if the use of certain Internet functions is associated with negative consequences of use, correlations were calculated between the use of the Internet functions and scores on the CERI and CERM. A multiple analysis of variance for gender and year of questionnaire administration was run to test the effects of these two factors on both the CERM and CERI scores.

3. Results

3.1. Results of the Present Study

The frequencies of Internet uses are shown in Table 1. The most frequent activities on the Internet were checking emails and sending messages, participating in social networks, and listening to music. The least frequent uses were gambling and visiting adult pages. Significant gender differences were found between all uses except for online purchases, viewing of TV series, movies or videos, and administrative tasks.

Table 1. Most frequent Internet uses by university students.

Internet Uses	Men	Women	Total	t	p
	M (SD)	M (SD)	M (SD)		
Phone calls and videoconferences	2.70 (1.12)	2.98 (1.11)	2.91 (1.27)	2.97	0.003
Email/Chat	4.60 (0.66)	4.76 (0.54)	4.72 (0.58)	3.38	0.001
Social networking	4.11 (1.06)	4.34 (0.97)	4.28 (0.99)	2.69	0.007
General information	3.80 (1.04)	3.33 (1.07)	3.44 (1.08)	5.18	0.007
Shopping	2.11 (1.00)	2.11 (1.06)	2.11 (1.05)	0.01	0.987
Videogames	2.32 (1.29)	1.63 (0.93)	1.79 (1.07)	7.94	0.000
Gambling/betting	1.44 (0.85)	1.06 (0.32)	1.15 (0.53)	8.96	0.000
Videos/TV series	3.43 (1.24)	3.52 (1.21)	3.50 (1.22)	0.90	0.364
Listening to music	3.99 (1.10)	4.24 (1.01)	4.18 (1.04)	2.90	0.004
Administrative tasks	2.58 (1.20)	2.60 (1.25)	2.59 (1.24)	0.21	0.832
Adult content	2.34 (1.08)	1.25 (0.65)	1.51 (0.90)	16.07	0.000
Academic activities	3.66 (1.01)	4.03 (0.99)	3.95 (1.01)	4.46	0.000

Descriptive statistics of the Cuestionario de Experiencias Relacionadas al Móvil (CERM) and Cuestionario de Experiencias Relacionadas con Internet (CERI) scores are presented in Table 2. The mean CERI score was 18.04 (SD = 4.50), and the mean CERM score was 15.77 (SD = 3.50) for the whole cohort 6. There were no significant differences either for gender or for the different major degrees ($F(4, 787) = 1.24$; $p = 0.291$ for the CERI, and $F(4, 787) = 1.85$; $p = 0.116$ for the CERM). To the question, "Do you think girls are more addicted to the Internet than boys?" 73.2% answered affirmatively.

Table 2. Means and standard deviations of the Cuestionario de Experiencias Relacionadas con Internet (CERI) and the Cuestionario de Experiencias Relacionadas al Móvil (CERM) scores for cohort 6.

Questionnaire	Men	Women	Total	t-Tests
	M (SD)	M (SD)	M (SD)	
Cuestionario de Experiencias Relacionadas al Móvil (CERM)	15.77 (3.55) $n = 186$	15.77 (3.50) $n = 606$	15.77 (3.50) $n = 792$	$t(790) = 0.012$ ($p = 0.884$)
Cuestionario de Experiencias Relacionadas con Internet (CERI)	18.09 (4.81) $n = 186$	18.04 (4.41) $n = 606$	18.04 (4.50) $n = 792$	$t(790) = 0.146$ ($p = 0.990$)

The correlation between the CERI and the CERM was high ($r = 0.76$, $p = 0.000$). The correlations between the different online functions and the CERI and CERM were mostly significant, but low, with the most relevant correlation being between social networks and both the CERI and the CERM (see Table 3).

Table 3. Correlations between CERI or CERM and the uses of the Internet.

Internet Uses	CERI	CERM
Email/Chat	0.15 **	0.14 **
Social networking	0.23 **	0.21 **
General information	0.14 **	0.08
Shopping	0.14 **	0.12 **
Videogames	0.10 **	0.11 **
Gambling/betting	0.15 **	0.17 **
Videos/TV series	0.12 **	0.11 **
Listen music	0.18 **	0.17 **
Administration	0.03	0.04
Adult content	0.12	0.13 **
Academic activities	0.07	0.01
Phone calls/videoconferences	0.04	0.08

Note: ** $p < 0.001$.

Regarding the self-assessment of whether they considered themselves addicted to the Internet, 375 students (47.4%) either agreed or agreed strongly with this statement. Table 4 shows that people who "strongly agreed" with the statement "I am addicted to the internet" obtained significantly higher results than the rest of the participants on both the CERI and the CERM. The correlations with the CERI and CERM were 0.38 ($p < 0.001$) and 0.34 ($p < 0.001$), respectively.

Table 4. Levels of agreement with the statement "I am addicted to the Internet" with the scores in CERI and CERM.

Level of Agreement	CERI	CERM
	M (SD)	*M* (SD)
Strongly agree	21.89 (4.71) $n = 93$	18.47 (4.22) $n = 93$
Agree	19.02 (4.20) $n = 282$	16.45 (3.25) $n = 282$
Neither agree nor disagree	16.82 (3.81) $n = 275$	14.89 (3.09) $n = 275$
Disagree	15.90 (3.58) $n = 123$	14.35 (2.71) $n = 123$
Totally disagree	16.47 (6.85) $n = 19$	14.47 (4.25) $n = 19$

3.2. Comparison between Present Study and Former Studies with CERM and CERI

As shown in Table 5, the scores in the CERI and CERM grew from 2005 to 2013 and remained stable during the 2013–2014 academic year. The number of students who showed problematic Internet use went from 1.5% in 2005 to 6.4% in 2017 and from 0.6% to 3.0%, in the case of problematic mobile phone use. The correlation between CERI and CERM was "moderate" up to 2014 and increased to "high" as of 2015.

Table 5. Scores of the university student cohorts in the CERI and the CERM.

Cohort	Year of Survey	CERI M (SD)	CERM M (SD)	Correlation CERI/CERM	Problematic Use CERI (%)	Problematic Use CERM (%)
Cohort 1	2006	14.44 (4.00)	13.07 (2.90)	0.439 **	2.2%	0.9%
Cohort 2	2013	19.65 (5.06)	17.83 (4.39)	0.530 **	9.2%	8.6%
Cohort 3	2015	18.64 (5.03)	18.38 (4.09)	0.692 **	7.5%	8.1%
Cohort 4	2015	17.05 (4.06)	16.68 (3.51)	0.734 **	2.0%	3.0%
Cohort 5	2015	17.98 (5.41)	17.88 (4.98)	0.851 **	9.3%	11.5%
Cohort 6	2017	18.04 (4.50)	15.77 (3.50)	0.760 **	6.4%	3.0%

Note: ** $p < 0.001$.

Descriptive statistics for the CERM and CERI scores, separately for males and females, over the period 2006–2017 are presented in Table 6. Cohorts 3, 4, and 5 were taken together, as data collection took place in the same year. The multivariate analysis of variance for the effects of gender and year of administration on CERI and CERM showed significant effects of both factors (except for gender on CERI) and also a combined effect (see Table 7).

Table 6. Means and standard deviations (in brackets) of the CERM and the CERI scores.

Questionnaire	2006		2013		2015		2017	
	Males	Females	Males	Females	Males	Females	Males	Females
CERM	12.28 (2.62)	13.42 (2.96)	16.95 (4.34)	19.01 (4.60)	17.39 (4.40)	17.86 (4.18)	15.77 (3.55)	15.77 (3.51)
CERI	14.35 (4.05)	14.49 (3.99)	19.11 (4.97)	19.97 (5.20)	18.19 (5.11)	17.89 (4.85)	18.09 (4.81)	18.04 (4.41)

Table 7. Tests of between-subjects effects for the factors gender and year in CERM and CERI.

Source	Variable	F	p	η^2
gender	CERM	24.16	<0.001	0.010
	CERI	0.508	0.476	0.000
year	CERM	140.99	<0.001	0.146
	CERI	70.99	<0.001	0.079
Gender × year	CERM	5.29	0.001	0.006
	CERI	1.21	0.304	0.001

Women presented higher scores in the CERM, but not in the CERI. Post-hoc Bonferroni pairwise comparisons showed that scores for both CERM and CERI had increased from the first survey: there was, in fact, a significant difference between the 2006 scores and those of all the following surveys ($p < 0.001$ in all comparisons). However, between the 2013 and the following years, there was no significant difference between 2013 and 2015, and there was even a decrease of the negative effects between 2015 and 2017 ($p < 0.001$) with respect to the CERM; for the CERI, there was a decrease from 2013 to 2015 and from 2015 to 2017 ($p < 0.001$ in both cases). Figures 1 and 2 present the results separately for CERM and CERI and for both sexes. As can be seen, the increase of the perception of problematic Internet and smartphone use was stronger for females than for males, but as of 2017, both sexes had lower scores and tended to present the same degree of negative effects.

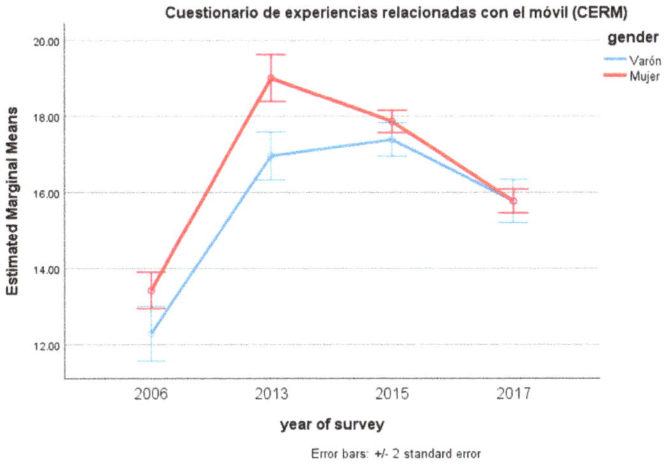

Figure 1. Estimated marginal means of Cuestionario de Experiencias Relacionadas al Móvil (CERM) for year of survey and gender.

Figure 2. Estimated marginal means of the Cuestionario de Experiencias Relacionadas con Internet (CERI) for year of survey and gender.

4. Discussion

The objective of this study was to evaluate the perception of problematic Internet and mobile phone use and compare these results with those of similar cohorts from up to a decade ago. One of the problems presented by studies evaluating technological addictions to the Internet, social networks, mobile phones, and video games is the absence of longitudinal studies. It is difficult for these studies to monitor a cohort in the medium or long term because the questionnaires are administered in person or online to: (i) cohorts obtained in the general population through social networks or the like; (ii) samples of high school or university students; (iii) video gamers identified in forums. Other factors that make longitudinal studies difficult are the need for respecting the participants' anonymity and the existence of time limitations (i.e., on funding for the research projects and doctoral theses).

To overcome these difficulties, this study compared data from different cohorts of university students assessed in different moments.

Our hypotheses were partially confirmed. Taken globally, the results of the study support the idea that the perception of a problematic use of the Internet and mobile phones exists and has increased over the last decade (H1). This perception seems to go along with the growth of Internet use and all kinds of electronic devices with a screen, with which our samples became familiar during their adolescence [44]. This problematic use of the Internet is specific and not general; that is, it depends on the concrete activity that is carried out [45–47]. This can also be affirmed for mobile phones [17,48,49]. Given that the Internet applications most widely used by university students are e-mail and messaging, participating in social networks, and listening to music, we infer that the increase in the perception of problematic use is associated with the use of online social networks. The activities in which the university students invest the least time are betting games and adult pages, like in other similar samples [17,45,46] and as is expected for university students. However, despite the existence of problematic use, it seems that the term 'addiction' is an inadequate construct when used as simply "Internet addiction", because: (i) the problematic use does not depend on the mobile phone or the Internet itself, but on the activities accessed on them; (ii) the problematic use can be the symptom of other disorders, not a primary disorder in itself [10,17]; (iii) there is a risk that labeling this problematic use an addiction means pathologizing the daily life [50]. The questionnaires used make it possible to detect a concern about certain technology-based behaviors but in no case to issue a clinical diagnosis. The term 'addiction' is probably adequate when related to specific types of use, such as addiction to gaming via the Internet or to pornography via the Internet.

It is difficult to compare our data on problematic mobile use with those of other prevalence studies because of the use of different measuring instruments. A preliminary comparison shows that the range of values for problematic users or addicts ranges between 0% and 35%, with 10–20% being the most frequent values [7,10,12], although there have been reports of 48% in university students [24]. In a recent research, the percentages ranged from 3.9% in Belgium to 1% in Poland, with 1.7% addicts identified in the Spanish sample [17].

It is also not surprising that the Internet uses of young men and women are quite different from each other as there are differences in behavior and attitudes between them in the real world which are perpetuated in the network (H2). Our results show that women use social networks and academic applications more and listen to more music than men. Men play more videogames and betting games and use more adult pages than women. In any case, women's problematic use is greater (H3), probably as a consequence of their using social networks more than males and of the role that those social networks play in communication and in creating and maintaining connections [17,40,51]. Some studies suggest that there may be different thresholds for males and females with respect to these negative effects [52].

The average scores in the CERI were higher than in the CERM, as has been the case with these instruments in other studies [18,20,53]. No differences were found between men and women in the problematic use of the mobile phone despite this being a frequent result in other investigations [15,17,40,53]. Although the Internet activities of men and women were different, there were no differences found in the problematic use of the Internet [53,54]. However, the perception of our students was that women are more addicted to the Internet than men probably because of the fact that using social networks is more common than using video games. Another possible explanation is that women are more vulnerable to this type of problem because it is related to communication practices such as establishing and actively maintaining relationships, which women engage in more than men [38,40].

Social network use is the only Internet use that moderately correlated with CERI and CERM. The other correlations, in line with other studies, were low or nonexistent [17]. In fact, online social networking is considered, along with video games, to be the use with the highest risk of becoming problematic [55,56] even though there is a lack of empirical confirmation [13]. Facebook and WhatsApp (as WhatsApp could be considered a social media) could be used as a key component to understand how young people socialize through these applications [57]. A large-scale tracking of online behavior showed that the use of WhatApp over the smartphone accounted for nearly 20% of all smartphone behavior, above Facebook use [58]. The low correlation with the different Internet uses can be explained because we are talking about a population that, as a whole, bets little and consumes little pornography, which leads us to think that we would obtain higher correlations if the CERI and the CERM were applied to cohorts of people extracted from the general population, who would be more likely to bet online, consume pornography, and/or be intensive video gamers.

The correlation between the CERI and CERM was high. In fact, we wonder if it is still convenient to use both the CERI and the CERM since, at the current time, both measures may be considered equivalent in the context of young people's technology use; young people use the mobile phone more and more frequently to access the Internet [8,17,40] and tend not to distinguish between the platforms (mobile vs. computer) and the program/application. The current study confirms how strong smartphone and Internet addiction and social media overlap [47,59] and advises that the evolution of technology forces us to change and update certain research questions. For example, when we designed the CERM, the mobile phones used by the participants did not have access to the Internet, whereas at present, there is not really a distinction between mobile phones and smartphones because they are considered synonymous. We will draw upon an anecdote to illustrate this situation. When, over a decade ago, the first papers expressing concern about mobile phone addiction were published [9], they were about mobile phones without Internet access. However, last year, when we showed a fourth-year psychology student a picture of an old Nokia phone from that era, one of them asked: "But... was it possible to be addicted to that?" This question reveals to what extent the diagnosis of technological addictions is influenced by time and social and cultural change.

From our point of view, the concern over mobile phone addiction came in two distinct waves. The first was focused on the non-smart mobile phone and was mainly due to two factors: the amount of phone bills that the use led to and the high use of text messages. Phone bills were a point of concern because a flat rate did not exist, and users needed a certain learning period to understand how to manage their use so that it remained within reasonable limits. Another point of concern was the number of text messages (i.e., for women: 11 or more calls or text messages per day, see Thomee, Dellve, Harenstam, Hagberg [6]). Several studies can serve as examples [6,9,34,36]. When, thanks to new billing structures, users managed to control their phone expenses and these worries seemed to dissipate, smartphones emerged on the market. At that point, a new wave of concern started in regard to this new device, because it allowed access to the Internet and Internet-based applications such as social networks and messaging services. Several studies can serve as examples [22–24,60,61]. Here again, the influence of context and culture is crucial.

Something similar is occurring with Internet addiction. In the last decade, we have learned that it is convenient to distinguish between behavioral addictions that take place on the Internet (for example, pathological gambling), the specific uses of the Internet that can become problematic (for example, videogames and social networks), and a possible generalized Internet addiction [45,46,49]. Davis [62] offered one of the first theoretical models, which differentiates between a generalized and a specific type of Internet addiction. Later, more sophisticated models to explain the different levels of the addiction process have been developed, such as the I-PACE model [27], which is useful for understanding the development and maintenance of specific Internet-use disorders. The students who responded to the CERI 10 years ago did so thinking primarily about their connection to the Internet via their computer, whereas now they access the Internet indiscriminately from their mobile phones, other handheld devices, and home or university computer. One example of how Internet use has changed in recent years is the application WhatsApp. WhatsApp is a telephone messaging service but it shares many features with social networks and, as of 2016, it can be accessed from the computer. Therefore, students can use WhatsApp on their mobile phones or from their laptop when they are in the classroom and they create class groups on both WhatsApp and Facebook indiscriminately. Another example of the merging of lines between phone applications and Internet applications is the difficulty in distinguishing between how much time is dedicated to each application or program, since it is so common to work in multi-screen mode. Students can write an academic paper, answer emails, and have a conversation on WhatsApp, all at the same time.

The results of the comparison between the present and former studies with the CERM and the CERI show that the negative effects of Internet and mobile phone use are considerably stronger now that at the first survey in 2006. However, this is due mainly to the first period, between 2006 and 2013. Apparently, there is a downward tendency in the perception of negative consequences in the recent years, which may correspond to a progressive normalization and integration of these new technologies into our daily life.

Although we have already mentioned the limitations of the CERI and CERM (they were created in the cell phone era before the existence of the smartphone and they are self-report measures), they are easy to use and to score, thereby inviting us to continue using them when possible in order to study the evolution of the perception of the problematic use of the Internet and mobile phones. They indicate, not so much the prevalence of an addiction, but the perception of a problem by the respondents. Although the correlation of the CERI and CERM with the statement "I am addicted to the internet" was moderate, the participants who were "very much in agreement" and "agreed" with this statement obtained significantly higher results than the rest in both the CERI and the CERM. This indicates that it might be possible to use this single question to detect the perception of problematic mobile phone use of individuals, as has already been suggested [46,63]. It is highlighted that our students seem to have a perception about their Internet addiction more in line with the data obtained through the questionnaires than the participants in the study done by Pontes, Szabo, and Griffiths [46], in which 51.9% of the participants identified themselves as Internet addicts.

This study is not without limitations. Firstly, university students have a higher than average level of academic development and their use of the Internet and mobile phones is not necessarily representative of the use that other young people engage in. Secondly, a problematic use of these technologies does not correspond to any diagnostic entity and may be a reflection of their social impact. Thirdly, both the CERI and the CERM should be updated, like other mobile addiction instruments (see, Kuss, Harkin, Kanjo and Billieux [64]), because a general use of the Internet is no longer conceivable; rather, a specific one is, and more so because of the new and expanded uses of smartphones. The terms "internet addiction" and "smartphone addiction" could be 'misnomers' [49] which we might not want to use anymore. Finally, it is possible that some differences found in this study were more influenced by the cultural differences between the samples than by the temporal differences.

5. Conclusions

Although men and women use the Internet differently, their problematic use of the Internet and mobile phones are quite similar. In university students, the use of social networks is the main factor responsible for the perception of problematic use; a casuistry that has increased in the decade 2006–2016. Despite the limitations of the CERI and CERM, estimating the prevalence at different points in time offers valuable information about the evolution of the perception of problematic Internet and mobile phone use. It is convenient to repeat the studies using the same instrument in order to understand the perception of problematic Internet and mobile phone use even if it lacks clinical significance. The degree of agreement with the statement "I am addicted to the Internet" might be used as a screening question for the problematic use. Young people are worried about the phenomenon, and it is convenient to keep track of their perception on the issue in order to design, if necessary, educational campaigns for an adequate use of these technologies.

Acknowledgments: We would like to thank Marta Beranuy, Tayana Panova, Diana-Ximena Puerta, Meritxell Puértolas, and Blanca Sánchez for the transfer of their databases.

Author Contributions: Xavier Carbonell designed the study. Beatriz Rodrigo and Mariona Prades acquired the data. Andres Chamarro, Ursula Oberst and Beatriz Rodrigo did the statistical analysis. Xavier Carbonell,

Andres Chamarro, and Ursula Oberst interpreted the results. Xavier Carbonell and Ursula Oberst wrote the initial draft of the article. All authors reviewed the initial draft, participated in the writing of the final draft, and approved the final version of the manuscript.

Conflicts of Interest: The authors declare no conflict of interest

References

1. Young, K.S. Internet Addiction: The emergence of a new clinical disorder. *Cyberpsychol. Behav.* **1998**, *1*, 237–244. [CrossRef]
2. Carbonell, X.; Guardiola, E.; Beranuy, M.; Bellés, A. A bibliometric analysis of the scientific literature on Internet, video games, and cell phone addiction. *J. Med. Libr. Assoc.* **2009**, *97*, 102–107. [CrossRef] [PubMed]
3. Griffiths, M.D. Technological addictions. *Clin. Psychol. Forum* **1995**, *76*, 14–19.
4. American Psychiatric Association. *Diagnostic and Statistical Manual of Mental Disorders (DSM-5)*; American Psychiatric Association: Arlington, VA, USA, 2013.
5. Lepp, A.; Li, J.; Barkley, J.E.; Salehi-Esfahani, S. Exploring the relationships between college students' cell phone use, personality and leisure. *Comput. Hum. Behav.* **2015**, *43*, 210–219. [CrossRef]
6. Thomée, S.; Härenstam, A.; Hagberg, M. Mobile phone use and stress, sleep disturbances, and symptoms of depression among young adults—A prospective cohort study. *BMC Public Health* **2011**, *11*, 66. [CrossRef] [PubMed]
7. Carbonell, X.; Fuster, H.; Chamarro, A.; Oberst, U. Internet and mobile phone addiction: A review of empirical Spanish studies. *Papeles Psicol.* **2012**, *33*, 82–89.
8. Fundación Telefónica. *La Sociedad de la Información en España 2016*; Fundación Telefónica: Barcelona, Spain, 2016.
9. Bianchi, A.; Phillips, J.G. Psychological predictors of problem mobile phone use. *Cyberpsychol. Behav.* **2005**, *8*, 39–51. [CrossRef] [PubMed]
10. Billieux, J.; Maurage, P.; Lopez-Fernandez, O.; Kuss, D.J.; Griffiths, M.D. Can disordered mobile phone use be considered a behavioral addiction? An update on current evidence and a comprehensive model for future research. *Curr. Addict. Rep.* **2015**, *2*, 156–162. [CrossRef]
11. Chóliz, M. Mobile phone addiction: A point of issue. *Addiction* **2010**, *105*, 373–374. [CrossRef] [PubMed]
12. Pedrero, E.J.; Rodríguez, M.T.; Ruiz, J.M. Adicción o abuso del teléfono móvil. Revisión de la literatura. *Adicciones* **2012**, *24*, 139–152. [CrossRef]
13. Carbonell, X.; Panova, T. A critical consideration of social networking sites' addiction potential. *Addict. Res. Theory* **2017**, *25*, 48–57. [CrossRef]
14. Lopez-Fernandez, O.; Honrubia-Serrano, M.L.; Gibson, W.; Griffiths, M.D. Problematic Internet use in British adolescents: An exploration of the addictive symptomatology. *Comput. Hum. Behav.* **2014**, *35*, 224–233. [CrossRef]
15. Lopez-Fernandez, O.; Losada-Lopez, J.L.; Honrubia-Serrano, M.L. Predictors of problematic Internet and mobile phone usage in adolescents. *Aloma* **2015**, *33*, 49–58.
16. Billieux, J.; Van der Linden, M.; Rochat, L. The role of impulsivity in actual and problematic use of the mobile phone. *Appl. Cogn. Psychol.* **2008**, *22*, 1195–1210. [CrossRef]
17. Lopez-Fernandez, O.; Kuss, D.; Romo, L.; Morvan, Y.; Kern, L.; Graziani, P.; Rousseau, A.; Rumpf, H.J.; Bischof, A.; Gässler, A.K.; et al. Self-reported dependence on mobile phones in young adults: A European cross-cultural empirical survey. *J. Behav. Addict.* **2017**, *6*, 168–177. [CrossRef] [PubMed]
18. Beranuy, M.; Chamarro, A.; Graner, C.; Carbonell, X. Validación de dos escalas breves para evaluar la adicción a Internet y el abuso de móvil. *Psicothema* **2009**, *21*, 480–485.
19. Muñoz-Miralles, R.; Ortega-González, R.; López-Morón, M.R.; Batalla-Martínez, C.; Manresa, J.M.; Montellà-Jordana, N.; Chamarro, A.; Carbonell, X.; Torán-Monserrat, P. The problematic use of Information and Communication Technologies (ICT) in adolescents by the cross sectional JOITIC study. *BMC Pediatr.* **2016**, *16*, 140. [CrossRef] [PubMed]
20. Panova, T.; Lleras, A. Avoidance or boredom: Negative mental health outcomes associated with use of Information and Communication Technologies depend on users' motivations. *Comput. Hum. Behav.* **2016**, *58*, 249–258. [CrossRef]

21. Panova, T.; Carbonell, X.; Chamarro, A. Uses and gratifications in smartphone use and its relationship with anxiety and depression: A multicultural perspective. *Comput. Hum. Behav.* **2018**. submitted.
22. Kwon, M.; Lee, J.Y.; Won, W.Y.; Park, J.W.; Min, J.A.; Hahn, C.; Gu, X.; Choi, J.H.; Kim, D.J. Development and validation of a Smartphone Addiction Scale (SAS). *PLoS ONE* **2013**, *8*, e56936. [CrossRef] [PubMed]
23. Lin, Y.H.; Chang, L.R.; Lee, Y.H.; Tseng, H.W.; Kuo, T.B.J.; Chen, S.H. Development and validation of the Smartphone Addiction Inventory (SPAI). *PLoS ONE* **2014**, *9*, e98312. [CrossRef] [PubMed]
24. Aljomaa, S.S.; Mohammad, M.F.; Albursan, I.S.; Bakhiet, S.F.; Abduljabbar, A.S. Smartphone addiction among university students in the light of some variables. *Comput. Hum. Behav.* **2016**, *61*, 155–164. [CrossRef]
25. Montag, C.; Walla, P. Carpe diem instead of losing your social mind: Beyond digital addiction and why we all suffer from digital overuse. *Cogent Psychol.* **2016**, *3*, 1157281. [CrossRef]
26. Anderson, E.L.; Steen, E.; Stavropoulos, V. Internet use and Problematic Internet Use: A systematic review of longitudinal research trends in adolescence and emergent adulthood. *Int. J. Adolesc. Youth* **2017**, *22*, 430–454. [CrossRef]
27. Brand, M.; Young, K.S.; Laier, C.; Wölfling, K.; Potenza, M.N. Integrating psychological and neurobiological considerations regarding the development and maintenance of specific Internet-use disorders: An Interaction of Person-Affect-Cognition-Execution (I-PACE) model. *Neurosci. Biobehav. Rev.* **2016**, *71*, 252–266. [CrossRef] [PubMed]
28. Brand, M.; Young, K.S.; Laier, C. Prefrontal control and internet addiction: A theoretical model and review of neuropsychological and neuroimaging findings. *Front. Hum. Neurosci.* **2014**, *8*, 357. [CrossRef] [PubMed]
29. Sayre, S.A.; Brod, C. *Technostress—The Human Cost of the Computer Revolution*; Addison-Wesley: Reading, MA, USA, 1984.
30. Butler, H. *Teens Pick "Smombie" as Hippest German Word*; The Local: Stockholm, Sweden, 2015.
31. Przybylski, A.K.; Murayama, K.; Dehaan, C.R.; Gladwell, V. Motivational, emotional, and behavioral correlates of fear of missing out. *Comput. Hum. Behav.* **2013**, *29*, 1841–1848. [CrossRef]
32. King, A.L.; Valença, A.M.; Nardi, A.E.; Valenca, A.M.; Nardi, A.E. Nomophobia: The mobile phone in panic disorder with agoraphobia: Reducing phobias or worsening of dependence? *Cogn. Behav. Neurol.* **2010**, *23*, 52–54. [CrossRef] [PubMed]
33. Ha, J.H.; Chin, B.; Park, D.-H.; Ryu, S.-H.; Yu, J. Characteristics of excessive cellular phone use in Korean adolescents. *Cyberpsychol. Behav.* **2008**, *11*, 783–784. [CrossRef] [PubMed]
34. Jenaro, C.; Flores, N.; Gómez-Vela, M.; González-Gil, F.; Caballo, C. Problematic internet and cell-phone use: Psychological, behavioral, and health correlates. *Addict. Res. Theory* **2007**, *15*, 309–320. [CrossRef]
35. Labrador, F.J.; Villadangos, S.M. Menores y nuevas tecnologías: Conductas indicadoras de posible problema de adicción. *Psicothema* **2010**, *22*, 180–188.
36. Sánchez-Martínez, M.; Otero, A. Factors associated with cell phone use in adolescents in the community of Madrid (Spain). *Cyberpsychol. Behav.* **2009**, *12*, 131–137. [CrossRef] [PubMed]
37. Bian, M.; Leung, L. Linking loneliness, shyness, smartphone addiction symptoms, and patterns of smartphone use to social capital. *Soc. Sci. Comput. Rev.* **2015**, *33*, 61–79. [CrossRef]
38. Gökçearslan, Ş.; Mumcu, F.K.; Haşlaman, T.; Çevik, Y.D. Modelling smartphone addiction: The role of smartphone usage, self-regulation, general self-efficacy and cyberloafing in university students. *Comput. Hum. Behav.* **2016**, *63*, 639–649. [CrossRef]
39. Lin, Y.-H.; Lin, Y.-C.; Lee, Y.-H.; Lin, P.H.; Lin, S.H.; Chang, L.R.; Tseng, H.W.; Yen, L.Y.; Yang, C.C.; Kuo, T.B. Time distortion associated with smartphone addiction: Identifying smartphone addiction via a mobile application (App). *J. Psychiatr. Res.* **2015**, *65*, 139–145. [CrossRef] [PubMed]
40. Van Deursen, A.J.A.M.; Bolle, C.L.; Hegner, S.M.; Kommers, P.A.M. Modeling habitual and addictive smartphone behavior. The role of smartphone usage types, emotional intelligence, social stress, self-regulation, age, and gender. *Comput. Hum. Behav.* **2015**, *45*, 411–420. [CrossRef]
41. Oberst, U.; Wegmann, E.; Stodt, B.; Brand, M.; Chamarro, A. Negative consequences from heavy social networking in adolescents: The mediating role of fear of missing out. *J. Adolesc.* **2017**, *55*, 51–60. [CrossRef] [PubMed]
42. Carbonell, X.; Chamarro, A.; Griffiths, M.D.; Oberst, U.; Cladellas, R.; Talarn, A. Problematic Internet and cell phone use in Spanish teenagers and young students. *Ann. Psychol.* **2012**, *28*, 789–796.
43. García-Oliva, C.; Piqueras, J.A. Experiential Avoidance and Technological Addictions in Adolescents. *J. Behav. Addict.* **2016**, *5*, 293–303. [CrossRef] [PubMed]

44. Martínez De Morentin, J.I.; Cortés, A.; Medrano, C.; Apodaca, P. Internet use and parental mediation: A cross-cultural study. *Comput. Educ.* **2014**, *70*, 212–221. [CrossRef]
45. Griffiths, M.D.; Szabo, A. Is excessive online usage a function of medium or activity? *J. Behav. Addict.* **2014**, *3*, 74–77. [CrossRef] [PubMed]
46. Pontes, H.M.; Szabo, A.; Griffiths, M.D. The impact of Internet-based specific activities on the perceptions of Internet addiction, quality of life, and excessive usage: A cross-sectional study. *Addict. Behav. Rep.* **2015**, *1*, 19–25. [CrossRef]
47. Montag, C.; Bey, K.; Sha, P.; Li, M.; Chen, Y.F.; Liu, W.Y.; Zhu, Y.K.; Li, C.B.; Markett, S.; Keiper, J. Is it meaningful to distinguish between generalized and specific Internet addiction? Evidence from a cross-cultural study from Germany, Sweden, Taiwan and China. *Asia-Pac. Psychiatry* **2015**, *7*, 20–26. [CrossRef] [PubMed]
48. Jeong, S.H.; Kim, H.; Yum, J.Y.; Hwang, Y. What type of content are smartphone users addicted to? SNS vs. games. *Comput. Hum. Behav.* **2016**, *54*, 10–17. [CrossRef]
49. Griffiths, M.D. Conceptual issues concerning Internet addiction and Internet gaming disorder: Further critique on Ryding and Kaye (2107). *Int. J. Ment. Health Addict.* **2018**, *16*, 233–239. [CrossRef] [PubMed]
50. Billieux, J.; Schimmenti, A.; Khazaal, Y.; Maurage, P.; Heeren, A. Are we overpathologizing everyday life? A tenable blueprint for behavioral addiction research. *J. Behav. Addict.* **2015**, *4*, 119–123. [CrossRef] [PubMed]
51. Sapacz, M.; Rockman, G.; Clark, J. Are we addicted to our cell phones? *Comput. Hum. Behav.* **2016**, *57*, 153–159. [CrossRef]
52. Lachmann, B.; Sariyska, R.; Kannen, C.; Cooper, A.; Montag, C. Life satisfaction and problematic Internet use: Evidence for gender specific effects. *Psychiatry Res.* **2016**, *238*, 363–367. [CrossRef] [PubMed]
53. Beranuy, M.; Oberst, U.; Carbonell, X.; Chamarro, A. Problematic Internet and mobile phone use and clinical symptoms in college students: The role of emotional intelligence. *Comput. Hum. Behav.* **2009**, *25*, 1182–1187. [CrossRef]
54. Casas, J.A.; Ruiz-Olivares, R.; Ortega-Ruiz, R. Validation of the Internet and Social Networking Experiences Questionnaire in Spanish adolescents. *Int. J. Clin. Health Psychol.* **2013**, *13*, 40–48. [CrossRef]
55. Kuss, D.J.; Griffiths, M.D. Online social networking and addiction—A review of the psychological literature. *Int. J. Environ. Res. Public Health* **2011**, *8*, 3528–3552. [CrossRef] [PubMed]
56. Ryan, T.; Chester, A.; Reece, J.; Xenos, S. The uses and abuses of Facebook: A review of Facebook addiction. *J. Behav. Addict.* **2014**, *3*, 133–148. [CrossRef] [PubMed]
57. O'Hara, K.P.; Massimi, M.; Harper, R.; Rubens, S.; Morris, J. Everyday dwelling with WhatsApp. In Proceedings of the 17th ACM Conference on Computer Supported Cooperative Work & Social Computing—CSCW'14, Baltimore, MD, USA, 15–19 February 2014; pp. 1131–1143.
58. Montag, C.; Błaszkiewicz, K.; Sariyska, R.; Lachmann, B.; Andone, I.; Trendafilov, B.; Eibes, M.; Markowetz, A. Smartphone usage in the 21st century: Who is active on WhatsApp? *BMC Res. Notes* **2015**, *8*, 331. [CrossRef] [PubMed]
59. Kittinger, R.; Correia, C.J.; Irons, J.G. Relationship between Facebook use and problematic Internet use among college students. *Cyberpsychol. Behav. Soc. Netw.* **2012**, *15*, 324–327. [CrossRef] [PubMed]
60. Haug, S.; Castro, R.P.; Kwon, M.; Filler, A.; Kowatsch, T.; Schaub, M.P. Smartphone use and smartphone addiction among young people in Switzerland. *J. Behav. Addict.* **2015**, *4*, 299–307. [CrossRef] [PubMed]
61. Darcin, A.E.; Kose, S.; Noyan, C.O.; Nurmedov, S.; Yılmaz, O.; Dilbaz, N. Smartphone addiction and its relationship with social anxiety and loneliness. *Behav. Inf. Technol.* **2016**, *35*, 520–525. [CrossRef]
62. Davis, R.A. A cognitive-behavioral model of pathological Internet use. *Comput. Hum. Behav.* **2001**, *17*, 187–195. [CrossRef]
63. Widyanto, L.; Griffiths, M.D.; Brunsden, V. A psychometric comparison of the Internet Addiction Test, the Internet-Related Problem Scale, and self-diagnosis. *Cyberpsychol. Behav. Soc. Netw.* **2011**, *14*, 141–149. [CrossRef] [PubMed]
64. Kuss, D.; Harkin, L.; Kanjo, E.; Billieux, J. Problematic smartphone use: Investigating contemporary experiences using a convergent design. *Int. J. Environ. Res. Public Health* **2018**, *15*, 142. [CrossRef] [PubMed]

International Journal of
Environmental Research and Public Health

MDPI

Brief Report

Mobile Phone Dependence, Social Support and Impulsivity in Chinese University Students

Songli Mei [1], Jingxin Chai [2,*], Shi-Bin Wang [3], Chee H. Ng [4], Gabor S. Ungvari [5,6] and Yu-Tao Xiang [7,*]

1 School of Public Health, Jilin University, Changchun 130021, China; meisongli@sina.com
2 Hepin Street, Dongcheng District, Beijing Center for Disease Prevention and Control, Beijing 100013, China
3 Guangdong Mental Health Center, Guangdong General Hospital and Guangdong Academy of Medical Sciences, Guangzhou 510120, China; spiriorwang@126.com
4 Department of Psychiatry, University of Melbourne, Melbourne, VIC 3050, Australia; cng@unimelb.edu.au
5 Division of Psychiatry, Notre Dame university Australia, Fremantle, WA 6160, Australia; sungvari@e.cuhk.edu.hk
6 Graylands Hospital, Claremont, WA 6010, Australia
7 Unit of Psychiatry, Faculty of Health Sciences, University of Macau, Avenida da Universidade, Taipa, Macau, China
* Correspondence: chaitong2008@163.com (J.C.); xyutly@gmail.com (Y.-T.X.); Tel./Fax: +86-10-65925223 (J.C.); Tel.: +853-8822-4223 (Y.-T.X.); Fax: +853-2288-2314 (Y.-T.X.)

Received: 12 January 2018; Accepted: 4 March 2018; Published: 13 March 2018

Abstract: This study examined the frequency of mobile phone dependence in Chinese university students and explored its association with social support and impulsivity. Altogether, 909 university students were consecutively recruited from a large university in China. Mobile phone use, mobile phone dependence, impulsivity, and social support were measured with standardized instruments. The frequency of possible mobile phone use and mobile phone dependence was 78.3% and 7.4%, respectively. Multinomial logistic regression analyses revealed that compared with no mobile phone dependence, possible mobile phone dependence was significantly associated with being male ($p = 0.04$, OR = 0.7, 95% CI: 0.4–0.98), excessive mobile phone use ($p < 0.001$, OR = 1.2, 95% CI: 1.09–1.2), and impulsivity ($p < 0.001$, OR = 1.05, 95% CI: 1.03–1.06), while mobile phone dependence was associated with length of weekly phone use ($p = 0.01$, OR = 2.5, 95% CI: 1.2–5.0), excessive mobile phone use ($p < 0.001$, OR = 1.3, 95% CI: 1.2–1.4), and impulsivity ($p < 0.001$, OR = 1.08, 95% CI: 1.05–1.1). The frequency of possible mobile phone dependence and mobile phone dependence was high in this sample of Chinese university students. A significant positive association with impulsivity was found, but not with social support.

Keywords: mobile phone dependence; mobile phone use; impulsivity; China

1. Introduction

Mobile phone dependence, defined as inapt use of a mobile phone, is broadly viewed as a subset of behavioral or technological addiction [1] which could lead to significant social and emotional impairment [2]. Excessive mobile phone use is common among young people and is negatively associated with academic performance [3], interpersonal relationships, self-esteem, self-regulation, and life satisfaction [4].

High level of impulsivity appears to be closely associated with excessive mobile phone use and dependence [5,6]. For example, Smetaniuk [7] reported that mobile phone dependence could be considered as an impulse control disorder. Those with a high level of impulsivity often prefer mobile phone use as a choice of gratification without fully thinking through consequences of the action [6].

In addition, social extraversion has been shown to have a positive association with mobile phone addiction, while self-esteem has a negative association [8]. Furthermore, social support is negatively associated with mobile phone use [9], but frequent social networking could increase the risk of mobile phone addiction [10].

China has become the largest mobile phone market worldwide and mobile phone use has increased dramatically in Chinese university students in recent years [11]. Better understanding of the frequency of mobile phone dependence and its associated factors is important to develop effective strategies to reduce its harmful impact. To date, few studies have examined the association of mobile phone dependence with impulsivity and social support. In this study, we aimed to examine the frequency of mobile phone dependence in Chinese university students and explore its association with demographic variables, impulsivity, and social support.

2. Materials and Methods

2.1. Study Design and Participants

This cross-sectional study was conducted in a large university in Jilin province, China between 1 September and 30 November 2013. Students who fulfilled the following inclusion criteria were recruited: (1) being registered as undergraduate students; (2) having a mobile phone; (3) having the ability to complete the assessment; (4) having the willingness to participate in the study and provide informed consent. The study protocol was approved by the Human Ethical Committee of the Jilin University School of Public Health (BSERE-APP002-FHS).

Three of the six campuses in the University were randomly selected and students who were enrolled in the selected campuses were consecutively screened for eligibility during the study period following a detailed explanation about the study protocol. The questionnaires were distributed to all eligible students in person by the research team members after they provided informed consent. Completed questionnaires were collected on a voluntary basis on the same day. The survey was completed anonymously and confidentially.

2.2. Assessment Instruments

Basic demographic characteristics were recorded on a standard study form. At the time of the study, no validated standardized questionnaires on mobile phone use were available in China. Therefore a self-reported questionnaire entitled "Mobile Phone Use Questionnaire (MPUQ)" was developed based on the recommendation of the China Internet Network Information Center [12]. The MPUQ consists of two parts. In the first part, respondents were screened for smart phone use by asking: "Do you own and use a smart phone?" If the answer was "yes", then the respondent was required to complete the second part. In the second part, the frequency of mobile phone use was assessed in two domains: (1) "use for networking purposes", including voice calling, sending short messages, using communication programs, watching internet news and blogs; (2) "use for entertainment purposes", including listening to music, watching videos, playing online games, and reading internet novels. Each domain includes four items with each item scoring from 1 (never) to 4 (frequently). The MPUQ total score ranges from 8 to 32, with a higher score indicating more frequent mobile phone use. The details of MPUQ items are presented in Supplementary Materials Table S1. The internal consistency of the MPUQ was fair: Cronbach's $\alpha = 0.66$ for the whole scale, and 0.68 and 0.58 for the two parts, respectively.

The Chinese version of the Barratt Impulsiveness Scale (BIS-11) was used to measure impulsivity trait [13]. The BIS-11 is a 30-item self-reported questionnaire with three subscales: attention impulsivity, motor impulsivity, and non-planning impulsivity. Each item is scored from 1 (never) to 5 (always). The Chinese version of the BIS-11 has been validated with good reliability and validity in China [14]. The internal consistency of the BIS-11 was satisfactory (Cronbach's $\alpha = 0.87$) in this study.

The factor structure of Chinese version of the BIS-11 was tested using the confirmatory factor analysis (CFA) with the AMOS 18.0 program. The model was acceptable if the value of χ^2/df was less than 8, the goodness of fit index (GFI) was greater than 0.90, the comparative fit index (CFI) was greater than 0.90, or the root mean square of approximation (RMSEA) was less than 0.08 [15]. The factor structure of Chinese version of the BIS-11 was found to be acceptable (χ^2/df = 7.42, *GFI* = 0.95, *CFI* = 0.89, *RMSEA* = 0.08).

The multidimensional scale of perceived social support (MSPSS) is a self-reported questionnaire measuring perceived social support. There are three subscales with a total of 12 items: family members' support, friends' support, and others' support. The items is scored from 1 (strongly disagree) to 7 (strongly agree). The Chinese version of the MSPSS has been validated in China [16]. The internal consistency of the MSPSS in this study was satisfactory (Cronbach's α = 0.83). The factor structure of Chinese version of the MSPSS was found to be acceptable (χ^2/df = 7.81, *GFI* = 0.93, *CFI* = 0.94, *RMSEA* = 0.08).

Mobile phone addiction was assessed using the mobile phone addiction scale for college students (MPAS) [17]. The MPAS comprises 16 items in four areas: withdrawal symptoms, salience, social comfort, and mood changes. Each item is rated on a five-point (1–5) scale with higher scores indicating higher level of mobile phone addiction. A total score of 16–31 indicates "no mobile phone addiction", while the score of 32–56 indicates "possible mobile phone addiction", and 57 or above indicates "mobile phone addiction" [9]. This definition of mobile phone addiction was also used in a previous study [16]. The factor structure of the MPAS was found to be acceptable (χ^2/df = 2.92, *NFI* = 0.94, *CFI* = 0.96, *RMSEA* = 0.07).

2.3. Statistical Analysis

The database was established using EpiData 3.2 (Epidata Assoc., Odense, Denmark). The data were analyzed with SPSS 18.0 (SPSS, Chicago, IL, USA) for Windows. Comparisons between no dependence, possible dependence, and mobile phone dependence individuals with respect to demographic and mental variables were conducted with chi-square and analysis of variance (ANOVA), as appropriate. Multinomial logistic regression analysis was used to determine the independent correlates significantly associated with mobile phone dependence. Mobile phone dependence was the dependent variable, while the demographic and clinical characteristics that significantly differed in the univariate analyses were entered as independent variables. Statistical significance was set at 0.05 (two-tailed).

3. Results

Out of 946 students who were approached, 909 students consented to the study and completed the questionnaire, giving a response rate of 96.1%. The frequency of having no mobile phone dependence, possible mobile phone dependence, and mobile phone dependence was 14.3% (n = 130), 78.3% (n = 712), and 7.4% (n = 67), respectively.

Table 1 presents the socio-demographic and clinical characteristics of the whole sample and separately by groups of mobile phone dependence. Univariate analyses revealed that being a single child, weekly mobile phone use time, and the total scores of the MPUQ and BIS were significantly different between the three groups.

Table 1. Comparison of socio-demographic and clinical characteristics between different groups of phone dependence.

	Whole Sample (n = 909)		No Phone Dependence (n = 130)		Possible Phone Dependence (n = 712)		Phone Dependence (n = 67)		Statistics [a]		
	n	%	n	%	n	%	n	%	χ^2	df	p
Male	404	44.4	70	53.8	302	42.5	31	46.3	5.9	2	0.053
Grade											
UG1-UG4	644	70.8	88	67.7	508	71.3	48	71.6	3.7	4	0.4
UG5	132	14.5	24	18.5	96	13.5	12	17.9			
PG	133	14.6	18	13.8	108	15.2	7	10.4			
Urban	426	46.9	63	48.5	330	46.3	33	49.3	0.4	2	0.8
Single child	485	53.4	66	50.8	373	52.4	46	68.7	6.9	2	**0.03**
Length of phone use >3 years	647	71.2	95	73.1	503	70.6	49	73.1	0.45	2	0.8
Length of weekly phone use >24 h	512	56.4	55	42.3	407	57.2	50	74.6	19.8	2	**<0.001**
	Mean	**SD**	**Mean**	**SD**	**Mean**	**SD**	**Mean**	**SD**	f	df	p
MPUQ total	26.0	5.1	23.0	5.2	26.2	4.8	29.7	5.6	44.7	2	**<0.001**
BIS total	73.8	12.9	67.7	13.2	74.4	12.7	78.8	11.5	21.0	2	**<0.001**
PSS total	64.4	11.1	64.0	13.1	64.2	10.8	66.8	9.9	1.8	2	0.2

Bolded values: <0.05; [a] comparison between no, possible and phone dependence; BIS = Barratt Impulsiveness Scale; PSS = perceived social support; PG = postgraduate; MPUQ = Mobile Phone Use Questionnaire; UG = under graduate.

Multinomial logistic regression analyses revealed that compared with no mobile phone dependence, possible mobile phone dependence was significantly associated with male gender ($p = 0.04$, OR = 0.7, 95% CI: 0.4–0.98) and MPUQ ($p < 0.001$, OR = 1.2, 95% CI: 1.09–1.2) and BIS total scores ($p < 0.001$, OR = 1.05, 95% CI: 1.03–1.06), while mobile phone dependence was associated with length of weekly phone use ($p = 0.01$, OR = 2.5, 95% CI: 1.2–5.0) and MPUQ ($p < 0.001$, OR = 1.3, 95% CI: 1.2–1.4) and BIS total scores ($p < 0.001$, OR = 1.08, 95% CI: 1.05–1.1) (see Table 2). Supplementary Materials Table S1 shows the details regarding the purpose of mobile phone use.

Table 2. Factors independently associated with phone dependence (multinomial logistic regression analysis with no dependence as the reference group).

	Possible Phone Dependence vs. No Dependence			Phone Dependence vs. No Dependence		
	p	OR	95% CI	p	OR	95% CI
Male	**0.04**	**0.7**	**0.4, 0.98**	0.6	0.8	0.5, 1.6
Single child	0.9	1.0	0.7, 1.5	0.2	1.6	0.8, 3.2
Length of weekly phone use >24 h	0.1	1.4	0.9, 2.1	**0.01**	**2.5**	**1.2, 5.0**
MPUQ total	**<0.001**	**1.2**	**1.09, 1.20**	**<0.001**	**1.3**	**1.2, 1.4**
BIS total	**<0.001**	**1.05**	**1.03, 1.06**	**<0.001**	**1.08**	**1.05, 1.1**

Bolded value: <0.05; BIS = Barratt Impulsiveness Scale; MPUQ = Mobile Phone Use Questionnaire; OR = odds ratio.

4. Discussion

In this study, 78.3% and 7.4% of university students reported possible mobile phone dependence and mobile phone dependence, respectively, which is inconsistent with previous findings using the MPAS in Chinese young adults (possible mobile phone dependence: 56.4%; mobile phone dependence: 36.4%) [9] and British and American college students (mobile phone dependence 10–25%) [7,18]. The discrepancy in the results across studies could be due to different socioeconomic and cultural factors, such as age group [7], type of university [11], and religion [19]. The major purpose of mobile phone use in this sample was voice calling and sending short messages, followed by using communication programs, watching internet news, reading blogs, listening music, watching video, playing online games, and reading internet novels.

Compared to male students, female students reported more frequent possible mobile phone dependence, but there was no significant gender difference in mobile phone dependence, which is partly supported by earlier findings in other settings [20,21]. This could be related to gender preferences in the use of mobile phones for maintaining interpersonal relationship [19]. Similar to previous studies [19,22] although different measures were used, we found that excessive use of mobile phones was positively associated with mobile phone dependence.

We found that impulsivity was positively associated with both possible mobile phone dependence and mobile phone dependence, which is in line with previous studies. For example, Smetaniuk [7] found a positive relationship between problematic mobile phone use and impulse control. Roberts et al. [23] found that attention impulsivity is predictive of mobile phone addiction. Further qualitative and empirical studies are warranted to explore the association between mobile phone behavior and impulsivity.

A mobile phone is a communication tool that can enhance social support. Therefore, greater family support could increase the likelihood of excessive mobile phone use [24], which may increase the risk of mobile phone dependence. For example, Crosswhite et al. [25] found that young adults prefer texting to engaging in general conversation with their parents. On the other hand, mobile phone dependence could negatively affect social relationships which in turn may have a negative impact on academic performance [26]. However, our study found no significant association between social support and mobile phone dependence in this study.

There are several limitations in this study. First, this was a cross-sectional study, therefore the causality between mobile phone dependence and other variables could be not examined. Second, the study was conducted only in one university so the findings could not be generalized to all Chinese university students. Multicenter studies involving different regions across China are needed. Third, other important information related to mobile phone dependence—such as depressive and anxiety symptoms—were not recorded. Fourth, second-digital divide and third-digital divide which may affect individual behavior including impulsiveness, was not examined in this study. Fifth, the mediating effects of various purposes of mobile phone use on impulsiveness could not be examined due to the current study design. Finally, the Cronbach's α of the MPUQ was relatively low.

5. Conclusions

In conclusion, the frequency of possible mobile phone dependence and mobile phone dependence in this sample of Chinese university students was relatively high. A significant positive association between mobile phone dependence and impulsivity was found. Further longitudinal research is needed to explore the relationship between mobile phone dependence and impulsivity, and how the different purposes for mobile phone use could mediate their interaction.

Supplementary Materials: The following are available online at http://www.mdpi.com/1660-4601/15/3/504/s1, Table S1: Factor loadings of mobile phone use questionnaire (MPUQ).

Acknowledgments: The authors are grateful to all the participants who were involved in this study. This study was supported by the Science and Technology International Cooperation Project of Jilin province (20160414035GH).

Author Contributions: Study design: Songli Mei, Jingxin Chai, and Yu-Tao Xiang Data collection, analysis, and interpretation of data: Jingxin Chai and Songli Mei Drafting of the manuscript: Jingxin Chai, Shi-Bin Wang, and Yu-Tao Xiang Critical revision of the manuscript: Chee H. Ng and Gabor S. Ungvari. Approval of the final version for publication: all co-authors.

Conflicts of Interest: The authors declare no conflict of interest.

References

1. Billieux, J. Problematic Use of the Mobile Phone: A Literature Review and a Pathways Model. *Curr. Psychiatry Rev.* **2012**, *8*, 299–307. [CrossRef]
2. Yen, C.F.; Tang, T.C.; Yen, J.Y.; Lin, H.C.; Huang, C.F.; Liu, S.C.; Ko, C.H. Symptoms of problematic cellular phone use, functional impairment and its association with depression among adolescents in Southern Taiwan. *J. Adolesc.* **2009**, *32*, 863–873. [CrossRef] [PubMed]
3. Lepp, A.; Barkley, J.E.; Karpinski, A.C. The Relationship Between Cell Phone Use and Academic Performance in a Sample of U.S. College Students. *SAGE Open* **2015**, *5*. [CrossRef]
4. Lepp, A.; Barkley, J.E.; Karpinski, A.C. The relationship between cell phone use, academic performance, anxiety, and Satisfaction with Life in college students. *Comput. Hum. Behav.* **2014**, *31*, 343–350. [CrossRef]
5. Billieux, J.; Van der Linden, M.; Rochat, L. The role of impulsivity in actual and problematic use of the mobile phone. *Appl. Cogn. Psychol.* **2008**, *22*, 1195–1210. [CrossRef]
6. Mitchell, M.R.; Potenza, M.N. Addictions and Personality Traits: Impulsivity and Related Constructs. *Curr. Behav. Neurosci. Rep.* **2014**, *1*, 1–12. [CrossRef] [PubMed]
7. Smetaniuk, P. A preliminary investigation into the prevalence and prediction of problematic cell phone use. *J. Behav. Addict.* **2014**, *3*, 41–53. [CrossRef] [PubMed]
8. Hong, F.-Y.; Chiu, S.-I.; Huang, D.-H. A model of the relationship between psychological characteristics, mobile phone addiction and use of mobile phones by Taiwanese university female students. *Comput. Hum. Behav.* **2012**, *28*, 2152–2159. [CrossRef]
9. Hoffner, C.A.; Lee, S. Mobile Phone Use, Emotion Regulation, and Well-Being. *Cyberpsychol. Behav. Soc. Netw.* **2015**, *18*, 411–416. [CrossRef] [PubMed]
10. Salehan, M.; Negahban, A. Social networking on smartphones: When mobile phones become addictive. *Comput. Hum. Behav.* **2013**, *29*, 2632–2639. [CrossRef]
11. Chen, L.; Yan, Z.; Tang, W.; Yang, F.; Xie, X.; He, J. Mobile phone addiction levels and negative emotions among Chinese young adults: The mediating role of interpersonal problems. *Comput. Hum. Behav.* **2016**, *55*, 856–866. [CrossRef]
12. China Internet Network Information Center (CNNIC). *The 2013 Statistics Report the Development of China Internet Network*; China Internet Network Information Center (CNNIC): Beijing, China, 2013; Available online: http://www.cnnic.net.cn/hlwfzyj/hlwxzbg/hlwtjbg/201307/P020130717505343100851.pdf (accessed on 17 July 2017). (In Chinese)
13. Patton, J.M.; Stanford, M.S.; Barratt, E.S. Factor structure of the Barratt Impulsiveness Scale. *J. Clin. Psychol.* **1995**, *51*, 768–774. [CrossRef]
14. Li, X.Y.; Phillips, M.R.; Xu, D.; Zhang, Y.L.; Yang, S.J.; Tong, Y.S.; Wang, Z.Q.; Niu, Y.J. Reliability and validity of an adapted Chinese version of Barratt Impulsiveness Scale. *Chin. Ment. Health J.* **2011**, *25*, 610–615.
15. Bentler, P.M. On the fit of models to covariances and methodology to the Bulletin. *Psychol. Bull.* **1992**, *112*, 400–404. [CrossRef] [PubMed]
16. Wang, X.D.; Jiang, C.Q.; Ma, H. *Rating Scales for Mental Health*; Mental Health in Chinese Press: Beijing, China, 1999.
17. Xiong, J.; Zhou, Z.K.; Chen, W.; You, Z.Q.; Zhai, Z.Y. Development of the Mobile Phone Addiction Tendency Scale for College Students. *Chin. Ment. Health J.* **2012**, *26*, 222–225.
18. Lopez-Fernandez, O.; Honrubia-Serrano, L.; Freixa-Blanxart, M.; Gibson, W. Prevalence of problematic mobile phone use in British adolescents. *Cyberpsychol. Behav. Soc. Netw.* **2014**, *17*, 91–98. [CrossRef] [PubMed]
19. Jiang, Z.; Zhao, X. Self-control and problematic mobile phone use in Chinese college students: The mediating role of mobile phone use patterns. *BMC Psychiatry* **2016**, *16*, 416. [CrossRef] [PubMed]
20. Lee, Y.-K.; Chang, C.-T.; Lin, Y.; Cheng, Z.-H. The dark side of smartphone usage: Psychological traits, compulsive behavior and technostress. *Comput. Hum. Behav.* **2014**, *31*, 373–383. [CrossRef]
21. Toda, M.; Ezoe, S. Multifactorial study of mobile phone dependence in medical students: Relationship to health-related lifestyle, Type A behavior, and depressive state. *Open J. Prev. Med.* **2013**, *3*, 99–103. [CrossRef]
22. Liu, H.; Yu, H.-L. The Relationship among University Students' Mobile Phone Addiction and Mobile Phone Motive, Loneliness. *J. Psychol. Sci.* **2011**, *34*, 1453–1457.
23. Roberts, J.A.; Pullig, C.; Manolis, C. I need my smartphone: A hierarchical model of personality and cell-phone addiction. *Personal. Individ. Differ.* **2015**, *79*, 13–19. [CrossRef]

24. Pourrazavi, S.; Allahverdipour, H.; Jafarabadi, M.A.; Matlabi, H. A socio-cognitive inquiry of excessive mobile phone use. *Asian J. Psychiatry* **2014**, *10*, 84–89. [CrossRef] [PubMed]
25. Crosswhite, J.M.; Rice, D.; Asay, S.M. Texting among United States young adults: An exploratory study on texting and its use within families. *Soc. Sci. J.* **2014**, *51*, 70–78. [CrossRef]
26. Dong, G.S.; Park, Y.; Min, K.K.; Park, J. Mobile phone dependency and its impacts on adolescents' social and academic behaviors. *Comput. Hum. Behav.* **2016**, *63*, 282–292.

International Journal of
Environmental Research and Public Health

MDPI

Article

Investigating the Effect of Personality, Internet Literacy, and Use Expectancies in Internet-Use Disorder: A Comparative Study between China and Germany

Benjamin Stodt [1], Matthias Brand [1,2], Cornelia Sindermann [3], Elisa Wegmann [1], Mei Li [4], Min Zhou [5], Peng Sha [6] and Christian Montag [3,7,*]

1 General Psychology: Cognition and Center for Behavioral Addiction Research (CeBAR), University of Duisburg-Essen, 47057 Duisburg, Germany; benjamin.stodt@uni-due.de (B.S.); matthias.brand@uni-due.de (M.B.); elisa.wegmann@uni-due.de (E.W.)

2 Erwin L. Hahn Institute for Magnetic Resonance Imaging, 45141 Essen, Germany

3 Department of Molecular Psychology, Institute of Psychology and Education, Ulm University, 89081 Ulm, Germany; cornelia.sindermann@uni-ulm.de

4 Student Counselling Centre, Beijing University of Civil Engineering and Architecture, Beijing 100037, China; amorelm415@gmail.com

5 Institute of Medical Statistics, Informatics and Epidemiology, University of Cologne, 50923 Cologne, Germany; min.zhou@hotmail.de

6 School of Journalism and Communication, Southwest University, Chongqing 400716, China; a2352893@gmail.com

7 Key Laboratory for NeuroInformation/Center for Information in Medicine, School of Life Science and Technology, University of Electronic Science and Technology of China, Chengdu 611731, China

* Correspondence: christian.montag@uni-ulm.de; Tel.: +49-731-502-6550

Received: 18 January 2018; Accepted: 19 March 2018; Published: 23 March 2018

Abstract: Research on Internet-use Disorder (IUD) has increased rapidly, indicating its clinical and global importance. Past studies suggested cultural diversity regarding the prevalence of an IUD, e.g., between Asian and European countries. Additionally, it was found that personality factors, Internet-related cognitions and specific competences seem to influence IUD tendencies, but research lacks in cultural comparative studies regarding these mechanisms. This study focuses on differences between Germany and China regarding the above-mentioned characteristics. German ($n = 411$; $M = 20.70$ years, $SD = 3.34$ years) and Chinese participants ($n = 410$; $M = 20.72$ years, $SD = 2.65$ years) answered the short Internet Addiction Test, Big Five Inventories, the Internet-use Expectancies Scale, as well as the Internet Literacy Questionnaire. The results revealed higher occurrence of IUD symptoms in China. Furthermore, Chinese participants scored significantly higher on neuroticism and agreeableness, whereas German participants scored higher on extraversion and openness. Compared to German participants, Chinese showed higher expectancies to avoid negative feelings online and to be positively reinforced. Regarding Internet literacy, German participants indicated higher skills concerning the reflection and critical analysis of online content, whereas Chinese showed higher expertise in producing and interacting online. Further, simple slope analyses indicated that certain Internet literacy domains were related differentially to IUD symptoms in Germany and China. While Chinese participants with higher reflective skills indicated highest IUD symptoms, reflective skills revealed no effect in Germany. Additionally, higher self-regulative skills correlated with lower IUD symptoms in the German, but not in the Chinese sample. The results give a hint to potential cultural differences regarding IUD, especially on the predictive and protective role of Internet literacy domains.

Keywords: Internet addiction; Internet-use disorder; Internet literacy; expectancies; personality; cultural differences

1. Introduction

Since the late 90s, the phenomenon of Internet addiction or Internet-use disorder (IUD) is focused more and more in the scientific literature (based on the term Internet-gaming disorder, the term Internet-use disorder is used in the following as a synonym for Internet addiction or pathological Internet use). Further, the number of theoretical assumptions about its classification as well as empirical studies, which give insights into the underlying mechanisms of its development, increase constantly [1–3]. Past research differentiates between unspecific and specific IUDs [4]. While an unspecific IUD goes along with an excessive and uncontrolled use of different Internet applications and websites without a clear preference [1], a specific IUD describes the addictive use of one specific application or certain type of website, such as games, shopping sites, or Internet communication applications (for an overview, see [2]). Not just since Internet-gaming disorder has been added to section III of the DSM-5 [5], there is an undisputable clinical relevance of IUDs. Nevertheless, further research is needed to conduct and clarify potential diagnostic criteria for an unspecified IUD or further specific forms, such as Internet-pornography-use disorder or Internet-communication disorder (ICD).

1.1. Prevalence of an IUD

According to a meta-analysis by Cheng and Li [6], who reviewed studies from 31 nations across seven world regions, the global occurrence of an IUD is estimated at 6%, illustrating it as a global phenomenon and universal issue. Past research on the epidemiology and cultural diversity of an IUD showed that prevalence rates strongly differ between different countries and cultures [7]. Cheng and Li [6] reported an occurrence of IUD symptoms in 2.6% of European individuals, which confirms the results by Spada [8], who identified 1 to 9% of the European population showing symptoms of a pathological Internet use. In a German representative population sample, a prevalence rate of 1% was estimated [9]. While prevalence rates in the Middle East range between 1% and 12%, highest prevalence rates up to 38% are reported for Asian countries [7,8,10]. Confirming the quality of real life hypothesis, Cheng and Li [6] identified several environmental characteristics associated with a higher occurrence of an IUD, like perception of less life satisfaction, greater traffic commute time consumption, as well as lower national income. Other reasons for cultural differences, such as Hofstede's [11] assumptions on power-distance, collectivism, and individualism, have also been discussed [12]. Besides, the broad range of reported prevalence rates can also be due to the absence of standardized diagnostic tools. Former studies used various questionnaires to measure symptoms of an IUD, which are grounded on different diagnostic criteria and cut-off scores [7]. Consequently, cross-cultural differences in the prevalence of an IUD found in previous studies still require a cautious interpretation.

1.2. Cultural Issues of an IUD: Results from Germany and China

Along with research on the occurrence of an IUD, former studies already focused on specific psychological aspects which are associated with an excessive and addictive use of the Internet in different nations and cultures [7,13,14]. However, there are just a few studies concentrating on cultural issues regarding the occurrence of an IUD as well as the effect of personality traits and Internet-related state variables on its development. In order to better understand the development and maintenance of an IUD within different national backgrounds, cross-cultural studies are important [13]. Besides, Ko and Yao [15] emphasized that cultures psychologically differ in a variety of individual characteristics, such as personality, emotions, and cognitions. Therefore, nationally differing effects of personal characteristics on the development of an IUD can be expected.

Within the scope of cultural-comparative studies, Asian and Western samples have been compared most frequently regarding specific individual characteristics as well as economic and infrastructural background [15]. Regarding digitization and Internet penetration rates, Germany and China display two key nations of their continents [16,17]. In Germany, 89.6% of the population is online (which are 72.3 million people) [16]. This corresponds to 11.1% of all European Internet users and is the second highest rate after Russia (16.6%). Nevertheless, only 76.4% of the Russian population is online [16]. In China, the Internet penetration rate is at 54.6% (772 million users), which displays the highest amount of Internet users in Asia with 38.1% in terms of total user numbers [16]. The highest Internet penetration rate for Asian nations is found in Japan (93.3%) [16]. However, this rate corresponds to just 5.9% of the Asian population. Due to these facts, we chose Germany and China as two representatives for well-digitized European and Asian nations.

Within Germany, IUD displays a serious health problem. In a representative study, 1% of the German population has been identified as pathological Internet users [9]. Higher prevalence rates were observed within the age of 14 to 24 (2.4%) and 14 to 16 years (4%) [9]. Another German study reported a prevalence rate of 2.1%, where the group of people suffering from IUD reported significant psychosocial and health consequences due to their excessive Internet use [18]. Further, male gender and social factors were significantly associated with IUD symptoms [18]. Further studies from Germany highlighted significant correlations of tendencies towards an unspecific and specific IUD with psychopathological symptoms (e.g., depression and interpersonal sensitivity [19–21]), higher impulsivity [19], younger age [22] and male gender [18,19].

Even if IUD is a growing topic of research in Germany, IUD has been studied more frequently in China. There, IUD has been recognized as a nationwide health problem with prevalence rates ranging from 2.4 to 6.4% [23,24]. Within China, Internet addicted individuals have been characterized with higher impulsivity [25], lower prosocial behaviors [23], higher depressive symptoms [24,26], shyness [27], lower self-esteem [26], younger age [24,28], and male gender [28].

In the following, results on the effect of specific personality traits and Internet-related state variables on IUD symptoms are summarized. Here, we focus on investigations carried out in Germany and China, supplemented by results from Western and Asian countries.

1.3. Predictors of an IUD: Trait Variables

Beyond epidemiological studies, many studies focus on individual factors influencing the development and maintenance of IUDs. Here it has to be distinguished between personal predispositions [29,30], which persist over a long period of time (traits), and factors which are situation-dependent, such as expectancies or actual mood (states) [31,32]. Regarding trait variables, depression, attention-deficit/hyperactivity disorder (ADHD), as well as social anxiety disorders have been verified as three main comorbid personal predispositions of IUDs [32–34]. Next to these predispositions, further person's core characteristics could play an important role in the addiction process. One popular concept of one person's core personality is the Big Five personality model, including the personality traits of neuroticism, extraversion, openness to experience, conscientiousness, and agreeableness [35,36]. In their theoretical model, called I-PACE, Brand et al. [2] describe different stages of the addiction process of specific Internet-use disorders, including the interaction of state and trait variables as well as different mediation effects. In this model and their previous literature review, the authors highlighted the relevance of the Big Five personality traits, especially low conscientiousness, low extraversion, and high neuroticism, as predispositions in persons who are vulnerable to use specific Internet applications excessively and uncontrolled [2].

According to cross-cultural studies, differences regarding the Big Five personality traits between East Asian and Western cultures exist. A meta-analysis revealed that especially East Asian countries score significant lower on the traits of extraversion, agreeableness, conscientiousness, and openness, but higher on neuroticism than other countries and cultures [37]. Moreover, mainly higher neuroticism and lower extraversion were found as main correlates of an IUD across different countries [38–40]. However, openness, conscientiousness and agreeableness also indicated significant relations to IUD symptoms within a sample of Chinese adolescents [28]. Despite these general cultural differences concerning Big Five personality traits and the correlations with IUD symptoms found within country-specific samples, it remains under-uninvestigated whether the strength of impact regarding these traits on the development and maintenance of an IUD significantly differs between different countries and cultures.

1.4. Predictors of an IUD: State Variables

Next to personal predispositions and traits, researchers focused on state variables, which are considered as situation-dependent factors, influencing and triggering human behavior [31]. In the context of an IUD, such factors are especially motivational factors, coping strategies, and cognitive biases [2,3]. It is assumed that these factors interact with trait variables and core characteristics in the developmental process of an IUD and in turn reinforce and stabilize cognitive and affective responses [2]. In the following, Internet-use expectancies and specific Internet literacy domains are discussed as two exemplary predictors of an IUD.

1.4.1. Internet-Use Expectancies

According to Lee et al. [41], person's expectancies towards the use of the Internet affect his or her attitudes regarding specific online applications. Furthermore, Internet users' expectancies (e.g., to gratify a certain need) lead to the decision to use the Internet or a specific application to experience gratification, as shown in the I-PACE model by Brand et al. [2] (also see their article for a more comprehensive overview of this decision-making process). One step further, symptoms of an IUD increase, when positive expectancies and metacognitions towards the Internet exist [42,43]. Brand et al. [44] already showed that Internet-use expectancies for positive reinforcement as well as avoidance expectancies are correlated with psychopathological symptoms and further personality aspects within a German sample. In turn, certain use expectancies and dysfunctional coping mechanisms are correlated with symptoms of an unspecified IUD. Also within a German sample, Wegmann et al. [21] could replicate these findings, as they found associations between psychopathological symptoms, Internet-use expectancies, self-regulation abilities, and symptoms of an excessive use of Internet-communication applications. According to a Taiwanese study by Lee et al. [41], Internet-use expectancies positively predicted students' attitudes towards and the use of specific Internet applications. In another Taiwanese study, Wu et al. [45] found that having expectancies that the use of the Internet helps to relieve negative emotions increases the probability of IUD symptoms in persons with borderline personality disorder. The authors stated that it is important to teach individuals to develop alternative coping strategies and not to use the Internet for escaping from negative feelings.

1.4.2. Internet Literacy

As an additional state variable, specific skills and competences in using the Internet and its applications could also play a key role in the development and maintenance of IUDs [22]. Today, there is still no common term summarizing such skills, whereas researchers stated that more empirical research in this area is required [46]. Some of these concepts are termed e.g., digital literacy [47], new media literacy [48,49], information literacy [50], or Internet literacy [22,51]. Fraillon et al. [52] defined digital literacy as "an individual's ability to use computers to investigate, create, and communicate in order to participate effectively at home, at school, in the workplace, and in society" (p. 18). Stodt, Wegmann, and Brand [22] recently published a four-dimensional concept of Internet literacy which they had also operationalized with a self-assessment questionnaire. Beyond the pure technical expertise in handling Internet applications and relevant software, the concept of Internet literacy by Stodt et al. [22] specifically focuses on a rather deliberate, conscious and reflective use of the Internet and certain applications (see also [51]). Besides that, the increasing possibilities through interactive participation brought by social media cause that Internet users need to critically evaluate online content and information. This is also crucial to become an inherent part of online social networks as an active producer of content [53,54]. As a last skill, it is necessary for individuals to control and self-regulate their own online behavior [55].

Some empirical studies found that specific domains of an Internet literacy could be both positively and negatively related to symptoms of an IUD or ICD in Asian as well as European samples, e.g., [22,51]. More detailed, the high impact of self-regulative skills in IUD and ICD was already highlighted in Western cultures, e.g., [21,56]. Here, the authors assumed that people with less self-regulative skills are more vulnerable to develop an IUD, also due to reduced cognitive control. Stodt et al. [22] demonstrated that higher technical and productive skills increase the risk to suffer from IUD symptoms, whereas higher reflective and self-regulative skills were negatively correlated with IUD symptoms within a German sample. Therefore, the latter domains of an Internet literacy can be seen as preventive factors, which are not stable but learnable. Nevertheless, the preventive values of specific Internet literacy domains and their mediating or moderating role in the addiction process is an important topic of research. In another German study, Wegmann et al. [21] showed that both Internet-use expectancies and self-regulation as one domain of Internet literacy mediate the effect between psychopathological predispositions and ICD symptoms and especially productive and self-regulative skills are correlated with expectancies towards the Internet. More detailed, higher productive and interactive skills as well as lower self-regulation abilities go along with higher expectancies regarding positive reinforcement and the avoidance of negative feelings.

1.5. Aims and Hypotheses

Recent German studies declared IUD as a nationwide problem, e.g., [9]. Its symptoms are correlated with lower extraversion and higher neuroticism as well as the expectancies someone has towards using the Internet, and certain Internet literacy dimensions [20–22,44,57]. Besides this, most epidemiological studies revealed high prevalence rates in China and adjacent nations, e.g., [6,10]. Addressing personality, also higher neuroticism and lower extraversion are mostly reported as significant predictors of an IUD in China, even if some studies indicated correlations with all traits of the Big Five [28]. Regarding Internet literacy, just a few findings have been reported for Chinese samples [51], while from our knowledge no research has been done on Internet-related specific cognitions.

The first aim of our study was to investigate whether previous findings regarding the relationship of personality, Internet-use expectancies, different Internet literacy domains and IUD symptoms could be replicated across different nations and cultures. We expect that (i) primarily lower extraversion and higher neuroticism should be linked to higher tendencies towards IUD, (ii) higher Internet use expectancies should be accompanied with IUD symptoms and (iii) primarily higher productive and interactive as well as lower self-regulative skills should be correlated with higher tendencies towards an IUD. Secondly, we wanted to have a closer look at potential differences and interaction effects between the above-mentioned factors among two countries which differ in their cultural background and most likely in their Internet-use behavior.

2. Materials and Methods

2.1. Participants

Overall, 821 participants (334 female, 487 male) from Germany and China took part in a multi-parted data acquisition. The mean age of the whole sample was $M = 20.71$ years ($SD = 3.01$) with a range from 16 to 30.

The German convenience subsample consisted of 411 participants (179 female, 232 male; age: $M = 20.70$, $SD = 3.34$, range: 16–29 years). German participants were recruited from all over Germany by advertisements in university, e-mail lists, announcements on social networking sites, as well as word-of-mouth recommendation. Most of the German participants were university students (33.5%), of which 72.3% named the Abitur (German's secondary school leaving diploma) and 13.1% the

Fachabitur (vocational baccalaureate diploma) to be their highest school educational level. Additionally, 10.2% of the university students already received their bachelor's degree and 2.2% received their master's degree or a comparable graduation. Further, 25.2% of the German sample were pupils, 22.2% trainees, 10.3% were employees (of which 50.1% had a university degree), and 8.8% of the subsample practiced other professions.

The Chinese convenience subsample consisted of 410 participants (155 female, 255 male; age: $M = 20.72$, $SD = 2.65$, range: 16–30 years). Chinese participants were recruited via announcements and postings in university as well as advertisements in university online forums. Overall, 202 Chinese participants took part in a data collection carried out in Beijing. Further 151 participants took part in the *Chengdu Gene Brain Behavior Project (CGBBP)* and have been recruited in Chengdu. An additional subsample of 57 participants took part in another data acquisition in Chengdu. The data collection was performed in these two Chinese cities due to location-dependent reasons (e.g., the studies were carried out in laboratories in these cities' universities). Within all surveys, most of the Chinese participants were undergraduate students (74.6%), followed by graduate students (23.4%) and high school graduates (2.0%). The corresponding local ethics committees approved the study.

2.2. Measures

Online surveys were used to gather participants' information of the below mentioned measurements in Germany and China. The short Internet Addiction Test, the Internet-use Expectancies Scale, and the Internet Literacy Questionnaire were already available in German language and professionally translated into Chinese. Due to methodological issues, which are described later, two different personality questionnaires were used in the two countries. In the German subsample, the 10-Item Big Five Inventory (BFI-10) was used to assess the Big Five personality traits, whereas the NEO Five-Factor Inventory was used in the Chinese subsample. Both questionnaires were already available in the respective languages. In former studies, all of the used questionnaires showed good psychometric properties and validity. In this study's sample, all questionnaires and its subscales revealed a good internal consistency with Cronbach's $\alpha > 0.700$, except for the subscale *openness* and *agreeableness* of the Chinese NEO Five-Factor Inventory, which however were still in the acceptable range. All reliabilities (Cronbach's α) of the used questionnaires are presented in Table 1. Cronbach's α reliabilities of the BFI-10 are not reported because this measure is not appropriate for calculating Cronbach's α as only two items are building each scale [58].

Table 1. Overview on reliabilities (Cronbach's α) of the NEO Five-Factor Inventory (NEO-FFI) *, Internet-use Expectancies Scale (IUES), Internet Literacy Questionnaire (ILQ), short Internet Addiction Test (s-IAT) and their subscales in the German and Chinese samples.

	Germany	China
NEO-FFI—*Neuroticism*	-	0.820
NEO-FFI—*Extraversion*	-	0.732
NEO-FFI—*Openness*	-	0.602
NEO-FFI—*Conscientiousness*	-	0.747
NEO-FFI—*Agreeableness*	-	0.698
IUES—*Positive reinforcement*	0.855	0.775
IUES—*Avoidance expectancies*	0.784	0.810
s-IAT—*Total score*	0.869	0.919
s-IAT—*Loss of control/time management*	0.819	0.862
s-IAT—*Craving/social problems*	0.809	0.870
ILQ—*Technical expertise*	0.805	0.775
ILQ—*Production and interaction*	0.759	0.834
ILQ—*Reflection and critical analysis*	0.707	0.723
ILQ—*Self-regulation*	0.772	0.810

* Note that in the German sample the BFI-10 was used, where no reliabilities can be presented.

We further compared the factorial structure of the short Internet Addiction Test, the Internet-use Expectancies Scale, and Internet Literacy Questionnaire using mean structure analyses with nation as grouping variable. Here, the model fits revealed no good fit with the data indicating that the factorial structure of the scales seem to be different between the nations. Nevertheless, all items scored significantly on their respective factor/latent dimension with acceptable to good factor loadings in both nations (weakest factor loading = 0.350, highest factor loading = 0.896). This issue will be further discussed in the limitations section.

2.2.1. Short Internet Addiction Test (s-IAT)

To measure symptoms of unspecified IUD, the short Internet Addiction Test (s-IAT; [19]) was used. It is a 12-item short version of Young's original Internet Addiction Test (IAT; [59]), an internationally esteemed questionnaire to cover IUD symptoms. The s-IAT has already been validated and frequently used in past research. Within this questionnaire, participants have to evaluate their subjective complaints and experienced negative consequences due to excessive online activities in everyday life. Besides the total sum score, the questionnaire consists of the two subscales *loss of control/time management* and *craving/social problems*, each consisting of six items. Items were rated on a 5-point Likert scale from 1 (*never*) to 5 (*very often*). The total sum score ranges from 12 to 60, in which a higher sum score of the s-IAT indicates higher symptoms of an IUD. To evaluate the strength of IUD symptoms, the cut-off for the total score was >30 to indicate a problematic Internet use and >37 to indicate a pathological Internet use [19]. The questionnaire in English, German, and Chinese language is reported in Table A1.

2.2.2. Big Five Inventory-10 and NEO Five-Factor Inventory

Because of methodological issues, we used two different questionnaires to measure the Big Five personality traits. These two questionnaires represent reliable and valid instruments for the measurement of the above-mentioned traits in the respective country. For the German data collection, we preferred to use the Big Five Inventory-10 (BFI-10; [60]), which (despite its shortness) represents a well-evaluated and valid instrument in Western countries. While the validity of the BFI-10 in Chinese language still has to be further tested [61], we decided to use the 60-item version of the NEO Five-Factor Inventory (NEO-FFI; [62]), which has been used in several Asian studies before and revealed a good validity, e.g., [63]. Both questionnaires were used to measure the participants' level of *neuroticism*, *extraversion*, *openness*, *conscientiousness*, and *agreeableness*. In both questionnaires, higher mean scores reveal a higher level of the respective characteristic.

The BFI-10 consists of 10 items with two items on each trait. All of the items were rated on a 5-point Likert scale ranging from 1 (*strongly disagree*) to 5 (*strongly agree*).

The NEO-FFI comprises overall 60 statements, which have to be evaluated on a 5-point Likert scale from 1 (*strongly disagree*) to 5 (*strongly agree*).

2.2.3. Internet-Use Expectancies Scale

To measure expectancies and motivating factors for using the Internet in general, the Internet-use Expectancies Scale (IUES; [44]) was used. The scale consists of the two dimensions *positive reinforcement* and *avoidance expectancies* (each having four items). The dimension positive expectancies measures positive feelings, which are accompanied by using the Internet, whereas the dimension avoidance expectancies indicated whether the Internet is used to avoid negative feelings. Every item was answered on a 6-point Likert scale with a range from 1 (*completely disagree*) to 6 (*completely agree*). Mean scores for each subscale have been calculated. Although the IUES is a relatively new developed questionnaire, it has already been used in several studies [20,21,44] and has also been adapted for specific Internet-use patterns, like Internet-communication, e.g., [57]. The questionnaire in English, German, and Chinese language is shown in Table A2.

2.2.4. Internet Literacy Questionnaire

The Internet Literacy Questionnaire (ILQ; [22]) was applied to measure the individuals' competent and adequate dealing with the Internet on the following four dimensions: *technical expertise, reflection and critical analysis, production and interaction,* and *self-regulation*. Depending on a new calculated exploratory factor analysis, a more economical version of the ILQ was used. The shortened version includes 18 items on the four above-mentioned dimensions, compared to the original 24-item version. According to Stodt et al. [22], the first dimension *technical expertise* measures "the individual's expertise in handling computer hard- and software as well as Internet applications" ([22], p. 31) on four items. The dimension *production and interaction* covers five items measuring "how and why an individual uses the Internet to create own content and to interact with others" ([22], p. 31). The third dimension *reflection and critical analysis* covers "the individual's ability to evaluate the credibility of online content and the behavior of others as well as critically reflecting one's own activities on the Internet" ([22], p. 31) on four items. The dimension *self-regulation* consists of five items covering "the individuals' ability to regulate their own Internet use to prevent negative consequences for daily life" ([22], p. 31). Every item had to be answered on a 6-point Likert scale ranging from 0 (*strongly disagree*) to 5 (*strongly agree*). Mean scores were calculated for each dimension.

Due to translation problems, item number 8 from the Chinese version of the ILQ had to be removed. Therefore, the subscale *reflection and critical analysis* consists of 3 items in the Chinese version. The subscales' internal consistency was still in a good range (Cronbach's $\alpha = 0.723$). Nevertheless, we provide readers with the complete questionnaire in English, German, and Chinese language as presented in Table A3, including a revised version of item number 8.

2.3. Statistical Analysis Section

Aside from the presentation of descriptive statistics, inferential statistics are provided in the following result section with respect to both the German and Chinese sample. *t*-Tests were used to examine differences between both nations regarding the manifold Internet variables, whereas Pearson's correlations were used to provide insights into correlation patterns between Internet variables in both countries. In addition, moderated regression analyses were used to consider possible interaction effects between nationality and Internet literacy within the development and maintenance of an IUD. The statistical analyses were performed by using IBM SPSS 24. MPlus version 6.12 (IBM, Armonk, NY, USA) was used to carry out the mean structure analyses. Reported effect sizes are in accordance with Cohen's d (small effect: $d = 0.30$; medium effect: $d = 0.50$; large effect: $d = 0.80$) for t-tests and r (small effect: $r = 0.10$; medium effect: $r = 0.30$; large effect: $r = 0.50$) for Pearson correlations [64].

3. Results

3.1. Descriptive Results

Table 2 contains the whole sample's (German and Chinese collapsed) descriptive statistics of the different questionnaires including average scores, standard deviations, and range.

Table 2. Descriptive statistics of the s-IAT, IUES, ILQ, BFI-10, and NEO-FFI.

Domain/Variable	*M*	*SD*	*Min–Max* [1]
Internet-use disorder (s-IAT)			
Total score	27.76	9.23	12.00–60.00
Loss of control/time management	15.62	4.98	6.00–30.00
Craving/social problems	12.14	4.97	6.00–30.00
Internet-use expectancies (IUES)			
Positive reinforcement	4.00	1.11	1.00–6.00
Avoidance expectancies	2.93	1.23	1.00–6.00

Table 2. *Cont.*

Domain/Variable	*M*	*SD*	*Min–Max* [1]
Internet literacy (ILQ)			
Technical expertise	2.99	1.16	0.00–5.00
Production and interaction	2.52	1.72	0.00–5.00
Reflection and critical analysis	3.11	0.96	0.00–5.00
Self-regulation	3.00	1.00	0.00–5.00
Big Five personality (BFI-10) [2]			
Neuroticism	2.66	0.95	1.00–5.00
Extraversion	3.53	0.96	1.00–5.00
Openness	3.46	1.08	1.00–5.00
Conscientiousness	3.30	0.90	1.00–5.00
Agreeableness	3.03	0.81	1.00–5.00
Big Five personality (NEO-FFI) [3]			
Neuroticism	2.84	0.58	1.00–4.50
Extraversion	3.20	0.47	1.50–4.67
Openness	3.28	0.43	2.17–4.58
Conscientiousness	3.34	0.45	2.25–5.00
Agreeableness	3.38	0.46	2.00–4.83

[1] Actual range of each scale, including the lowest and highest values measured in both subsamples. [2] Used in the German subsample (n = 411). [3] Used in the Chinese subsample (n = 410).

3.1.1. Occurrence of IUD

Based on the suggested cut-off scores of the s-IAT by Pawlikowski et al. [19], 55 participants of the German sample indicated a problematic use and 25 participants reported symptoms of a pathological use of the Internet. In China, 115 of the participants reported a problematic, 95 a pathological use. Figure 1 illustrates the percentage distribution of IUD tendencies in both countries.

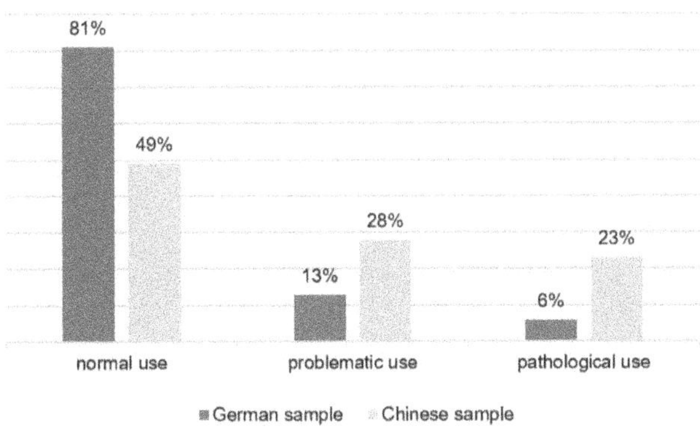

Figure 1. Internet-use disorder symptoms in the German and Chinese sample.

3.2. Inferential Statistical Analyses

3.2.1. Testing Both the Chinese and German Sample for Differences in Age and Gender Ratios

Both samples did not differ with respect to age (t = 0.12, p = 0.901) and gender ratios (χ^2 = 2.81, df = 1, p = 0.094). However, since both subsamples were tried to be parallelized as good as possible regarding sample size and age, no age differences were expected. In terms of gender ratios, both samples show merely slight descriptive differences. We further tested for significant age and gender differences in tendencies towards IUD, Internet-use expectancies, Internet literacy, and personality in both samples. Here, age was significantly associated with IUD in the Chinese ($r = 0.142$, $p = 0.004$), but not in the German sample ($r = -0.072$, $p = 0.145$). Regarding gender effects, male participants in the German sample indicated higher scores in the s-IAT compared with females ($t = 1.98$, $p = 0.049$; $M = 24.81$, $SD = 7.48$ versus $M = 23.35$, $SD = 7.38$). No significant gender differences in the Chinese sample could be observed ($t = 0.46$, $p = 0.644$; $M = 31.53$, $SD = 8.89$ versus $M = 31.08$, $SD = 10.30$). While investigating the hypothesized cultural differences, we additionally controlled for the variables of age and gender. Calculated ANCOVAs revealed that the significant cultural effects on the observed variables reported below are not caused by age and gender effects. Nevertheless, some significant effects of age and gender on the measured Internet and personality variables in both countries could be observed. For an overview please see Tables S1 and S2 in the Supplementary Materials.

3.2.2. Comparative Analyses

In a first step, *t*-tests for independent samples including effect sizes by Cohen [64] were calculated to compare the different scores of the s-IAT, ILQ, IUES, and the personality questionnaires of both countries. The analyses revealed significant differences between Germany and China regarding the s-IAT total score and both subscores with medium to high effect sizes, in which the Chinese sample indicated higher scores. Concerning Internet-use expectancies, also the Chinese sample showed higher scores regarding the *positive reinforcement* and *avoidance expectancies* with medium to high effect sizes. Regarding the self-evaluated Internet literacy, Chinese participants indicated significant higher scores in the dimension *production and interaction* (medium effect size), whereas the German sample scored higher on the dimension *reflection and critical analysis* (small effect size). No significant differences were found regarding the dimensions *technical expertise* and *self-regulation*. With regard to personality differences, the Chinese sample reported significantly higher scores of *neuroticism* and *agreeableness* than the German sample (small effect size). For *extraversion* and *openness*, German participants indicated higher scores (small to medium effect sizes). No significant differences were found regarding the trait *conscientiousness* (see Table 3).

Table 3. Differences between the German and Chinese samples regarding the s-IAT, IUES, ILQ, and Big Five personality traits.

Domain/Variable	Germany		China		*t*	*df*	*p*	*d*
	M	*SD*	*M*	*SD*				
Internet-use disorder (s-IAT)								
Total score	24.17	7.46	31.36	9.44	−12.10	776.84	<0.001	0.85
Loss of control/time management	14.14	4.56	17.10	4.96	−8.89	819.00	<0.001	0.62
Craving/social problems	10.03	3.81	14.26	5.10	−13.47	757.47	<0.001	0.94
Internet-use expectancies (IUES)								
Positive reinforcement	3.70	1.20	4.31	0.92	−8.27	765.55	<0.001	0.57
Avoidance expectancies	2.47	1.14	3.40	1.13	−11.77	819.00	<0.001	0.82
Internet literacy (ILQ)								
Technical expertise	2.94	1.23	3.03	1.10	−1.06	808.87	0.291	0.08
Production and interaction	2.12	1.11	2.93	1.08	−10.59	819.00	<0.001	0.74
Reflection and critical analysis [1]	3.18	0.88	3.03	1.09	2.29	798.88	0.022	0.15
Self-regulation	3.06	1.02	2.94	0.99	1.73	819.00	0.085	0.12
Big Five personality [2]								
Neuroticism	2.66	0.95	2.84	0.58	−3.22	679.50	0.001	0.23
Extraversion	3.53	0.96	3.20	0.47	6.34	595.08	<0.001	0.44
Openness	3.46	1.08	3.28	0.43	3.10	536.97	0.002	0.22
Conscientiousness	3.30	0.90	3.34	0.45	−0.93	607.75	0.354	0.06
Agreeableness	3.03	0.81	3.38	0.46	7.57	650.46	<0.001	0.53

[1] As already mentioned in the method section, the mean score of this dimension was calculated by using three instead of four items in the Chinese subsample. [2] German sample ($n = 411$): measured with BFI-10; Chinese sample ($n = 410$): measured with NEO-FFI.

3.2.3. Correlations

Table 4 shows bivariate correlations between the s-IAT and the Big Five personality traits, divided by country. In the German sample, *neuroticism* (positive), *extraversion*, and *conscientiousness* (both negative) indicated a significant relationship to the s-IAT with low effects. For the Chinese sample, all personality traits of the Big Five showed significant relations with the s-IAT. *Neuroticism*, *conscientiousness*, and *agreeableness* indicated the highest effect sizes. Except for *neuroticism*, all scores of the Big Five were negatively related to the scores of the s-IAT. According to Fisher's z comparisons, the correlations between *neuroticism*, *extraversion*, *conscientiousness*, *agreeableness*, and the s-IAT were significantly different between both countries (see Table 4).

Table 4. Correlations between s-IAT and Big Five personality traits [1] (Pearson correlations) including Fisher's z comparison.

Domain/ Variable [2]	Germany			China			Fisher's z		
	s-IAT Total Score	s-IAT LoC/TM [3]	s-IAT C/SP [3]	s-IAT Total Score	s-IAT LoC/TM [3]	s-IAT C/SP [3]	s-IAT Total Score	s-IAT LoC/TM [3]	s-IAT C/SP [3]
N	0.219 **	0.180 **	0.213 **	0.502 **	0.438 **	0.504 **	−4.70 **	−4.11 **	−4.83 **
E	−0.242 **	−0.183 **	−0.254 **	−0.186 **	−0.155 **	−0.195 **	−0.84	−0.41	−0.89
O	−0.012	−0.056	0.043	−0.160 **	−0.105 *	−0.193 **	2.13 *	0.70	3.40 **
C	−0.192 **	−0.204 **	−0.131 **	−0.365 **	−0.377 **	−0.308 **	2.69 **	2.71 **	2.66 **
A	−0.030	0.023	−0.085	−0.386 **	−0.275 **	−0.447 **	5.38 **	4.36 **	5.65 **

[1] German sample (n = 411): measured with BFI-10; Chinese sample (n = 410): measured with NEO-FFI. [2] N = neuroticism, E = extraversion, O = openness, C = conscientiousness, A = agreeableness. [3] LoC/TM = Loss of control/time management, C/SP = Craving/social problems. * $p \leq 0.05$, ** $p \leq 0.01$.

Table 5 includes the bivariate correlations between the s-IAT, ILQ and IUES scores, divided by country. For both countries, Internet-use expectancies correlated positively with the s-IAT total score and both subscores with medium to high effect sizes. The ILQ dimension *technical expertise* showed small significant correlations with the s-IAT scores in both samples. Here, the correlation between *technical expertise* and the s-IAT subscore *craving/social problems* differs significantly between Germany and China. The ILQ dimension *production and interaction* correlated positively with symptoms of an IUD in both countries. Regarding the domain *reflection and critical analysis*, the correlations just revealed significant positive and small effects with the s-IAT total score and the subscore *craving/social problems* in the Chinese sample. Here, Fisher's z revealed significant differences between the countries. Lastly, the ILQ score for *self-regulation* showed significant negative correlations with IUD symptoms in the German but not in the Chinese sample (except for one s-IAT subdimension), resulting in a significant Fisher's z comparison.

Table 5. Correlations between s-IAT and Internet literacy as well as Internet-use expectancies (Pearson correlations) including Fisher's z comparison.

Domain/Variable [1]	Germany			China			Fisher's z		
	s-IAT Total Score	s-IAT LoC/TM [2]	s-IAT C/SP [2]	s-IAT Total Score	s-IAT LoC/TM [2]	s-IAT C/SP [2]	s-IAT Total Score	s-IAT LoC/TM [2]	s-IAT C/SP [2]
Internet-use expectancies									
PR	0.390 **	0.352 **	0.344 **	0.369 **	0.338 **	0.356 **	0.35	0.23	−0.20
AE	0.567 **	0.507 **	0.505 **	0.556 **	0.480 **	0.564 **	0.23	0.51	−1.18
Internet literacy									
TE	0.105 *	0.120 *	0.062	0.153 *	0.069	0.215 **	−0.70	0.74	−2.23 *
PI	0.317 **	0.236 **	0.339 **	0.346 **	0.298 **	0.350 **	−0.47	−0.95	−0.18
RCA	−0.074	−0.064	−0.068	0.165 **	0.086	0.221 **	−3.44 **	−2.15 *	−4.18 **
SR	−0.516 **	−0.531 **	−0.375 **	−0.088	−0.185 **	0.018	−6.89 **	−5.77 **	−5.88 **

[1] PR = positive reinforcement, AE = avoidance expectancies, TE = technical expertise, PI = production and interaction, RCA = reflection and critical analysis, SR = self-regulation. [2] LoC/TM = Loss of control/time management, C/SP = Craving/social problems. * $p \leq 0.05$, ** $p \leq 0.01$.

3.2.4. Analyses of Interaction Effects

Based on the relationships found between certain Internet literacy domains and IUD symptoms as well as the different correlation effects between both samples, we calculated further moderated regression analyses to consider possible interaction effects between Internet literacy and the participants' cultural background explaining the level of IUD symptoms. Because particularly the two domains *reflection and critical analysis* as well as *self-regulation* revealed high cross-cultural differences regarding their association with IUD symptoms, we used these two domains as the predictors in the upcoming analyses.

In the first analysis, the ILQ score for *reflection and critical analysis* explained no significant amount of variance in the s-IAT total score ($R^2 = 0.001$, $F(1, 819) = 0.83$, $p = 0.361$). Nevertheless, the country ($\Delta R^2 = 0.155$, $\Delta F(1, 818) = 149.94$, $p < 0.001$) and the interaction term of both variables explained a significant part of variance ($\Delta R^2 = 0.012$, $\Delta F(1, 817) = 11.75$, $p = 0.001$). The whole moderation model explained 16.8% of the s-IAT score's variance ($R^2 = 0.168$, $F(3, 817) = 54.89$, $p < 0.001$; see Table 6, model 1 for beta-coefficients).

For further illustration of this interaction effect, we calculated simple slope analysis (see Figure 2). The slope illustrating the Chinese sample was significantly different from zero ($t = 3.73$, $p < 0.001$). The slope for Germany was not significant ($t = 1.32$, $p = 0.187$). These results indicate that there was an effect of reflective skills on the level of IUD symptoms in the Chinese sample. More detailed, Chinese participants with higher reflective skills tend to indicate higher s-IAT total scores in contrast to participants with lower skills. In the German sample the domain of *reflection and critical analysis* revealed no significant effect on IUD symptoms.

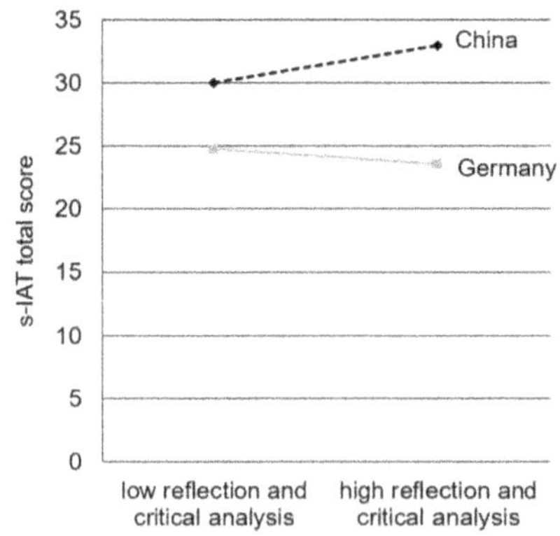

Figure 2. Simple slopes to illustrate the interaction effect between reflection and critical analysis and country on the s-IAT total score.

Secondly, we calculated a moderated regression analysis where we added the ILQ domain *self-regulation* as the predictor. Here, the score for *self-regulation* ($R^2 = 0.078$, $F(1, 819) = 68.83$, $p < 0.001$), the country ($\Delta R^2 = 0.140$, $\Delta F(1, 818) = 145.76$, $p < 0.001$), and the interaction term of both variables explained a significant part of variance ($\Delta R^2 = 0.026$, $\Delta F(1, 817) = 27.90$, $p < 0.001$). All variables explained 24.3% of the s-IAT score's variance ($R^2 = 0.243$, $F(3, 817) = 87.37$, $p < 0.001$; see Table 6, model 2 for beta-coefficients).

Simple slope analysis revealed significant slopes for the Chinese ($t = 2.08$, $p = 0.038$) and the German sample ($t = 9.69$, $p < 0.001$). However, while the effect of self-regulating skills is relatively low in the Chinese sample, in the German sample higher self-regulative skills reduce the score for IUD symptoms considerably in contrast to the Chinese sample (see Figure 3).

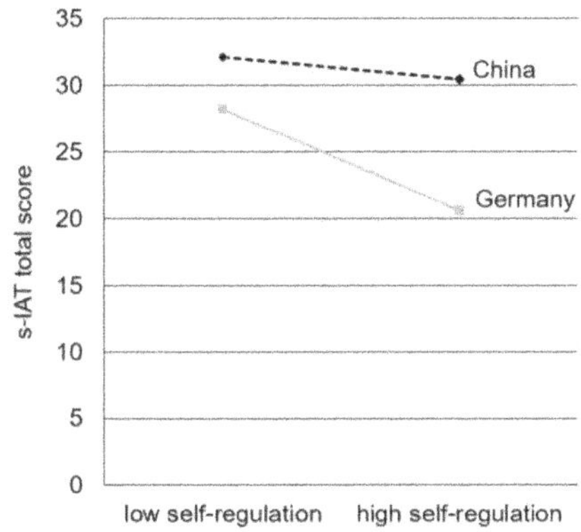

Figure 3. Simple slopes to illustrate the interaction effect between self-regulation and country on the s-IAT total score.

Table 6. Regression coefficients of the moderated regression analyses with the s-IAT total score as dependent variable.

	B	*SE (B)*	*β*	*t*	*p*
Model 1					
Reflection and critical analysis	0.44	0.31	0.046	1.42	0.156
Country	7.26	0.59	0.393	12.28	<0.001
Interaction	2.14	0.63	0.111	3.42	0.001
Model 2					
Self-regulation	−2.32	0.28	−0.252	−8.27	<0.001
Country	−6.91	0.56	−0.374	−12.278	<0.001
Interaction	−2.96	0.56	−0.161	−5.28	<0.001

4. Discussion

This study is the first one investigating a cultural comparison of the relationship between personal core characteristics, specific Internet-related cognitions and skills, as well as symptoms of an IUD. The present study's first aim was to verify whether previous findings on the effect of personality traits as well as state variables like Internet-use expectancies and Internet literacy on the development and maintenance of an IUD are stable across different cultures and nations. Another purpose was to point out potential differences in the relationship of these variables between two countries from different cultural areas.

4.1. IUD Symptoms

By using the same questionnaire across both countries, we found significant higher IUD symptoms in the Chinese sample, where the average total score of the s-IAT was almost at the cut-off score for a problematic Internet use. The significant cultural difference is reflected by the prevalence rates found in both countries: Whereas 6% of the German sample indicated a pathological Internet use, which value is in line with former epidemiological studies [7,65], 23% of the Chinese sample indicated an excessive and pathological use of the Internet. This rate is much higher than the prevalence reported for China in recent studies, where the rates range between 8 and 15 percent [66–68]. A different age range as well as different locations of recruitment in this study may explain the higher prevalence compared to previous investigations. Further, the year of publication could explain varying IUD frequencies, since prevalence rates could have been grown in the recent five years. Higher prevalence rates of IUD in Asian countries with respect to Western cultures have also been reported in the past [7,14]. However, because of the missing clinical evidence and differentiating diagnostic criteria of the methods used, reported prevalence rates generally have to be compared with caution.

4.2. Theoretical Integration of the Results

In the following, we want to integrate the results of this study into a recently published theoretical framework, called the I-PACE model (I-PACE stands for Interaction of Person-Affect-Cognition-Execution). In this empirically based model, Brand et al. [2] illustrate different stages of the addiction process of specific Internet-use disorders. The model illustrates the interaction of person's core characteristics, such as personality traits, social cognitions, and psychopathological symptoms (P-component) and possible mediating or moderating state variables like coping styles, Internet-related cognitive biases, as well as affective and cognitive responses (A/C-component). According to the authors, the interaction of these variables lead to the decision to use a certain application (E-component). The later experienced gratification could positively reinforce such mediating and moderating effects, resulting in an addictive use.

4.2.1. Personality

In accordance with the addiction process described in the I-PACE model, we start to discuss our findings regarding the Big Five personality traits. The found cultural differences in the traits extraversion, openness, and neuroticism are in line with a worldwide comparative study, where East Asian countries indicated the lowest scores of extraversion and openness as well as the highest scores in neuroticism [37]. The results on conscientiousness (no significant difference) and agreeableness (higher scores in China), are in line with typical national stereotypes. Both German and Chinese citizens are described as working hard to reach their goals and having a strong will power [69–72]. This cultural commonality might be reflected by a comparable score in the conscientiousness scale of both samples. In addition, China is considered as a nation tending to hierarchical structures and following social norms [69]. This could be reflected by higher levels of agreeableness of the Chinese sample compared to the German sample. However, the agreeableness finding speaks against observations, where Chinese samples scored significantly lower on agreeableness compared to a German sample [63,73]. It remains difficult to interpret and ascribe the found personality differences solely to cultural differences. Therefore, the results of this study have to be interpreted cautiously. Previous German and Chinese studies on the association between personality, general Internet use and IUDs revealed controversial results, but most consistent relations have been found between IUDs, low conscientiousness and high neuroticism [2,19,74–76]. Following this, we could show similar findings by indicating significant correlations between high neuroticism, low conscientiousness and IUD symptoms in both countries. Both traits revealed higher effect sizes in the Chinese sample, meaning that higher levels of neuroticism and lower levels of conscientiousness are more strongly associated with IUD symptoms in China than in Germany. Our results also demonstrate negative correlations between openness, agreeableness, and IUD symptoms in the Chinese sample. Regarding extraversion, significant relationships in both countries were found, which do not significantly differ. While interpreting these results, it must be considered that some personality factors have not been extracted consistently over several cultural regions [37]. Furthermore, the Big Five personality model mostly displays typical traits of people living in Western individualistic cultures but not of citizens in Asian cultures, which are more collectivistic [37]. This could lead to the different effects of some of the Big Five personality traits on IUD symptoms in both samples.

4.2.2. Internet-Related Cognitions and Skills

The next level of the I-PACE model comprises mediating and moderating state variables between the P-component and the decision to use a certain application. In this study, this component is represented by Internet-use expectancies and Internet literacy. Regarding Internet use expectancies, we found that Chinese participants indicated higher expectancies towards the Internet as a useful tool to reduce stress or to feel confident compared with the German sample. This is in line with Li and Kirkup [13], who found cultural differences between British and Chinese students regarding Internet experiences and attitudes. Although higher use expectancies were accompanied with higher IUD scores in both countries, no significant cultural differences regarding the strength of this relationship were found. There are already some empirical German studies underlining Internet-use expectancies as mediating variables, reinforcing the effect between personal predispositions and IUD symptoms [20,21,44,57]. Our study's results suggest similar effects in China, but further research with mean structure analyses is needed to investigate the role of Internet-use expectancies in the addiction process in a cross-cultural comparison.

Even though Internet literacy covers no inherent part in the I-PACE model, previous studies already indicated that Internet-related skills play an important role in the addiction process, e.g., [21,22,51]. Further, the dimensions of Internet literacy postulated by Stodt et al. [22] share similarities with Internet-related cognitive biases, as productive and interactive skills are a result of the subjective awareness of today's technological possibilities. In addition, the possession of self-regulative skills is theoretically accompanied with addiction-related concepts like cue-reactivity and craving as well as coping mechanisms. This study indicates less productive and interactive skills as well as higher reflective and analyzing skills of the German compared to the Chinese sample. Besides that, our analyses revealed two further interesting findings on the interaction between Internet literacy and IUD symptoms across both countries. Firstly, reflective and analytical skills seem to play no role in the development and maintenance of an IUD in German individuals. In contrast, we found a positive relationship between reflective skills and IUD symptoms in the Chinese sample. Simple slope analyses illustrated that Chinese participants show higher symptoms of an IUD compared to the German sample, but surprisingly indicate the highest symptomatology, when reflective skills are well-marked. At first site, this result appears to be counterintuitive, since reflective skills have been emphasized as preventive factors in past studies, e.g., [22]. However, it can be assumed that culture affects in what way an effect of reflective skills on IUD symptoms can occur. People living in collectivistic cultures like China are described as being more concerned about the effect of one's own behavior on others [11]. It can be assumed that the possession of reflective skills is necessary to evaluate this effect. In fact that in online environments such as games or social networks the interaction with others displays an essential feature, it could be that especially collectivistic thinking people with high reflective skills tend to use such applications more frequently and intense while neglecting personal needs and duties, leading to higher IUD symptoms. Past studies already investigated that people living in collectivistic countries are associated with a preference for online social interaction and appreciate their online social network [77,78].

Secondly, the ability to self-regulate one's own Internet use had just a low negative effect on the level of IUD symptoms among Chinese participants. On the other hand, the effect of self-regulation was much higher in the German sample, where individuals who cannot self-regulate their Internet-use behavior showed higher IUD symptoms compared to the people who have good self-regulation skills. The latter finding is in line with former research, where people with symptoms of an IUD reported problems in regulating and controlling their own Internet-use [21,22]. Regarding Cloninger [79], higher self-regulation abilities (in the offline context) are associated with higher self-directedness. Furthermore, people scoring high on self-directedness indicate high levels of self-esteem, will-power, and are self-satisfied with their own personality and problem-solving. As a result the authors emphasized self-directedness as an important personality dimension affecting personality disorders. Further, less self-directed people indicate higher symptoms of an IUD [29,30,80–83]. In a cross-cultural study by Sariyska et al. [30], however, the relation between self-directedness and IUD was slightly different across countries. Here, the effect sizes in two Chinese samples were lower than in the German sample. Our study found that even though online-focused self-regulation is not different across both investigated countries, it has no significant impact on IUD in China. Putting it together with previous research, these findings suggest the existence of heterogeneous mechanisms and effects regarding the interaction of online-focused self-regulation skills, self-directedness, and IUD symptoms in different countries and cultures.

4.3. Further Theoretical Integration

In the following, further reasons regarding the described cultural differences are discussed. The continuum of collectivism and individualism represents a frequently studied dimension of cultural variability, showing high differences such as between European and Asian countries, e.g., [12,84,85]. While European population indicates a higher degree of individualism, resulting from being more concerned about consequences of one's own behavior, needs, and goals, Asian population displays higher degrees of collectivism, represented by being more concerned of one's own behavior on others, collective needs, and group goals [11,12]. Further, individuals living in collectivistic societies show greater levels of a need to belong, while members of individualistic cultures show a greater need for self-presentation [11,84,86]. Also comparing a German and a Chinese sample, a recent study by Montag et al. [12] pointed out that low acceptance of power distance as a predictive factor for IUD symptoms, especially in males and Chinese individuals. The authors assume that an individual's greater acceptance of power distance and the higher exerted power on themselves leads to the desire to find a way and place to relieve from stress, which could be the Internet.

Furthermore, China displays the world's fastest growing Internet population [87] with an increasing online gaming industry [88]. Especially in today's social, mobile, or online role-playing games, it is necessary for users to play together with others to reach certain group goals. It can be assumed that the collectivistic orientation within the Chinese society contribute to a high involvement in online gaming as well as social networking due to its collaborative features. Further, the competitive character of online games as well as further social factors (e.g., socializing, giving support, teamwork) display important motivational factors for playing online games [89,90], which could be more distinct in collectivistic cultures. Specific using motives have been assumed as significant predictors of an addictive behavior [2]. Further studies should examine potential effects of collectivism/individualism in different countries and its impact on (excessive) Internet use.

4.4. Practical Implications for Both Countries

Although both investigated samples are not representative for its respective population, we think that this study's results could give a contribution to prevention and intervention programs in both nations, which are mentioned in the following.

4.4.1. Practical Implications for Germany

The results of the current study point IUD out as a significant problem within the investigated German sample, where especially two Internet literacy domains are related to higher symptoms: higher productive/interactive as well as lower self-regulative skills. Besides, no considerable effect of technical and reflective skills on tendencies towards an IUD have been found. These correlations strengthen the findings by Stodt et al. [22], who stated that pure technical skills do not counteract a dysfunctional Internet use. As demonstrated in the I-PACE model and relying on empirical investigations, Internet-related cognitions and skills play a key role in the process of developing a specific Internet-use disorder [2,20,21]. Such state factors could be addressed in specific training programs and probably diminish the effect of dysfunctional personality factors and psychopathological symptoms on the development of an IUD. Currently, the improvement of self-regulative skills (without disregarding technical knowledge) still does not find its way into today's curricula of media and/or Internet literacy in Germany. Based on the current findings, we recommend to include the improvement of self-regulative skills in current curricula of media competence in Germany.

4.4.2. Practical Implications for China

Statistics revealed high occurrence of a pathological use of the Internet as well as a high number of people who indicated a problematic use in the investigated sample, mostly consisting of university students. Taking the pathological and the problematic groups together, more than half of the Chinese sample indicated at least a problematic Internet use, pointing towards a serious health problem in these Chinese students. Previous studies from China showed that especially university freshmen suffer from higher IUD symptoms, often associated with psychopathological symptoms, such as depressive symptoms, social anxiety, or higher levels of loneliness [24]. These findings emphasize the importance of prevention and intervention programs not only in middle school, but also in university. Such programs should screen for high-risk Internet users and focus on the improvement of Internet-related cognitions and skills that are related to IUD symptoms, as the current study shows. More detailed, Chinese participants indicated to have high expectancies towards the use of the Internet. Those people who think that the Internet is helpful to avoid negative feelings, like stress and loneliness, or to gain positive emotions to a high degree tend to use the Internet more excessively. That is why prevention and intervention programs in China should focus on the outcomes people expect by using certain Internet applications and should demonstrate alternative coping strategies. Besides that, self-regulation skills are not correlated with the s-IAT total score in the Chinese students, which is different from the German sample. Nevertheless, based on the results of the current study that specific skills play a significant role in the development and maintenance of an IUD, we propose to include such skills in current prevention programs. Future studies should address how these skills could be systematically integrated in upcoming prevention programs.

4.5. *Limitations*

Some limitations of this study need to be addressed in the following. First, the use of two different questionnaires for measuring the Big Five personality traits, the BFI-10, which was used in the German sample, and the NEO-FFI, which was used in the Chinese sample, could be a reason for the different results regarding the personality traits and their impact on IUD symptoms in both samples. Nevertheless, the use of these two questionnaires was carefully considered during planning the study. In previous studies, the BFI-10 displayed a valid and reliable instrument with good psychometric properties in German samples. Further, its 10-item version is very efficient and advantageous for the use in online studies, where time is limited. For the best of our knowledge, there is no valid and evaluated short version of a Big Five questionnaire in Chinese language (to be more precise: please see that the BFI-10 has been published in Mandarin, recently, but a problem concerning one introversion/extraversion item is discussed in Lachmann et al. [61]). That is why we decided to use the 60-item version of the NEO-FFI, which has been used in several Asian studies before and therefore displays a well-evaluated instrument. Although, the NEO-FFI is also evaluated sufficiently in Germany and other Western countries, we used the BFI-10 in the German data collection due to practical reasons and its above mentioned advantages.

One further limitation is the removal of one item out of the subscale *reflection and critical analysis* of the ILQ in the Chinese sample. We had to remove this item directly after data collection because of a problematic translation, which could lead to different possible understandings by the participants. Additionally, the removal should ensure the subscale's content validity. Regarding the subscale's internal consistency, Cronbach's α was still in the acceptable range. With the removal of one item, we do not expect different results because the remaining items substantially represent the same theoretical facet. For the German sample, we additionally calculated all analyses including the subscale *reflection and critical analysis* again with the same three items as used in the Chinese sample. The results did not differ between the three- and four-item solutions.

As a further limitation, the different factorial structures of the s-IAT, IUES, and ILQ in both nations, which are based on weak model fits in mean structure analyses, have to be mentioned. Even if the analyses revealed non-acceptable model fits, we had a closer look on the factor loadings within both nations. Here, every item loaded significantly on its theoretically suggested and empirically approved factor with just slightly varying effects between both nations. Therefore, we assume that the results of the current study are quite comparable. Further, former studies already demonstrated the applicability of the s-IAT and the IUES in a cross-cultural setting, e.g., [57]. However, further studies are needed to test if these questionnaires are suitable for investigating cross-cultural issues.

Furthermore, limitations regarding the representativeness of both investigated samples exist. First, in China and Germany data was collected from convenience samples. Whereas German participants came from all over Germany and were not recruited in a specific region, Chinese data was collected in two certain cities exclusively due to location-dependent reasons, which may have led to region-based effects. Second, the Chinese sample almost consisted of university students, whereas the German sample also contains participants who are employed. Strictly speaking, these facts reduce the level of generalizability and interpretation of the findings in terms of underlying cultural differences. Nevertheless, the German sample was predominately well educated with a high number of participants holding a university degree. Furthermore, both samples did not differ in age and gender. At this point, we want to clarify that this study did not have the intention to examine cultural differences with two nationally representative samples. Notwithstanding, although several limitations exist that have to be beared in mind, this study indicated several similarities between both nations (e.g., level of Internet-use expectancies and personality correlations) which are theoretically grounded and indicate global effects. On the other hand, the differences found between both nations (influence of reflective and regulative skills) have to be interpreted with caution. Here, further replications are necessary to check the meaningfulness of these results. Hence, this study's findings provide evidence that cultural differences between Germany and China regarding symptoms of an IUD and influencing trait and state variables could exist. However, these differences need to be replicated and further investigated in upcoming research explicitly.

4.6. Future Directions

This study suggests some directions for upcoming research. Future studies should address possible mediating and moderating effects between personality traits, Internet-related cognitions and skills, as well as IUD symptoms across different countries such as by using structural equation modelling and mean structure analyses. Based on the findings of this study, examining certain parts and interaction effects of the I-PACE model with representative samples from different countries would be helpful promoting a better understanding of cultural differences in IUD. Additionally, further empirical studies should focus on the role of one person's knowledge and skills, which could represent additional mediating or moderating factors in the addiction process but have not been a part of previous theoretical models. Furthermore, there are just a few longitudinal studies covering IUD. The investigation of long-term effects in different countries and cultures would be helpful to shed light onto the reinforcing or protective value of Internet literacy and further Internet-related cognitive biases on the development and maintenance of an IUD and respective cultural differences. Especially, longitudinal designs would be helpful to develop specific prevention and intervention programs.

5. Conclusions

The present study strengthens previous findings on the effect of personality traits as well as state variables such as Internet-use expectancies and Internet literacy domains on the development and maintenance of an IUD within two samples from different countries and cultural areas. In addition, some interesting country-specific differences regarding these variables and their relationship to symptoms of an IUD were pointed out. Next to a higher occurrence of IUD symptoms in the Chinese sample compared to the German sample, Chinese participants scored significantly higher on the Big Five personality traits of neuroticism and agreeableness, whereas German participants scored higher on the traits of extraversion and openness. However, please note that different inventories to measure the Big Five personality traits were applied in each country. Furthermore, Chinese individuals showed higher Internet-use expectancies to avoid negative feelings or to be positively reinforced. Further analyses indicated that Chinese participants with higher reflective skills regarding their Internet behavior indicated highest IUD symptoms, whereas such skills indicated no effect within the German sample. Moreover, higher skills in self-regulating one's own Internet use were accompanied with lower IUD symptoms in the German, but not in the Chinese sample. Despite the missing representativeness of both samples and some limitations of the current study, several theoretically grounded findings have to be underlined, strengthening the global applicability of this study and providing evidence that cultural differences between Germany and China regarding IUD symptoms and influencing trait and state variables could exist. Because Internet literacy domains seem to play a significant role in the development and maintenance of an IUD, we propose to encourage skills such as reflective and self-regulating skills in current or future IUD prevention programs, which should be individually matched to each countries conditions. Upcoming research should address the replication of this study's results to check its meaningfulness.

Supplementary Materials: The following are available online at http://www.mdpi.com/1660-4601/15/4/579/s1, Table S1: Gender differences in the s-IAT, IUES, ILQ, and Big Five personality traits in Germany and China (ANOVA), Table S2: Correlations between age and the observed Internet and personality (Pearson correlations) including Fisher's z comparison between Germany and China.

Acknowledgments: The position of Christian Montag is funded by a Heisenberg grant awarded to him by the German Research Foundation (DFG, MO 2363/3-2). Cornelia Sindermann is a stipend of the German Academic Scholarship Foundation (Studienstiftung des deutschen Volkes). Moreover, the study has been funded by a grant to study Internet-gaming disorder and Internet addiction by the German Research Foundation (DFG, MO 2363/2-1).

Author Contributions: Christian Montag and Matthias Brand designed the present study. Benjamin Stodt wrote the manuscript and carried out the statistical analysis. Elisa Wegmann, Matthias Brand and Benjamin Stodt collected the German data. Cornelia Sindermann, Christian Montag and Mei Li collected the Chinese data. Min Zhou and Peng Sha were responsible for the translation process of the inventories into the Chinese language. All authors critically revised the manuscript and approved the final version.

Conflicts of Interest: The authors declare no conflict of interest.

Appendix A

Table A1. Short Internet Addiction Test (s-IAT) in English, German, and Chinese language.

	English	German	Chinese
1.	How often do you find that you stay online longer than you intended?	Wie oft stellen Sie fest, dass Sie länger als beabsichtigt im Internet waren?	你经常会发现自己在网上呆的时间比你预期的要长么?
2.	How often do you neglect household chores to spend more time online?	Wie oft vernachlässigen Sie alltägliche Pflichten, um mehr Zeit online zu verbringen?	你经常会为了在网上多呆一会儿而忽略家务琐事么?
3.	How often do your grades or school work suffer because of the amount of time you spend online?	Wie häufig leidet Ihre schulische, Ihre universitäre oder Ihre berufliche Arbeit darunter, dass Sie so viel Zeit online verbringen?	你经常会因为上网用了很多时间而耽误了成绩或者大学的功课么?
4.	How often do you become defensive or secretive when anyone asks you what you do online?	Wie häufig reagieren Sie ausweichend oder verteidigend, wenn Sie jemand fragt, was Sie online tun?	你经常会在别人问起你在网上做什么的时候变得防御或者遮遮掩掩么?
5.	How often do you snap, yell, or act annoyed if someone bothers you while you are online?	Wie oft reagieren Sie patzig, schimpfen oder sind genervt, wenn Sie jemand stört, während Sie online sind?	你经常会因为别人打扰你上网而呵斥、大叫或者作出生气的行为么?
6.	How often do you lose sleep due to being online late at night?	Wie oft fehlt Ihnen Schlaf, weil Sie spät nachts noch online sind?	你经常会因为上网到半夜而失眠么?
7.	How often do you feel preoccupied with the Internet when offline, or fantasize about being online?	Wie oft denken Sie ans Internet, wenn Sie offline sind oder stellen sich vor, online zu sein?	你经常会在线下的时候总想着上网的事情或者想象着在网络上么?
8.	How often do you find yourself saying "just a few more minutes" when online?	Wie oft ertappen Sie sich dabei zu sagen: "Nur noch ein paar Minuten", während Sie online sind?	你经常会发现在上网时你对自己说"再多几分钟"么?
9.	How often do you try to cut down the amount of time you spend online and fail?	Wie häufig versuchen Sie weniger Zeit im Internet zu verbringen und schaffen es nicht?	你经常试图减少上网的时间但是却以失败告终么?
10.	How often do you try to hide how long you've been online?	Wie häufig versuchen Sie zu verbergen, wie lange Sie online waren?	你经常会尝试隐瞒实际在网上呆的时间么?
11.	How often do you choose to spend more time online over going out with others?	Wie oft kommt es vor, dass Sie lieber mehr Zeit online verbringen als mit Anderen etwas zu unternehmen?	你经常会在网络上花费的时间比与他人一起外出的时间更多么?
12.	How often do you feel depressed, moody, or nervous when you are offline, which goes away once you are back online?	Wie oft fühlen Sie sich deprimiert, verstimmt oder nervös, wenn Sie offline sind – was sich ändert, wenn Sie wieder online sind?	你经常会因为下线而感到郁闷、不高兴或焦虑，而一旦上网这些感受就烟消云散了么?

Items must be answered on a 5-point Likert scale with the following response levels: 1 = never/nie/从不, 2 = rarely/selten/很少, 3 = sometimes/manchmal/有时, 4 = often/oft/经常, 5 = very often/sehr oft/非常常见. Scoring of the s-IAT scales: *Total score* is based on the sum score of all 12 items; *Loss of control/time management* is based on the sum score of the following items: 1, 2, 3, 6, 8, 9; *Craving/social problems* is based on the sum score of the following items: 4, 5, 7, 10, 11, 12.

Table A2. Internet-Use Expectancies Scale (IUES) in English, German, and Chinese language.

	English	German	Chinese
	I use the Internet because it makes possible/facilitates . . .	Ich nutze das Internet, weil es mir ermöglicht/erleichtert . . .	我使用网络是因为它在如下方面创造了可能性／便利性:
1.	to experience pleasure.	Freude zu erleben.	为了体验快乐。
2.	to distract from problems.	Problemen aus dem Weg zu gehen.	为了从问题（麻烦）中转移注意力。
3.	to have fun.	Spaß zu haben.	为了玩得愉快。
4.	to avoid loneliness.	Gefühle der Einsamkeit zu vermeiden.	为了避免孤独。
5.	to feel good.	mich gut zu fühlen.	为了感觉好。
6.	to escape from reality.	vor der Realität zu flüchten.	为了逃避现实。
7.	to gain positive emotions.	positive Gefühle zu erreichen.	为了获得正性情感。
8.	to avoid annoying duties.	lästige Aufgaben zu vermeiden.	为了回避恼人的责任。

Items must be answered on a 6-point Likert scale with the following response levels: 1 = completely disagree/stimme gar nicht zu/完全不同意, 2 = disagree/stimme nicht zu/不同意, 3 = rather disagree/stimme eher nicht zu/有一点不同意, 4 = rather agree/stimme eher zu/有点同意, 5 = agree/stimme zu/同意, 6 = completely agree/stimme voll und ganz zu/完全同意. Scoring of the IUES scales: *Positive reinforcement* is based on the mean score of the following items: 1, 3, 5, 7; *Avoidance expectancies* is based on the mean score of the following items: 2, 4, 6, 8.

Table A3. Internet Literacy Questionnaire (ILQ) in English, German, and Chinese language.

	English	German	Chinese
1.	I can quickly familiarize myself with new sites on the Internet or applications I have not used before.	Ich kann mich schnell mit neuen Internetseiten oder -anwendungen vertraut machen, die ich zuvor noch nicht benutzt habe.	我能够很快熟悉那些我从未使用过的网站和应用。
2.	It is easy for me to limit my Internet use when I notice that it has a negative effect on my private life.	Es fällt mir leicht, meinen Internetkonsum einzuschränken, wenn ich merke, dass er sich negativ auf mein Privatleben auswirkt.	当我察觉网络使用已经对我的个人生活产生负面影响时，我可以很容易地控制自己使用网络。
3.	It is easier to maintain social contacts on the Internet than offline.	Es ist einfacher im Internet soziale Kontakte zu pflegen als offline.	在网络中保持社交接触比线下更容易。
4.	I make sure to use the Internet to an extent that is appropriate for me.	Ich achte darauf, das Internet in einem für mich angemessenen Umfang zu nutzen.	我确保在一个对我适度的范围内使用网络。
5.	I can easily assess the credibility of information on websites even when I am not familiar with the topic.	Ich kann Informationen auf Internetseiten mühelos hinsichtlich ihrer Glaubwürdigkeit einschätzen, auch wenn mir das Thema nicht vertraut ist.	即便在不熟悉的领域里，我也可以比较容易地确定网站上的信息的可信程度。

Table A3. *Cont.*

	English	German	Chinese
6.	I inform myself regularly about developments regarding the Internet.	Ich informiere mich regelmäßig über Entwicklungen im Bereich des Internets.	我经常关注互联网发展方面的信息。
7.	When I am online, I make sure that my Internet use does not negatively affect my private life.	Wenn ich online bin, achte ich darauf, dass sich meine Internetnutzung nicht negativ auf mein Privatleben auswirkt.	当我在线的时候，我会注意我的网络使用情况不会对我的个人生活造成负面的影响。
8.	I notice very quickly when my counterpart on the Internet is pretending.	Ich merke sehr schnell, wenn sich mein Gegenüber im Internet verstellt.	当对方在网络上假意的时候，我很快可以注意到。
9.	Friends and acquaintances ask me for advice when they have problems with Internet applications.	Freunde und Bekannte fragen mich nach Rat, wenn sie Probleme mit Internetanwendungen haben.	当朋友们或熟人们遇到网络运用方面的问题时会请教我。
10.	Causally exchanging views with other people on the Internet is easier than offline.	Im Internet kann man sich zwangloser mit anderen Personen austauschen als offline.	在线上和他人随意地交换观点要比在线下更加容易。
11.	It is easier for me to make casual contact with another person online than offline.	Es fällt mir online leichter, unverbindlichen Kontakt mit einer anderen Person zu knüpfen als offline.	对我来说，与别人建立一般的联系在线上要比在线下更容易。
12.	I can easily assess what others on the Internet want to achieve with their behavior.	Ich kann gut einschätzen, was Andere im Internet mit ihrem Verhalten bezwecken wollen.	在网络中，我能比较容易地从他人的行为中推断出他们想要达到的目的。
13.	I can tell the difference between credible and untrustworthy content on the Internet.	Ich kann zwischen glaubwürdigen und unglaubwürdigen Inhalten im Internet unterscheiden.	我能在网络中分辨清楚可信和不可信的内容。
14.	It is easier for me to be creative online than offline.	Es fällt mir online leichter kreativ zu sein als offline.	对我来说，在线上比线下更容易有创造力。
15.	I go offline when I feel I have done or found everything on the Internet that is relevant at the time.	Ich gehe offline, wenn ich das Gefühl habe, alles gerade Relevante im Internet erledigt oder gefunden zu haben.	当我认为已经找到相关的信息或处理完相关的事情后，我就会下线。
16.	When I am on the Internet, I make sure not to be online longer than planned.	Wenn ich im Internet bin, achte ich darauf, dass ich nicht länger als beabsichtigt online bin.	当我使用网络时，我会确保不要超过计划的时间。
17.	I can deal with errors in computer software and apps and can fix them myself.	Ich kann mit auftretenden Fehlern in Computerprogrammen/Apps umgehen und sie eigenständig beheben.	我会自己处理和修理电脑软件和应用软件里的错误。
18.	I can more easily express myself on a topic on the Internet than offline.	Im Internet kann ich mich leichter zu einem Thema äußern als offline.	与线下相比，在网络上我更容易针对某一个话题发表自己的观点。

Items must be answered on a Likert scale of six response levels from 0 (=strongly disagree/stimme überhaupt nicht zu/非常不同意) to 5 (=strongly agree/stimme vollkommen zu/非常同意). Scoring of the ILQ scales: *Technical expertise* is based on the mean score of the following items: 1, 6, 9, 17; *Production & interaction* is based on the mean score of the following items: 3, 10, 11, 14, 18; *Reflection & critical analysis* is based on the mean score of the following items: 5, 8, 12, 13; *Self-regulation* is based on the mean score of the following items: 2, 4, 7, 15, 16.

References

1. Davis, R.A. A cognitive-behavioral model of pathological Internet use. *Comput. Hum. Behav.* **2001**, *17*, 187–195. [CrossRef]
2. Brand, M.; Young, K.; Laier, C.; Wölfling, K.; Potenza, M. Integrating psychological and neurobiological considerations regarding the development and maintenance of specific Internet-use disorders: An Interaction of Person-Affect-Cognition-Execution (I-PACE) model. *Neurosci. Biobehav. Rev.* **2016**, *71*, 252–266. [CrossRef] [PubMed]
3. Brand, M.; Young, K.S.; Laier, C. Prefrontal control and Internet addiction: A theoretical model and review of neuropsychological and neuroimaging findings. *Front. Hum. Neurosci.* **2014**, *8*, 375. [CrossRef] [PubMed]
4. Montag, C.; Bey, K.; Sha, P.; Li, M.; Chen, Y.-F.; Liu, W.-Y.; Zhu, Y.-K.; Li, C.-B.; Markett, S.; Keiper, J.; et al. Is it meaningful to distinguish between generalized and specific Internet addiction? Evidence from a cross-cultural study from Germany, Sweden, Taiwan and China. *Asia Pac. Psychiatry* **2015**, *7*, 20–26. [CrossRef] [PubMed]
5. American Psychiatric Association. *Diagnostic and Statistical Manual of Mental Disorders*, 5th ed.; American Psychiatric Publishing: Washington, DC, USA, 2013; ISBN 8123923791.
6. Cheng, C.; Li, A.Y.-L. Internet addiction prevalence and quality of (real) life: A meta-analysis of 31 nations across seven world regions. *Cyberpsychol. Behav. Soc. Netw.* **2014**, *17*, 755–760. [CrossRef] [PubMed]
7. Kuss, D.J.; Griffiths, M.; Karila, M.; Billieux, J. Internet addiction: A systematic review of epidemiological research for the last decade. *Curr. Pharm. Des.* **2014**, *20*, 4026–4052. [CrossRef] [PubMed]
8. Spada, M.M. An overview of problematic Internet use. *Addict. Behav.* **2014**, *39*, 3–6. [CrossRef] [PubMed]
9. Rumpf, H.-J.; Meyer, C.; Kreuzer, A.; John, U. Prävalenz der Internetabhängigkeit (PINTA). Bericht an das Bundesministerium für Gesundheit (Prevalence of Internet addiction (PINTA). Report to the Federal Ministry of Health). Available online: https://www.bundesgesundheitsministerium.de/fileadmin/Dateien/5_Publikationen/Drogen_und_Sucht/Berichte/Forschungsbericht/Studie_Praevalenz_der_Internetabhaengigkeit__PINTA_.pdf (accessed on 18 January 2018).
10. Ko, C.H.; Yen, J.Y.; Yen, C.F.; Chen, C.S.; Chen, C.C. The association between Internet addiction and psychiatric disorder: A review of the literature. *Eur. Psychiatry* **2012**, *27*, 1–8. [CrossRef] [PubMed]
11. Hofstede, G. *Culture's Consequences: International Differences in Work-related Values*; Sage Publications: Newbury Park, CA, USA, 1984; ISBN 0803913060.
12. Montag, C.; Duke, E.; Sha, P.; Zhou, M.; Sindermann, C.; Li, M. Does acceptance of power distance influence propensities for problematic Internet use? Evidence from a cross-cultural study. *Asia Pac. Psychiatry* **2016**, *8*, 296–301. [CrossRef] [PubMed]
13. Li, N.; Kirkup, G. Gender and cultural differences in Internet use: A study of China and the UK. *Comput. Educ.* **2007**, *48*, 301–317. [CrossRef]
14. Zhang, L.; Amos, C.; Mcdowell, W.C. A comparative study of Internet addiction between the United States and China. *Cyberpsychol. Behav.* **2008**, *11*, 727–729. [CrossRef] [PubMed]
15. Ko, D.; Yao, M. Internet addiction: An cross-cultural perspective. In *The Psychology of Social Networking Vol. 2: Identity and Relationships in Online Communities*; Riva, G., Wiederhold, B.K., Cipresso, P., Eds.; De Gruyter: Warsaw, Poland; Berlin, Germany, 2016; pp. 141–158, ISBN 3110473844.
16. Internet World Stats. World Internet Users and 2017 Population Stats. Available online: http://www.internetworldstats.com/stats.htm (accessed on 26 January 2018).
17. Internet Live Stats. Internet Users by Country. Available online: http://www.internetlivestats.com/internet-users-by-country/ (accessed on 26 January 2018).
18. Müller, K.W.; Glaesmer, H.; Brähler, E.; Wölfling, K.; Beutel, M.E. Prevalence of Internet addiction in the general population: Results from a German population-based survey. *Behav. Inf. Technol.* **2014**, *33*, 757–766. [CrossRef]
19. Pawlikowski, M.; Altstötter-Gleich, C.; Brand, M. Validation and psychometric properties of a short version of Young's Internet Addiction Test. *Comput. Human Behav.* **2013**, *29*, 1212–1223. [CrossRef]
20. Wegmann, E.; Brand, M. Internet-communication disorder: It's a matter of social aspects, coping, and Internet-use expectancies. *Front. Psychol.* **2016**, *7*, 1747. [CrossRef] [PubMed]

21. Wegmann, E.; Stodt, B.; Brand, M. Addictive use of social networking sites can be explained by the interaction of Internet use expectancies, Internet literacy, and psychopathological symptoms. *J. Behav. Addict.* **2015**, *4*, 155–162. [CrossRef] [PubMed]
22. Stodt, B.; Wegmann, E.; Brand, M. Predicting dysfunctional Internet use: The role of age, conscientiousness, and Internet literacy in Internet addiction and cyberbullying. *IJCBPL* **2016**, *6*, 28–43. [CrossRef]
23. Cao, F.; Su, L. Internet addiction among Chinese adolescents: Prevalence and psychological features. *Child Care Health Dev.* **2006**, *33*, 275–282. [CrossRef] [PubMed]
24. Ni, X.; Yan, H.; Chen, S.; Liu, Z. Factors influencing Internet addiction in a sample of freshmen university students in China. *Cyberpsychol. Behav.* **2009**, *12*, 327–330. [CrossRef] [PubMed]
25. Cao, F.; Su, L.; Liu, T.; Gao, X. The relationship between impulsivity and Internet addiction in a sample of Chinese adolescents. *Eur. Psychiatry* **2007**, *22*, 466–471. [CrossRef] [PubMed]
26. Nie, J.; Zhang, W.; Liu, Y. Exploring depression, self-esteem and verbal fluency with different degrees of Internet addiction among Chinese college students. *Compr. Psychiatry* **2017**, *72*, 114–120. [CrossRef] [PubMed]
27. Chak, K.; Leung, L. Shyness and locus of control as predictors of Internet addiction and Internet use. *Cyberpsychol. Behav.* **2004**, *7*, 559–570. [CrossRef] [PubMed]
28. Zhou, Y.; Li, D.; Li, X.; Wang, Y.; Zhao, L. Big five personality and adolescent Internet addiction: The mediating role of coping style. *Addict. Behav.* **2017**, *64*, 42–48. [CrossRef] [PubMed]
29. Montag, C.; Jurkiewicz, M.; Reuter, M. Low self-directedness is a better predictor for problematic Internet use than high neuroticism. *Comput. Hum. Behav.* **2010**, *26*, 1531–1535. [CrossRef]
30. Sariyska, R.; Reuter, M.; Bey, K.; Sha, P.; Li, M.; Chen, Y.F.; Liu, W.Y.; Zhu, Y.K.; Li, C.B.; Suárez-Rivillas, A.; et al. Self-esteem, personality and Internet addiction: A cross-cultural comparison study. *Pers. Individ. Differ.* **2014**, *61–62*, 28–33. [CrossRef]
31. Chaplin, W.F.; John, O.P.; Goldberg, L.R. Conceptions of states and traits: Dimensional attributes with ideals as prototypes. *J. Pers. Soc. Psychol.* **1988**, *54*, 541–557. [CrossRef] [PubMed]
32. Steyer, R.; Mayer, A.; Geiser, C.; Cole, D.A. A theory of states and traits - revised. *Annu. Rev. Clin. Psychol.* **2015**, *11*, 71–98. [CrossRef] [PubMed]
33. Sariyska, R.; Reuter, M.; Lachmann, B.; Montag, C. Attention deficit/hyperactivity disorder is a better predictor for problematic Internet use than depression: Evidence from Germany. *J. Addict. Res. Ther.* **2015**, *6*, 1–6. [CrossRef]
34. Prizant-Passal, S.; Shechner, T.; Aderka, I.M. Social anxiety and Internet use—A meta-analysis: What do we know? What are we missing? *Comput. Hum. Behav.* **2016**, *62*, 221–229. [CrossRef]
35. Goldberg, L.R. An alternative "description of personality": The Big-Five factor structure. *J. Pers. Soc. Psychol.* **1990**, *59*, 1216–1229. [CrossRef] [PubMed]
36. Soto, C.J.; John, O.P.; Gosling, S.D.; Potter, J. Age differences in personality traits from 10 to 65: Big Five domains and facets in a large cross-sectional sample. *J. Pers. Soc. Psychol.* **2011**, *100*, 330–348. [CrossRef] [PubMed]
37. Schmitt, D.P.; McCrae, R.R.; Bennett, K.L.; Grammer, K. The geographic distribution of Big Five personality traits—Patterns and profiles of human self-description across 56 nations. *J. Cross Cult. Psychol.* **2007**, *38*, 173–212. [CrossRef]
38. Hardie, E.; Tee, M.Y. Excessive Internet use: The role of personality, loneliness, and social support networks in Internet addiction. *AJETS* **2007**, *5*, 34–47.
39. Tsai, H.F.; Cheng, S.H.; Yeh, T.L.; Shih, C.-C.; Chen, K.C.; Yang, Y.C.; Yang, Y.K. The risk factors of Internet addiction—A survey of university freshmen. *Psychiatry Res.* **2009**, *167*, 294–299. [CrossRef] [PubMed]
40. Montag, C.; Reuter, M. Molecular genetics, personality and Internet addiction. In *Internet Addiction: Neuroscientific Approaches and Therapeutical Interventions*; Springer International Publishing: Heidelberg, Germany, 2015; pp. 93–109, ISBN 9781611226164.
41. Lee, Y.H.; Ko, C.H.; Chou, C. Re-visiting Internet addiction among Taiwanese students: A cross-sectional comparison of students' expectations, online gaming, and online social interaction. *J. Abnorm. Child Psychol.* **2015**, *43*, 589–599. [CrossRef] [PubMed]
42. Lin, M.-P.; Ko, H.-C.; Wu, J.Y.-W. The role of positive/negative outcome expectancy and refusal self-efficacy of Internet use on Internet addiction among college students in Taiwan. *Cyberpsychol. Behav.* **2008**, *11*, 451–457. [CrossRef] [PubMed]

43. Casale, S.; Caplan, S.E.; Fioravanti, G. Positive metacognitions about Internet use: The mediating role in the relationship between emotional dysregulation and problematic use. *Addict. Behav.* **2016**, *59*, 84–88. [CrossRef] [PubMed]
44. Brand, M.; Laier, C.; Young, K.S. Internet addiction: Coping styles, expectancies, and treatment implications. *Front. Psychol.* **2014**, *5*, 1256. [CrossRef] [PubMed]
45. Wu, J.Y.W.; Ko, H.C.; Tung, Y.Y.; Li, C.C. Internet use expectancy for tension reduction and disinhibition mediates the relationship between borderline personality disorder features and Internet addiction among college students—One-year follow-up. *Comput. Hum. Behav.* **2016**, *55*, 851–855. [CrossRef]
46. Iordache, C.; Mariën, I.; Baelden, D. Developing digital skills and competences: A quick-scan analysis of 13 digital literacy models. *IJSE* **2017**, *9*, 6–30. [CrossRef]
47. Martin, A.; Madigan, D. *Digital Literacies for Learning*; Facet: London, UK, 2006; ISBN 1856045633.
48. Jenkins, H. *Confronting the Challenges of Participatory Culture: Media Education for the 21st Century*; MIT Press: Cambridge, MA, USA, 2009; ISBN 978-0-262-51362-3.
49. Lin, T.-B.; Jen-Yi, L.; Deng, F.; Lee, L. Understanding new media literacy: An explorative theoretical framework. *J. Educ. Technol. Soc.* **2013**, *16*, 160–170.
50. Katz, I.R.; Elliot, N. Information literacy in digital environments: Construct mediation, construct modeling, and validation processes. In *Information Literacy: Research and Collaboration across Disciplines*; D'Angelo, B.J., Jamieson, S., Maid, B., Walker, J.R., Eds.; The WAC Clearinghouse: Fort Collins, CO, USA, 2016; pp. 97–116.
51. Leung, L.; Lee, P.S.N. The influences of information literacy, Internet addiction and parenting styles on Internet risks. *New Media Soc.* **2011**, *14*, 117–136. [CrossRef]
52. Fraillon, J.; Schulz, W.; Ainley, J. *International Computer and Information Literacy Study: Assessment Framework*; International Association for the Evaluation of Educational Achievement (IEA): Amsterdam, The Netherlands, 2013.
53. Johnson, G.M. Functional Internet literacy: Required cognitive skills with implications for instruction. *E-Learn. Digit. Media* **2007**, *4*, 433–441. [CrossRef]
54. Livingstone, S.; Helsper, E. Balancing opportunities and risks in teenagers' use of the Internet: The role of online skills and Internet self-efficacy. *New Media Soc.* **2009**, *12*, 309–329. [CrossRef]
55. LaRose, R.; Lin, C.A.; Eastin, M.S. Unregulated Internet usage: Addiction, habit, or deficient self-regulation? *Media Psychol.* **2003**, *5*, 225–253. [CrossRef]
56. Hahn, E.; Reuter, M.; Spinath, F.M.; Montag, C. Internet addiction and its facets: The role of genetics and the relation to self-directedness. *Addict. Behav.* **2017**, *65*, 137–146. [CrossRef] [PubMed]
57. Wegmann, E.; Oberst, U.; Stodt, B.; Brand, M. Online-specific fear of missing out and Internet-use expectancies contribute to symptoms of Internet-communication disorder. *Addict. Behav. Rep.* **2017**, *5*, 33–42. [CrossRef] [PubMed]
58. Eisinga, R.; Te Grotenhuis, M.; Pelzer, B. The reliability of a two-item scale: Pearson, Cronbach, or Spearman-Brown? *Int. J. Public Health* **2013**, *58*, 637–642. [CrossRef] [PubMed]
59. Young, K.S. *Caught in the Net: How to Recognize the Signs of Internet Addiction—And a Winning Strategy for Recovery*; John Wiley & Sons: New York, NY, USA, 1998; ISBN 0471191590.
60. Rammstedt, B.; John, O.P. Measuring personality in one minute or less: A 10-item short version of the Big Five Inventory in English and German. *J. Res. Pers.* **2007**, *41*, 203–212. [CrossRef]
61. Lachmann, B.; Sariyska, R.; Kannen, C.; Błaszkiewicz, K.; Trendafilov, B.; Andone, I.; Eibes, M.; Markowetz, A.; Li, M.; Kendrick, K.M.; et al. Contributing to overall life satisfaction: Personality traits versus life satisfaction variables revisited—Is replication impossible? *Behav. Sci.* **2018**, *8*, 1. [CrossRef] [PubMed]
62. Costa, P.T.; McCrae, R.R. *NEO Five-Factor Inventory (NEO-FFI): Professional Manual*; Psychological Assessment Resources: Odessa, FL, USA, 1989.
63. Melchers, M.C.; Li, M.; Haas, B.W.; Reuter, M.; Bischoff, L.; Montag, C. Similar personality patterns are associated with empathy in four different countries. *Front. Psychol.* **2016**, *7*, 290. [CrossRef] [PubMed]
64. Cohen, J. *Statistical Power Analysis for the Behavioral Sciences*; Lawrence Earlbaum Associates: Hillsdale, NJ, USA, 1988.
65. Durkee, T.; Kaess, M.; Carli, V.; Parzer, P.; Wasserman, C.; Floderus, B.; Apter, A.; Balazs, J.; Barzilay, S.; Bobes, J.; et al. Prevalence of pathological Internet use among adolescents in Europe: Demographic and social factors. *Addiction* **2012**, *107*, 2210–2222. [CrossRef] [PubMed]

66. Cao, H.; Sun, Y.; Wan, Y.; Hao, J.; Tao, F. Problematic Internet use in Chinese adolescents and its relation to psychosomatic symptoms and life satisfaction. *BMC Public Health* **2011**, *11*, 802. [CrossRef] [PubMed]
67. Liu, Q.-X.; Fang, X.-Y.; Deng, L.-Y.; Zhang, J.-T. Parent–adolescent communication, parental Internet use and Internet-specific norms and pathological Internet use among Chinese adolescents. *Comput. Hum. Behav.* **2012**, *28*, 1269–1275. [CrossRef]
68. Chi, X.; Lin, L.; Zhang, P. Internet addiction among college students in China: Prevalence and psychosocial correlates. *Cyberpsychol. Behav. Soc. Netw.* **2016**, *19*, 567–573. [CrossRef] [PubMed]
69. Bond, M.H. *The Oxford Handbook of Chinese Psychology*; Oxford Library of Psychology: Oxford, NY, USA, 2010; ISBN 019954185X.
70. Noesselt, N.; Schüller, M.; Schüler-Zhou, Y. Deutschland und China—Wahrnehmung und Realität—Die Huawei-Studie 2014 (Germany and China—Perception and Reality—The Huawei Study 2014). Available online: www.huawei-studie.de/downloads/Huawei-Studie-2014-DE.pdf (accessed on 24 January 2018).
71. Noesselt, N.; Schüller, M.; Schüler-Zhou, Y. Deutschland und China—Wahrnehmung und Realität—Die Huawei-Studie 2016 (Germany and China—Perception and Reality—The Huawei Study 2016). Available online: http://www.huawei-studie.de/downloads/Huawei-Studie-2016-DE.pdf (accessed on 24 January 2018).
72. Schroll-Machl, S. *Doing Business with Germans: Their Perception, Our Perception*; Vandenhoeck & Ruprecht: Göttingen, Germany, 2008; ISBN 3525461674.
73. Sindermann, C.; Luo, R.; Zhao, Z.; Li, M.; Kendrick, K.; Becker, B.; Panksepp, J.; Montag, C. High ANGER and low agreeableness predict vengefulness in German and Chinese participants. *Pers. Individ. Differ.* **2018**, *121*, 184–192. [CrossRef]
74. Müller, K.W.; Beutel, M.E.; Egloff, B.; Wölfling, K. Investigating risk factors for Internet gaming disorder: A comparison of patients with addictive gaming, pathological gamblers and healthy controls regarding the Big Five personality traits. *Eur. Addict. Res.* **2014**, *20*, 129–136. [CrossRef] [PubMed]
75. Hong, F.-Y.; Huang, D.-H.; Lin, H.-Y.; Chiu, S.-L. Analysis of the psychological traits, Facebook usage, and Facebook addiction model of Taiwanese university students. *Telemat. Inform.* **2014**, *31*, 597–606. [CrossRef]
76. Wang, C.-W.; Ho, R.T.H.; Chan, C.L.W.; Tse, S. Exploring personality characteristics of Chinese adolescents with Internet-related addictive behaviors: Trait differences for gaming addiction and social networking addiction. *Addict. Behav.* **2015**, *42*, 32–35. [CrossRef] [PubMed]
77. La Ferle, C.; Kim, H.-J. Cultural influences on Internet motivations and communication styles: A comparison of Korean and US consumers. *IJIMA* **2006**, *3*, 142–157. [CrossRef]
78. Xu-Priour, D.-L.; Truong, Y.; Klink, R.R. The effects of collectivism and polychronic time orientation on online social interaction and shopping behavior: A comparative study between China and France. *Technol. Forecast. Soc. Chang.* **2014**, *88*, 265–275. [CrossRef]
79. Cloninger, C.R.; Svrakic, D.M.; Przybeck, T.R. A psychobiological model of temperament and character. *Arch. Gen. Psychiatry* **1993**, *50*, 975–990. [CrossRef] [PubMed]
80. Ha, J.H.; Kim, S.Y.; Bae, S.C.; Bae, S.; Kim, H.; Sim, M.; Lyoo, I.K.; Cho, S.C. Depression and Internet addiction in adolescents. *Psychopathology* **2007**, *40*, 424–430. [CrossRef] [PubMed]
81. June, K.J.; Sohn, S.Y.; So, A.Y.; Yi, G.M.; Park, S.H. A study of factors that influence Internet addiction, smoking, and drinking in high school students. *J. Korean Acad. Nurs.* **2007**, *37*, 872–882. [CrossRef]
82. Stieger, S.; Burger, C. Implicit and explicit self-esteem in the context of Internet addiction. *Cyberpsychol. Behav. Soc. Netw.* **2010**, *13*, 681–688. [CrossRef] [PubMed]
83. Montag, C.; Flierl, M.; Markett, S.; Walter, N.; Jurkiewicz, M.; Reuter, M. Internet addiction and personality in first-person-shooter video gamers. *J. Media Psychol.* **2011**, *23*, 163–173. [CrossRef]
84. Triandis, H.C. The self and social behavior in differing cultural contexts. *Psychol. Rev.* **1989**, *96*, 506–520. [CrossRef]
85. Wheeler, L.; Reis, H.T.; Bond, M.H. Collectivism-individualism in everyday social life: The middle kingdom and the melting pot. *J. Pers. Soc. Psychol.* **1989**, *57*, 79–86. [CrossRef]
86. Nadkarni, A.; Hofmann, S.G. Why do people use Facebook? *Pers. Individ. Differ.* **2012**, *52*, 243–249. [CrossRef] [PubMed]
87. Weber, I.; Jia, L. Internet and self-regulation in China: The cultural logic of controlled commodification. *Media Cult. Soc.* **2007**, *29*, 772–789. [CrossRef]

88. Kshetri, N. The evolution of the Chinese online gaming industry. *JTMC* **2009**, *4*, 158–179. [CrossRef]
89. Yee, N. Motivations for play in online games. *Cyberpsychol. Behav.* **2006**, *9*, 772–775. [CrossRef] [PubMed]
90. Demetrovics, Z.; Urbán, R.; Nagygyörgy, K.; Farkas, J.; Zilahy, D.; Mervó, B.; Reindl, A.; Ágoston, C.; Kertész, A.; Harmath, E. Why do you play? The development of the motives for online gaming questionnaire (MOGQ). *Behav. Res. Methods* **2011**, *43*, 814–825. [CrossRef] [PubMed]

International Journal of
Environmental Research and Public Health

MDPI

Review

Association between Internet Gaming Disorder or Pathological Video-Game Use and Comorbid Psychopathology: A Comprehensive Review

Vega González-Bueso [1,†], Juan José Santamaría [1,*,†], Daniel Fernández [2,3], Laura Merino [1], Elena Montero [1] and Joan Ribas [1]

1 Atención e Investigación en Socioadicciones (AIS), Mental Health and Addictions Network, Generalitat de Catalunya (XHUB), C/Forn-7-9 Local, 08014 Barcelona, Spain; vgonzalez@ais-info.org (V.G.-B.); lmerino@ais-info.org (L.M.); emontero@ais-info.org (E.M.); 38039jrs@comb.cat (J.R.)

2 Research and Development Unit, Parc Sanitari Sant Joan de Déu, Fundació Sant Joan de Déu, CIBERSAM, Dr. Antoni Pujadas, 42, Sant Boi de Llobregat, 08830 Barcelona, Spain; df.martinez@pssjd.org

3 School of Mathematics and Statistics, Victoria University of Wellington, Wellington 6140, New Zealand

* Correspondence: jsantamaria@ais-info.org; Tel.: +34-93-301-3024

† These authors contributed equally to this work.

Received: 27 February 2018; Accepted: 31 March 2018; Published: 3 April 2018

Abstract: The addictive use of video games is recognized as a problem with clinical relevance and is included in international diagnostic manuals and classifications of diseases. The association between "Internet addiction" and mental health has been well documented across a range of investigations. However, a major drawback of these studies is that no controls have been placed on the type of Internet use investigated. The aim of this study is to review systematically the current literature in order to explore the association between Internet Gaming Disorder (IGD) and psychopathology. An electronic literature search was conducted using PubMed, PsychINFO, ScienceDirect, Web of Science and Google Scholar (r.n. CRD42018082398). The effect sizes for the observed correlations were identified or computed. Twenty-four articles met the eligibility criteria. The studies included comprised 21 cross-sectional and three prospective designs. Most of the research was conducted in Europe. The significant correlations reported comprised: 92% between IGD and anxiety, 89% with depression, 85% with symptoms of attention deficit hyperactivity disorder (ADHD), and 75% with social phobia/anxiety and obsessive-compulsive symptoms. Most of the studies reported higher rates of IGD in males. The lack of longitudinal studies and the contradictory results obtained prevent detection of the directionality of the associations and, furthermore, show the complex relationship between both phenomena.

Keywords: pathological video-game use; Internet Gaming Disorder; comorbid psychopathology; review

1. Introduction

The problematic use of video games is recognized by mental health professionals as an addictive behavior with clinical relevance. This is due to the negative consequences it may have for affected people in several functional areas such as relationship conflicts, sleep problems or occupational functioning [1,2]. However, in the current literature, the terms "Internet addiction" (IA) and "pathological Internet use" (PIU) have commonly been used to refer to all sorts of activities including, but not limited to, the use of video games. All these activities are derived from the excessive use of devices connected to the Internet (i.e., computers, smartphones and other devices to play on and navigate). This classification has frequently been criticized as being too broad and not distinguishing between problematic activities and the medium itself on which they take place [3,4], despite the fact that persons engaged in these activities have different sociodemographic characteristics and motivations [5]. For example, the Internet preference activities for males are those related to entertainment and leisure, whereas women tend to choose activities related to interpersonal communication and educational assistance; additionally, these differences may be mediated by age [6].

The non-inclusion of IA as a diagnosis, and the inclusion of "Internet Video-Game Disorder" (Internet Gaming Disorder, IGD) in Section III of the diagnostic manual DSM-5 [7] as a condition that requires further study, seems to support considering both disorders as different problems. Likewise, the most recent inclusion of Gaming Disorder in the beta version of the ICD-11 (International Classification of Diseases) of the World Health Organization [8] seems to confirm this trend. In this document, the problem is defined as "a pattern of persistent or recurrent gaming behavior ('digital gaming' or 'video-gaming'), which may be online (i.e., over the Internet) or offline, manifested by: (1) impaired control over gaming (e.g., onset, frequency, intensity, duration, termination, context); (2) increasing priority given to gaming to the extent that gaming takes precedence over other life interests and daily activities; and (3) continuation or escalation of gaming despite the occurrence of negative consequences. The behavior pattern is of sufficient severity to result in significant impairment in personal, family, social, educational, occupational or other important areas of functioning. The pattern of gaming behavior may be either continuous or, on the other hand, episodic and recurrent. The gaming behavior and other features are normally evident over a period of at least 12 months for a diagnosis to be assigned, although the required duration may be shortened if all diagnostic requirements are met and symptoms are severe".

The psychopathology associated with addictive behaviors, with or without substance, can result from a problem or, alternatively, lead to further issues [9,10]. If the association between two disorders is higher than expected by chance, it is likely that there are mechanisms contributing to that association. Four general models of increased comorbidity have been described [11–13]: common factor models, secondary substance-use disorder models, secondary psychiatric disorder models, and bidirectional models. In the first instance, both disorders share risk factors and the higher comorbidity is the result. In the second case, the addictive disorder contributes to other psychiatric disorders. In the third condition, the psychiatric disorder precipitates the addictive behavior. Finally, either disorder can increase vulnerability to the other disorder; in such cases the higher comorbidity reported may be due to inappropriate sampling, assessment, study design or other biases in the published studies.

In the case of behavioral addictions, the temporal linearity of that relationship remains unclear. Associations between IA or PIU and various psychiatric symptoms have been reported in the literature. Specifically, they have been related to depression, attention deficit hyperactivity disorder (ADHD), anxiety, obsessive-compulsive symptoms, and hostility or aggression [14]. Depression seems to be the most common comorbidity in all age groups (adolescents, adults and the general population). However, the designs used to explore these relationships are not sufficiently comprehensive or complex to confirm the hypothesis for the above models. It is possible that a specific psychiatric problem might have an influence on developing an IA, or that a person with an IA diagnosis, due to various negative consequences, will later develop a comorbid psychiatric disorder. It is also possible that both problems share biological, sociodemographic or psychological underlying mechanisms that make people vulnerable to both pathologies; these may thus become evident at the same time [15]. A major drawback of these studies is that, in most, the type of Internet use is not controlled or, alternatively, the results are not separated by use. In many studies, playing video games is the most common activity among people with IA [16–19]; still, the results have been analyzed without taking this aspect into account.

Therefore, some interesting questions remain. One is whether IGD has similar comorbidities to IA or, rather, the comorbidities are different. In the latter case, one may wonder if other Internet-based issues are affecting in some way the results of studies focused on IA in general. An additional question pertains to the directionality of both conditions (IGD and psychopathology).

The aim of this study is to review systematically the current literature to elicit epidemiological evidence supporting or refuting the association between Internet gaming addiction and psychopathology. An additional objective is to explore the relationship between these conditions. Such results can furnish clinicians with updated information and provide a direction for future investigative endeavors.

2. Materials and Methods

This systematic review was conducted in accordance with the Preferred Reporting Items for Systematic Reviews and Meta-Analyses-P 2015 statement for systematic review and meta-analysis protocols [20]. The databases reviewed between October and December 2017 were PubMed, PsychINFO, ScienceDirect, Web of Science, and Google Scholar, using the following search terms and logic: "(Internet OR online) gaming addiction AND (psychopathology OR comorbidity)". Without considering the results in Google Scholar, these database search parameters yielded a total of 688 results, including the following results in each database: PubMed (54 results), PsychINFO (354 results), and ScienceDirect (280 results). Due to the large number of results provided by Google Scholar (more than 17,500 results), we reviewed only the first 30 pages of results. Additional articles were identified through searching the citations in the literature selected.

The studies were systematically and independently reviewed by the authors (Vega González-Bueso and Juan José Santamaría); paying attention to the study type, study population, methodology, outcome measures, effect sizes and interpretation of results. In cases of discrepancies, these were resolved through consensus or referral to a third reviewer (Laura Merino). The inclusion criteria were: (i) the inclusion of empirically collected data; (ii) IGD assessed by standardized questionnaires or other proposed criteria based on international disease classifications; (iii) psychiatric comorbidity assessed by standardized questionnaires; (iv) availability of the full text; (v) published after the year 2010 (this allowed us to review the most recent research in a field where the subject of addiction evolves rapidly); (vi) written in English or in Spanish (the two languages known by the authors); and (vii) article published in a peer-reviewed journal.

Studies were also included if the object of research was IA, only if it was specified that the Internet was used to play video games, and/or the results were separated according to Internet use and whether video games were one of those activities.

The exclusion criteria were: (i) articles containing only anecdotal evidence on psychopathology associated with IGD; (ii) authors not providing a specific definition or criteria for IGD; (iii) case reports and case series; (iv) studies only reporting results on phenomenons such as motivation to play video games, decision-making, stress, lifestyle, impulsivity and sexual attitude, without reporting other psychiatric comorbidity.

A review protocol exists at the PROSPERO International prospective register of systematic reviews [21] registration number CRD42018082398.

In order to facilitate the comparisons with pathological Internet use, the reviewing method applied by Carli et al. in 2013 [14] was followed: the effect sizes of the associations between IGD and psychopathology were identified by the reviewed publications or calculated using the data provided by the authors, when available. In order to compare the different associations, the effect sizes d and R^2 were stated as small, moderate, or large, according to Cohen [22]; OR were converted into these groups according to Chinn [23]. The effect sizes were interpreted accordingly: small ($d = 0.2$, $R^2 = 0.01$, OR = 1.45), moderate ($d = 0.5$, $R^2 = 0.06$, OR = 2.50), and large ($d = 0.8$, $R^2 = 0.14$, OR = 4.25). Full association was considered when a correlation was found for both genders after multivariate analyses. If a correlation was identified for only one gender, it was classified as a partial association. The geographical distribution of studies was also mapped (Figure 1).

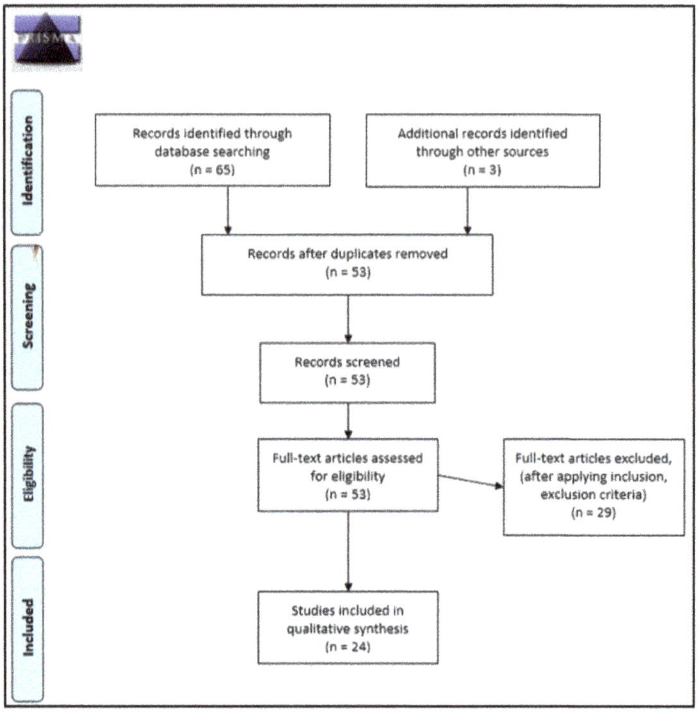

Figure 1. PRISMA 2009 protocols flow diagram.

3. Results

After deleting duplicate studies, a total of 68 articles were screened and identified through the present systematic search. After applying inclusion and exclusion criteria, a total of 24 studies were selected and included. Table 1 shows a summary of the main characteristics of the studies examining the relationship between IGD and comorbid psychopathology, including effect sizes.

Table 1. Studies examining the relationship between Internet Gaming Disorder (IGD) and comorbid psychopathology, including effect sizes.

Source	Study Type	N	Population Age [a]	Sex	Country	IGD Measures	Psychopathology Measures	Psychopathology Outcome	Association	Effect Size	95% CI of d
Baer et al., 2011 [24]	cross-sectional	102	adolescents 13.7 ± 1.9	M/F	Canada	Computer/Gaming-station Addiction Scale (CGAS)	Strengths and Difficulties Questionnaire	Emotional problems	full	$R^2 = 0.29$	–
								Hyperactivity	full	$R^2 = 0.18$	–
Cole & Hooley, 2013 [25]	cross-sectional	163	general population 27.3 ± 9.1	M/F	USA	Generalized Problematic Internet Use Scale (GPIUS)	State-Trait Anxiety Inventory (STAI)	Anxiety state	full	d = 0.26	−0.05–0.57
								Anxiety trait	full	d = 1.07	0.74–1.40
							Social Phobia Scale	Social phobia	full	d = 1.17	0.83–1.50
Jiménez-Murcia et al., 2014 [26]	cross-sectional	193	adults with GD 42.4 ± 13.4	M/F	Spain	Video-game Dependency Test (VDT)	Symptom Checklist 90-revision	Somatization	full	d = 0.57	0.16–0.983
								Obsessive-Compulsive	full	d = 0.84	0.424–1.257
								Interpersonal Sensitivity	full	d = 0.76	0.341–1.169
								Depression	full	d = 0.58	0.17–0.991
								Anxiety	full	d = 0.64	0.216–1.064
								Hostility	full	d = 0.68	0.255–1.106
								Phobic-Anxiety	full	d = 0.55	0.127–0.973
								Paranoid Ideation	full	d = 0.83	0.402–1.259
								Psychoticism	full	d = 0.56	0.137–0.983
Kim et al., 2016 [27]	cross-sectional	3041	adults 20–49	M/F	South Korea	IGD diagnostic criteria in DSM-5	Brief Symptom Inventory (BSI)	Somatization	full	d = 1.59	1.481–1.703
								Obsessive-Compulsive	full	d = 1.67	1.557–1.78
								Interpersonal Sensitivity	full	d = 1.61	1.499–1.721
								Depression	full	d = 1.75	1.642–1.867
								Anxiety	full	d = 1.75	1.642–1.866
								Hostility	full	d = 1.72	1.61–1.834
								Phobic-Anxiety	full	d = 1.82	1.705–1.928
								Paranoid Ideation	full	d = 1.74	1.623–1.847
								Psychoticism	full	d = 1.76	1.646–1.87
King et al., 2013 [28]	cross-sectional	1287	adolescents 12–18	M/F	Australia	Pathological Technology Use (PTU)	Revised Children's Anxiety and Depression Scale	Depression	none	–	

Table 1. *Cont.*

Source	Study Type	*N*	Population Age [a]	Sex	Country	IGD Measures	Psychopathology Measures	Psychopathology Outcome	Association	Effect Size	95% CI of d
								Obsessive-Compulsive Disorder (OCD)	none	–	
								Anxiety	none	–	
King & Delfabbro, 2016 [29]	cross-sectional	824	adolescents 14.1 ± 1.5	M/F	Australia	IGD Diagnostic criteria in DSM-5	Depression Anxiety Stress Scales, 21-item version	Depression	full *	d = 0.62	0.087–1.155
								Anxiety	full *	d = 0.50	−0.035–1.025
Laconi et al., 2017 [30]	cross-sectional	418	adults 21.9 ± 3	M/F	France	Internet Gaming Disorder Test-10 (IGDT-10)	Center for Epidemiologic Studies, Depression Scale-10	Depression	full	d = 2.687	1.969–3.405
Männikkö et al., 2015 [31]	cross-sectional	293	general population 18.7 ± 3.4	M/F	Finland	Gaming Addiction Scale (GAS)	School Health Promotion (SHP)	Depression	full	$R^2 = 0.17$	-
								Anxiety	full	$R^2 = 0.11$	-
Mentzoni, et al., 2011 [32]	cross-sectional	816	general population 15–40	M/F	Norway	Gaming Addiction Scale for Adolescents (GASA)	Hospital Anxiety and Depression Scale (HADS)	Depression	full	n/a	-
								Anxiety	full	n/a	-
Müller et al., 2015 [33]	cross-sectional	12,938	adolescents 15.8 ± 0.7	M/F	Germany	Assessment of Internet and Computer Game Addiction (AICGA)	Youth Self-Report	Anxious-Depression	full	d = 0.34	0.183–0.496
								Withdrawn-Depression	full	d = 0.35	0.347–0.507
Na et al., 2017 [34]	cross-sectional	1819	adults 20–49	M/F	South Korea	IGD diagnostic criteria in DSM-5	Symptom Checklist 90-revision	Depression	full	n/a	-
								Anxiety	full	n/a	-
Starcevic et al., 2011 [35]	cross-sectional	1945	general population over 14	M/F	Australia	Video-Game Use Questionnaire (VGUQ)	Symptom Checklist 90	Somatization	partial>	d = 1.02	0.854–1.187
								Obsessive-Compulsive	partial	d = 1.365	1.196–1.534
								Interpersonal Sensitivity	partial	d = 1.228	1.059–1.396
								Depression	partial	d = 1.264	1.096–1.433
								Anxiety	partial>	d = 1.149	0.981–1.317
								Hostility	partial>	d = 1.276	1.108–1.445

Table 1. *Cont.*

Source	Study Type	*N*	Population Age [a]	Sex	Country	IGD Measures	Psychopathology Measures	Psychopathology Outcome	Association	Effect Size	95% CI of d
								Phobic-Anxiety	partial>	d = 1.131	0.964–1.299
								Paranoid Ideation	partial>	d = 1.203	1.035–1.371
								Psychoticism	partial>	d = 1.368	1.199–1.537
Stetina et al., 2011 [36]	cross-sectional	468	general population 11–67	M/F	Austria	Problematic Internet use scale (ISS-20)	Questionnaire for depression diagnostics (FDD for DSM-IV)	Depression	none	-	-
Strittmatter et al., 2015 [37]	cross-sectional	9758	adolescents 15.0 ± 1.3	M/F	Germany	Young Diagnostic Questionnaire (YDQ)	Beck Depression Inventory II	Depression	full	d = 0.58	0.449–0.702
							Strengths and Difficulties Questionnaire (SDQ)	Hyperactivity	full	d = 0.53	0.399–0.652
Vadlin et al., 2016 [38]	cross-sectional	*N1* (1868) *N2* (242)	adolescents 12–18	M/F	Sweden	Gaming Addiction Identification (GAIT)	Depression Self-Rating Scale (DSRS-A)	Depression	full	OR 2.47 (1.44–4.25)	-
							Spence Children's Anxiety Scale (SCAS)	Anxiety	full	OR 2.06 (1.27–3.33)	-
							Adult ADHD Self-Report Scale (ASRS-A)	Attention Deficit Hyperactivity Disorder (ADHD)	full	OR 2.43 (1.44–4.11)	-
							Psychotic-like experiences (PLEs)	Psychoticism	none	-	-
Wang et al., 2018 [39]	cross-sectional	7200	general population 14–39	M/F	South Korea	IGD diagnostic criteria in DSM-5	Patient Health Questionnaire9 (PHQ9)	Depression	full	n/a	-
							Generalized Anxiety Disorder Scale (GAD-7)	Anxiety	n/a	-	-
Wartberg et al., 2017 [40]	cross-sectional	1095	adolescents 13.0 ± 0.82	M/F	Germany	Internet Gaming Disorder Scale (IGDS)	Reynolds Adolescent Adjustment Screening Inventory	Depression and anxiety	full	OR 1.09 (1.02–1.17)	-
								Hyperactivity	full	OR 1.27 (1.16–1.39)	-
Wei et al., 2012 [41]	cross-sectional	722	general population 21.8 ± 4.9	M/F	Taiwan	Chen's Internet Addiction Scale (CIAS)	Depression and Somatic Symptoms Scale (DSSS)	Depression	full	$R^2 = 0.298$	-

Table 1. *Cont.*

Source	Study Type	*N*	Population Age [a]	Sex	Country	IGD Measures	Psychopathology Measures	Psychopathology Outcome	Association	Effect Size	95% CI of d
							Social Phobia Inventory (SPIN)	Social phobia	full	n/a	-
Panagiotidi, 2017 [42]	cross-sectional	205	adults 27.4 ± 10	M/F	United Kingdom	Problem Video-Game Playing Test (PVGT)	ADHD Self-Report Scale (ASRS)	ADHD	full	$R^2 = 0.22$	-
Gentile et al., 2011 [43]	Longitudinal	3034	children, adolescents 11.2 ± 2.06	M/F	Singapore	Pathological Technology Use (PTU)	Asian Adolescent Depression Scale (AADS)	Depression	full	$R^2 = 0.49$	-
							Child Anxiety-Related Emotional Disorders (SCARED)	Anxiety	full	$R^2 = 0.29$	-
							Adult ADHD Self-Report Scale (ASRS-A)	ADHD	none	-	-
							Social Phobia Inventory (SPIN)	Social phobia	full	$R^2 = 0.20$	-
Van Rooij et al., 2011 [44]	Longitudinal	T1 (1572) T2 (1476)	children 13–16	M/F	Deutschland	Compulsive Internet Use Scale (CIUS)	Depressive Mood List	T1: Depression T2: Depression	none full #	n/a	-
							Revised Social Anxiety Scale for Children	T1: Social anxiety T2: Social anxiety	none none	-	-
Hyun et al., 2015 [45]	case-control	308	general population 21.0 ± 5.9	M/F	South Korea	Young Internet Addiction Scale (YIAS)	Beck Depressive Inventory (BDI)	Depression	full	d = 1.09	0.88–1.305
							Beck Anxiety Scale (BAI)	Anxiety	full	d = 0.64	0.437–0.845
							Dupaul's ADHD scale (K-ARS)	ADHD	full	d = 1.05	0.838–1.262
Yen et al., 2016 [46]	case-control	174	adults 23.29 ± 2.34 23.38 ± 2.40	M/F	Taiwan	Semi-structured interview with the DSM-5 IGD criteria	ADHD DSM-IV-TR criteria diagnosis for adult and childhood	ADHD	full	OR 13.51 (4.49–40.64)	-
Brunborg et al., 2014 [47]	cohort	1928	adolescents 13–17	M/F	Norway	Game Addiction Scale for Adolescents (GASA)	Hopkins Symptom Checklist	Depression	T1: full other time: none	$R^2 = 0.25$	- -

[a] Age is presented in years as a range or mean with standard deviation (SD). M/F = both males and females analyzed together. * Low severity symptoms. n/a Non-enough data provided to calculate the effect size or not applicable. # When non-addicted heavy gamers and addicted heavy gamers compared. > A difference was found between IGD subjects and non IGD subjects but the psychopathology scores on both groups were not clinical.

3.1. Design of the Included Studies

Nineteen of the 24 articles included were cross-sectional studies [24–42], the rest were two longitudinal studies [43,44], two case-control studies [45,46], and a cohort study [47]. The research was performed, in descending order, in South Korea (4), Australia (3), Germany (3), Norway (2), Taiwan (2), Canada (1), USA (1), Singapore (1), Spain (1), United Kingdom (1), France (1), Finland (1), Deutschland (1) Austria (1) and Sweden (1). Most of the studies were performed in European countries (12).

3.2. Characteristics of the Used Samples

The 24 studies had a total of 53,889 participants. All studies examined both genders. The number of participants in each study ranged from 102 to 12,938 (M = 2155.56; standard deviation (SD): 3176.05). Nine of the studies in this review [24,28,29,33,37,38,40,43,47] targeted adolescent groups, six studies [26,27,30,34,42,46] targeted adults, one [44] targeted children and eight studies [25,31,32,35,36,39,41,45] were carried out in the general population. A total of three studies were conducted in clinical populations, using people in outpatient treatment for IGD [45] or other mental health problems, namely Gambling Disorder [26] and other unspecified psychiatric problems [38].

3.3. Methods of Assessing Internet Gaming Disorder (IGD)

Since 2013, the DSM-5 includes a proposal of diagnostic criteria for IGD. However, only five out of 15 of the reviewed articles published after this year used these criteria [27,29,34,39,46]; three use psychometric questionnaires based on them [30,38,40] to assess the problem.

These diagnostic criteria pertain to repetitive use of Internet-based games, often with other players, that leads to significant issues with functioning. Five of the following criteria must be met within one year: "(i) Preoccupation or obsession with Internet games. (ii) Withdrawal symptoms when not playing Internet games. (iii) A build-up of tolerance (i.e., more time needs to be spent playing the games). (iv) The person has tried to stop or curb playing Internet games but has failed to do so. (v) The person has had a loss of interest in other life activities, such as hobbies. (vi) A person has had continued overuse of Internet games even with awareness of how much they impact a person's life. (vii) The person has lied to others about his or her Internet game usage. (viii) The person uses Internet games to relieve anxiety or guilt (i.e., it is a way to escape). (ix) The person has lost or put at risk opportunities or relationships because of Internet games".

The questionnaires based on these criteria were the Internet Gaming Disorder Test-10 (IGDT-10) [48]; the Gaming Addiction Identification (GAIT) [49] and the Internet Gaming Disorder Scale (IGDS) [50].

The IGDT-10 includes the nine diagnostic criteria of the DSM-5. Each criterion was operationalized using a single item, except for the last criterion referring to "jeopardy or losing a significant relationship, job, or educational or career opportunity because of participation in Internet games." This criterion was operationalized with two items, given its complexity and description of more than one construct.

The GAIT is a screening instrument used to identify addictive factors related to gaming addiction in adolescents. Primarily developed based on items from the AUDIT Alcohol Consumption Questions (AUDIT-C) [51], and the criteria for gambling disorder suggested by the DSM-5, GAIT covers seven of the nine criteria in the proposed IGD criteria. These items are: preoccupation, withdrawal, tolerance, unsuccessful attempts to control the behavior, loss of interests, harm, and loss of a significant relationship or educational opportunity due to gaming. Questions regarding lying/deception to hide the gaming, and escape/mood modification, are not included.

Finally, the IGDS measures each of the nine DSM-5 definitions with three items, either through separating core aspects of a criterion into different items or by applying changes in phrasing or synonyms. Furthermore, the proposed terms "Internet gaming" or "Internet games" were replaced with "gaming" or "games."

The remaining studies employed either measures based on the DSM-IV Gambling Disorder criteria (Pathological Technology Use (PTU), Gaming Addiction Scale (GAS)) or based on DSM-IV Addiction criteria (Gaming Addiction Scale for Adolescents (GASA), Video-game Dependency Test (VDT), Assessment of Internet and Computer Game Addiction (AICGA), Video-Game Use Questionnaire (VGUQ)), or questionnaires used to measure IA problems (Computer/Gaming-station Addiction Scale (CGAS), Generalized Problematic Internet Use Scale (GPIUS), Young Internet Addiction Scale (YIAS), Compulsive Internet Use Scale (CIUS), Problematic Internet use scale (ISS-20), Young Diagnostic Questionnaire (YDQ), Chen's Internet Addiction Scale (CIAS), and Problem Video-Game Playing Test (PVGT)).

3.4. Methods Assessing Psychopathology

Different psychometric assessments were used in the reviewed articles to measure psychopathology.

Depression was measured using various assessment tools, i.e., the Hopkins Symptom Checklist [52], the Asian Adolescent Depression Scale [53], the Beck Depressive Inventory [54], the Beck Depressive Inventory-II [55], the Center for Epidemiologic Studies-Depression Scale-10 [56], the Depressive Mood List [57], the Questionnaire for Depression Diagnostics [58], the Depression Self-Rating Scale [59], the Patient Health Questionnaire-9 [60] and the Depression and Somatic Symptoms Scale [61].

To assess anxiety, in each study, different measures were used, these are the State-Trait Anxiety Inventory [62], the Screen for Child Anxiety-Related Emotional Disorders [63], the Beck Anxiety Scale [64], the Spence Children's Anxiety Scale [65], and the Generalized Anxiety Disorder Scale-7 [66]. In addition, some authors used questionnaires evaluating both depression and anxiety, the Revised Children's Anxiety and Depression Scale [67], the School Health Promotion [68], the Hospital Anxiety and Depression Scale [69], the Youth Self-Report [70] and the Reynolds Adolescent Adjustment Screening Inventory [71].

To measure ADHD symptoms or hyperactivity, three authors [38,42,43] used the ADHD Self-Report Scale [72], two authors [24,37] used the Strengths and Difficulties Questionnaire [73], one author [45] used the Dupaul's ADHD scale [74], and one author [46] used the ADHD DSM-IV-TR criteria diagnosis for adult and childhood [75].

To assess social phobia and social anxiety, two studies [41,43] used the Social Phobia Inventory [76], one study [25] used the Social Phobia Scale [77], and one study [44] used the Revised Social Anxiety Scale for Children [78].

Several studies used questionnaires to assess multiple conditions: in three articles [26,34,35] the Symptom Checklist 90-Revision [79] was employed to assess several conditions (somatization, obsessive-compulsive disorder, interpersonal sensitivity, depression, anxiety, hostility, phobic anxiety, paranoid ideation, and psychoticism), and one study [27] used the Brief Symptoms Inventory [80] to measure the same psychopathologies. Another study [28] evaluated depression, anxiety and obsessive-compulsive disorder through the Revised Children's Anxiety and Depression Scale [67]. Finally, one article [24] assessed emotional problems and hyperactivity using the Strengths and Difficulties Questionnaire [73].

Finally, in one study [38] the association between IGD and psychoticism was explored through the Psychotic-like Experiences Test [81].

3.5. Effect Size of the Associations of Psychopathology with IGD

Regarding the associations between the analyzed mental disorders and IGD, the effect sizes reported in the reviewed papers comprised different levels of association: 35 large [24–27,30,31,41–43,45–47,82], 13 moderate [26,29,31,37,45], eight small [25,33,38,40], and seven non-association [36,38,43,44,83]. In order to summarize these results, Table 2 shows the observed associations identified between IGD and psychopathology only for the main four outcomes. The largest correlations were identified between IGD and anxiety and depression and ADHD, whereas the weakest were observed between IGD and obsessive-compulsive disorder.

Table 2. Number of observed associations identified between IGD and psychopathology stratified by effect size for the four main outcomes.

Effect Size	Depression	Anxiety	ADHD/Hyper-Activity	Social Phobia/Anxiety
Small [a]	2	2	2	0
Moderate [b]	3	5	1	0
Large [c]	8	2	4	2
None	2	1	1	1
Total	15	10	8	3

[a] d = 0.2, R^2 = 0.01, OR = 1.45. [b] d = 0.5, R^2 = 0.06, OR = 2.50. [c] d = 0.8, R^2 = 0.14, OR = 4.25.

3.6. Psychopathology, IGD and Sample Characteristics (Age, Gender)

Twenty-one studies were conducted in healthy populations; only three analyzed clinical populations (IGD or other mental health problems).

Regarding age, the analyzed studies included in the present review focused on three age groups as target populations: general population, adolescents and adults.

Eight articles examined groups of general population formed by children, adolescents and adults together, exploring the association between IGD and 1 depression and anxiety [31,32,39,45], depression [36,41], anxiety [25], social phobia [25,41], ADHD [45] and several psychiatric symptoms using the SCL-90-R [35]. One of these studies focused on a clinical sample of IGD patients [45]. All studies found a large effect size in the correlation between IGD in the general population and depression, except for one that found a non-correlation between both disorders. Large correlation effects with IGD in the general population were also found with ADHD and social phobia. Two studies analyzing anxiety found large effect sizes and two found moderate effect sizes. Large effect sizes were also found with the remaining SCL-90-R scales.

Six studies were focused on adults, analyzing the association between IGD and depression and anxiety [34], depression [30], ADHD [42,46] and several psychiatric symptoms [26,27]; here the SCL-90-R and the Brief Symptom Inventory (BSI) questionnaires were used. One of these studies focused on a clinical sample of pathological gamblers [26]. The authors identified correlations between IGD and depression and anxiety with large and moderate effect sizes, large effect sizes with ADHD, paranoid ideation and obsessive-compulsive symptoms, and finally, large and moderate effect sizes with the remaining SCL-90-R scales.

Adolescent participant groups were used in the remaining 10 studies. One of these studies [38] focused on adolescents with unspecified psychiatric problems. An association between depression and IGD in adolescents was found in seven articles and non-association in one; the effect sizes varied between large (2), moderate (2) and small (2) and no association (1). Anxiety correlated with IGD in adolescents in four of the five studies exploring this relationship; the sizes of the effects varied between large (1), moderate (1) and small (2). The association with ADHD was found in four out of five studies, with effect sizes: large (1), moderate (1) and small (2). Social phobia or social anxiety showed a large association and no association in two studies. Finally, non-association was found with obsessive-compulsive disorder (OCD) and psychoticism in the adolescent population.

With respect to gender, all studies reported higher video-game use among males. Seventeen studies [25–28,30,32–35,37–40,42,43,45,46] found higher rates of IGD among males. Two [24,29] reported no gender differences. The association between psychopathology and IGD was found for both sexes in all the articles (full association), except one [35] that only analyzed the relationship between males.

3.7. IGD and Depression

Nineteen of the 21 studies examined some form of depression as a comorbid symptom. Thirteen studies found a full association [26,27,29–31,37,38,40,41,43,45,47,82], and two [28,36] found no association. Specifically, King et al. [28] reported association with depression in PIU groups, demonstrating significantly more severe depression and anxiety symptoms than either the non-problematic user's group or the pathological video gamers group. In contrast, the pathological video gamers group scores did not differ significantly from the non-problematic users group.

Four studies were not cross-sectional, there were two longitudinal studies [43,44], one cohort study [47], and one case-control [45]. The results of these studies showed large effect size associations with depression. In the case of the longitudinal studies, Gentile et al. [43] reported elevated depressive symptoms after the pathological video-gaming problems started and these symptoms persisted and increased only if the pathological abuse persisted, while Van Rooij et al. [44], in their longitudinal study exploring two different times (years 2008 and 2009) found correlations with depressive mood only in Time 2 when comparing addicted heavy gamers with non-addicted heavy gamers. In the cohort study, the authors reported a correlation between video game addiction and depression with a large effect size only in Time 1, but they did not find any significant correlation between these two variables two years later. Among the rest of correlations detected, the effect sizes for the association with depression comprised eight large [27,30,31,41,43,45,47,82], three moderate [26,29,37], and two small [38,40] observed effects.

3.8. IGD and Anxiety

Regarding the correlation between IGD and anxiety, 11 studies found a full association, one study found a partial association, and one study found no association. The studies finding full association were: a longitudinal study [43] identifying a large effect size; a case-control study [45] identifying a moderate effect size; a cross-sectional study [25], where the authors reported a large effect size in the correlation with the anxiety trait, but a small effect size with anxiety state; and eight cross-sectional studies [26,27,29,31,32,34,38,40] identifying large effect sizes (1), moderate effect sizes (3), and small effect sizes (2). Just as in the case of depression, in the longitudinal study carried out by Gentile et al. [43], the anxiety symptoms appeared after pathological video-gaming problems. A partial association only in males was found in a study [35] and here there was a moderate effect size. Finally, no association with anxiety was found in one cross-sectional study [28].

3.9. IGD and Attention Deficit Hyperactivity Disorder (ADHD)

The relationship between IGD and ADHD and hyperactivity symptoms were analyzed in eight studies. Seven of them reported full association, with four finding large [24,42,45,46], two finding small [38,40], and one reporting moderate, effect sizes [37]. The studies comprised two case-control, five cross-sectional and one longitudinal design; the latter found no association between the two variables [43].

3.10. IGD and Social Phobia and Social Anxiety

Four studies included social phobia or social anxiety as a comorbid symptom in their studies. These studies comprised two longitudinal [43,44] and three cross-sectional designs [25,41,44]. One longitudinal and two cross-sectional studies found full association with IGD, reporting large effect sizes. Furthermore, the longitudinal study, similar to the results found regarding anxiety and depression, found that social phobia symptoms worsen after a youth becomes a pathological gamer, and improve if an individual stops this activity. In the remaining longitudinal study, no association was found between social anxiety and IGD.

3.11. IGD and Obsessive-Compulsive Symptoms

Four studies examined obsessive-compulsive symptoms as a comorbid problem. Three studies [26,27,35] found a full association with large effect sizes, and one [28] found no association.

3.12. Publication Bias

In order to detect possible publication bias, a funnel plot was conducted for depression and anxiety, as there was only a sufficient number of studies reporting results for these two pathologies (according to Grading of Recommendations, Assessment, Development and Evaluation Working Group (GRADE guides), a minimum of five to 10 studies with the same statistic reported are needed). A total of seven studies analyzing the relationship between depression and IGD, and a total of five analyzing anxiety and IGD, reported d values or data to calculate them. Figure 2 depicts the distribution of the reported or calculated correlations for depression and anxiety. The x-axis and y-axis represent the reported d values and the inverse of the sample size, respectively.

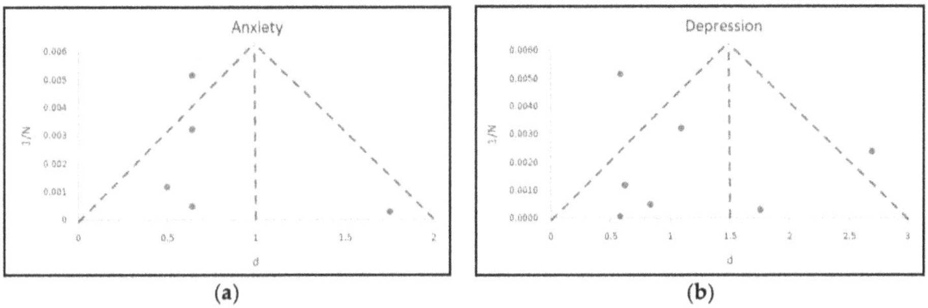

Figure 2. Funnel plots with pseudo-95% confidence limits: (**a**) anxiety, (**b**) depression.

The location of the studies shows a bias towards the left side of the funnel plot, i.e., low values of d, indicating a possible publication bias. Even so, we would like to remark that the number of studies is very small to conclude with definitive results in both psychopathologies [84,85], and thus this information must be interpreted cautiously.

4. Discussion

The main purpose of this review was to explore the state of current literature about the relationship between IGD and comorbid psychopathologies, as this knowledge is crucial to the positioning of the disorder as a behavioral addiction. A secondary aim was to analyze the effect size of these correlations and the potential effect of publication bias. In the reviewed papers on IGD and comorbid psychological pathologies, 92% of the studies describe significant correlations with anxiety, 89% with depression, 87% with ADHD or hyperactivity symptoms, and 75% with social phobia/anxiety and obsessive-compulsive symptoms. However, the potential publication bias detected in the preliminary analysis demands caution in interpretation of the results. Notwithstanding this, it should be noted that despite the inclusion of IGD in Section III of the Diagnostic and Statistical Manual DSM-5 [7] and in the beta version of the ICD-11 (International Classification of Diseases) [8], only a marginally small number of publications were centered on IGD in the literature, and several authors continue analyzing IA or PIU as a whole, without distinguishing the different possible problematic activities that users experience with this medium.

With regard to the main purpose, IGD showed strong correlations with most of the analyzed psychopathologies, in comparison with PIU, where the strongest association was found with depression [14]. The effect sizes examined indicated that the strongest associations were found with anxiety, depression, and ADHD or hyperactivity symptoms and social phobia/anxiety.

The fact as to whether the addictive behaviors (with or without substance use) may be a consequence or a trigger of psychopathology [15] cannot be unraveled yet. The lack of longitudinal studies analyzing the temporal linearity of these events in AI or PIU precludes clarifying whether a specific psychiatric problem helps to develop an AI or, alternatively, a person with a diagnosis of AI —due to negative consequences stemming from it—later developed a comorbid psychiatric disorder. A third possibility is that both problems share underlying biological, sociodemographic or psychological mechanisms that make people vulnerable to both pathologies (which manifest at the same time). In the case of this review, two longitudinal studies and one cohort study required data on whether IGD was the cause or consequence of psychopathological problems; as a result, contradictory results were obtained. On the one hand, the results of the longitudinal study performed by Gentile et al. [43] showed that the adolescents who became and stayed pathological gamers during the study period, in the last time measured, ended up with increased levels of depression, anxiety and social phobia, while those who were pathological at the start but stopped being pathological, ended up with reduced levels of depression, anxiety and social phobia. These results seem to demonstrate that gaming predicts other mental health disorders longitudinally, rather than simply being correlated with them. On the other hand, van Rooij et al. [44] found a relation between addicted heavy gamers and depression in the second year, but no correlation with social anxiety at any time. Finally, Brunborg et al. [47] only found a correlation between depression and IGD at Time 1, but not at other times.

These ambiguous results show the complex relationship between the intrinsic characteristics of online video games, the consequences of their abuse, and associated psychopathologies. The literature shows that adolescents with high scores in IGD also have negative consequences at the psychosocial level: fewer recreational activities, fewer social activities and contacts, and diminished academic performance [86,87]. These abnormalities in "real-world" social support can affect people with different personality profiles in different ways. Generally, each online video game has an associated players' community. This may lead players to find people online with similar interests and, thus, expand or replace their "real-life" social network. As these online relationships spend more and more time, "real-world" social relations will tend to deteriorate or disappear and this lack of "real-life" social support can lead some players to develop symptomatology. But in other cases, establishing this type of online relationships can help alleviate the psychological distress of some players, helping the person to establish social relationships through the Internet and build their lives around it. Some authors provide evidence that personality characteristics (e.g., extraversion, introversion) affect the choice of online or offline options for relationships [88].

Finally, age could be another key factor influencing comorbid psychopathology. In the present review, the strongest associations were found in the adult population. Results focusing on other behavioral addictions (i.e., Gambling Disorder), shows that younger adults, as opposed to older patients, only experience the symptoms of the addiction as psychological discomfort [89], without another comorbid psychopathology. One possible explanation is that older gamblers have experienced the negative consequences of the disorder for a longer period, and this has led them to develop comorbid psychopathology. It is also possible that the psychological symptoms associated with IGD require a longer time period to appear in certain subjects. Another hypothesis is that, first, children and adolescents tend to underestimate the long-term negative consequences of risky or prejudicial behaviors; and second, compared with adults, when making decisions adolescents tend to give more weight to short-term rewards compared with attendant risks [90]. Future research should analyze the differences in the perception of the negative consequences caused by IGD among adults and adolescents.

In relation to gender differences, similar to IA results all the reviewed studies reported higher video-game use among males, and most of the articles found a higher prevalence of IGD in males. Other authors have found that female respondents report less frequent play and less orientation to game genres featuring competition and three-dimensional rotation [91,92]. These characteristics in women players may be a protective factor against IGD. Regarding the amount of time spent playing, although contradictory results have been found regarding the relationship between this factor and IGD [31,42], some authors suggest that its control could be a protective factor in its appearance [93]. With respect to the type of video game chosen, it is likely that both the competitive factor and the immersive factor (in this case favored by a three-dimensional environment) of the online games, characteristics that women do not usually choose, may influence the development of IGD [94–96].

In order to clarify these points, future studies should focus on an analysis of the relationships between the personality of the affected people, the video-game preferences (e.g., massively multiplayer online role-playing game or MMORPG, multiplayer online battle arena or MOBA, first-person shooter gamers), the perception of the negative consequences generated by the problematic use, and the associated psychopathology.

The geographical distribution of the research in IGD seems to be more homogeneous than in IA; 50% of the included studies were developed in Europe and 50% were conducted in the rest of the world (29% in Asia, 30% in Australia, and 8% in North America). The prevalence of the problem and its correlation with psychopathology has been reported in all countries; therefore, it seems that it is a global problem and independent of cultural variation.

In contrast to IA, where there is a lack of common diagnostic criteria [14], in the case of IGD there are several questionnaires available based on the proposed diagnostic criteria for the disorder in the DSM-5. Despite this inclusion, the debate about the adequacy of these criteria and the emphasis upon online gaming rather than "general" gaming addiction is still active [97,98]. Therefore, although there is no gold standard questionnaire for IGD, the authors have a diagnostic base in which to frame their research. In the present review, of the 15 included articles published after the appearance of the DSM-5, only eight authors used these criteria or questionnaires based on them. The rest of the published research is based on measures for IA problems or questionnaires adapted from Gambling Disorder and general addiction criteria. This variability in evaluation methods, and basing the division of the comparison groups (IGD problems vs. no IGD problems) exclusively in the results of auto-administered data, could in part explain the variability found between IGD and comorbid psychopathology.

A consensus on the evaluation method of the problem is critical; in addition, studies focused on clinical populations with a diagnosis confirmed by professionals are needed. The data based on self-reports may not be accurate and may be limited in how they diagnose people [99]; therefore, in future research it would be helpful to complement the results of self-report questionnaires with clinical interviews (at least for the positive cases).

5. Limitations

The results of this review should be interpreted with several limitations in mind. First (as noted), some of the studies were published before the inclusion of IGD as a diagnostic category in the DSM-5. Thus, inconsistencies in clinical definitions and evaluations should be expected. Second, restrictions applied to the language of the articles, and heterogeneity in the nomenclature surrounding IGD across the different studies, suggests a potential risk that a relevant article was missed. However, articles written in other languages (with abstracts in English) were included in the review process; furthermore, a search in the citations of the selected literature was carried out. Third, reviewing only the first 30 pages of results in Google Scholar may have produced some bias; however, this method has been shown to be commonly used [100] and seems not to influence the results of the reviews. In addition, searches in other search engines and citations of included articles may have reduced that risk.

6. Conclusions

The present review included 24 studies analyzing the association between IGD and psychopathology. Compared with IA (which showed strong correlations only with depression), IGD showed strong correlations with anxiety, depression, ADHD or hyperactivity symptoms, social phobia/anxiety, and obsessive-compulsive symptoms. The lack of longitudinal studies and the contradictory results obtained makes it difficult to detect the directionality of these associations and shows the existing complexity of the relationship between IGD and psychopathology. In addition, due to a possible publication bias, the results should be interpreted with caution.

For future research, it would be helpful to investigate the relationships between personality styles, type of video-game problem, negative consequences, and associated psychopathology. It is also necessary to reach a consensus on the diagnostic criteria of IGD and on psychometric instruments used to research the subject. Studies centered in the clinical population, with diagnostic interviews that confirm the presence of the disorder, are critically needed.

Acknowledgments: This work was funded by an AIS (Atención e InvestigaciónenSocioaddiciones) intramural research program. This research has been partially supported by the Marsden grant number E2987-3648 (Royal Society of New Zealand). This partial funder had no role in the study design, data collection, and analysis, decision to publish, or preparation of the manuscript.

Author Contributions: Vega González-Bueso and Juan José Santamaría conceived and planned the review. Vega González-Bueso and Juan José Santamaría carried out the search and revision of the literature. Juan José Santamaría and Daniel Fernández analyzed the data. Vega González-Bueso and Juan José Santamaría drafted the study. All authors (Vega González-Bueso, Juan José Santamaría, Daniel Fernández, Laura Merino, Elena Montero and Joan Ribas) revised the article critically for important intellectual content. All authors (Vega González-Bueso, Juan José Santamaría, Daniel Fernández, Laura Merino, Elena Montero and Joan Ribas) commented on and approved the final manuscript and are accountable for all aspects of the work.

Conflicts of Interest: The authors declare no conflict of interest.

References

1. King, D.L.; Haagsma, M.C.; Delfabbro, P.H.; Gradisar, M.; Griffiths, M.D. Toward a consensus definition of pathological video-gaming: A systematic review of psychometric assessment tools. *Clin. Psychol. Rev.* **2013**, *33*, 331–342. [CrossRef] [PubMed]
2. Sim, T.; Gentile, D.A.; Bricolo, F.; Serpelloni, G.; Gulamoydeen, F. A Conceptual Review of Research on the Pathological Use of Computers, Video Games, and the Internet. *Int. J. Ment. Health Addict.* **2012**, *10*, 748–769. [CrossRef]
3. King, D.L.; Delfabbro, P.H. Issues for DSM-5: Video-gaming disorder? *Aust. N. Z. J. Psychiatry* **2013**, *47*, 20–22. [CrossRef] [PubMed]
4. Starcevic, V.; Aboujaoude, E. Internet addiction: Reappraisal of an increasingly inadequate concept. *CNS Spectr.* **2017**, *22*, 7–13. [CrossRef] [PubMed]
5. Thompson, T. Demographic and motivation variables associated with Internet usage activities. *Internet Res.* **2001**, *11*, 125–137. [CrossRef]
6. Weiser, E.B. Gender Differences in Internet Use Patterns and Internet Application Preferences: A Two-Sample Comparison. *CyberPsychol. Behav.* **2000**, *3*, 167–178. [CrossRef]
7. American Psychiatric Association. *Diagnostic and Statistical Manual of Mental Disorders (DSM-5)*, 5th ed.; American Psychiatric Association: Washington, DC, USA, 2013.
8. ICD-11 Beta Draft—Mortality and Morbidity Statistics. Available online: https://icd.who.int/dev11/l-m/en (accessed on 17 November 2017).
9. Saban, A.; Flisher, A.J. The Association between Psychopathology and Substance Use in Young People: A Review of the Literature. *J. Psychoact. Drugs* **2010**, *42*, 37–47. [CrossRef] [PubMed]
10. Petit, A.; Karila, L.; Chalmin, F.; Lejoyeux, M. Methamphetamine Addiction: A Review of the Literature. *J. Addict. Res. Ther.* **2012**, *1*, 2–7. [CrossRef]
11. Anthony, J.C. Epidemiology of drug dependence and illicit drug use. *Curr. Opin. Psychiatry* **1991**, *4*, 435–439. [CrossRef]
12. Kosten, T.R.; Ziedonis, D.M. Substance abuse and schizophrenia: Editors' introduction. *Schizophr. Bull.* **1997**, *23*, 181–186. [CrossRef] [PubMed]
13. Lehman, A.F.; Myers, C.P.; Corty, E. Assessment and classification of patients with psychiatric and substance abuse syndromes. *Hosp. Community Psychiatry* **1989**, *40*, 1019–1025. [CrossRef] [PubMed]

14. Carli, V.; Durkee, T.; Wasserman, D.; Hadlaczky, G.; Despalins, R.; Kramarz, E.; Wasserman, C.; Sarchiapone, M.; Hoven, C.W.; Brunner, R.; et al. The association between pathological internet use and comorbid psychopathology: A systematic review. *Psychopathology* **2013**, *46*, 1–13. [CrossRef] [PubMed]
15. Dong, G.; Lu, Q.; Zhou, H.; Zhao, X. Precursor or Sequela: Pathological Disorders in People with Internet Addiction Disorder. *PLoS ONE* **2011**, *6*, e14703. [CrossRef] [PubMed]
16. Floros, G.; Siomos, K.; Stogiannidou, A.; Giouzepas, I.; Garyfallos, G. Comorbidity of psychiatric disorders with Internet addiction in a clinical sample: The effect of personality, defense style and psychopathology. *Addict. Behav.* **2014**, *39*, 1839–1845. [CrossRef] [PubMed]
17. Young, K.S. Cognitive behavior therapy with Internet addicts: Treatment outcomes and implications. *Cyberpsychol. Behav.* **2007**, *10*, 671–679. [CrossRef] [PubMed]
18. Chang, F.-C.; Chiu, C.-H.; Lee, C.-M.; Chen, P.-H.; Miao, N.-F. Predictors of the initiation and persistence of Internet addiction among adolescents in Taiwan. *Addict. Behav.* **2014**, *39*, 1434–1440. [CrossRef] [PubMed]
19. Durkee, T.; Kaess, M.; Carli, V.; Parzer, P.; Wasserman, C.; Floderus, B.; Apter, A.; Balazs, J.; Barzilay, S.; Bobes, J.; et al. Prevalence of pathological internet use among adolescents in Europe: Demographic and social factors. *Addiction* **2012**, *107*, 2210–2222. [CrossRef] [PubMed]
20. Moher, D.; Shamseer, L.; Clarke, M.; Ghersi, D.; Liberati, A.; Petticrew, M.; Shekelle, P.; Stewart, L.A. PRISMA-P Group Preferred reporting items for systematic review and meta-analysis protocols (PRISMA-P) 2015 statement. *Syst. Rev.* **2015**, *4*, 1. [CrossRef] [PubMed]
21. Available online: https://www.crd.york.ac.uk/PROSPERO/ (accessed on 17 November 2017).
22. Cohen, J. *Statistical Power Analysis for the Behavioral Sciences*, 2nd ed.; L. Erlbaum Associates: Hillsdale, NJ, USA, 1988; ISBN 9780805802832.
23. Chinn, S. A simple method for converting an odds ratio to effect size for use in meta-analysis. *Stat. Med.* **2000**, *19*, 3127–3131. [CrossRef]
24. Baer, S.; Bogusz, E.; Green, D.A. Stuck on screens: Patterns of computer and gaming station use in youth seen in a psychiatric clinic. *J. Can. Acad. Child Adolesc. Psychiatry* **2011**, *20*, 86–94. [PubMed]
25. Cole, S.H.; Hooley, J.M. Clinical and Personality Correlates of MMO Gaming. *Soc. Sci. Comput. Rev.* **2013**, *31*, 424–436. [CrossRef]
26. Jiménez-Murcia, S.; Fernández-Aranda, F.; Granero, R.; Chóliz, M.; La Verde, M.; Aguglia, E.; Signorelli, M.S.; Sá, G.M.; Aymamí, N.; Gómez-Peña, M.; et al. Video game addiction in gambling disorder: Clinical, psychopathological, and personality correlates. *Biomed. Res. Int.* **2014**, *7*, 105–110. [CrossRef] [PubMed]
27. Kim, N.R.; Hwang, S.S.-H.; Choi, J.-S.; Kim, D.-J.; Demetrovics, Z.; Király, O.; Nagygyörgy, K.; Griffiths, M.D.; Hyun, S.Y.; Youn, H.C.; et al. Characteristics and Psychiatric Symptoms of Internet Gaming Disorder among Adults Using Self-Reported DSM-5 Criteria. *Psychiatry Investig.* **2016**, *13*, 58. [CrossRef] [PubMed]
28. King, D.L.; Delfabbro, P.H.; Zwaans, T.; Kaptsis, D. Clinical features and axis I comorbidity of Australian adolescent pathological Internet and video game users. *Aust. N. Z. J. Psychiatry* **2013**, *47*, 1058–1067. [CrossRef] [PubMed]
29. King, D.L.; Delfabbro, P.H. The Cognitive Psychopathology of Internet Gaming Disorder in Adolescence. *J. Abnorm. Child Psychol.* **2016**, *44*, 1635–1645. [CrossRef] [PubMed]
30. Laconi, S.; Pirès, S.; Chabrol, H. Internet gaming disorder, motives, game genres and psychopathology. *Comput. Hum. Behav.* **2017**, *75*, 652–659. [CrossRef]
31. Männikkö, N.; Billieux, J.; Kääriäinen, M. Problematic digital gaming behavior and its relation to the psychological, social and physical health of Finnish adolescents and young adults. *J. Behav. Addict.* **2015**, *4*, 281–288. [CrossRef] [PubMed]
32. Mentzoni, R.A.; Brunborg, G.S.; Molde, H.; Myrseth, H.; Skouverøe, K.J.M.; Hetland, J.; Pallesen, S. Problematic Video Game Use: Estimated Prevalence and Associations with Mental and Physical Health. *Cyberpsychol. Behav. Soc. Netw.* **2011**, *14*, 591–596. [CrossRef] [PubMed]
33. Müller, K.W.; Janikian, M.; Dreier, M.; Wölfling, K.; Beutel, M.E.; Tzavara, C.; Richardson, C.; Tsitsika, A. Regular gaming behavior and internet gaming disorder in European adolescents: Results from a cross-national representative survey of prevalence, predictors, and psychopathological correlates. *Eur. Child Adolesc. Psychiatry* **2015**, *24*, 565–574. [CrossRef] [PubMed]
34. Na, E.; Lee, H.; Choi, I.; Kim, D.-J. Comorbidity of Internet gaming disorder and alcohol use disorder: A focus on clinical characteristics and gaming patterns. *Am. J. Addict.* **2017**, *26*, 326–334. [CrossRef] [PubMed]

35. Starcevic, V.; Berle, D.; Porter, G.; Fenech, P. Problem Video Game Use and Dimensions of Psychopathology. *Int. J. Ment. Health Addict.* **2011**, *9*, 248–256. [CrossRef]
36. Stetina, B.U.; Kothgassner, O.D.; Lehenbauer, M.; Kryspin-Exner, I. Beyond the fascination of online-games: Probing addictive behavior and depression in the world of online-gaming. *Comput. Hum. Behav.* **2011**, *27*, 473–479. [CrossRef]
37. Strittmatter, E.; Kaess, M.; Parzer, P.; Fischer, G.; Carli, V.; Hoven, C.W.; Wasserman, C.; Sarchiapone, M.; Durkee, T.; Apter, A.; et al. Pathological Internet use among adolescents: Comparing gamers and non-gamers. *Psychiatry Res.* **2015**, *228*, 128–135. [CrossRef] [PubMed]
38. Vadlin, S.; Åslund, C.; Hellström, C.; Nilsson, K.W. Associations between problematic gaming and psychiatric symptoms among adolescents in two samples. *Addict. Behav.* **2016**, *61*, 8–15. [CrossRef] [PubMed]
39. Wang, H.R.; Cho, H.; Kim, D.-J. Prevalence and correlates of comorbid depression in a nonclinical online sample with DSM-5 internet gaming disorder. *J. Affect. Disord.* **2018**, *226*, 1–5. [CrossRef] [PubMed]
40. Wartberg, L.; Kriston, L.; Kramer, M.; Schwedler, A.; Lincoln, T.M.; Kammerl, R. Internet gaming disorder in early adolescence: Associations with parental and adolescent mental health. *Eur. Psychiatry* **2017**, *43*, 14–18. [CrossRef] [PubMed]
41. Wei, H.-T.; Chen, M.-H.; Huang, P.-C.; Bai, Y.-M. The association between online gaming, social phobia, and depression: An internet survey. *BMC Psychiatry* **2012**, *12*, 92. [CrossRef] [PubMed]
42. Panagiotidi, M. Problematic Video Game Play and ADHD Traits in an Adult Population. *Cyberpsychol. Behav. Soc. Netw.* **2017**, *20*, 292–295. [CrossRef] [PubMed]
43. Gentile, D.A.; Choo, H.; Liau, A.; Sim, T.; Li, D.; Fung, D.; Khoo, A. Pathological video game use among youths: A two-year longitudinal study. *Pediatrics* **2011**, *127*, e319–e329. [CrossRef] [PubMed]
44. Van Rooij, A.J.; Schoenmakers, T.M.; Vermulst, A.A.; Van Den Eijnden, R.J.J.M.; Van De Mheen, D. Online video game addiction: Identification of addicted adolescent gamers. *Addiction* **2011**, *106*, 205–212. [CrossRef] [PubMed]
45. Hyun, G.J.; Han, D.H.; Lee, Y.S.; Kang, K.D.; Yoo, S.K.; Chung, U.-S.; Renshaw, P.F. Risk factors associated with online game addiction: A hierarchical model. *Comput. Hum. Behav.* **2015**, *48*, 706–713. [CrossRef]
46. Yen, J.-Y.; Liu, T.-L.; Wang, P.-W.; Chen, C.-S.; Yen, C.-F.; Ko, C.-H. Association between Internet gaming disorder and adult attention deficit and hyperactivity disorder and their correlates: Impulsivity and hostility. *Addict. Behav.* **2017**, *64*, 308–313. [CrossRef] [PubMed]
47. Brunborg, G.S.; Mentzoni, R.A.; Frøyland, L.R. Is video gaming, or video game addiction, associated with depression, academic achievement, heavy episodic drinking, or conduct problems? *J. Behav. Addict.* **2014**, *3*, 27–32. [CrossRef] [PubMed]
48. Király, O.; Sleczka, P.; Pontes, H.M.; Urbán, R.; Griffiths, M.D.; Demetrovics, Z. Validation of the Ten-Item Internet Gaming Disorder Test (IGDT-10) and evaluation of the nine DSM-5 Internet Gaming Disorder criteria. *Addict. Behav.* **2017**, *64*, 253–260. [CrossRef] [PubMed]
49. Vadlin, S.; Åslund, C.; Nilsson, K.W. Development and content validity of a screening instrument for gamingaddiction in adolescents: The Gaming Addiction Identification Test (GAIT). *Scand. J. Psychol.* **2015**, *56*, 458–466. [CrossRef] [PubMed]
50. Lemmens, J.S.; Valkenburg, P.M.; Gentile, D.A. The Internet Gaming Disorder Scale. *Psychol. Assess.* **2015**, *27*, 567–582. [CrossRef] [PubMed]
51. Bush, K.; Kivlahan, D.R.; McDonell, M.B.; Fihn, S.D.; Bradley, K.A. The AUDIT alcohol consumption questions (AUDIT-C): An effective brief screening test for problem drinking. Ambulatory Care Quality Improvement Project (ACQUIP). Alcohol Use Disorders Identification Test. *Arch. Intern. Med.* **1998**, *158*, 1789–1795. [CrossRef] [PubMed]
52. Derogatis, L.R.; Lipman, R.S.; Rickels, K.; Uhlenhuth, E.H.; Covi, L. The Hopkins Symptom Checklist (HSCL): A self-report symptom inventory. *Behav. Sci.* **1974**, *19*, 1–15. [CrossRef] [PubMed]
53. Woo, B.S.C.; Chang, W.C.; Fung, D.S.S.; Koh, J.B.K.; Leong, J.S.F.; Kee, C.H.Y.; Seah, C.K.F. Development and validation of a depression scale for Asian adolescents. *J. Adolesc.* **2004**, *27*, 677–689. [CrossRef] [PubMed]
54. Beck, A.T.; Ward, C.H.; Mendelson, M.; Mock, J.; Erbaugh, J. An inventory for measuring depression. *Arch. Gen. Psychiatry* **1961**, *4*, 561–571. [CrossRef] [PubMed]
55. Beck, A.T.; Steer, R.A.; Ball, R.; Ranieri, W.F. Comparison of Beck Depression Inventories-IA and-II in Psychiatric Outpatients. *J. Pers. Assess.* **1996**, *67*, 588–597. [CrossRef] [PubMed]

56. Andresen, E.M.; Malmgren, J.A.; Carter, W.B.; Patrick, D.L. Screening for depression in well older adults: Evaluation of a short form of the CES-D (Center for Epidemiologic Studies Depression Scale). *Am. J. Prev. Med.* **1994**, *10*, 77–84. [CrossRef]
57. Kandel, D.B.; Davies, M. Epidemiology of depressive mood in adolescents: An empirical study. *Arch. Gen. Psychiatry* **1982**, *39*, 1205–1212. [CrossRef] [PubMed]
58. Kühner, C. *Fragebogen zur Depressionsdiagnostik nach DSM-IV (FDD-DSMIV)*; Hogrefe: Göttingen, Germany, 1997.
59. Svanborg, P.; Ekselius, L. Self-assessment of DSM-IV criteria for major depression in psychiatric out- and inpatients. *Nord. J. Psychiatry* **2003**, *57*, 291–296. [CrossRef] [PubMed]
60. Spitzer, R.L.; Kroenke, K.; Williams, J.B. Validation and utility of a self-report version of PRIME-MD: The PHQ primary care study. Primary Care Evaluation of Mental Disorders. Patient Health Questionnaire. *JAMA* **1999**, *282*, 1737–1744. [CrossRef] [PubMed]
61. Hung, C.-I.; Wang, S.-J.; Liu, C.-Y. Validation of the Depression and Somatic Symptoms Scale by comparison with the Short Form 36 scale among psychiatric outpatients with major depressive disorder. *Depress. Anxiety* **2009**, *26*, 583–591. [CrossRef] [PubMed]
62. Spielberger, C.; Gorsuch, R.L.; Lushene, R.E. *Manual for the State/Trait Anxiety Inventory*; Consulting Psychologists Press: Palo Alto, CA, USA, 1970.
63. Birmaher, B.; Khetarpal, S.; Brent, D.; Cully, M.; Balach, L.; Kaufman, J.; Neer, S.M. The Screen for Child Anxiety Related Emotional Disorders (SCARED): Scale construction and psychometric characteristics. *J. Am. Acad. Child Adolesc. Psychiatry* **1997**, *36*, 545–553. [CrossRef] [PubMed]
64. Beck, A.T.; Epstein, N.; Brown, G.; Steer, R.A. An inventory for measuring clinical anxiety: Psychometric properties. *J. Consult. Clin. Psychol.* **1988**, *56*, 893–897. [CrossRef] [PubMed]
65. Spence, S.H. A measure of anxiety symptoms among children. *Behav. Res. Ther.* **1998**, *36*, 545–566. [CrossRef]
66. Spitzer, R.L.; Kroenke, K.; Williams, J.B.W.; Löwe, B. A Brief Measure for Assessing Generalized Anxiety Disorder. *Arch. Intern. Med.* **2006**, *166*, 1092. [CrossRef] [PubMed]
67. Chorpita, B.F.; Yim, L.; Moffitt, C.; Umemoto, L.A.; Francis, S.E. Assessment of symptoms of DSM-IV anxiety and depression in children: A revised child anxiety and depression scale. *Behav. Res. Ther.* **2000**, *38*, 835–855. [CrossRef]
68. Kunttu, K.; Pesonen, T. *Student Health Survey 2012: A National Survey among Finnish University Students*; Finnish Student Health Service; Ylioppilaiden Terveydenhoitosäätiön Tutkimuksia 47: Helsinki, Finland, 2013.
69. Zigmond, A.S.; Snaith, R.P. The hospital anxiety and depression scale. *Acta Psychiatr. Scand.* **1983**, *67*, 361–370. [CrossRef] [PubMed]
70. Achenbach, T. *Manual for the Youth Self-Report and 1991 Profile*; Department of Psychiatry, University of Vermont: Burlington, VT, USA, 1999.
71. Reynolds, W. *Reynolds Adolescent Adjustment Screening InventoryTM (RAASITM): Professional Manual*; Psychological Assessment Resources: Lutz, FL, USA, 2001.
72. Kessler, R.C.; Adler, L.; Ames, M.; Demler, O.; Faraone, S.; Hiripi, E.; Howes, M.J.; Jin, R.; Secnik, K.; Spencer, T.; et al. The World Health Organization Adult ADHD Self-Report Scale (ASRS): A short screening scale for use in the general population. *Psychol. Med.* **2005**, *35*, 245–256. [CrossRef] [PubMed]
73. Goodman, R. The Strengths and Difficulties Questionnaire: A research note. *J. Child Psychol. Psychiatry* **1997**, *38*, 581–586. [CrossRef] [PubMed]
74. DuPaul, G.J. Parent and Teacher Ratings of ADHD Symptoms: Psychometric Properties in a Community-Based Sample. *J. Clin. Child Psychol.* **1991**, *20*, 245–253. [CrossRef]
75. American Psychiatric Association. *Diagnostic and Statistical Manual of Mental Disorders*, 4th ed.; American Psychiatric Association Text Revision: Arlington, TA, USA, 2000.
76. Connor, K.M.; Davidson, J.R.; Churchill, L.E.; Sherwood, A.; Foa, E.; Weisler, R.H. Psychometric properties of the Social Phobia Inventory (SPIN). New self-rating scale. *Br. J. Psychiatry* **2000**, *176*, 379–386. [CrossRef] [PubMed]
77. Mattick, R.P.; Clarke, J.C. Development and validation of measures of social phobia scrutiny fear and social interaction anxiety. *Behav. Res. Ther.* **1998**, *36*, 455–470. [CrossRef]
78. La Greca, A.M.; Stone, W.L. Social Anxiety Scale for Children-Revised: Factor Structure and Concurrent Validity. *J. Clin. Child Psychol.* **1993**, *22*, 17–27. [CrossRef]

79. Derogatis, L.R. *SCL-90-R. Administration, Scoring and Procedures Manual*; Clinical Psychometric Research Inc.: Baltimore, MD, USA, 1990.
80. Derogatis, L.R.; Melisaratos, N. The Brief Symptom Inventory: An introductory report. *Psychol. Med.* **1983**, *13*, 595–605. [CrossRef] [PubMed]
81. Laurens, K.; Hodgins, S.; MaughanN, B.; Murray, R.; Rutter, M.; Taylor, E. Community screening for psychotic-like experiences and other putative antecedents of schizophrenia in children aged 9–12 years. *Schizophr. Res.* **2007**, *90*, 130–146. [CrossRef] [PubMed]
82. STARCEVIC, V. Problematic Internet use: A distinct disorder, a manifestation of an underlying psychopathology, or a troublesome behaviour? *World Psychiatry* **2010**, *9*, 92–93. [CrossRef] [PubMed]
83. King, D.L.; Delfabbro, P.H.; Griffiths, M.D. Trajectories of Problem Video Gaming Among Adult Regular Gamers: An 18-Month Longitudinal Study. *Cyberpsychol. Behav. Soc. Netw.* **2013**, *16*, 72–76. [CrossRef] [PubMed]
84. Monroe, J. *Meta-Analysis for Observational Studies: Statistical Methods for Heterogeneity, Publication Bias and Combining Studies: Statistics*; University of California: Los Angeles, CA, USA, 2007.
85. Sutton, A.J. *Methods for Meta-Analysis in Medical Research*; J. Wiley: Chichester, UK, 2000; ISBN 9780471490661.
86. Beutel, M.E.; Hoch, C.; Wölfling, K.; Müller, K.W. Clinical characteristics of computer game and internet addiction in persons seeking treatment in an outpatient clinic for computer game addiction. *Z. Psychosom. Med. Psychother.* **2011**, *57*, 77–90. [CrossRef] [PubMed]
87. Batthyány, D.; Müller, K.W.; Benker, F.; Wölfling, K. Computer game playing: Clinical characteristics of dependence and abuse among adolescents. *Wien. Klin. Wochenschr.* **2009**, *121*, 502–509. [CrossRef] [PubMed]
88. Goby, V.P. Personality and Online/Offline Choices: MBTI Profiles and Favored Communication Modes in a Singapore Study. *Cyberpsychol. Behav.* **2006**, *9*, 5–13. [CrossRef] [PubMed]
89. González-Ibáñez, A.; Mora, M.; Gutiérrez-Maldonado, J.; Ariza, A.; Lourido-Ferreira, M.R. Pathological gambling and age: Differences in personality, psychopathology, and response to treatment variables. *Addict. Behav.* **2005**, *30*, 383–388. [CrossRef] [PubMed]
90. Halpern-Felsher, B.L.; Cauffman, E. Costs and benefits of a decision: Decision-making competence in adolescents and adults. *J. Appl. Dev. Psychol.* **2001**, *22*, 257–273. [CrossRef]
91. Lucas, K.; Sherry, J.L. Sex Differences in Video Game Play. *Commun. Res.* **2004**, *31*, 499–523. [CrossRef]
92. Greenberg, B.S.; Sherry, J.; Lachlan, K.; Lucas, K.; Holmstrom, A. Orientations to Video Games Among Gender and Age Groups. *Simul. Gaming* **2010**, *41*, 238–259. [CrossRef]
93. Young, K. Understanding Online Gaming Addiction and Treatment Issues for Adolescents. *Am. J. Fam. Ther.* **2009**, *37*, 355–372. [CrossRef]
94. Floros, G.; Siomos, K. Patterns of Choices on Video Game Genres and Internet Addiction. *Cyberpsychol. Behav. Soc. Netw.* **2012**, *15*, 417–424. [CrossRef] [PubMed]
95. Elliott, L.; Golub, A.; Ream, G.; Dunlap, E. Video Game Genre as a Predictor of Problem Use. *Cyberpsychol. Behav. Soc. Netw.* **2012**, *15*, 155–161. [CrossRef] [PubMed]
96. King, D.; Delfabbro, P.; Griffiths, M. Video Game Structural Characteristics: A New Psychological Taxonomy. *Int. J. Ment. Health Addict.* **2010**, *8*, 90–106. [CrossRef]
97. Petry, N.M.; Rehbein, F.; Gentile, D.A.; Lemmens, J.S.; Rumpf, H.-J.; Mößle, T.; Bischof, G.; Tao, R.; Fung, D.S.S.; Borges, G.; et al. An international consensus for assessing internet gaming disorder using the new DSM-5 approach. *Addiction* **2014**, *109*, 1399–1406. [CrossRef] [PubMed]
98. Kuss, D.J.; Griffiths, M.D.; Pontes, H.M. Chaos and confusion in DSM-5 diagnosis of Internet Gaming Disorder: Issues, concerns, and recommendations for clarity in the field. *J. Behav. Addict.* **2017**, *6*, 103–109. [CrossRef] [PubMed]
99. Bhandari, A.; Wagner, T. Self-Reported Utilization of Health Care Services: Improving Measurement and Accuracy. *Med. Care Res. Rev.* **2006**, *63*, 217–235. [CrossRef] [PubMed]
100. Kaptsis, D.; King, D.; Delfabbro, P.; Gradisar, M. Withdrawal symptoms in internet gaming disorder: A systematic review. *Clin. Psychol. Rev.* **2016**, *43*, 58–66. [CrossRef] [PubMed]

International Journal of
Environmental Research and Public Health

MDPI

Article

A Phenotype Classification of Internet Use Disorder in a Large-Scale High-School Study

Katajun Lindenberg [1,*], Katharina Halasy [1], Carolin Szász-Janocha [1] and Lutz Wartberg [2]

1 Institute for Psychology, University of Education Heidelberg, 69120 Heidelberg, Germany; halasy@ph-heidelberg.de (K.H.); szaszc@ph-heidelberg.de (C.S.-J.)
2 German Center for Addiction Research in Childhood and Adolescence, University Medical Center Hamburg-Eppendorf, 20246 Hamburg, Germany; lwartberg@uke.de
* Correspondence: lindenberg@ph-heidelberg.de; Tel.: +49-6221-477159

Received: 28 February 2018; Accepted: 10 April 2018; Published: 12 April 2018

Abstract: Internet Use Disorder (IUD) affects numerous adolescents worldwide, and (Internet) Gaming Disorder, a specific subtype of IUD, has recently been included in DSM-5 and ICD-11. Epidemiological studies have identified prevalence rates up to 5.7% among adolescents in Germany. However, little is known about the risk development during adolescence and its association to education. The aim of this study was to: (a) identify a clinically relevant latent profile in a large-scale high-school sample; (b) estimate prevalence rates of IUD for distinct age groups and (c) investigate associations to gender and education. N = 5387 adolescents out of 41 schools in Germany aged 11–21 were assessed using the Compulsive Internet Use Scale (CIUS). Latent profile analyses showed five profile groups with differences in CIUS response pattern, age and school type. IUD was found in 6.1% and high-risk Internet use in 13.9% of the total sample. Two peaks were found in prevalence rates indicating the highest risk of IUD in age groups 15–16 and 19–21. Prevalence did not differ significantly between boys and girls. High-level education schools showed the lowest (4.9%) and vocational secondary schools the highest prevalence rate (7.8%). The differences between school types could not be explained by academic level.

Keywords: Internet Use Disorder; prevalence; epidemiology; adolescence; latent profile analysis

1. Introduction

Internet Use Disorder (IUD) describes mental disorders due the problematic use of the Internet. Internet Gaming Disorder (IGD), a specific subtype of IUD characterized by the persistent and recurrent use of the Internet to engage in games, has been included in the DSM-5 section "conditions for further studies" [1] because of its significant health importance. The World Health Organization stated recently that Gaming disorder (GD) will be listed as independent diagnosis in the upcoming ICD-11 and defined GD as "a pattern of recurrent video-gaming" (online or offline), manifested by impaired control over gaming, increasing priority of gaming over other life interests and daily activities and continuation of gaming despite the occurrence of negative consequences, which results in significant impairment in personal, family, social, educational, occupational or other important areas of functioning and should be evident over a period of at least 12 months [2].

Non-gaming subtypes, such as problematic social media use, were not included in DSM-5 due to lack of empirical research. In ICD-11, however, other (non-gaming) Internet-related disorders can be classified as other specific disorders due to addictive behaviors. It is discussed that gaming and non-gaming IUD should be defined using the same criteria. Facing the multiple severe comorbidities and impairments which are associated with IUD, the early diagnosis is of particular interest [3–5]. Moreover, it is essential not only to identify but also to differentiate between adolescents with IUD and high-risk Internet use in order to tailor preventive interventions and treatments for those individuals most in need.

The DSM-5 group has noted that the majority of studies lack of a standard definition from which to derive prevalence data and called for further research [1]. The use of various conceptualizations, assessment tools and cut-off scores for estimating the prevalence of IUD led to widely heterogeneous findings of prevalence rates. International prevalence estimates for IUD in adolescents vary between 0.8% in Italy [6] and 26.7% in China [7]. A multinational meta-analysis of 31 nations revealed a global prevalence of 6.0% [8]. The estimate was based on the Young Diagnostic Questionnaire (YDQ) [9] and the Internet Addiction Test (IAT) [10]. The highest prevalence rates were reported in the Middle East and the lowest prevalence rates in Northern and Western Europe. Recent studies investigating IGD in Germany reported prevalence rates of 1.2% in adolescents aged 11–18 years [11] to 5.7% in adolescents and young adults aged 12–25 years [12].

In Europe, several studies have used the Compulsive Internet Use Scale (CIUS) to estimate the prevalence of IUD [13–15], which resulted in more homogeneous findings. The CIUS is a widely used questionnaire for assessing IUD [16,17] and for the German version of the instrument good psychometric properties were reported [18]. However, a sound cut-off score of the CIUS has not yet been identified to classify IUD and high-risk Internet users (HR-IU). The authors have suggested a cut-off score of 28 to identify IUD [17]. A study comparing the CIUS with the IAT proposed a cut-off value of 18 for case finding and 21 for prevalence estimates [19]. Furthermore, empirical data of a representative German sample suggested a cut-off score of 30 for prevalence estimates and a cut-off score of 24 for the detection of HR-IU [13]. As a methodological superior alternative to cut-off scores, previous research [14,15,20,21] has suggested to empirically identify distinct groups based on latent profiles to estimate the prevalence of clinically relevant IUD. Latent profile analysis (LPA) is a powerful method to classify phenotypes based on latent mixture modeling. It allows identification of underlying (or latent) groups of individuals on the basis of their response pattern across a set of indicators for IUD and thus, identifying the prevalence of the group characterized by clinically relevant IUD symptoms.

Applying LPA to CIUS data, Rumpf et al. [14] found a prevalence rate of 4.0% in German adolescents aged 14–16 years in a general population sample. In the total sample (14 to 64 years) the prevalence rate was 1.0%. In line with these results, a similar rate of 4.7% was classified as IUD based on latent profiles indicated by the CIUS [22]. In this large school survey, the latent profile classifying individuals with IUD showed a mean sum score of 30. Another empirically driven LPA using the CIUS revealed a prevalence of 3.2% in adolescents aged 14 to 17 years and a CIUS mean score of 31.93 in the profile group of adolescents with IUD [15]. These prevalence rates were confirmed by another representative study investigating prevalence based on external ratings by parents [20].

Findings about sociodemographic variables associated with IUD are inconsistent. Some studies assessing the relationship between gender and IUD reported higher prevalence rates in males [6,23–27]. Despite these, other studies found no difference between males and females e.g., [28] or higher prevalence rates in females e.g., [22]. Moreover, empirical research shows a higher association between males and gaming related IUD and females and non-gaming IUD including the use of social networking and other applications of the Internet [13,22,29]. Rehbein and Mößle [22] investigated gaming related IUD and non-gaming IUD separately and emphasized the need for differentiation in a number of ways. They reported a female preponderance of non-gaming IUD and a male preponderance of gaming related IUD. Furthermore, in students of grades 7 to 10, higher risk of non-gaming IUD was found in older students and higher risk of gaming related IUD in younger students. Although prevalence rates indicate that compared to adults adolescents seem to be particularly affected by IUD e.g., [14], currently it remains unclear how the risk for IUD develops (e.g., decreases or increases) over the course of adolescence. According to Cerniglia et al. [30] concerning this matter the specific neuro-developmental plasticity in adolescence is an important aspect.

Bakken et al. [31] conducted a prevalence study in a large representative Norwegian sample aged 16 to 74 years. They found highest prevalence rates among males aged 16–29 years. Furthermore, achieved high education level (university level vs. senior high school and junior high school) was positively associated with Internet addiction. In line with these results, other studies [27,32,33] showed a positive relationship between the current academic level (lower vs. higher degree of current education) and IUD. However, parental education level has shown to be negatively correlated with IUD [34].

Studies assessing the effects of age on IUD show inconsistent findings. A recent longitudinal study [26] revealed a decrease of IUD symptoms between 16 and 18 years. Another study [24] conducted in Croatia, Finland and Poland surveyed an adolescent sample of n = 1078 aged 11 to 18 years and reported the lowest risk of IUD among the 11–12 year old participants and the highest risk among the 15–16 year old participants. It was discussed that this peak was driven by a greater level of independence and less parental control over free time and social activities.

In sum, research assessing IUD, i.e., non-gaming IUD and (I)GD, show inconsistent findings with regard to epidemiology and sociodemographic risk factors. Therefore, it is a public health challenge to understand risk factors and upholding conditions which are associated with IUD in order to develop effective prevention programs and bring them to broader dissemination. Although research supported first evidence for an increase of risk in middle adolescence with a peak at the ages of 15 and 16, little is known about the development in late adolescence until the age of 21.

The purpose of this study was to: (a) empirically derive a clinically relevant latent profile characterized by an IUD phenotype in a large epidemiological sample of children and adolescents using LPA; (b) to estimate prevalence of IUD and HR-IU in adolescence; (c) to estimate the differential risk of IUD for distinct age groups from 11 to 21 years separately; (d) to identify effects of gender on risk of IUD and (e) to assess the associations between academic career and risk of IUD. To the best of our knowledge, this is the first large-scale epidemiological study investigating the prevalence rate for distinct age groups from early to late adolescence. We explored the following research questions: (1) How many different profile groups of adolescent Internet users can be identified in our sample by a LPA? (2) Is the number of profile groups comparable with the findings in previous studies? (3) What is the prevalence estimate of Internet Use Disorder (IUD) and high-risk Internet use (HR-IU) in our sample? (4) Are there gender differences in the prevalence of IUD and HR-IU? (5) Are there differences in the prevalence rates of IUD or HR-IU between different age groups? (6) Are there differences in the prevalence rates of IUD or HR-IU between different school types?

2. Materials and Methods

2.1. Data Collection

Data were collected within the PROTECT study (ClinicalTrials.gov: NCT02907658). Ethical approval was obtained from the University of Education Heidelberg Research Ethics Committee on 3 September 2015 (Az.: 7741.35-13). Approval from the Regierungspräsidium Karlsruhe was obtained on 19 October 2015 (Az.: 71c2-6499.25) for school-wide screenings and thus, individual written consent was not necessary. The study was supported by the Dietmar Hopp foundation, thus, in line with the funding priority we focused on schools in the Rhine-Neckar region and adjoining areas. Data collection was conducted in schools during regular school hours between September 2015 and February 2017. Target groups were students aged 11–21 from 41 secondary schools (convenience sample of schools). All types of schools within the German school system were included, i.e., low education level (Werkrealschule), middle education level (Realschule), high education level (Gymnasium), comprehensive schools (Gemeinschaftsschule), vocational school (Berufsschule) and vocational upper secondary school (Berufliches Gymnasium). Data collection was conducted by a team of trained psychologists. Teachers were present in the classroom during completion of the survey, but were not involved in the data collection. Total time for introduction and data collection in each class took approximately 20 min.

2.2. Measures

We used the German version of the Compulsive Internet Use Scale [18] for investigating the severity of IUD. The questionnaire consists of 14 items with a 5-level Likert-scale ranging from 0 = never to 4 = very often. Hence 0 to 56 points can be obtained in the total score. Internal consistency lies between Cronbach's $\alpha = 0.89$–0.90 [17]. The questionnaire has been translated to various languages and examined empirically. The German version was psychometrically validated in a representative sample of adolescents [18]. The majority of research supports a one dimensional structure of the scale [18,35,36]. We applied the same model as Meerkerk et al. [17], correlating the error variances of items 1 and 2, items 6 and 7, items 8 and 9, items 10 and 11 as well as items 12 and 13, and could confirm the one-dimensional structure of the instrument in our sample (RMSEA = 0.047, CFI = 0.962, TLI = 0.952) with standardized factor loadings between 0.33 (Item 8) and 0.64 (Item 14). Cronbach's α was 0.87 in our sample.

According to the authors, the CIUS assesses the symptoms loss of control, withdrawal symptoms, coping with unpleasant mood, mental and behavioral preoccupation as well as inter- and intrapersonal conflicts CIUS (see Table 1; [17,35,36]).

Table 1. Symptoms assessed by CIUS items.

	Symptoms Assessed by CIUS Items
1	Loss of control (LOC)
	1. How often do you find it difficult to stop using the Internet when you are online? 2. How often do you continue to use the Internet despite your intention to stop? 5. How often are you short of sleep because of the Internet? 9. How often have you unsuccessfully tried to spend less time on the Internet?
2	Withdrawal symptoms (WS)
	14. How often do you feel restless, frustrated, or irritated when you cannot use the Internet?
3	Coping with unpleasant mood (C)
	12. How often do you go on the Internet when you are feeling down? 13. How often do you use the Internet to escape from your sorrows or get relief from negative feelings?
4	Mental and behavioral preoccupation (MBP)
	4. How often do you prefer to use the Internet instead of spending time with others (e.g., partner, children, parents, friends *)? 6. How often do you think about the Internet, even when not online? 7. How often do you look forward to your next Internet session?
5	Inter- and intrapersonal conflicts (IIC)
	3. How often do others (e.g., partner, children, parents, friends *) say you should use the Internet less? 8. How often do you think you should use the Internet less often? 10. How often do you rush through your (home) work in order to go on the Internet? 11. How often do you neglect your daily obligations (work, school, or family life) because you prefer to go on the Internet?

Note: * "friends" was not mentioned in the original version by Meerkerk et al. [17] but was included in the German translation by Gürtler et al. [35] and Peukert et al. [36].

The five symptom scales were calculated by the means of respective variables (range 0–4). In addition to the CIUS, we recorded sociodemographic data consisting of age, gender, grade and school type.

2.3. Sample

Overall, 5549 students from 41 schools participated in the study (range 11–52 years). The high range in age was due to the inclusion of vocational schools. We excluded all students older than 21 years ($n = 162$), because our focus was to study IUD in adolescence. Remaining sample size was 5387 with a mean age of 14.72 ($SD = 1.96$). Gender ratio was 51.4% male, 47.3% female and 1.3% without information on their gender ($n = 68$). The age distribution was 613 11- to 12-year olds (11.4% of the sample), 2185 13- to 14-year olds (40.6%), 1526 15- to 16-year olds (28.3%), 865 17- to 18-year olds (16.1%), 198 19- to 21-year olds (3.7%). Overall, 11.6% of the pupils attended the "Werkrealschule" (low educational level), 11.4% "Realschule" (middle educational level), 40.4% "Gymnasium" (high educational level), 5.8% "Gesamtschule" (comprehensive school), 17.6% "Berufsschule" (vocational school) and 13.1% attended "berufliches Gymnasium" (vocational upper secondary school).

2.4. Statistical Analyses

All statistical analyses were conducted with SPSS Statistics for Windows version 24.0 (IBM, Armonk, NY, USA) and MPlus version 8 (Muthén & Muthén, Los Angeles, CA, USA). Because our aim was to empirically classify phenotypes based on the underlying latent profile, we chose LPA as statistical method. LPA is a person-centered approach which allows " ... to classify individuals into distinct groups or categories based on individual response patterns so that individuals within a group are more similar than individuals between groups" [37] (p. 309). We tested six models including 1–6 latent profiles and used the Akaike information criterion (AIC), Bayesian information criterion, BIC, the entropy value as well as the Lo-Mendell-Rubin likelihood ratio test [38] as fit indices to compare the models. The BIC was given priority among the criteria as it has been found to be superior [39]. In line with prior studies [15], we allowed within-class correlations in the LPA between the CIUS items because of the given one-dimensionality [17], which we confirmed in our data. However, within-class correlations were restricted to be equal across classes. This procedure was equal to the study conducted by Wartberg et al. [15]. For comparison of the different profile groups, we used the Kruskal Wallis test due to disparity of variances and chi-square test for comparison of frequencies.

3. Results

3.1. Phenotype Classification Using LPA

LPA models varying the number of latent profiles from 1–6 were evaluated. BIC was lowest for a 5-profile solution, thus, the 5-profile solution was found to be the best-fitting model (see Table 2). The Lo-Mendell-Rubin likelihood ratio test for 4 (H0) versus 5 indicated that five profiles resulted in a significantly better model fit that four profiles ($p < 0.001$), whereas the Lo-Mendell-Rubin likelihood ratio test for five (H0) versus six profiles was no longer significant ($p = 0.51$), i.e., the inclusion of a sixth profile did not result in a better model fit. Of the total sample, $n = 134$ (2.5%) were members of latent profile group (LPG) 1, $n = 2483$ (46.1%) were members of LPG2, $n = 1695$ (31.5%) were members of LPG3, $n = 748$ (13.9%) were members of LPG4 and $n = 327$ (6.1%) were members of LPG5.

Table 2. Fit indices of the LPA.

Profile No.	AIC	BIC	Entropy	p (LMR-LRT)	Persons Per Profile Group, n	Relative Frequency
1	203,864.875	204,649.292			5387	
2	202,484.369	203,367.662	0.869	0.0000	4112 1275	0.76 0.24
3	201,610.314	202,592.484	0.948	0.0000	2510 2373 504	0.47 0.44 0.09
4	192,946.834	194,027.880	1.000	0.0000	2617 748 1695 327	0.49 0.14 0.31 0.06
5	192,753.95	193,933.868	0.975	0.0000	134 2483 1695 748 327	0.02 0.46 0.31 0.14 0.06
6	192,716.38	193,995.181	0.951	0.5069	2401 132 748 83 1696 327	0.45 0.02 0.14 0.02 0.31 0.06

Note: AIC = Akaike information criterion; BIC = Bayesian information criterion; LMR-LRT = Lo-Mendell-Rubin likelihood ratio test.

3.2. Comparison of Latent Profile Groups

A comparison of the CIUS total scores showed significant differences ($H = 1968.65$, $df = 4$, $p = 0.000$) between the 5 LPGs. Post hoc tests, with Bonferroni correction for multiple tests, showed significant differences between each of the five groups. The largest profile (LPG2) was characterized by a CIUS total score of $M = 12.07$ ($SD = 6.72$), loss of control $M = 0.96$ ($SD = 0.61$), withdrawal symptoms $M = 0.57$ ($SD = 8.84$), coping with unpleasant mood $M = 1.18$ ($SD = 1.02$), mental and behavioral preoccupation $M = 0.81$ ($SD = 0.68$), inter- and intrapersonal conflicts $M = 0.73$ ($SD = 0.53$). The differences in symptom severity across profile groups can be found in Table 3.

LPG5 was characterized by the highest CIUS total scores ($M = 32.16$, $SD = 9.01$). The values of all symptom scales were significantly higher in LPG5 as compared to all other profiles (loss of control $M = 2.28$, $SD = 0.78$; withdrawal symptoms $M = 2.06$, $SD = 1.33$; coping with unpleasant mood $M = 2.57$, $SD = 1.26$; mental and behavioral preoccupation $M = 2.11$, $SD = 0.97$; inter- and intrapersonal conflicts $M = 2.40$, $SD = 0.67$).

The LGP with the second highest burden was LPG4 (CIUS total score $M = 24.64$, $SD = 7.55$; loss of control $M = 1.84$, $SD = 0.70$; withdrawal symptoms $M = 1.39$, $SD = 1.14$; coping with unpleasant mood $M = 2.08$, $SD = 1.16$; mental and behavioral preoccupation $M = 1.55$, $SD = 0.80$; inter- and intrapersonal conflicts $M = 1.79$, $SD = 0.58$).

3.3. Prevalence Estimates of IUD and HR-IU

Based on the profile characteristics, LPG5 was found to be the group that comprises individuals with IUD. It consisted of 327 adolescents. Thus, the prevalence of IUD was found to be 6.1% in the total sample. LPG4 was defined as HR-IU group. It included 748 adolescents and accounted for 13.9% of the total sample. Clinically relevant IUD symptoms were absent in LPG1, LPG2 and LPG3, thus, these groups are defined as unproblematic Internet users.

Table 3. Symptom severity across profile groups: mean CIUS symptom scales and total score.

Parameter	TS	PG1	PG2	PG3	PG4 HR-IU	PG5 IUD	*H*	*df*	*p*
N	5387	134	2483	1695	748	327			
Mean age	14.72 (1.96)	14.31 (1.95)	14.48 (1.95)	14.89 (1.93)	15.06 (1.98)	15.14 (1.91)	108.80	4	0.000
CIUS total score	17.24 (9.27)	20.60 (6.87)	12.07 (6.72)	18.40 (7.35)	24.64 (7.55)	32.16 (9.01)	1968.65	4	0.000
Loss of control (LOC)	1.34 (0.78)	1.99 (0.72)	0.96 (0.61)	1.43 (0.69)	1.84 (0.70)	2.28 (0.78)	1410.05	4	0.000
Withdrawal symptoms (WS)	0.90 (1.07)	0.93 (1.09)	0.57 (0.84)	0.96 (1.03)	1.39 (1.14)	2.06 (1.33)	698.56	4	0.000
Coping with unpleasant mood (C)	1.53 (1.15)	1.74 (1.09)	1.18 (1.02)	1.60 (1.08)	2.08 (1.16)	2.57 (1.26)	615.27	4	0.000
Mental and behavioral preoccupation (MBP)	1.11 (0.81)	1.08 (0.71)	0.81 (0.68)	1.17 (0.72)	1.55 (0.80)	2.11 (0.97)	937.37	4	0.000
Inter- and intrapersonal conflicts (IIC)	1.16 (0.74)	1.28 (0.59)	0.73 (0.53)	1.27 (0.56)	1.79 (0.58)	2.40 (0.67)	2127.43	4	0.000

Notes: Profiles were compared using Kruskal-Wallis *H*-test. Values are presented as means (with *SD* in parentheses). TS = total sample, LPG = latent profile group, HR-IU = high risk Internet use, IUD = Internet Use Disorder, *H* = Kruskal Wallis test statistic.

3.4. Differential Risk for Distinct Age Groups

We found significant difference in the distribution of age across the five profile groups ($H = 108.80$, $df = 4$, $p = 0.000$). Post hoc tests showed significant differences between profile group 1 compared to profile groups 3, 4, 5 ($p = 0.002$, $p = 0.000$, $p = 0.000$), profile group 2 and profile groups 3, 4, 5 ($p = 0.000$ respectively). Relative frequencies of LPG 5 (IUD) increased with rising age, with prevalence peaks at the age groups of 15- to 16-year-olds and 19- to 21-year-olds (see Figure 1). The prevalence of IUD was 2.8% for the 11- to 12-year-olds (17 of 613), 5.5% for the 13- to 14-year-olds (121 of 2185), 7.6% for the 15- to 16-year-olds (116 of 1526), 6.4% for the 17- to 18-year-olds (55 of 865) and 9.1% for the 19- to 21-year-olds (18 of 198) adolescents. The same pattern was found in LGP4 (HR-IU). Relative frequencies increased with rising age, with prevalence peaks at the age groups of 15- to 16-year-olds and 19- to 21-year-olds. The prevalence of HR-IU was 10.0% for the 11- to 12-year-olds (61 of 613),

12.4% for the 13- to 14-year-olds (272 of 2185), 16.1% for the 15- to 16-year-olds (246 of 1526), 14.6% for the 17- to 18-year-olds (126 of 865) and 21.7% for the 19- to 21-year-olds (43 of 198) adolescents.

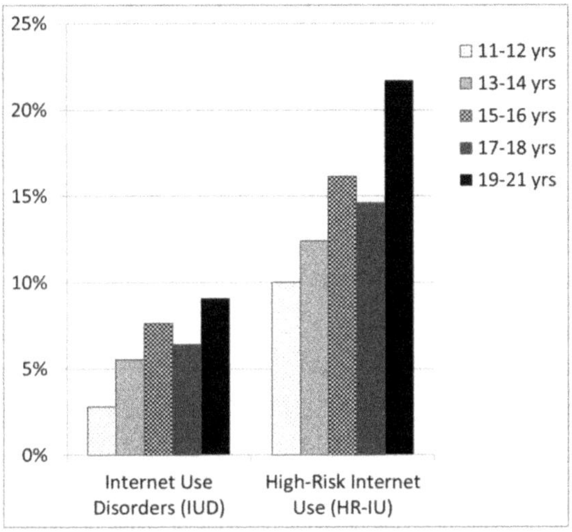

Figure 1. Changes in prevalence rates from early to late adolescence in IUD: two peaks at 15–16 and 19–21 years.

3.5. Gender Effects

We did not find significant gender effects in profile group membership (LPG1 2.2% males vs. 2.8% females; LPG2 45.8% males vs. 46.5% females; LPG3 32.8% males vs. 30.2% females; LPG4 13.4% males vs. 14.4% females; LPG5 5.9% males vs. 6.2% females; χ^2 (4) = 6.27, p = 0.18).

3.6. Associations between Education and IUD

Because age groups were not equally distributed across school types in the German school system, we analyzed associations between education and IUD or HR-IU for different age groups. As presented in Table 4, prevalence rates of IUD and HR-IU differed between school types. Within the same school type, risk increased with age.

Table 4. Associations of school type (separated by age group) and prevalence of IUD/HR-IU.

		11–12 Years	13–14 Years	15–16 Years	17–18 Years	19–21 Years
Low Educational Level	IUD	1.2%	7.9%	7.9%	-	-
	HR-IU	19.5%	14.2%	15.8%	-	-
Middle Educational Level	IUD	5.7%	6.7%	11.4%	-	-
	HR-IU	13.8%	13.1%	18.6%	-	-
High Educational Level	IUD	2.4%	4.8%	6.5%	-	-
	HR-IU	7.1%	12.6%	18.5%	-	-
Comprehensive School	IUD	-	4.8%	-	-	-
	HR-IU	-	9.6%	-	-	-
Vocational School	IUD	-	6.2%	6.3%	5.8%	7.4%
	HR-IU	-	4.6%	11.9%	9.8%	18.9%
Vocational Upper Secondary School	IUD	-	-	9.4%	7.0%	14.3%
	HR-IU	-	-	15.1%	18.0%	30.6%

Notes: Cell counts below n = 40 within age group per school type are not reported.

4. Discussion

The primary objective of this study was to use a phenotype classification approach to estimate prevalence rates of clinically relevant IUD in a large-scale high-school sample including both gaming and non-gaming subtypes. Moreover, we aimed at identifying the differences in phenotypes between problematic and unproblematic Internet users. The LPA was based on the response pattern of the CIUS, indicating the severity of the symptoms (1) loss of control; (2) withdrawal symptoms; (3) coping with unpleasant mood; (4) mental and behavioral preoccupation; and (5) inter- and intrapersonal conflicts. These symptoms cover most of the diagnostics criteria as proposed by DSM-5 and ICD-11 for (I)GD.

Specifically, we identified distinct profile groups characterized by IUD and HR-IU as compared to unproblematic users. The prevalence estimate was 6.1% for IUD in adolescents aged 11–21. The impairment of individuals classified to the IUD group (CIUS total $M = 32.16$) was comparable to the clinically relevant groups identified by Rumpf et al. ($M > 28.00$) [21], Rumpf et al. ($M = 34.24$) [14] and Wartberg et al. ($M = 31.93$) [15]. Thus, the definition of illness underlying the prevalence estimates is comparable across the previous and our studies. As a second clinically relevant group, we identified 13.9% of all adolescents characterized by high-risk behavior (LPG4: HR-IU). The impairment of this sub-threshold pathology group (CIUS total $M = 24.64$) was also comparable to the sub-clinical group identified by Rumpf et al. ($M = 23.09$) [14]. We observed no significant gender effect on profile group membership for IUD and HR-IU. This finding is in line with several studies reporting no differences in the estimated IUD prevalence between German female and male adolescents in representative samples [14,15,20].

Comparing the phenotypes of profile members, the largest differences between IUD and functional Internet use can be observed for the symptom inter- and intrapersonal conflicts (see Table 4), followed by loss of control, mental and behavioral preoccupation coping with unpleasant mood and withdrawal symptoms. In other words, having inter- and intrapersonal conflicts due to recurrent Internet use seems to be the best factor to differentiate between functional and dysfunctional Internet use. This is an important finding in the discussion about the specificity of symptoms defining the disorder.

Another important result was that risk of IUD and HR-IU was significantly associated with age. Distribution of age showed an increase for the IUD and HR-IU groups from 11 to 21. We identified two peaks, showing the largest prevalence rates at the ages of 15–16 (IUD = 7.6%, HR-IU = 16.1%) and 19–21 years (IUD = 9.1%, HR-IU = 21.7%). Few studies have found similar results showing an increased risk at the ages of 15–16 [24,26]. However, the age group of 19–21, which was found to be the group of highest risk in our study, was not investigated separately. Most studies that investigated prevalence rates in adolescents either summarized the findings over age groups because of fewer sample sizes [21,22,31] or investigated fewer age groups that were analyzed separately [24,26].

One finding that requires further attention is the significant difference of risk across school types. The findings must be interpreted carefully, because school types and age groups are intercorrelated. Specific school types exclude higher age groups (e.g., low level education and middle level education schools), whereas other school types exclude lower age groups (e.g., vocational school and vocational upper secondary school). Therefore, we analyzed the effects of school type separately for each age group. It was notable that even within each school type, prevalence increased with age. However, the base rates were highly different across schools and the differences could not be explained by the level of education per se. More research is needed to investigate, if specific school characteristics (i.e., humanistic vs. vocational track) might be associated with IUD. e.g., in the age group 15–16, risk of IUD seems to be highest in middle education level (11.4%) followed by vocational upper secondary school (9.4%), low educational level (7.9%), high educational level (6.5%) and vocational school (6.3%). A quite similar pattern was found for HR-IU, with the highest risk in middle educational level (18.6%) followed by high educational level (18.5%), low educational level (15.8%), vocational upper secondary school (15.1%) and vocational school (11.9%).

Strengths of the present study are the large number of cases and the broad age range of the sample, but of course, the survey has also several limitations. Although we used the same instrument (CIUS) and methods as Rumpf et al. [14] and Wartberg et al. [15], our prevalence rate (6.1%) was higher than previously reported (i.e., 4.0% [14] and 3.2% [15]) in representative samples of adolescents. This might be due to several reasons: First, it is assumable that prevalence rates have increased over time. Second, prevalence estimates have been found to be very sensitive to effects of age, school type and probably some more factors which have not yet been identified. Although our sample size was larger than most of the studies that have been conducted in this age group, it was based on an ad-hoc sample of schools that agreed to participate within a specific region and the distribution of school types (e.g., regarding Gymnasium and Gesamtschule) was not representative for the Rhine-Neckar region or for Germany. Further limitations of our study were the focus on self-report measures, the absence of external ratings of parents, teachers or peers and the absence of diagnostic interviews ("gold standard"). Nonetheless, to our knowledge this is the first study that estimates prevalence rates of IUD and HR-IU in a large adolescent sample for distinct age groups ranging from early to late adolescence.

5. Conclusions

IUD was found to affect 6.1% of German adolescents aged 11–21. The risk of IUD increased with age, ranging from 2.8% to 9.1%. The course of prevalence rates over adolescence showed two peaks at age groups 15–16 and 19–21. Gender was equally distributed across all phenotype profiles. Inter- and intrapersonal conflicts were identified to be the symptom to differentiate most specifically between functional and dysfunctional Internet use.

One finding which needs further attention was the fact prevalence rates differed significantly between school types, which could not be explained by the level of education. Future research on school characteristics and education is needed to identify factors that explain these considerable differences in risk of illness and academic career. Longitudinal designs are needed to investigate causality.

Our results indicate that IUD is a major health challenge that increases over the course of adolescence and it should be a public health policy priority to identify individuals most in need that might benefit from effective preventive approaches and treatments.

Acknowledgments: The data were collected within the PROTECT study (ClinicalTrials.gov: NCT02907658), which is supported by the Dietmar Hopp Foundation. No conditions were imposed by the financing. We thank all students who participated in the study and the headmaster, teacher and social worker which supported us at the schools.

Author Contributions: K.L. was principle investigator, conceived and designed the study, analyzed the data and wrote the paper. K.H. was involved in data collection, data analysis and writing. C.S.-J. was involved was involved in data collection, data analysis and writing. L.W. was involved in data analysis and writing.

Conflicts of Interest: The authors declare no conflict of interest.

References

1. American Psychiatric Association. *Diagnostic and Statistical Manual of Mental Disorders: DSM-5*, 5th ed.; American Psychiatric Association: Washington, DC, USA, 2013.
2. ICD-11 Beta Draft. Available online: https://icd.who.int/dev11/l-m/en (accessed on 27 February 2018).
3. Ho, R.C.; Zhang, M.W.B.; Tsang, T.Y.; Toh, A.H.; Pan, F.; Lu, Y.; Cheng, C.; Yip, P.S.; Lam, L.T.; Lai, C.-M.; et al. The association between internet addiction and psychiatric co-morbidity: A meta-analysis. *BMC Psychiatry* **2014**, *14*, 183. [CrossRef] [PubMed]
4. Ko, C.H.; Yen, J.Y.; Yen, C.F.; Chen, C.S.; Chen, C.C. The association between Internet addiction and psychiatric disorder: A review of the literature. *Eur. Psychiatry* **2012**, *27*, 1–8. [CrossRef] [PubMed]
5. Carli, V.; Durkee, T.; Wasserman, D.; Hadlaczky, G.; Despalins, R.; Kramarz, E.; Wasserman, C.; Sarchiapone, M.; Hoven, C.W.; Brunner, R.; et al. The association between pathological internet use and comorbid psychopathology: A systematic review. *Psychopathology* **2013**, *46*, 1–13. [CrossRef] [PubMed]
6. Poli, R.; Agrimi, E. Internet addiction disorder: Prevalence in an Italian student population. *Nordic J. Psychiatry* **2012**, *66*, 55–59. [CrossRef] [PubMed]
7. Shek, D.T.L.; Yu, L. Internet addiction phenomenon in early adolescents in Hong Kong. *Sci. World J.* **2012**, *2012*, 104304. [CrossRef] [PubMed]

8. Cheng, C.; Li, A.Y.L. Internet addiction prevalence and quality of (real) life: A meta-analysis of 31 nations across seven world regions. *Cyberpsychol. Behav. Soc. Netw.* **2014**, *17*, 755–760. [CrossRef] [PubMed]
9. Young, K.S. Internet Addiction: The Emergence of a New Clinical Disorder. *Cyber Psychol. Behav.* **1998**, *1*, 237–244. [CrossRef]
10. Young, K.S. *Caught in the Net: How to Recognize the Signs of Internet Addiction—And a Winning Strategy for Recovery*; John Wiley & Sons: New York, NY, USA, 1998.
11. Rehbein, F.; Kliem, S.; Baier, D.; Mößle, T.; Petry, N.M. Prevalence of Internet gaming disorder in German adolescents: Diagnostic contribution of the nine DSM-5 criteria in a state-wide representative sample. *Addict. Abingd. Engl.* **2015**, *110*, 842–851. [CrossRef] [PubMed]
12. Wartberg, L.; Kriston, L.; Thomasius, R. The Prevalence and Psychosocial Correlates of Internet Gaming Disorder. *Dtsch. Arzteblatt Int.* **2017**, *114*, 419–424. [CrossRef]
13. Bischof, G.; Bischof, A.; Meyer, C.; John, U.; Rumpf, H.J. *Prävalenz der Internetabhängigkeit—Diagnostik und Risikoprofile (PINTA-DIARI)*; Report to the German Federal Ministry of Health: Lübeck, Germany, 2013.
14. Rumpf, H.J.; Vermulst, A.A.; Bischof, A.; Kastirke, N.; Gürtler, D.; Bischof, G.; Meerkerk, G.J.; John, U.; Meyer, C. Occurence of internet addiction in a general population sample: a latent class analysis. *Eur. Addict. Res.* **2014**, *20*, 159–166. [CrossRef] [PubMed]
15. Wartberg, L.; Kriston, L.; Kammerl, R.; Petersen, K.U.; Thomasius, R. Prevalence of pathological Internet use in a representative German sample of adolescents: Results of a latent profile analysis. *Psychopathology* **2015**, *48*, 25–30. [CrossRef] [PubMed]
16. Laconi, S.; Rodgers, R.F.; Chabrol, H. The measurement of Internet addiction: A critical review of existing scales and their psychometric properties. *Comput. Hum. Behav.* **2014**, *41*, 190–202. [CrossRef]
17. Meerkerk, G.-J.; Van Den Eijnden, R.J.J.M.; Vermulst, A.A.; Garretsen, H.F.L. The Compulsive Internet Use Scale (CIUS): Some psychometric properties. *Cyber Psychol. Behav.* **2009**, *12*, 1–6. [CrossRef] [PubMed]
18. Wartberg, L.; Petersen, K.-U.; Kammerl, R.; Rosenkranz, M.; Thomasius, R. Psychometric Validation of a German Version of the Compulsive Internet Use Scale. *Cyberpsychol. Behav. Soc. Netw.* **2014**, *17*, 99–103. [CrossRef] [PubMed]
19. Guertler, D.; Rumpf, H.J.; Bischof, A.; Kastirke, N.; Petersen, K.U.; John, U.; Meyer, C. Assessment of problematic internet use by the Compulsive Internet Use Scale and the Internet Addiction Test: A sample of problematic and pathological gamblers. *Eur. Addict. Res.* **2014**, *20*, 75–81. [CrossRef] [PubMed]
20. Wartberg, L.; Kriston, L.; Bröning, S.; Kegel, K.; Thomasius, R. Adolescent problematic Internet use: Is a parental rating suitable to estimate prevalence and identify familial correlates? *Comput. Hum. Behav.* **2017**, *67*, 233–239. [CrossRef]
21. Rumpf, H.J.; Meyer, C.; Kreuzer, A.; John, U. *Prävalenz der Internetabhängigkeit (PINTA)*; Report to the German Federal Ministry of Health: Lübeck, Germany, 2011.
22. Rehbein, F.; Mößle, T. Video Game and Internet Addiction: Is there a Need for Differentiation? *SUCHT* **2013**, *59*, 129–142. [CrossRef]
23. Anderson, E.L.; Steen, E.; Stavropoulos, V. Internet use and Problematic Internet Use: A systematic review of longitudinal research trends in adolescence and emergent adulthood. *Int. J. Adolesc. Youth* **2016**, *22*, 430–454. [CrossRef]
24. Karacic, S.; Oreskovic, S. Internet Addiction through the Phase of Adolescence: A Questionnaire Study. *JMIR Mental Health* **2017**, *4*, e11. [CrossRef] [PubMed]
25. Ostovar, S.; Allahyar, N.; Aminpoor, H.; Moafian, F.; Nor, M.B.M.; Griffiths, M.D. Internet Addiction and its Psychosocial Risks (Depression, Anxiety, Stress and Loneliness) among Iranian Adolescents and Young Adults: A Structural Equation Model in a Cross-Sectional Study. *Int. J. Ment. Health Addict.* **2016**, *14*, 257–267. [CrossRef]
26. Stavropoulos, V.; Griffiths, M.D.; Burleigh, T.L.; Kuss, D.J.; Doh, Y.Y.; Gomez, R. Flow on the Internet: A longitudinal study of Internet addiction symptoms during adolescence. *Behav. Inf. Technol.* **2018**, 1–14. [CrossRef]
27. Xu, J.; Shen, L.X.; Yan, C.H.; Hu, H.; Yang, F.; Wang, L.; Kotha, S.R.; Zhang, L.N.; Liao, X.P.; Zhang, J.; et al. Personal characteristics related to the risk of adolescent internet addiction: A survey in Shanghai, China. *BMC Public Health* **2012**, *12*, 1106. [CrossRef] [PubMed]

28. Jackson, L.A.; von Eye, A.; Biocca, F.A.; Barbatsis, G.; Fitzgerald, H.E.; Zhao, Y. Personality, cognitive style, demographic characteristics and Internet use—Findings from the HomeNetToo project. *Swiss J. Psychol.* **2003**, *62*, 79–90. [CrossRef]
29. Riedl, D.; Stöckl, A.; Nussbaumer, C.; Rumpold, G.; Sevecke, K.; Fuchs, M. Nutzungsmuster von Internet und Computerspielen. *Neuropsychiatrie* **2016**, *30*, 181–190. [CrossRef] [PubMed]
30. Cerniglia, L.; Zoratto, F.; Cimino, S.; Laviola, G.; Ammaniti, M.; Adriani, W. Internet Addiction in adolescence: Neurobiological, psychosocial and clinical issues. *Neurosci. Biobehav. Rev.* **2017**, *76*, 174–184. [CrossRef] [PubMed]
31. Bakken, I.J.; Wenzel, H.G.; Götestam, K.G.; Johansson, A.; Oren, A. Internet addiction among Norwegian adults: A stratified probability sample study. *Scand. J. Psychol.* **2009**, *50*, 121–127. [CrossRef] [PubMed]
32. Kuss, D.J.; van Rooij, A.J.; Shorter, G.W.; Griffiths, M.D.; van de Mheen, D. Internet addiction in adolescents: Prevalence and risk factors. *Comput. Hum. Behav.* **2013**, *29*, 1987–1996. [CrossRef]
33. Stavropoulos, V.; Alexandraki, K.; Motti-Stefanidi, F. Recognizing internet addiction: Prevalence and relationship to academic achievement in adolescents enrolled in urban and rural Greek high schools. *J. Adolesc.* **2013**, *36*, 565–576. [CrossRef] [PubMed]
34. Tsitsika, A.; Janikian, M.; Schoenmakers, T.M.; Tzavela, E.C.; Olafsson, K.; Wójcik, S.; Macarie, G.F.; Tzavara, C.; Richardson, C. Internet addictive behavior in adolescence: A cross-sectional study in seven European countries. *Cyberpsychol. Behav. Soc. Netw.* **2014**, *17*, 528–535. [CrossRef] [PubMed]
35. Gürtler, D.; Rumpf, H.-J.; Bischof, A.; Kastirke, N.; Meerkerk, G.J.; John, U.; Meyer, C. Psychometrische Eigenschaften und Normierung der deutschen Version der Compulsive Internet Use Scale (CIUS). *Diagnostica* **2014**, *61*, 210–221. [CrossRef]
36. Peukert, P.; Steffen, S.; ElKasmi, J.; Barth, G.M.; Meerkerk, G.J.; Batra, A. Faktorielle Struktur der deutschen Version der Compulsive Internet Use Scale (CIUS) nach konfirmatorischer Faktorenanalyse. *Z. Klin. Psychol. Psychother.* **2012**, *41*, 101–108. [CrossRef]
37. Jung, T.; Wickrama, K.A.S. An Introduction to Latent Class Growth Analysis and Growth Mixture Modeling. *Soc. Personal. Psychol. Compass* **2008**, *2*, 302–317. [CrossRef]
38. Lo, Y. Testing the number of components in a normal mixture. *Biometrika* **2001**, *88*, 767–778. [CrossRef]
39. Nylund, K.L.; Asparouhov, T.; Muthen, B.O. Deciding on the number of classes in latent class analysis and growth mixture modeling: A Monte Carlo simulation study. *Struct. Equ. Model.* **2007**, *14*, 535–569. [CrossRef]

International Journal of
Environmental Research and Public Health

MDPI

Article

Effects of Internet and Smartphone Addictions on Depression and Anxiety Based on Propensity Score Matching Analysis

Yeon-Jin Kim [1], Hye Min Jang [2], Youngjo Lee [3], Donghwan Lee [2,*,†] and Dai-Jin Kim [4,*,†]

1 Department of Psychiatry, SMG-SNU Boramae Medical Center, Seoul 07061, Korea; loveaj1220@gmail.com
2 Department of Statistics, Ewha Womans University, Seoul 03760, Korea; janghyemin23@naver.com
3 Department of Statistics, Seoul National University, Seoul 08826, Korea; youngjo@snu.ac.kr
4 Department of Psychiatry, Seoul St. Mary's Hospital, The Catholic University of Korea College of Medicine, Seoul 06591, Korea
* Correspondence: donghwan.lee@ewha.ac.kr (D.L.); Kdj922@catholic.ac.kr (D.-J.K.); Tel.: +82-2-3277-2300 (D.L.); +82-2-2258-7546 (D.-J.K.)
† These authors contributed equally to this work.

Received: 23 February 2018; Accepted: 21 April 2018; Published: 25 April 2018

Abstract: The associations of Internet addiction (IA) and smartphone addiction (SA) with mental health problems have been widely studied. We investigated the effects of IA and SA on depression and anxiety while adjusting for sociodemographic variables. In this study, 4854 participants completed a cross-sectional web-based survey including socio-demographic items, the Korean Scale for Internet Addiction, the Smartphone Addiction Proneness Scale, and the subscales of the Symptom Checklist 90 Items-Revised. The participants were classified into IA, SA, and normal use (NU) groups. To reduce sampling bias, we applied the propensity score matching method based on genetics matching. The IA group showed an increased risk of depression (relative risk 1.207; $p < 0.001$) and anxiety (relative risk 1.264; $p < 0.001$) compared to NUs. The SA group also showed an increased risk of depression (relative risk 1.337; $p < 0.001$) and anxiety (relative risk 1.402; $p < 0.001$) compared to NCs. These findings show that both, IA and SA, exerted significant effects on depression and anxiety. Moreover, our findings showed that SA has a stronger relationship with depression and anxiety, stronger than IA, and emphasized the need for prevention and management policy of the excessive smartphone use.

Keywords: anxiety; depression; Internet addiction; smartphone addiction; propensity score

1. Introduction

With the increasing use and convenience of the Internet and smartphones in daily life, the accumulated research has shown the negative effects of the excessive Internet and smartphone use in the realm of mental health [1].

The smartphone user rate in South Korean population is approximately 85%, the highest worldwide [2]. However, the excessive smartphone use is strongly associated with a number of mental health issues, including stress and an increased risk of abnormal anxiety [3,4]. Smartphone addiction (SA) has emerged as a new form of addiction along with Internet addictions (IA), and the clinical characteristic of the SA have received attention in recent years [5]. For example, there are some differences regarding the nature of the devices, such as the easy portability, real-time Internet access and direct communication features of smartphones [6]. Similarities and differences between IA and SA have been reported with respect to demographic variables and the motivational aspects of media use [1,6].

From the environmental aspect, a lack of alternative activities is associated with IA [7]. Additionally, being single has been reported to be strongly associated with both a social network and online gaming [8]. As to the educational level and monthly income dimensions, a recent study in people with SA found significant differences in the health dimension in favor of those who had a lower income and a lower degree of education [9]. Consistent with this finding, a systematic review reported significant correlation between academic performance and severity of IA [10]. With regard to age, a recent review found that problematic Internet use is most relevant to both adolescent and emerging adults (19 years and older) [10], while smartphone addiction is more prevalent in younger adolescents compared with emerging adults (19 years and older) [11]. A recent study showed that women tend to have a higher average of daily usage times and dependency scores for smartphones, compared to men [4]. Choi et al. (2015) reported that the male gender has a relevant risk factor for IA, and the female gender for SA [1]. Regarding the purpose of use, social networking showed to be more strongly related to a high smartphone dependence, compared to other mobile telephone-related functions [11]. In individuals with IA, Anderson et al. (2016) reported that male gender was significantly associated with online PC gaming [10].

With regard to psychological aspects, the positive associations of IA and SA with depression and anxiety have been widely reported [12,13]. Recent studies have suggested that addiction to the Internet and smartphones may arise by user's individual cognitive-emotional and behavioral profile rather than the medium itself [14–16]. A recent study observed the role of empathy and life satisfaction in both IA and SA [17]. With regard to psychopathology, several studies reported a positive correlation between IA, depression, and anxiety [18–20], while a recent study reported a relationship between smartphone use and severity, depression, and anxiety [13]. Therefore, the interrelationship between IA, SA, and mental health problems needs to be precisely delineated. Moreover, given both the overlap and differences between IA and SA [16], then the question that arises is to what extent IA and SA are linked to the increased in the level of depression and anxiety after adjusting the confounding demographic and socioeconomic factors?

It remains unclear whether mental health problems are causes or consequences of excessive reliance on the Internet and smartphones. Cross-sectional studies have employed multiple regression analyses to investigate the relationships between mental health problems, IA, and SA in people [21]. However, in observational studies, which lack randomization, multiple regression analysis has limitations, such as the possibility of overestimation and a poor standard error when numerous covariates are present, in addition to the selection bias [22]. Thus, estimating the effects of addiction by simply examination of a particular outcome, such as depression and anxiety, would be biased by the imbalance of the demographic and socioeconomic factors associated with IA and SA. Moreover, no studies have yet investigated the differential effects according to the characteristics of Internet and smartphone users, including environmental contexts and users' psychological profiles, of IA and SA on depression and anxiety. Propensity score matching (PSM) has become a popular approach to reduce the selection bias in observational studies [23,24]. In this paper, we applied PSM analysis to investigate the effects of IA and SA on depression and anxiety, in order to reduce the selection bias in our data. We chose sex, age, education, marital status, and income as confounding variable, considering the association of these sociodemographic variables with IA and SA in our study [9,25].

The primary aim of this study is to examine the interrelationships between IA, SA, and mood status, that is depression and anxiety, using propensity score matching analysis. Second, we seek to discover how the effects of depression and anxiety differ between IA and SA.

2. Materials and Methods

2.1. Study Participants

The data consisted of the online anonymous self-diagnosis survey responses of 5003 Korean adults (aged 19–49 years), conducted by the Catholic University of Korea, Seoul; and St. Mary's Hospital in December 2014 [26]. The study was conducted in accordance with the Declaration of Helsinki. The institutional review boards of the Catholic University of Korea, Seoul; and St. Mary's Hospital approved this study. All participants were informed about the study and provided written informed consent. The survey participants were recruited by a panel of a research company and self-report questionnaires were administered through Internet without any compensation. Only 149 respondents, who did not use smartphones, were excluded. Finally, we analyzed the data of 4854 participants. In the final sample, the ages were classified into three categories: Below 30 (33.19%), 30–39 (43.94%), and 40–49 (22.87%). There were 2573 males (53.01%) and 2281 females (46.99%). The additional demographic variables of participants considered were education, marital status, and income.

2.2. Measures

2.2.1. Measurement of Internet Addiction

The Korean Scale for Internet Addiction (K-scale) was developed in Korea to assess IA and has been validated in the Korean population with a high reliability of internal consistency [27]. The Cronbach's alpha coefficient for the K-Scale was 0.91 [28]. It has seven subscales and 40 items, measuring daily life disturbance, disturbance of reality testing, automatic addictive thoughts, virtual interpersonal relationships, deviant behavior, withdrawal, and tolerance. This Likert type scale has been set from 1 (not at all) to 4 (always). According to the previous report using this scale, the participants were sorted into three groups: normal, potential risk, and high-risk [29]. The high-risk group was defined as having a standardized score of 70 or higher, in daily life disturbance, automatic addictive thoughts, tolerance factors, or at least 70 in total. The potential risk group was defined as a score of 62 or higher in daily life disturbance, automatic addictive thoughts, tolerance factors, or at least 63 in total. The normal use group contained those scores below these numbers. In this study, IA groups were made up of the potential risk and high-risk groups.

2.2.2. Measurement of Smartphone Addiction

The Smartphone Addiction Proneness Scale (K-SAS) has been validated and widely used to screen for SA [30]. It consists of 15 items rated in a four-point Likert type scale of distress from 1 (not at all) to 4 (always). The questions examined three factors: daily life disturbance, automatic addictive thoughts, and tolerance. The Cronbach's alpha coefficient for the K-SAS was 0.880 [5].

Based on a previous report using this scale, we used the scores to classify the participants into three groups: Normal, potential risk, and high-risk [30]. The high-risk group was defined as having a score of 44 or more in total, or having a subscore of 15 or more in daily life disturbance along with subscores of 13 or more, in both automatic addictive thoughts and tolerance. The potential risk group was defined as having 41 or more in the total score, or 15 or more in the daily life disturbance factor. The normal use group contained those scores below these numbers [30]. In this study, the smartphone-addicted group were made up of high-risk and potential risk groups.

2.2.3. Measurement of Mental Health Problems: Depression and Anxiety

The SCL-90-R is a multidimensional questionnaire developed to screen a range of psychological and psychopathological features of 9 subscales: Somatization, obsessive–compulsive, interpersonal sensitivity, depression, anxiety, hostility, phobic anxiety, paranoid ideation, and psychoticism [31]. The SCL-90 contains 90 items rated in a 5-point scale of distress from 0 (none) to 4 (extreme). The test–retest reliability of the SCL-90-R in the Korean language was 0.76 for depression and 0.77 for anxiety. The internal consistency was 0.89 for depression and 0.86 for anxiety [31]. Depression and anxiety have been reported to be the psychiatric symptoms most strongly associated with IA and SA [12,13]. The specific dimensions of interest to screen in this study included the SCL-90-R subscales for Depression and Anxiety.

2.3. Data Analysis

2.3.1. Statistical Definition

Let Z_i be a binary addiction indicator for the ith subject; that is, $Z_i = 1$ if the ith subject is addicted (IA or SA), and $Z_i = 0$ otherwise. The potential outcome of a mental problem (depression or anxiety) is defined as $Y_i(Z_i)$. Note that only one of the potential outcomes is observed at the same time for each subject, so direct computation of $Y_i(1) - Y_i(0)$ is impossible. Instead of the individual effect, the primary parameter of interest is the expected addiction effect on the addicted population

$$\tau = E(Y_i(1) - Y_i(0)|Z_i = 1) = E(Y_i(1)|Z_i = 1) - E(Y_i(0)|Z_i = 1) \quad (1)$$

However, the estimation of τ still has a problem because $E(Y_i(0)|Z_i = 1)$ cannot be directly estimated. Of course, in randomized experiments, $E(Y_i(0)|Z_i = 1) = E(Y_i(0)|Z_i = 0)$ is satisfied, so τ can easily be estimated. However, in an observation study, the naïve estimation of τ can be biased because $E(Y_i(0)|Z_i = 1) \neq E(Y_i(0)|Z_i = 0)$. To adjust this selection bias, we assume that we can observe the covariates X_i that are not affect by any addiction, and for a given covariates X_i, the potential outcomes $Y_i(1)$, $Y_i(0)$ are conditionally independent of addiction indicator Z_i. Furthermore, if potential outcomes are independent of the addiction conditional on covariates X_i, they are also independent of the addiction conditional in the propensity score $P(X_i) = P(Z_i = 1|X_i)$ [19]. The PSM estimator for τ becomes

$$\tau^{PSM} = E_{P(X)|Z=1}[E(Y_i(1)|Z_i = 1, P(X_i)) - E(Y_i(0)|Z_i = 0, P(X_i))] \quad (2)$$

2.3.2. Estimating the Propensity Score

Propensity scores are calculated using logistic regression, a model used to predict the probability that an addiction occurs

$$\log \frac{P(Z_i = 1|X_i)}{1 - P(Z_i = 1|X_i)} = \alpha + \beta^T X_i \quad (3)$$

In this paper, as the covariates for X_i, we consider five categorical covariates: sex (1 = male and 2 = female), age (1 = 20–29, 2 = 30–39, and 3 = 40–49), education (1 = middle school, 2 = high school, and 3 = university or above), marital status (1 = single, 2 = cohabitation, 3 = married, 4 = divorced, and 5 = bereaved), and income (1 = low, 2 = mid-low, 3 = middle, 4 = mid-high, and 5 = high). In Section 1, these covariates may influence simultaneously the outcomes (depression or anxiety) and addictions. Thus, for each subject, we estimated the propensity scores; that is, the conditional probability of being addicted given the observed covariates [32].

2.3.3. Matching Methods Based on the Estimated Propensity Score

Once the propensity scores are estimated, matching can be used to estimate the treatment effect after adjusting to the differences between the two groups [33]. The goal of matching is to produce a matched sample that balances the distribution of a study's patient and matched the covariates of the control groups observed. This adjusting method allows us to control the confounding variables. In this study, we adopted two widely used matching methods, the optimal and genetic matching [34].

2.3.4. Estimation of the Relative Risks of Addiction on Mental Health Problems after Propensity Score Matching

After propensity score matching by using the observed covariates (age, gender, marriage, income, and education), we have a more balanced dataset. To model the mental health problem (depression or anxiety), we applied generalized linear models (GLMs) to the matched sample. Because the mental health scores are positive and biased, the gamma distribution with log link is fitted. Let Y_i be an outcome of interest (an score of depression or anxiety) with mean μ_i, we can use the Gamma GLM framework with covariates X_i:

$$\log \mu_i = \gamma^T X_i$$

Through modeling, we estimated e^γ as the relative risks (as an expected mean difference between groups) of IA and SA for each covariate.

3. Results

In addition to the 4854 participants, 126 (2.60%) were included in the IA group and 652 (13.43%) were included in the SA group. Table 1 shows the descriptive statistics of the depression and anxiety scores. The mean scores of depression and anxiety of IA and SA groups are larger than those of the normal use (NU) group.

Table 1. Descriptive statistics of Depression and Anxiety scores.

Outcome	Statistics	Total (n = 4854)	Internet: NU (n = 4728)	IA (n = 126)	Smartphone: NU (n = 4202)	SA (n = 652)
Depression	Mean	26.69	26.52	33.01	25.49	34.42
	SD	10.3	10.23	10.7	9.55	11.48
	Min	13	13	13	13	13
	Max	65	65	62	65	65
	Skewness	0.74	0.76	0.29	0.74	0.34
Anxiety	Mean	18.47	18.33	23.75	17.51	24.67
	SD	7.79	7.7	9.21	7.04	9.37
	Min	10	10	10	10	10
	Max	50	50	50	50	50
	Skewness	1.02	1.03	0.56	1	0.47

Abbreviations: SD, standard deviation; NU, normal use; IA, Internet addiction; SA, smartphone addiction.

3.1. Matching Quality of the Propensity Score Matching Method

Although we condition only a few of the covariates in the questionnaires of this study, via the propensity score, we found that the matching procedure was sufficient to balance the distribution of each covariate, Tables 2 and 3. We assessed the distances in the marginal distributions of X_i. For each covariate, we computed the bias; that is, the difference in sample averages of the addicted and normal samples. Before applying the propensity score matching, the biases were not ignored. However, after propensity score matching, the addiction and normal subsamples had a very similar marginal distribution for all covariates.

Table 2. Comparison of the mean percentage of baseline characteristics between IA and normal use groups, in the original sample and the propensity score matched sample, using the genetic and optimal matching.

	Before PSM			After PSM (Genetic)			After PSM (Optimal)		
	Normal (n = 4728)	IA (n = 126)	Bias	Normal (n = 3722)	IA (n = 124)	Bias	Normal (n = 126)	IA (n = 126)	Bias
Sex (male)	53.51	34.13	19.38	33.87	33.87	0	34.92	34.13	0.79
Sex (female)	46.49	65.87	−19.38	66.13	66.13	0	65.08	65.87	−0.79
Age (19–29)	33.1	36.51	−3.41	36.29	36.29	0	32.54	36.51	−3.97
Age (30–39)	43.99	42.06	1.93	42.74	42.74	0	44.44	42.06	2.38
Age (40–49)	22.91	21.43	1.48	20.97	20.97	0	23.02	21.43	1.59
Education (middle school)	0.59	0	0.59	0	0	0	0	0	0
Education (high school)	27.33	30.16	−2.83	30.65	30.65	0	37.3	30.16	7.14
Education (university or above)	72.08	69.84	2.24	69.35	69.35	0	62.7	69.84	−7.14

Table 2. *Cont.*

	Before PSM			After PSM (Genetic)			After PSM (Optimal)		
	Normal (n = 4728)	IA (n = 126)	Bias	Normal (n = 3722)	IA (n = 124)	Bias	Normal (n = 126)	IA (n = 126)	Bias
Marriage (single)	48.1	50	−1.9	50.81	50.81	0	43.65	50	−6.35
Marriage (cohabitation)	0.8	0.79	0.01	0	0	0	0.79	0.79	0
Marriage (married)	49.34	46.03	3.31	46.77	46.77	0	51.59	46.03	5.56
Marriage (divorced)	1.61	3.17	−1.56	2.42	2.42	0	3.97	3.17	0.8
Marriage (bereaved)	0.15	0	0.15	0	0	0	0	0	0
Income (low)	11.84	11.9	−0.06	12.1	12.1	0	14.29	11.9	2.39
Income (mid-low)	31.58	33.33	−1.75	33.06	33.06	0	34.13	33.33	0.8
Income (middle)	44.35	45.24	−0.89	45.97	45.97	0	42.06	45.24	−3.18
Income (mid-high)	10.89	7.14	3.75	6.45	6.45	0	7.14	7.14	0
Income (high)	1.33	2.38	−1.05	2.42	2.42	0	2.38	2.38	0

Abbreviations: PSM, propensity score matching; IA, internet addiction.

Table 3. Comparison of the mean percentage of baseline characteristics between SA and normal groups, in the original sample and the propensity score matched sample, using the genetic and optimal matching.

	Before PSM			After PSM (Genetic)			After PSM (Optimal)		
	Normal (n = 4202)	SA (n = 652)	Bias	Normal (n = 3873)	SA (n = 643)	Bias	Normal (n = 652)	SA (n = 652)	Bias
Sex (male)	55.45	37.27	18.18	36.86	36.86	0	36.5	37.27	−0.77
Sex (female)	44.55	62.73	−18.18	63.14	63.14	0	63.5	62.73	0.77
Age (19–29)	32.2	39.57	−7.37	39.5	39.5	0	39.26	39.57	−0.31
Age (30–39)	43.69	45.55	−1.86	45.72	45.72	0	45.86	45.55	0.31
Age (40–49)	24.11	14.88	9.23	14.77	14.77	0	14.88	14.88	0
Education (middle school)	0.55	0.77	−0.22	0.16	0.16	0	0.92	0.77	0.15
Education (high school)	27.44	27.15	0.29	27.22	27.22	0	28.68	27.15	1.53
Education (university or above)	72.01	72.09	−0.08	72.63	72.63	0	70.4	72.09	−1.69
Marriage (single)	47.52	52.15	−4.63	52.57	52.57	0	52.45	52.15	0.3
Marriage (cohabitation)	0.74	1.23	−0.49	0.93	0.93	0	0.61	1.23	−0.62
Marriage (married)	49.9	45.09	4.81	45.41	45.41	0	45.86	45.09	0.77
Marriage (divorced)	1.67	1.53	0.14	1.09	1.09	0	1.07	1.53	−0.46
Marriage (bereaved)	0.17	0	0.17	0	0	0	0	0	0
Income (low)	11.66	13.04	−1.38	12.75	12.75	0	13.5	13.04	0.46
Income (mid-low)	31.89	29.91	1.98	29.86	29.86	0	28.68	29.91	−1.23
Income (middle)	44.17	45.71	−1.54	46.35	46.35	0	46.01	45.71	0.3
Income (mid-high)	10.97	9.66	1.31	9.64	9.64	0	10.28	9.66	0.62
Income (high)	1.31	1.69	−0.38	1.4	1.4	0	1.53	1.69	−0.16

Abbreviations: PSM, propensity score matching; SA, smartphone addiction.

3.2. *Effects of the Internet Addiction on Depression and Anxiety*

The effects of IA on depression and anxiety obtained using propensity score matching are reported in Table 4. Through genetic matching, 3846 samples were selected. The IA was related to a greater risk of depression (relative risk 1.207, 95% confidence interval 1.128–1.292, and $p < 0.001$) and anxiety (relative risk 1.264, 95% confidence interval 1.173–1.362, and $p < 0.001$). All these relative risk ratios are significant because the confidence interval does not contain the 1. Through optimal matching, 252 samples were selected. The IA was related to a greater depression (relative risk 1.243, 95% confidence interval 1.145–1.348, and $p < 0.001$) and anxiety (relative risk 1.308, 95% confidence interval 1.192–1.435, and $p < 0.001$). Similar to the genetic matching, the relative risk ratios on both, depression and anxiety, are significantly larger than 1.

Table 4. Effects of the internet and smartphone addiction on depression and anxiety, based on propensity score matching.

		Internet Addiction			Smartphone Addiction		
Outcome	**Type of PSM**	*n*	**RR**	**CI**	*n*	**RR**	**CI**
Depression	Optimal	252	1.243	1.145–1.348	1304	1.386	1.334–1.440
	Genetic	3846	1.207	1.128–1.292	4516	1.337	1.296–1.378
Anxiety	Optimal	252	1.308	1.192–1.435	1304	1.44	1.380–1.503
	Genetic	3846	1.264	1.173–1.362	4516	1.402	1.355–1.450

Abbreviations: RR. relative risk; CI, confidence interval.

3.3. Effects of the Smartphone Addiction on Depression and Anxiety

The effects of SA on depression and anxiety using propensity score matching are reported in Table 4. Through genetic matching, 4516 samples were selected. The SA was related to a greater risk of depression (relative risk 1.337, 95% confidence interval 1.296–1.378, and $p < 0.001$) and anxiety (relative risk 1.402, 95% confidence interval 1.355–1.450, and $p < 0.001$). Through optimal matching, 1304 samples were selected. The SA was related to a greater risk of depression (relative risk 1.386, 95% confidence interval 1.334–1.440, and $p < 0.001$) and anxiety (relative risk 1.440, 95% confidence interval 1.380–1.503, and $p < 0.001$). All these relative risk ratios are significant.

3.4. Differences in Effects of the Internet and Smartphone Addiction on Depression and Anxiety

The relative risk ratios for depression and anxiety, from both genetic and optimal matching, were 10% higher for SA than for IA. This means that SA has a greater risk for depression and anxiety than IA. Those confidence intervals do not contain the 1, so we can say that SA is 34–44% more likely to cause a mental disorder.

4. Discussion

Our findings are that both IA and SA exert significant effects on depression and anxiety, even after controlling the confounders using propensity score matching. Epidemiological studies have estimated a higher prevalence of depression in IA [35,36]. A number of cross-sectional studies have reported that individuals with IA or SA showed higher levels of depression and anxiety than normal users [13,37]. In the present study, our results show the roles of IA and SA in developing depression and anxiety. There are some possible explanations for the current findings. First, addictive use of internet and smartphones can increase interpersonal problems, which is related to depression and anxiety, such as family conflicts, lack of off-line relationships, and a heightened need for approval in cyberspace. Second, withdrawal symptoms are proposed as psychopathological patterns in IA and SA, comparable to substance abuse disorders [5]. When they do not have access to a PC or smartphone, the individuals with IA or SA may become anxious, and then desire to use the Internet or a smartphone in to escape such negative feelings [38]. Another possible explanation is that unlike other addictive substances, such as alcohol and nicotine, internet and smartphones over-users may have little insight about their excessive use in daily life because of free and flexible access to the devices [3], making them experience their excessive use as an annoyance rather than as a sign of problematic behavior [39]. Another interesting finding was that SA exerted stronger effects on depression and anxiety than IA. This leads us to speculate that IA and SA have different influences on mental health problems. There could be several possible explanations for this finding. First, considering the media characteristics, it is easier for the excessive smartphone use develops through habit-forming nature of the device, because of its higher accessibility to the wireless network and 24 h of frequent notifications [39]. Second, with regard to environmental factors, this finding may reflect the current radical change of daily life average from PCs to smartphones. People may use the PC internet for complicated work and carry out the other daily tasks with smartphones, leading to a decrease in labor productivity and a higher level of stress [40]. Finally, individuals with SA may use smartphones to maintain relationships and a sense of connectedness with the online social network [41], leading to the fear of missing out and the fear of loss of connection, while triggering a higher smartphone use [42].

This study has several limitations to generalize findings to the entire population, such as the cross-sectional nature of the data limits and the interpretation of causal inference between the Internet and smartphone addiction, depression, and anxiety. Propensity matching also has limitations and requirements. The major limitation is that propensity scores can only control by observed confounders [43]. The possibility of unobserved confounders may remain, limiting the study finding for generalization. Furthermore, because of all observed confounders in this study were collected as categorical variables, there may be information loss when building PSM model. Therefore, our findings should be interpreted with caution. However, to get the robust results of matching, we considered two matching methods, genetic matching and optimal matching. Especially, genetic matching uses a genetic search algorithm, so its process can find a good matching solution with less loss of information [44]. Lastly, assessment of the depression and anxiety symptom was conducted by self-report psychological symptom measure using SCL-90-R. To evaluate mental health problems more accurately and consistently. A structured interview by clinician should be conducted in further studies.

5. Conclusions

In this study, we investigated how IA and SA influence mental health problems, depression and anxiety. To the best of our knowledge, this is the first study to estimate the association between IA, SA and psychopathology using propensity matching score method from cross-sectional data, and to investigate the differential effect in the psychopathology between IA and SA. In conclusion, our findings reveal that both IA and SA increase the risk of depression and anxiety. In addition, SA showed a stronger relationship with depression and anxiety compared to IA.

An implication of these findings is that individuals with a problematic smartphone use should be closely monitored for mental health problems, highlighting the need to establish prevention and management policies aimed at the pre-clinical level of SA. Further prospective studies should investigate the causal directions of the relationships among IA, SA, and mental health problems and should identify the discriminative factors of IA and SA.

Author Contributions: D.-J.K. and D.L. conceived and designed the experiments; H.M.J. analyzed the data; Y.-J.K. wrote the paper. Y.L. supervised the data collection. All authors contributed to the development of the manuscript, revised it critically, and approved the final manuscript.

Acknowledgments: This work was supported by a grant from the National Research Foundation of Korea (Grant No. 2014M3C7A1062894, 2014M3C7A1062896).

Conflicts of Interest: The authors declare no conflict of interest.

References

1. Choi, S.-W.; Kim, D.-J.; Choi, J.-S.; Ahn, H.; Choi, E.-J.; Song, W.-Y.; Kim, S.; Youn, H. Comparison of risk and protective factors associated with smartphone addiction and Internet addiction. *J. Behav. Addict.* **2015**, *4*, 308–314. [CrossRef] [PubMed]
2. *2016 the Survey on Internet Overdependence*; Ministry of Science, ICT and Future Planning: Seoul, Korea, 2017.
3. Lee, Y.-K.; Chang, C.-T.; Lin, Y.; Cheng, Z.-H. The dark side of smartphone usage: Psychological traits, compulsive behavior and technostress. *Comput. Hum. Behav.* **2014**, *31*, 373–383. [CrossRef]
4. Lee, K.E.; Kim, S.-H.; Ha, T.-Y.; Yoo, Y.-M.; Han, J.-J.; Jung, J.-H.; Jang, J.-Y. Dependency on smartphone use and its association with anxiety in Korea. *Public Health Rep.* **2016**, *131*, 411–419. [CrossRef] [PubMed]
5. Kim, D.; Chung, Y.; Lee, J.; Kim, M.; Lee, Y.; Kang, E.; Keum, C.; Nam, J. Development of smartphone addiction proneness scale for adults: Self-report. *Korean J. Couns.* **2012**, *13*, 629–644.
6. Kwon, M.; Lee, J.-Y.; Won, W.-Y.; Park, J.-W.; Min, J.-A.; Hahn, C.; Gu, X.; Choi, J.-H.; Kim, D.-J. Development and validation of a smartphone addiction scale (SAS). *PLoS ONE* **2013**, *8*, e56936. [CrossRef] [PubMed]

7. Kuss, D.J.; Griffiths, M.D.; Karila, L.; Billieux, J. Internet addiction: A systematic review of epidemiological research for the last decade. *Curr. Pharm. Des.* **2014**, *20*, 4026–4052. [CrossRef] [PubMed]
8. Andreassen, C.S.; Billieux, J.; Griffiths, M.D.; Kuss, D.J.; Demetrovics, Z.; Mazzoni, E.; Pallesen, S. The relationship between addictive use of social media and video games and symptoms of psychiatric disorders: A large-scale cross-sectional study. *Psychol. Addict. Behav.* **2016**, *30*, 252. [CrossRef] [PubMed]
9. Aljomaa, S.S.; Qudah, M.F.A.; Albursan, I.S.; Bakhiet, S.F.; Abduljabbar, A.S. Smartphone addiction among university students in the light of some variables. *Comput. Hum. Behav.* **2016**, *61*, 155–164. [CrossRef]
10. Anderson, E.L.; Steen, E.; Stavropoulos, V. Internet use and Problematic Internet Use: A systematic review of longitudinal research trends in adolescence and emergent adulthood. *Int. J. Adolesc. Youth* **2017**, *22*, 430–454. [CrossRef]
11. Haug, S.; Castro, R.P.; Kwon, M.; Filler, A.; Kowatsch, T.; Schaub, M.P. Smartphone use and smartphone addiction among young people in Switzerland. *J. Behav. Addict.* **2015**, *4*, 299–307. [CrossRef] [PubMed]
12. Ko, C.-H.; Yen, J.-Y.; Yen, C.-F.; Chen, C.-S.; Chen, C.-C. The association between Internet addiction and psychiatric disorder: A review of the literature. *Eur. Psychiatry* **2012**, *27*, 1–8. [CrossRef] [PubMed]
13. Demirci, K.; Akgönül, M.; Akpinar, A. Relationship of smartphone use severity with sleep quality, depression, and anxiety in university students. *J. Behav. Addict.* **2015**, *4*, 85–92. [CrossRef] [PubMed]
14. Brand, M.; Young, K.S.; Laier, C.; Wölfling, K.; Potenza, M.N. Integrating psychological and neurobiological considerations regarding the development and maintenance of specific Internet-use disorders: An Interaction of Person-Affect-Cognition-Execution (I-PACE) model. *Neurosci. Biobehav. Rev.* **2016**, *71*, 252–266. [CrossRef] [PubMed]
15. Kim, Y.-J.; Kim, D.-J.; Choi, J. The cognitive dysregulation of Internet addiction and its neurobiological correlates. *Front. Biosci (Elite ed.)* **2017**, *9*, 307–320.
16. Lachmann, B.; Duke, É.; Sariyska, R.; Montag, C. Who's Addicted to the Smartphone and/or the Internet? *Psychol. Pop. Media Cult.* **2017**. [CrossRef]
17. Lachmann, B.; Sindermann, C.; Sariyska, R.Y.; Luo, R.; Melchers, M.C.; Becker, B.; Cooper, A.J.; Montag, C. The Role of Empathy and Life Satisfaction in Internet and Smartphone Use Disorder. *Front. Psychol.* **2018**, *9*, 398. [CrossRef] [PubMed]
18. Banjanin, N.; Banjanin, N.; Dimitrijevic, I.; Pantic, I. Relationship between internet use and depression: Focus on physiological mood oscillations, social networking and online addictive behavior. *Comput. Hum. Behav.* **2015**, *43*, 308–312. [CrossRef]
19. Akin, A.; Iskender, M. Internet addiction and depression, anxiety and stress. *Int. Online J. Educ. Sci.* **2011**, *3*, 138–148.
20. Ostovar, S.; Allahyar, N.; Aminpoor, H.; Moafian, F.; Nor, M.B.M.; Griffiths, M.D. Internet addiction and its psychosocial risks (depression, anxiety, stress and loneliness) among Iranian adolescents and young adults: A structural equation model in a cross-sectional study. *Int. J. Ment. Health Addict.* **2016**, *14*, 257–267. [CrossRef]
21. Cheung, L.M.; Wong, W.S. The effects of insomnia and internet addiction on depression in Hong Kong Chinese adolescents: An exploratory cross-sectional analysis. *J. Sleep Res.* **2011**, *20*, 311–317. [CrossRef] [PubMed]
22. Cepeda, M.S.; Boston, R.; Farrar, J.T.; Strom, B.L. Comparison of logistic regression versus propensity score when the number of events is low and there are multiple confounders. *Am. J. Epidemiol.* **2003**, *158*, 280–287. [CrossRef] [PubMed]
23. Austin, P.C. A critical appraisal of propensity-score matching in the medical literature between 1996 and 2003. *Stat. Med.* **2008**, *27*, 2037–2049. [CrossRef] [PubMed]
24. Austin, P.C.; Grootendorst, P.; Anderson, G.M. A comparison of the ability of different propensity score models to balance measured variables between treated and untreated subjects: A Monte Carlo study. *Stat. Med.* **2007**, *26*, 734–753. [CrossRef] [PubMed]
25. Müller, K.W.; Glaesmer, H.; Brähler, E.; Woelfling, K.; Beutel, M.E. Prevalence of internet addiction in the general population: Results from a German population-based survey. *Behav. Inf. Technol.* **2014**, *33*, 757–766. [CrossRef]
26. Rho, M.J.; Lee, H.; Lee, T.-H.; Cho, H.; Jung, D.; Kim, D.-J.; Choi, I.Y. Risk Factors for Internet Gaming Disorder: Psychological Factors and Internet Gaming Characteristics. *Int. J. Environ. Res. Public Health* **2018**, *15*, 40. [CrossRef] [PubMed]

27. National Information Service Agency. *A Study of Internet Addiction Proneness Scale for Adults*; National Information Service Agency: Seoul, Korea, 2005.
28. Kim, D. *The Follow up Study of Internet Addiction Proneness Scale*; Korea Agency for Digital Opportunity and Promotion: Seoul, Korea, 2008; Available online: http://www.nia.or.kr/site/nia_kor/ex/bbs/View.do?cbIdx=39485&bcIdx=277&parentSeq=277 (accessed on 8 May 2008).
29. Kim, D.-I.; Chung, Y.-J.; Lee, E.-A.; Kim, D.-M.; Cho, Y.-M. Development of internet addiction proneness scale-short form (KS scale). *Korean J. Couns.* **2008**, *9*, 1703–1722.
30. National Information Service Agency. *Development of Korean Smartphone Addiction Proness Scale for Youth and Adults*; National Information Service Agency: Seoul, Korea, 2011; pp. 85–86.
31. Kim, K-I.; Kim, J-W. The standardizaion study of symptom checklist-90-R in Korea III. *Ment. Health Res.* **1984**, *2*, 278–311.
32. Heckman, J.; Smith, J. Assessing the Case for Social Experiments. *J. Econ. Perspect.* **1995**, *9*, 85–110. [CrossRef]
33. Caliendo, M.; Kopeinig, S. Some practical guidance for the implementation of propensity score matching. *J. Econ. Surv.* **2008**, *22*, 31–72. [CrossRef]
34. Sekhon, J.S.; Diamond, A. Genetic Matching for Estimating Causal Effects, unpublished Manuscript. Presented at the Annual Meeting of the Political Methodology, Tallahassee, FL, USA, July 2005.
35. Ghassemzadeh, L.; Shahraray, M.; Moradi, A. Prevalence of Internet addiction and comparison of Internet addicts and non-addicts in Iranian high schools. *Cyberpsychol. Behav.* **2008**, *11*, 731–733. [CrossRef] [PubMed]
36. Yen, J.-Y.; Ko, C.-H.; Yen, C.-F.; Wu, H.-Y.; Yang, M.-J. The comorbid psychiatric symptoms of Internet addiction: Attention deficit and hyperactivity disorder (ADHD), depression, social phobia, and hostility. *J. Adolesc. Health* **2007**, *41*, 93–98. [CrossRef] [PubMed]
37. Tonioni, F.; Mazza, M.; Autullo, G.; Cappelluti, R.; Catalano, V.; Marano, G.; Fiumana, V.; Moschetti, C.; Alimonti, F.; Luciani, M. Is Internet addiction a psychopathological condition distinct from pathological gambling? *J. Addict. Behav.* **2014**, *39*, 1052–1056. [CrossRef] [PubMed]
38. Kuss, D.J.; Griffiths, M.D. Online social networking and addiction—A review of the psychological literature. *Int. J. Environ. Res. Public Health* **2011**, *8*, 3528–3552. [CrossRef] [PubMed]
39. Oulasvirta, A.; Rattenbury, T.; Ma, L.; Raita, E. Habits make smartphone use more pervasive. *Pers. Ubiquitous Comput.* **2012**, *16*, 105–114. [CrossRef]
40. Duke, É.; Montag, C. Smartphone addiction, daily interruptions and self-reported productivity. *Addict. Behav. Rep.* **2017**, *6*, 90–95. [CrossRef] [PubMed]
41. Kuss, D.J.; Griffiths, M.D. Social networking sites and addiction: Ten lessons learned. *Int. J. Environ. Res. Public Health* **2017**, *14*, 311. [CrossRef] [PubMed]
42. Oberst, U.; Wegmann, E.; Stodt, B.; Brand, M.; Chamarro, A. Negative consequences from heavy social networking in adolescents: The mediating role of fear of missing out. *J. Adolesc.* **2017**, *55*, 51–60. [CrossRef] [PubMed]
43. Joffe, M.M.; Rosenbaum, P.R. Invited commentary: Propensity scores. *Am. J. Epidemiol.* **1999**, *150*, 327–333. [CrossRef] [PubMed]
44. Diamond, A.; Sekon, J. Genetic matching for estimating causal effects: A new method of achieving balance in observational studies. *Rev. Econ. Stat.* **2013**, *95*, 932–945. [CrossRef]

International Journal of
Environmental Research and Public Health

MDPI

Article

Personality Traits, Strategies for Coping with Stress and the Level of Internet Addiction—A Study of Polish Secondary-School Students

Joanna Chwaszcz [1,*], Bernadeta Lelonek-Kuleta [2,*], Michał Wiechetek [1], Iwona Niewiadomska [1] and Agnieszka Palacz-Chrisidis [1]

[1] Institute of Psychology, The John Paul II Catholic University of Lublin, Aleje Racławickie 14, 20-950 Lublin, Poland; wiechetek@kul.pl (M.W.); iwona.niewiadomska@kul.pl (I.N.); agnieszka.palacz@kul.pl (A.P.-C.)

[2] Institute of Family Studies, The John Paul II Catholic University of Lublin, Aleje Racławickie 14, 20-950 Lublin, Poland

* Corresponding author: chwaszcz@kul.pl (J.C.); bernadetalelonek@kul.lublin.pl (B.L.-K.)

Received: 27 February 2018; Accepted: 4 May 2018; Published: 14 May 2018

Abstract: Among the many contributing factors in addictions there are also those describing the individual characteristics and ways of dealing with various life challenges. Despite numerous studies in this area, there is still no unambiguous data on the nature and specificity of this relationship in different age groups. The aim of the study was to assess the relationship between personality dimensions and strategies of coping with stress and the level of Internet addiction. The study was funded by the Ministry of Health under grant no. 93/HM/2015. The study was carried out in a group of 383 persons aged 15 to 19 ($M = 16.6$, $SD = 0.77$) attending secondary schools. The following research tools were used: Ten Item Personality Measure, Brief Cope and Internet Addiction Test. Both specific personality traits and styles of coping with stress are related to the addiction to the analysed medium. The personality traits most strongly associated with the risky Internet use were conscientiousness and emotional stability. An association was demonstrated between Internet addiction and the use of coping strategies, such as disengagement, substance use and self-blame. The results obtained demonstrate a major role of personality-related factors in the development of Internet addiction. The attitude to difficulties seems to be the key issue. The findings presented also make it possible to delineate the areas for improvement (e.g., through psychoeducational interventions) to protect young people from the risk of developing the addiction.

Keywords: Internet addiction; coping strategies; personality traits; young people

1. Introduction

The results of studies on the severity of Internet addiction among young people are alarming but at the same time are characterised by significant variance, which is most likely due to the diverse assessment tools and research methodologies applied, as well as to differences between study groups [1–4]. For instance, in Africa, 3.3% of young people meet the criteria for Internet addiction [5]. In China, moderate Internet addiction is found in 10.4% teenagers, while its serious levels are observed in 0.2% of the study population [6]. In Hong Kong, Internet addiction among young people ranges between 17% and 26.8% [7].

In India, it is estimated that approx. 0.7% of teenagers are addicted to the Internet [8]. The proportion of Internet addicts among Dutch teenagers is 3.7% [9]. Polish studies show that 1.3% of young people are addicted to the Internet and 12.1% are at risk of developing the addiction, i.e., in total there are 13.3% of problem Internet users [10]. The results of studies into the risk factors of pathological Internet use have shown that up to 89.9% of young people in the study group (the sample size was 11,931 female and male adolescents) display a positive correlation between pathological Internet use and repeated risky behaviours [11].

It is important to note that currently (2017), Internet addiction is yet to be recognised by the World Health Organisation or the American Psychiatric Association as an addiction or any other disease class (1992, 2013). The latest revision of the International Classification of Diseases, ICD-11, to be published in 2018, is likely to include only the associated gaming disorder [12]. However, the term has become so well-established in specialist language that it is commonly used in literature. In this paper, too, we will be using the term 'Internet addiction' to refer to various disorders associated with excessive use of the Internet. Excessive Internet use can be defined as the inability to control one's Internet use, leading to psychological, social, school-related and/or professional problems [13].

Many studies on excessive Internet use or Internet addiction have confirmed the relationship between personality traits and excessive online activity. Significant correlates of Internet addiction, as identified in specialist literature, include life satisfaction [14] and self-esteem [15,16], as well as negative valence (demonstrated by demanding, needy attitudes, and eagerness to impress), and attractiveness (demonstrated by care about one's looks, being well groomed, neat and efficient, and highly motivated) [17]. Studies conducted in 2014 in a sample of 3568 Korean Internet game players verified the risk factors for the development of Internet addiction, including: high impulsivity (odds ratio: 1.138), believed self-control (odds ratio: 1.034), anxiety (odds ratio: 1.086), pursuit of desired appetitive goals (odds ratio: 1.105), money spent on gaming (odds ratio: 1.005), weekday game time (odds ratio: 1.081), offline community meeting attendance (odds ratio: 2.060), game community membership (odds ratio: 1.393; $p < 0.05$ for all eight risk factors) [18].

A number of studies on personality-related correlates of excessive Internet use are based on the Big Five model. Meta-analyses of literature show, for instance, that there is a positive correlation between neuroticism and Internet addiction, and a negative one between the addiction and conscientiousness, agreeableness, extraversion and openness to experience in adults [19]. Multiple findings corroborate the existence of a positive correlation between neuroticism and Internet addiction [20–24].

A study by Montag et al., which, in addition to the Big Five personality dimensions, considered Eysenck's Personality Inventory measures and temperament and character traits, shows that low self-directedness is a better predictor for problematic Internet use than neuroticism [25]. A negative correlation with Internet addiction, on the other hand, is shown for conscientiousness, which acts as a protective factor against problematic Internet use in this context [26,27].

A similar relationship is observed between Internet addiction and agreeableness, which, as a personality trait, can also serve as a protective factor in relation to problematic Internet use [27–30].

A negative correlation with Internet addiction has also been demonstrated for extraversion, as confirmed by various scholars [21,24,31].

The findings related to openness to experience and its relationship with Internet addiction are, in turn, conflicting. Some scholars have confirmed a negative correlation between these two variables [29,32,33]. However, there are also studies which show a positive correlation between openness to experience and Internet addiction [21,30]. Still other studies show no relationship between openness to experience and Internet addiction [28,31].

Kuss et al. argue that findings related to adults should not be extrapolated to adolescents because of the unique character of this developmental period [34]. In their study, Kuss et al. show that the factors associated with Internet addiction in young people are high neuroticism, low conscientiousness, low agreeableness, and high openness to experience, with no confirmed relationship between extraversion and the addiction [34]. In turn, in a study by Zamani et al. [35] it was high neuroticism, low conscientiousness and low extraversion that increased the risk of developing Internet addiction, while agreeableness and openness to experience did not show any such correlation.

An interesting study on the impact of the Big Five personality traits on the development of Internet addiction in young people was conducted by Zhou et al. [36]. The study explored the mediating role of teenagers' coping styles in linking Internet addiction and personality traits. Its findings (for 998 study participants) showed that conscientiousness and agreeableness were negatively correlated with Internet addiction, while neuroticism, extraversion, and openness to experience had a positive correlation with the addiction. In-depth analyses indicated that conscientiousness indirectly influenced the development of Internet addiction in young people by reducing emotion-focused coping. Neuroticism, extraversion and openness to experience, on the other hand, indirectly influenced the development of Internet addiction in young people by strengthening emotion-focused coping. Problem-focused coping did not play any mediating role [36]. The study shows that personality traits can be correlated with Internet addiction as a result of additional mediating factors, which should be further explored by testing new Internet addiction models.

A study by Tang Jie et al. confirms the correlation between coping style and the addiction [37]. In the light of the study, a negative coping style can play a mediating role, increasing the risk of developing Internet addiction in young people who experience very stressful events in their lives. Internet addicts scored higher in negative coping, while non-addicts scored higher in positive coping [37].

A study into the mediating role of the Big Five personality traits in the development of Internet addiction in young people was also conducted by Niels van der Aa et al. [33]. Its findings indicated that daily Internet use was indirectly correlated with low well-being (i.e., loneliness, depressive mood, and low self-esteem), with compulsive Internet use having the mediating role. Moreover, daily Internet use was proven to be strongly correlated with compulsive Internet use in introverted young people with low agreeableness and emotional instability (increased neuroticism). In addition, compulsive Internet use was shown to be strongly linked to loneliness in introverted, emotionally less stable and less agreeable adolescents [38].

The above-mentioned examples corroborate the crucial role of the Big Five personality traits in the development of Internet addiction in both adolescents and adults. Nevertheless, it is important to note that unambiguous results are usually obtained in relation to neuroticism and conscientiousness. When it comes to the correlations between extraversion, agreeableness and openness to experience, and Internet addiction, findings are inconclusive. These differences in findings indicate that further research is needed to ascertain whether there are any such correlations.

In this paper, we raise questions about the relationships between the Big Five personality traits and the risk of Internet addiction, and about the mediating role of coping strategies in secondary-school students. We have assumed that there is a correlation between personality dimensions and coping strategies, and Internet addiction, and that coping strategies act as mediating variables between personality and addiction.

2. Materials and Methods

2.1. Participants

The study was conducted on a group of 383 people aged 15 to 19 (M = 16.60; SD = 0.77) attending Polish secondary schools of general education. The study group included 215 female and 168 male students. 30.3% of respondents lived in rural areas (n = 116), 14.1% in towns (n = 54), and 55.1% in cities (n = 211), while 0.5% did not disclose their place of residence (n = 2). Most respondents had full families (86.7%; n = 332), 12.8% had single-parent families (n = 47), and 0.5% did not disclose such information (n = 2). A substantial majority of respondents reported no addictions in their families (79.4%; n = 304). 19.3% (n = 74) of study participants had addicts in their families. Five persons (1.3%) did not answer this question. The most common addictions were alcohol- and nicotine-related.

2.2. Procedure

The presented findings are part of a larger research project on Internet addiction operated by the Chair of Social Psychoprevention at the John Paul II Catholic University of Lublin. The study was questionnaire-based and conducted in groups during classes. Before the study was commenced, we had obtained consent from head teachers. The study was anonymous. Students were assured that their individual answers would not be disclosed to anyone. The study was conducted by trained university students participating in a psychoprevention course at the Chair of Social Psychoprevention at the Catholic University of Lublin. Prior to the survey, interviewers had introduced themselves and described the general purpose of the study. Next, they assured participants of their anonymity and distributed the questionnaires. When respondents completed the survey, interviewers collected the forms and thanked respondents for their participation.

2.3. Measures

The study employed a personal information questionnaire and three research tools to measure such psychological variables as personality traits, coping strategies and the level of Internet addiction. Survey forms were printed and handed out to study participants. The forms included a general instruction and a brief description of each method used to examine key project variables.

2.3.1. Personality Traits

Personality traits were measured on the basis of the Big Five model, using the Ten Item Personality Measure [39]. This method comprises 10 items (adjectives) referring to specific individual characteristics, e.g., extraverted, enthusiastic. Respondents had to evaluate each item using a 7-point scale ranging from 1 = Disagree strongly, to 7 = Agree strongly. Each of the five personality dimensions (Extraversion, Agreeableness, Conscientiousness, Emotional Stability and Openness to Experience) was measured by two items. Scores for each dimension were averaged on the basis of the answers. Cronbach's alphas for specific dimensions of TIPI in the Polish sample range from 0.41 to 0.67 [40].

2.3.2. Coping Strategies

Coping was measured using the Brief Cope [41]. This tool helps determine how often the individual uses each of the 14 different strategies. When the Polish version of this tool was prepared, a factor analysis showed that these coping strategies could be grouped into four more complex classes, namely Active Coping, Helplessness, Use of Support, and Avoidance [42]. The Active Coping dimension includes a scale with the same name, i.e., Active coping, and Planning and Positive reframing scales. The Helplessness class comprises such scales as Substance use, Behavioural disengagement, and Self-blame. The Use of support is made up of such strategies as Use of emotional support and Use of instrumental support. Finally, the Avoidance dimension includes the following three strategies—Self-distraction, Denial and Venting. The remaining three strategies, i.e., Religion, Acceptance, and Humour constituted independent dimensions in the factor analysis.

In the Brief Cope, respondents are asked to describe how they usually cope with very difficult situations. This method comprises 28 items (e.g., I've been turning to work or other activities to take my mind off things). Each item is evaluated by respondents using a 4-point scale, from 0 = I haven't been doing this at all, to 3 = I've been doing this a lot. The score corresponds to the average answer to the statements related to the respective strategies. Each strategy is measured on the basis of two statements. Cronbach's alphas for specific coping strategies in the Polish sample range from 0.62 to 0.89 [42].

2.3.3. Internet Addiction

The level of Internet addiction was measured using The Problematic Internet Use Test by R. Poprawa [43], which is a Polish adaptation of the Internet Addiction Test by K. Young. The original version of the test includes 20 questions based on the addiction criteria specified in DSM IV. Respondents provide answers using a scale from 0 to 5, where 0 means not applicable, 1 = rarely, 2 = occasionally, 3 = frequently, 4 = often, and 5 = always. The Polish adaptation includes 3 additional items, which refer to problematic Internet use, and has 1 original item removed due to the lowest factor loading and the weakest discriminative power. Factor analysis showed that the tool had a one-factor structure with the factor score of 9.089 and 41% of the explained variance, with loadings between 0.41 and 0.72.

We assumed that the test score is the sum of answers for the 22 items, which could range from 0 to 110 points. The higher the score, the greater the risk of Internet addiction. Cronbach's alpha is .94 [44].

2.3.4. Data Analysis

The data was analysed using SPSS version 24 and IBM SPSS AMOS 24 (IBM Inc., Chicago, IL, USA). Descriptive statistics were used to illustrate the distribution of scores across individual variables, i.e., personality traits, coping strategies, and Internet addiction. Next, we used Pearson's r to evaluate the strength of the correlation between Internet addiction and personality dimensions and coping strategies. The last stage of the analysis was the development of a path model which would explain Internet addiction using personality traits and complex coping strategies. Coping was considered a mediating variable between Internet addiction and personality dimensions.

3. Results

The results obtained for each tested variable are presented in Table 1. Based on Polish standards, the overall score obtained in the Internet Addiction Test for all of the interviewed secondary-school students can be considered to fall among average values [43]. A detailed analysis of the score distribution in this test shows that 1.3% (n = 5) of the students scored very low, 11.7% (n = 45) scored low, 77.8% (n = 298) had an average score, 8.9% (n = 34) scored high, and only 0.3% (n = 1) had a very high score.

The assessment of personality traits in the study group shows that secondary-school students obtained relatively highest scores for openness to experience and extraversion. The lowest scores were recorded for emotional stability. The assessment of coping strategies shows that the studied secondary-school students prefer active strategies or ones which involve actions designed to use either instrumental or emotional support. They were the least likely to show behavioural disengagement, denial and substance use.

Our analysis of the relationships between Internet addiction and the studied variables showed a number of statistically significant correlations (c.f. Table 2). These suggest that Internet addiction among Polish secondary-school students is associated with all personality variables but extraversion. Relatively strongest negative correlations were observed between the addiction and conscientiousness and emotional stability. We also demonstrated a weak correlation between Internet addiction and selected coping strategies. The attitudes to difficult situations that were found to co-occur with Internet addiction, and which could aggravate it, are Self-distraction, Venting, Behavioural disengagement, Substance use and Self-blame. On the other hand, the strategies that seem to be associated with lower risk of addiction to online media include Active coping and Planning.

Table 1. Interplay between Internet addiction, personality traits and coping strategies (Pearson's correlation coefficient).

Variables	1. Internet Addiction	2.	3.	4.	5.	6.	7.	8.	9.	10.	11.	12.	13.	14.	15.	16.	17.	18.	19.	20.	21.	22.	23.
Personality traits																							
2. Extraversion	−0.08																						
3. Agreeableness	−0.12 *	0.03																					
4. Conscientiousness	−0.32 **	0.01	0.22 **																				
5. Emotional stability	−0.26 **	0.08	0.24 **	0.28 **																			
6. Openness to experience	−0.11 *	0.35 **	0.09	0.14 **	0.23 **																		
Coping strategies																							
7. Active coping	−0.19 **	0.21 **	0.14 **	0.30 **	0.28 **	0.29 **																	
8. Planning	−0.13 **	0.16 **	0.05	0.25 **	0.18 **	0.18 **	0.59 **																
9. Positive reframing	−0.04	0.13 **	0.15 **	0.10 *	0.24 **	0.23 **	0.30 **	0.36 **															
10. Acceptance	0.06	0.03	0.06	−0.01	0.04	0.10	0.21 **	0.25 **	0.33 **														
11. Humour	0.07	0.04	−0.05	−0.21 **	−0.07	0.09	−0.08	−0.04	0.28 **	0.17 **													
12. Religion	−0.01	−0.02	0.06	0.11 *	−0.01	0.05	0.12 *	0.19 **	0.21 **	0.10	−0.01												
13. Use of emotional support	−0.05	0.29 **	0.08	0.08	−0.01	0.16 **	0.26 **	0.16 **	0.14 **	0.02	−0.01	0.20 **											
14. Use of instrumental support	−0.01	0.23 **	0.11 *	0.08	−0.08	0.08	0.18 **	0.18 **	0.12 *	0.00	−0.06	0.22 **	0.78 **										
15. Self-distraction	0.12 *	−0.06	−0.05	−0.03	−0.16 **	−0.12 *	−0.14 **	−.05	0.13 *	0.11 *	0.18 **	0.12 *	0.10	0.12 *									
16. Denial	0.08	−0.02	−0.05	−0.08	−0.20 **	−0.08	−0.14 **	−0.10	−0.03	−0.10 *	0.21 **	0.04	0.02	0.05	0.16 **								
17. Venting	0.11 *	0.07	−0.10	−0.17 **	−0.35 **	−0.03	−0.08	0.01	0.12 *	0.03	0.17 **	0.05	0.26 **	0.28 **	0.25 **	0.21 **							
18. Substance use	0.13 *	0.09	−0.18 **	−0.16 **	−0.20 **	−0.02	−0.09	−0.04	0.02	−.01	0.21 **	−0.10	0.08	0.06	0.04	0.35 **	0.16 **						
19. Behavioural disengagement	0.29 **	−0.23 **	−0.08	−0.29 **	−0.30 **	−0.34 **	−0.43 **	−0.30 **	−0.19 **	−0.10 *	0.18 **	−0.10	−0.13 *	−0.06	0.14 **	0.36 **	0.15 **	0.31 **					
20. Self-blame	0.20 **	−0.11 *	−0.15 **	−0.16 **	−0.47 **	−0.27 **	−0.12 *	0.03	−0.13 *	0.03	0.07	0.02	−0.08	−0.03	0.19 **	0.22 **	0.26 **	0.09	0.31 **				
Complex coping strategies																							
21. Active coping	−0.15 **	0.21 **	0.15 **	0.27 **	0.29 **	0.29 **	0.78 **	0.83 **	0.73 **	0.34 **	0.08	0.22 **	0.24 **	0.20 **	−0.01	−0.11 *	0.03	−.04	−0.38 **	−0.09			
22. Helplessness	0.29 **	−0.13 *	−0.19 **	−0.29 **	−0.47 **	−0.31 **	−0.30 **	−0.14 **	−0.13 **	−0.04	0.21 **	−0.08	−0.06	−0.02	0.18 **	0.44 **	0.28 **	0.64 **	0.76 **	0.71 **	−0.24 **		
23. Use of support	−0.03	0.28 **	0.10	0.09	−0.04	0.13 *	0.23 **	0.18 **	0.14 **	0.01	−0.04	0.22 **	0.95 **	0.94 **	0.12 *	0.04	0.29 **	0.07	−0.10 *	−0.05	0.23 **	−0.05	
24. Avoidance	0.15 **	0.01	−0.09	−0.14 **	−0.34 **	−0.11 *	−0.17 **	−0.07	0.10 *	0.02	0.27 **	0.11 *	0.19 **	0.22 **	0.68 **	0.68 **	0.71 **	0.26 **	0.31 **	0.33 **	−0.05	0.43 **	0.21 **
M	27.73	4.86	4.78	**4.46**	**4.02**	**5.41**	**2.11**	**2.01**	**1.57**	**1.87**	**1.18**	**1.07**	**1.77**	**1.75**	**1.63**	**0.70**	**1.49**	**0.42**	**0.74**	**1.60**	**1.90**	**0.92**	**1.76**
SD	15.43	1.53	1.24	**1.54**	**1.60**	**1.29**	**.66**	**0.75**	**0.80**	**0.70**	**0.79**	**0.94**	**0.88**	**0.85**	**0.72**	**0.75**	**0.73**	**0.73**	**0.74**	**0.86**	**0.57**	**0.55**	**0.82**

* $p \leq 0.05$; ** $p \leq 0.01$.

Table 2. Overall, direct and indirect effect of personality traits on Internet Addiction mediated by Hopelessness.

Type of Effect:	Openness to Experience	Emotional Stability	Conscientiousness
Overall effect	−0.04 **	−0.17 **	−0.26 **
Direct effect	-	−0.11	−0.24 **
Indirect effect	−0.04 **	−0.07 **	−0.03 *

* $p < 0.05$; ** $p < 0.01$.

The final step in our analyses was to investigate the interplays between Internet addiction and personality traits and complex coping strategies. For this purpose, we used AMOS to develop a path model, in which complex coping strategies served as mediating variables between personality and Internet addiction (see Figure 1). We started from testing a saturated model in which Internet addiction was a dependent variable; specific personality traits (Extraversion, Agreeableness, Conscientiousness, Emotional stability, Openness to experience) were independent variables and complex coping strategies (Active coping, Helplessness, Use of support, Avoidance) were mediators. Then, insignificant paths were removed. A model so modified has satisfactory fit indices (RMSEA < 0.001; $\chi^2(1) = 0.001$; $p = 0.99$). In our model, the variables that best explained the Internet addiction level were conscientiousness, emotional stability, and openness to experience, and helplessness as a complex coping strategy. These variables accounted for 15% of the variance in Internet addiction. All the paths included in the model proved statistically significant. Low conscientiousness ($\beta = -0.24$) can directly contribute to the development of the addiction. In the case of openness to experience, this correlation is mediated by helplessness. As regards emotional stability, its impact on Internet addiction is both direct ($\beta = -0.11$) and mediated by helplessness. Low emotional stability contributes to problem Internet use, but can also make the individual more helpless, which, in turn, increases the risk of Internet addiction. It is worth emphasising that Helplessness is a partial mediator between Internet Addiction and Conscientiousness just as it is between Internet Addiction and Emotional stability. On the other hand, Helplessness is a full mediator between openness to experience and problematic Internet use. Detailed information about the overall, direct and indirect effect of analysed variables computed by bias-corrected percentile methods is reported in Table 2. It can be concluded from the analysis of the indirect to overall effect ratio that the indirect effect fully explains the overall effect of openness to experience. For emotional stability and conscientiousness, however, the ratios are relatively low at 0.41 and 0.12, respectively. This result suggests that research is needed to identify yet further mediators between personality and Internet addiction.

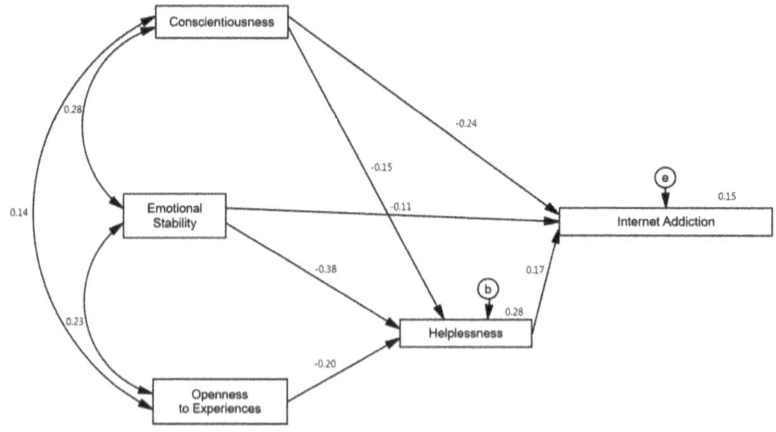

Figure 1. A path model explaining Internet addiction using personality traits and coping strategies. Note: e,b—residuals.

4. Discussion

Our findings are consistent with those of other scholars in relation to personality risk factors of Internet addiction.

The positive correlation between Internet addiction and emotional instability is confirmed by studies in both adults and adolescents [9,19–24,33,36].

A negative relationship between the risk of Internet addiction and conscientiousness was observed in studies by Montag et al., 2011 [26]; Zamani, 2011 [35] and Randler et al., 2014 [27]. A negative correlation between Internet addiction and agreeableness was found by Andreassen et al., 2013 [28]; Durak, Senol-Durak, 2014 [29]; Hwang et al., 2014 [30]; Randler et al., 2014 [27], Kuss et al., 2013 [9], and Zhou, 2017 [36].

Another trait associated with Internet addiction in our study in secondary-school students was openness to experience, which showed a negative correlation with the risk of the addiction. This was also confirmed by Durak, Senol-Durak, 2014 [29]; Servidio, 2014 [32]; Van der Aa et al., 2009 [33], and Kuss et al., 2013 [9].

Our study did not reveal any relationship between the risk of developing the addiction and extraversion. This is consistent with the findings reported by Kuss et al., 2013 [9]. A negative relationship between extraversion and Internet addiction was demonstrated by Buckner et al., 2012 [31]; Rahmani, Lavasani, 2011 [21]; Zamani, 2011 [35]; Yan, Li, Sui, 2014 [24], and Kayiş et al., 2016 [19].

In addition, our study showed a clear relationship between negative coping strategies and the risk of Internet addiction. The risk of developing the addiction is associated with the teenagers' use of such strategies as distraction, cessation of activities leading to problem resolution, release of emotions, self-blame, and substance use. Similar findings were reported by Tang Jie et al. [37], who suggested a relationship between negative coping styles and an increased risk of media addiction. Persons with developed Internet addiction display emotional coping strategies, which can additionally strengthen any depressive and autoaggressive tendencies in these individuals [38].

Zhou [36] observed an important role of emotion-focused coping style, which, when co-occurring with neuroticism, increased the risk of Internet addiction. Our study found a relationship between emotional instability and the risk of Internet addiction, which is reinforced when young people use emotion-focused strategies described as helplessness, namely substance use, behavioural disengagement and self-blame. In addition, the relationship between openness to experience and risk of Internet addiction was found to be mediated by helplessness.

Limitations

The research was conducted in one voivodeship, which is one of the poorest in Poland. It is worth to considerate to replicate it in other regions of the country. There is also a limitation in form of short study method to asses personality traits which has relatively low measures of reliability. It appears to be important to conduct similar research with application of more reliable methods. The last limitation of the research is the study group—high school students. It is worth to conduct the research on other age groups to asses whether noted results are similar.

5. Conclusions

To summarise, the study explored possible correlations between the Big Five personality traits and the risk of Internet addiction in young people and sought mediating variables for these correlations. Our findings are consistent with those of other scholars in relation to the correlations found between four out of five personality factors and the risk of Internet addiction. These correlations are observed between emotional stability, conscientiousness, agreeableness and openness to experience, and the risk of Internet addiction. Also, the correlations between Internet addiction and such traits as openness to experience and emotional stability are mediated by a factor referred to as helplessness. In our path model, no direct or indirect correlation was found between extraversion and the risk of Internet addiction. In correlation analyses, in turn, positive relationships were observed for such coping strategies as self-distraction, behavioural disengagement, venting, and self-blame, and the risk of developing the addiction, while negative correlations were established between active coping and planning, and the risk of Internet addiction.

A practical conclusion offered by the study is that some personality traits can act as protective factors, while others can serve as risk factors, in the context of Internet addiction. These are emotional stability, conscientiousness, agreeableness, and openness to experience. The strongest correlations are found between emotional stability and conscientiousness, and the risk of developing the addiction. Emotional stability and openness to experience, on the other hand, show a relationship between the risk of Internet addiction, which can be direct or mediated by helplessness-related strategies. The coping strategies which negatively correlate with Internet addiction are active coping and planning. A risk factor is the use of nonconstructive helplessness-related coping strategies, such as behavioural disengagement, substance use, self-blame, and venting.

As regards the correlation between personality traits and the risk of Internet addiction, it seems reasonable to argue that, given their basis, personality traits can be considered as contributing to the risk of addiction. Coping strategies can be both a cause and a consequence of excessive involvement in online activity. An interesting observation made in our study was that helplessness, defined as behavioural disengagement, substance use and self-blame, acted as a mediator of the relationship between emotional instability and openness to experience, and the risk of Internet addiction. Low emotional stability is associated with increased use of helplessness-related strategies, which, in turn, increases the risk of developing the addiction. Similarly, low openness to experience is associated with helplessness-related strategies, which increases the risk of Internet addiction. Given the constancy of the Big Five personality traits, and the fact that they are to a large extent inherited, prevention measures are not designed to induce a change in personality but to develop constructive and active coping strategies to minimise the risk of Internet addiction. Such measures are particularly important during adolescence, when children are already aware of the reasons for, and consequences of, their behaviour over time, but are still socially malleable, having no fixed cognitive or behavioural patterns which are characteristic of an addiction.

The exploration of the correlations between personality traits and the risk of addiction in practical terms is justified only when it addresses questions about the factors which mediate these correlations and can be changed in the process of socialisation. Our study has provided only a partial explanation of the reality, which must be explored further.

Author Contributions: J.C., B.L.-K., M.W., I.N. and A.P.-C. conceived and designed the studies, analysed the data and wrote the paper.

Acknowledgments: The studies supported by the Ministry of Health under grant no. 93/HM/2015. We received no funding to cover the cost to publish in open access.

Conflicts of Interest: The authors declare no conflict of interest. The funders had no role in the design of the study; in the collection, analyses, or interpretation of data; in the writing of the manuscript, and in the decision to publish the results.

References

1. Deryakulu, D.; Ursavas, Ö.F. Genetic and environmental influences on problematic Internet use: A twin study. *Comput. Hum. Behav.* **2014**, *39*, 331–338. [CrossRef]
2. Müller, K.W.; Dreier, M.; Duven, E.; Giralt, S.; Beutel M, E.; Wölfling, K. A hidden type of Internet addiction? Intense and addictive use of social networking sites in adolescents. *Comput. Hum. Behav.* **2016**, *55*, 172–177. [CrossRef]
3. Oberst, U.; Wegmann, E.; Stodt, B.; Brand, M.; Chamarro, A. Negative consequences from heavy social networking in adolescents: The mediating role of fear of missing out. *J. Adolesc.* **2017**, *55*, 51–60. [CrossRef] [PubMed]

4. Tokunaga, R.S. A meta-analysis of the relationships between psychosocial problems and internet habits: Synthesizing internet addiction, problematic internet use, and deficient self-regulation research. *Commun. Monogr.* **2017**, *84*, 423–446. [CrossRef]
5. Adiele, I.; Olatokun, W. Prevalence and determinants of Internet addiction among adolescents. *Comput. Hum. Behav.* **2014**, *31*, 100–110. [CrossRef]
6. Wua, X.-S.; Zhang, Z.-H.; Zhao, F.; Wang, W.-J.; Li, Y.-F.; Bi, L.; Qian, Z.-Z.; Lu, S.-S.; Feng, F.; Hu, C.-Y.; et al. Prevalence of Internet addiction and its association with social support and other related factors among adolescents in China. *J. Adolesc.* **2016**, *52*, 103–111. [CrossRef] [PubMed]
7. Shek, D.T.L.; Yu, L. Adolescent Internet Addiction in Hong Kong: Prevalence, Change, and Correlates. *J. Pediatr. Adolesc. Gynecol.* **2016**, *29* (Suppl. 1), S22–S30. [CrossRef] [PubMed]
8. Goel, D.; Subramanyam, A.; Kamath, R. A study on the prevalence of internet addiction and its association with psychopathology in Indian adolescents. *Indian J. Psychiatry* **2013**, *55*, 140–143. [CrossRef] [PubMed]
9. Kuss, D.J.; van Rooij, A.J.; Shorter, G.W.; Griffiths, M.D.; van de Mheen, D. Internet addiction in adolescents: Prevalence and risk factors. *Comput. Hum. Behav.* **2013**, *29*, 1987–1996. [CrossRef]
10. Makaruk, K.; Wójcik, S. *EU NET ADB Badanie Nadużywania Internetu Przez Młodzież w Polsce [A Study on Internet Overuse by Youngpeople in Poland]*; Fundacja Dzieci Niczyje: Warsaw, Poland, 2012.
11. Durkee, T.; Carli, V.; Floderus, B.; Wasserman, C.; Sarchiapone, M.; Apter, A.; Wasserman, D. Pathological Internet use and risk-behaviors among European adolescents. *Int. J. Environ. Res. Public Health* **2016**, *13*, 294. [CrossRef] [PubMed]
12. Van den Brink, W. ICD-11 Gaming Disorder: Needed and just in time or dangerous and much too early? Commentary on: Scholars' open debate paper on the World Health Organization ICD-11 Gaming Disorder proposal (Aarseth et al.). *J. Behav. Addict.* **2017**, *6*, 290–292. [CrossRef] [PubMed]
13. Spada, M.M. An overview of problematic Internet use. *Addict. Behav.* **2014**, *39*, 3–6. [CrossRef] [PubMed]
14. Wang, L.; Luo, J.; Bai, Y.; Kong, J.; Gao, W.; Sun, X. Internet addiction of adolescents in China: Prevalence, predictors, and association with well-being. *Addict. Res. Theory* **2013**, *21*, 62–69. [CrossRef]
15. Ayas, T.; Horzum, M.B. Relation between Depression, Loneliness, Self-Esteem and Internet Addiction. *Education* **2013**, *133*, 283–290.
16. Bozoglan, B.; Demirer, V.E.; Sahin, L. Loneliness, self-esteem, and life satisfaction as predictors of Internet addiction: A cross-sectional study among Turkish university students. *Scand. J. Psychol.* **2013**, *54*, 313–319. [CrossRef] [PubMed]
17. Charlton, J.P.; Danforth, I.D.W. Validating the distinction between computer addiction and engagement: Online game playing and personality. *Behav. Inf. Technol.* **2010**, *29*, 601–613. [CrossRef]
18. Rho, M.J.; Lee, H.; Lee, T.-H.; Cho, H.; Jung, D.J.; Kim, D.-J.; Choi, I.Y. Risk Factors for Internet Gaming Disorder: Psychological Factors and Internet Gaming Characteristics. *Int. J. Environ. Res. Public Health* **2018**, *15*, 40. [CrossRef] [PubMed]
19. Kayiş, A.R.; Satici, S.A.; Yilmaz, M.F.; Simsek, D.; Ceyhan, E.; Bakioglu, F. Big five-personality trait and internet addiction: A meta-analytic review. *Comput. Hum. Behav.* **2016**, *63*, 35–40. [CrossRef]
20. Hardie, E.; Tee, M.Y. Excessive internet use: The role of the personality, loneliness and social support networks in Internet addiction. *Aust. J. Emerg. Technol. Soc.* **2007**, *1*, 34–47.
21. Rahmani, S.; Lavasani, M.G. The relationship between internet dependency with sensation seeking and personality. *Procedia Soc. Behav. Sci.* **2011**, *30*, 272–277. [CrossRef]
22. Tsai, H.F.; Cheng, S.H.; Yeh, T.L.; Shih, C.C.; Chen, K.C.; Yang, Y.C.; Yang, Y.K. The risk factors of Internet addiction a survey of university freshmen. *Psychiatry Res.* **2009**, *167*, 294–299. [CrossRef] [PubMed]
23. Yao, M.Z.; He, J.; Ko, D.M.; Pang, K. The influence of personality, parental behaviors, and self-esteem on Internet addiction: A study of Chinese college students. *Cyberpsychol. Behav. Soc. Netw.* **2014**, *17*, 104–110. [CrossRef] [PubMed]
24. Yan, W.; Li, Y.; Sui, N. The relationship between recent stressful life events, personality traits, perceived family functioning and internet addiction among college students. *Stress Health* **2014**, *30*, 3–11. [CrossRef] [PubMed]
25. Montag, C.; Jurkiewicz, M.; Reuter, M. Low self-directedness is a better predictor for problematic internet use than high neuroticism. *Comput. Hum. Behav.* **2010**, *26*, 1531–1535. [CrossRef]
26. Montag, C.; Flierl, M.; Markett, S.; Walter, N.; Jurkiewicz, M.; Reuter, M. Internet addiction and personality in first-person shooter video gamers. *J. Media Psychol.* **2011**, *23*, 163–173. [CrossRef]

27. Randler, C.; Horzum, M.B.; Vollmer, C. Internet addiction and its relationship to chronotype and personality in a Turkish university student sample. *Soc. Sci. Comput. Rev.* **2014**, *32*, 484–485. [CrossRef]
28. Andreas44sen, C.; Griffiths, M.; Gjertsen, S.; Krossbakken, E.; Kvam, S.; Pallesen, S. The relationships between behavioral addictions and the five-factor model of personality. *J. Behav. Addict.* **2013**, *2*, 90–99. [CrossRef] [PubMed]
29. Durak, M.; Senol-Durak, E. Which personality traits are associated with cognitions related to problematic Internet use? *Asian J. Soc. Psychol.* **2014**, *17*, 206–218. [CrossRef]
30. Hwang, J.Y.; Choi, J.S.; Gwak, A.R.; Jung, D.; Choi, S.W.; Lee, J.; Lee, J.Y.; Jung, H.Y.; Kim, D.J. Shared psychological characteristics that are linked to aggression between patients with Internet addiction and those with alcohol dependence. *Ann. Gen. Psychiatry* **2014**, *13*, 6. [CrossRef] [PubMed]
31. Buckner, J.E.V.; Castille, C.M.; Sheets, T.L. The five factor model of personality and employees' excessive use of technology. *Comput. Hum. Behav.* **2012**, *28*, 1947–1953. [CrossRef]
32. Servidio, R. Exploring the effects of demographic factors, Internet usage and personality traits on Internet addiction in a sample of Italian university students. *Comput. Hum. Behav.* **2014**, *35*, 85–92. [CrossRef]
33. Van der Aa, N.; Overbeek, G.; Engels, R.C.; Scholte, R.H.; Meerkerk, G.J.; Van den Eijnden, R.J. Daily and Compulsive Internet Use and Well-Being in Adolescence: A Diathesis-Stress Model Based on Big Five Personality Traits. *J. Youth Adolesc.* **2009**, *38*, 765–776. [CrossRef] [PubMed]
34. Kuss, D.J.; Griffiths, M.D.; Karila, L.; Billieux, J. Internet addiction: A systematic review of epidemiological research for the last decade. *Curr. Pharm. Des.* **2014**, *20*, 4026–4052. [CrossRef] [PubMed]
35. Zamani, B.E.; Abedini, Y.; Kheradmand, A. Internet addiction based on personality characteristics of high school students in Kerman, Iran. *Addict. Health* **2011**, *3*, 85–91. [PubMed]
36. Zhou, Y.; Li, D.; Li, X.; Wang, Y.; Zhao, L. Big five personality and adolescent Internet addiction: The mediating role of coping style. *Addict. Behav.* **2017**, *64*, 42–48. [CrossRef] [PubMed]
37. Tang, J.; Yu, Y.; Du, Y.; Ma, Y.; Zhang, D.; Wang, J. Prevalence of internet addiction and its association with stressful life events and psychological symptoms among adolescent internet users. *Addict. Behav.* **2014**, *39*, 744–747. [CrossRef] [PubMed]
38. Yen, J.-Y.; Yeh, Y.-C.; Wang, P.-W.; Liu, T.-L.; Chen, Y.-Y.; Ko, C.-H. Emotional Regulation in Young Adults with Internet Gaming Disorder. *Int. J. Environ. Res. Public Health* **2018**, *15*, 30. [CrossRef] [PubMed]
39. Gosling, S.D.; Rentfrow, P.J.; Swann, W.B., Jr. A Very Brief Measure of the Big Five Personality Domains. *J. Res. Personal.* **2003**, *37*, 504–528. [CrossRef]
40. Łaguna, M.; Bąk, W.; Purc, E.; Mielniczuk, E.; Oleś, P. Short measure of personality TIPI-P in a Polish sample. *Rocz. Psychol.* **2014**, *17*, 421–437.
41. Carver, C.S. You want to measure coping but your protocol's too long: Consider the Brief COPE. *Int. J. Behav. Med.* **1997**, *4*, 92–100. [CrossRef] [PubMed]
42. Juczyński, Z.; Ogińska-Bulik, N. *Narzędziapomiarustresu i radzeniasobiezestresem [Tools for Measuring Stress and Coping]*; Pracownia Testów Psychologicznych: Warsaw, Poland, 2009.
43. Poprawa, R. Test problematycznego używania Internetu. Adaptacja i ocenapsychometryczna Internet Addiction Test K. Young [The Problematic Internet Use Test. Adaptation and psychometric assessment of Internet Addiction Test of K. Young]. *Rocz. Psychol.* **2011**, *54*, 193–216.
44. Poprawa, R. W poszukiwaniu psychologicznych mechanizmów problematycznego używania Internetu wśród polskich internautów [Searching for the psychologicalmechanisms of problem Internet useamongPolish Internet users]. In *Oblicza Internetu. Architektura Komunikacyjna Sieci [Faces of the Internet. Web Communication Architecture]*; Sokołowski, M., Ed.; IIS PWSZ: Elbląg, Poland, 2006; pp. 113–124.

International Journal of
Environmental Research and Public Health

MDPI

Article

Measurement Invariance of the Short Version of the Problematic Mobile Phone Use Questionnaire (PMPUQ-SV) across Eight Languages

Olatz Lopez-Fernandez [1,2,*], Daria J. Kuss [1], Halley M. Pontes [1], Mark D. Griffiths [1], Christopher Dawes [1], Lucy V. Justice [1], Niko Männikkö [3], Maria Kääriäinen [4], Hans-Jürgen Rumpf [5], Anja Bischof [5], Ann-Kathrin Gässler [5], Lucia Romo [6], Laurence Kern [7], Yannick Morvan [6], Amélie Rousseau [8], Pierluigi Graziani [9,10], Zsolt Demetrovics [11], Orsolya Király [11], Adriano Schimmenti [12], Alessia Passanisi [12], Bernadeta Lelonek-Kuleta [13], Joanna Chwaszcz [14], Mariano Chóliz [15], Juan José Zacarés [16], Emilia Serra [16], Magali Dufour [17], Lucien Rochat [18], Daniele Zullino [19,20], Sophia Achab [19,20], Nils Inge Landrø [21], Eva Suryani [22], Julia M. Hormes [23], Javier Ponce Terashima [24] and Joël Billieux [2,20,25]

1 International Gaming Research Unit, Psychology Department, Nottingham Trent University, Nottingham NG1 4FQ, UK; daria.kuss@ntu.ac.uk (D.J.K.); halleypontes@gmail.com (H.M.P.); mark.griffiths@ntu.ac.uk (M.D.G.); lpxcad@nottingham.ac.uk (C.D.); lucy.justice@ntu.ac.uk (L.V.J.)
2 Laboratory for Experimental Psychopathology, Psychological Sciences Research Institute, Université Catholique de Louvain, 1348 Louvain-la-Neuve, Belgium; joel.billieux@uni.lu
3 Department of Social Services and Rehabilitation, Oulu University of Applied Sciences, 90220 Oulu, Finland; niko.mannikko@oamk.fi
4 Research Unit of Nursing Science and Health Management, University of Oulu and Oulu University Hospital, 90014 Oulu, Finland; maria.kaariainen@oulu.fi
5 Department for Psychiatry and Psychotherapy, University of Lübeck, 23538 Lübeck, Germany; Hans-Juergen.Rumpf@uksh.de (H.-J.R.); Anja.Bischof@uksh.de (A.B.); akgaessler@outlook.de (A.-K.G.)
6 EA 4430 Clinique Psychanalyse Développement (CLIPSYD), Université Paris Nanterre, France; U894 Centre de Psychiatrie et Neurosciences, (CPN), Inserm, 92000 Paris, France; romodesprez@gmail.com (L.R.); ymorvan@parisnanterre.fr (Y.M.)
7 EA 2931, Centre de Recherches sur le Sport et le Mouvement (CESRM), Université Paris Nanterre, 92000 Nanterre, France; laurence.kern@gmail.com
8 Psychology Department, PSITEC EA 4074, Université Lille Nord de France, 59650 Villeneuve d'Ascq, France; amelie.rousseau@univ-lille3.fr
9 LPS EA 849, Aix-Marseille University, 13007 Marseille, France; pierluigi.graziani@free.fr
10 Psychologie, Langues, Lettres et Histoire Département, University of Nîmes, 30000 Nîmes, France; pierluigi.graziani@free.fr
11 Institute of Psychology, ELTE Eötvös Loránd University, 1064 Budapest, Hungary; demetrovics@t-online.hu (Z.D.); orsolya.papay@gmail.com or kiraly.orsolya@ppk.elte.hu (O.K.)
12 Faculty of Human and Social Sciences, UKE—Kore University of Enna, Cittadella Universitaria, 94100 Enna, Italy; adriano.schimmenti@unikore.it (A.S.); alessia.passanisi@unikore.it (A.P.)
13 Department of Family Science and Social Work, Katolicki Uniwersytet Lubelski Jana Pawła II, 20-950 Lublin, Poland; bernadetalelonek@kul.lublin.pl
14 Department of Psychology, Katolicki Uniwersytet Lubelski Jana Pawła II, 20-950 Lublin, Poland; chwaszcz@kul.pl
15 Department of Basic Psychology, University of Valencia, 46010 Valencia, Spain; Mariano.Choliz@uv.es
16 Department of Developmental and Educational Psychology, University of Valencia, 46010 Valencia, Spain; Juan.J.Zacares@uv.es (J.J.Z.); Emilia.Serra@uv.es (E.S.)
17 Service de Toxicomanie, Faculte de medicine Université de Sherbrooke, Longueuil, Qc, J4K 0A8, Canada; magali.dufour@usherbrooke.ca
18 Department of Psychology and Educational Sciences, University of Geneva, 1205 Geneva, Switzerland; Lucien.Rochat@unige.ch
19 Department of Psychiatry—Research Unit Addictive Disorders, University of Geneva, 1205 Geneva, Switzerland; Daniele.Zullino@hcuge.ch (D.Z.); Sophia.Achab@hcuge.ch (S.A.)

[20] Department of Mental Health and Psychiatry—Addiction Division, University Hospitals of Geneva, 1205 Geneva, Switzerland
[21] Clinical Neuroscience Research Group, Department of Psychology, University of Oslo, 0317 Oslo, Norway; n.i.landro@psykologi.uio.no
[22] Department Psychiatry and Behavior, School of Medicine and Health Science, Atma Jaya Catholic University of Indonesia, Jakarta 14440, Indonesia; eva.suryani@atmajaya.ac.id or amyeva511@gmail.com
[23] Department of Psychology, University at Albany State University of New York, Albany, NY, USA; jhormes@albany.edu
[24] University Hospitals Cleveland Medical Center/Case Western Reserve University, Cleveland, OH 44106, USA; javier@incaas.org
[25] Addictive and Compulsive Behaviours Lab (ACB-lab), Institute for Health and Behaviour, University of Luxembourg, 4366 Esch-sur-Alzette, Luxembourg
* Correspondence: olatz.lopez-fernandez@ntu.ac.uk or lopez.olatz@gmail.com; Tel.: +44-(0)-115-848-2977

Received: 14 May 2018; Accepted: 5 June 2018; Published: 8 June 2018

Abstract: The prevalence of mobile phone use across the world has increased greatly over the past two decades. Problematic Mobile Phone Use (PMPU) has been studied in relation to public health and comprises various behaviours, including dangerous, prohibited, and dependent use. These types of problematic mobile phone behaviours are typically assessed with the short version of the Problematic Mobile Phone Use Questionnaire (PMPUQ–SV). However, to date, no study has ever examined the degree to which the PMPU scale assesses the same construct across different languages. The aims of the present study were to (i) determine an optimal factor structure for the PMPUQ–SV among university populations using eight versions of the scale (i.e., French, German, Hungarian, English, Finnish, Italian, Polish, and Spanish); and (ii) simultaneously examine the measurement invariance (MI) of the PMPUQ–SV across all languages. The whole study sample comprised 3038 participants. Descriptive statistics, correlations, and Cronbach's alpha coefficients were extracted from the demographic and PMPUQ-SV items. Individual and multigroup confirmatory factor analyses alongside MI analyses were conducted. Results showed a similar pattern of PMPU across the translated scales. A three-factor model of the PMPUQ-SV fitted the data well and presented with good psychometric properties. Six languages were validated independently, and five were compared via measurement invariance for future cross-cultural comparisons. The present paper contributes to the assessment of problematic mobile phone use because it is the first study to provide a cross-cultural psychometric analysis of the PMPUQ-SV.

Keywords: mobile phone use; smartphone use; Problematic Mobile Phone Use; Problematic Mobile Phone Use Questionnaire; psychometric testing; measurement invariance

---

1. Introduction

Mobile phones have become a ubiquitous technology and their use is widespread internationally. However, there appear to be differences in terms of technology use across various geographical regions according to the International Telecommunication Union (ITU). Recently, ITU Facts and Figures 2017 [1] demonstrated that mobile phone use has experienced the largest growth compared with other technologies over the last two decades. More specifically, worldwide mobile phone subscriptions per 100 inhabitants were 15.5 in 2001, 76.6 in 2010, and 103.5 in 2017. At the same time, subscriptions for landline telephones were 16.6 in 2001, 17.8 in 2010, and 13 in 2017. According to a study by ProQuest [2], the number of scientific papers and reports published on this topic has grown markedly. The study examined 26 scientific databases simultaneously (e.g., PsycINFO) using the search terms "mobile phone" or "cell* phone" and "smartphone". It was reported that 490 academic outputs were published in 2001, 3225 in 2010, and 8224 in 2017 (these results referred to scholarly peer-reviewed journal articles, as well as trade journals, magazines, conference proceedings, and other reports).

Negative aspects related to mobile phone use are often conceptualised within the umbrella term of Problematic Mobile Phone Use (PMPU; [3,4]). According to Billieux and colleagues [4–7], PMPU can be understood as a heterogeneous and multidimensional construct involving the potential negative effects of mobile phone use. Accordingly, these authors formulated an integrative pathway model to account for the various types of problematic mobile phone use (i.e., dangerous, prohibited/antisocial, and dependent). Based on this model, each pathway to mobile phone overuse (i.e., extraversion pathway, reassurance-seeking pathway, impulsive pathway) is underlain by specific psychosocial factors and individual differences. Although maladaptive mobile phone use was initially considered a public health issue in child and adolescent populations [8–11], over the past decade, mobile phone use has been considered to involve potential risks for all populations across the different dimensions of problematic use, namely dangerous, prohibited, or dependent use [4].

Regarding general health issues traditionally associated with mobile phone use, several studies have shown significant associations between mobile phone use and users' lifestyles and wellbeing. For example, Ezoe and colleagues [11] found that PMPU among Japanese female college students was associated with poor sleep, low physical activity, decreased work performance, and skipping breakfast. Similarly, Gallimberti and colleagues [12] observed that reading books, higher school marks, and longer hours of sleep were associated with low PMPU in Italian adolescents. Conversely, and in line with previous studies, other authors have reported PMPU to be positively associated with stress, depression, sleep disturbances, extraversion, female gender, young age, and poor academic or professional competence or performance [13–22]. Furthermore, Yang and colleagues [13] investigated the health and psychological problems associated with mobile phone use in adolescent Southern Taiwanese students and found that PMPU was associated with aggression, insomnia, smoking, suicidal tendencies, and low self-esteem.

For instance, two studies analysing young Swedish adults' perceptions of the need of being available at all times via their mobile phones [14,15] reported that mobile phone use was positively associated with stress, depression, and sleep disorders. Similarly, a recent systematic review carried out by Elhai and colleagues [16] found that PMPU was usually related to depression, anxiety, chronic stress, and low self-esteem. However, only depression and anxiety were consistently related to this problematic use, with medium and small effect sizes, respectively. In another paper, the same authors even stated that while depression was inversely associated with social PMPU (e.g., social networking, messaging), anxiety was positively related to problematic use as a process or being consumption-based (e.g., news consumption, entertainment, relaxation) [17].

Associated behaviours, such as dependency and/or compulsiveness, have also been reported when individuals check their phone display, and even when not interacting with their mobile phone directly. This is because auditory and/or tactile notifications prompt thoughts that affect attention, and which negatively impact on performance [18] (a phenomenon coined as 'technoference'; such use of mobile phones results in conflicts in interpersonal relationships and decreased wellbeing [19]). In addition to this, physical reactions, such as headaches and heat sensations, have been reported. In the same vein, Bickham and colleagues [20] found associations between PMPU and depression in North American adolescents. In sum, the existing evidence on smartphone use suggests a clear association between PMPU and decreased wellbeing, especially in young populations worldwide.

In relation to dangerous mobile phone use, PMPU has initially been negatively associated with safety behaviours [1,2], such as using mobile phones when driving, cycling, or walking. The importance of this factor is supported by the development of specific policies and regulations related to mobile phone use (i.e., to prevent road accidents). A study conducted in China [21] assessed unintentional injuries (i.e., road traffic injuries, pedestrian collisions, and falls) due to mobile phone use and psychopathological symptoms in adolescence. The most prevalent injury was collisions (followed by falls and other injuries), where adolescents experienced PMPU, as well as negative emotional, behavioural, and social adaptation symptoms. Another study from the United States (US) [22] reviewed the associations between motor vehicle crashes and PMPU in adolescents because drivers between 16 and 19 years in the US are the most likely to die as a consequence of distractions caused by mobile phones. The review evidenced that half of all adolescents texted on their mobile phone while driving.

Prohibition of mobile phone use (or its regulation) is another specific aspect of PMPU, and is usually associated with legal or public regulations. However, some individuals do not abstain from using phones in such circumstances (i.e., public spaces, such as libraries, cinemas, or theatres). According to Takao and colleagues [23], personality traits may be associated with these types of behaviours, such as self-monitoring (i.e., traits related to the tendency to control and regulate the public self) and approval motivations (i.e., the need for favourable evaluations from others). Both are associated with an extraverted personality, as indicated by previous research [14] because individuals with the extraversion trait are sensitive to social cues and peer pressure, which involves being prone to risk behaviours when using mobile phones constantly, even when their use is banned. This aspect of problematic mobile phone use can also be related to the fact that individuals use mobile phones in a way that interferes with social situations. A prototypical example is the act of snubbing someone in a social setting by using one's mobile phone instead of interacting, a phenomenon referred to as "phubbing" [24,25].

The most studied type of negative outcome associated with mobile phone use is dependence, also conceptualised as a genuine addictive behaviour by some researchers [9,26]. The introduction of the internet and instant messaging (IM) on mobile phones (i.e., smartphones) has been associated with mobile phone dependence [21]. Moreover, it has also been associated with sociability levels of mobile phone users [27–29] and peer pressure [28]. However, studies examining peer pressure have reported slightly contradictory findings [29], where PMPU has not necessarily been associated with peer support or social acceptance. Therefore, it appears there is a potential association between mobile phone dependence (especially texting) and levels of sociability in adolescent and young adult populations [27,28,30]. Other factors usually associated with this type of problematic use include emotional symptoms (e.g., stress, anxiety, and depression [31–33]), reward seeking [26], and heightened impulsivity [2,26]. Moreover, specific mobile phone use patterns have also been associated with dependent use, except for some entertainment uses, such as downloading or playing mobile games [26,34,35], or using the mobile phone for travel bookings, online payments, and online shopping [34].

In sum, on the one hand, a few authors have claimed that the negative nature of dependent mobile phone use is not always severe, such as Chung [36], who argued that levels of dependence in South Korean female adolescent mobile phone users (i.e., withdrawal, maladjustment, tolerance, obsession, and flashiness) are associated with high levels of interpersonal solidarity (i.e., shared sentiments, intimacy, and similarities). Similarly, other scholars [37] have alerted researchers concerning the risk of overpathologizing everyday life behaviours in the context of behavioural addictions research, such as PMPU. On the other hand, Chóliz [38] has claimed that mobile phone addiction is a clinically relevant condition. Therefore, further research is warranted to assess the underlying motivations behind dependent use.

In relation to the cross-cultural assessment of PMPU, only a few studies have been conducted [39,40]. A number of different scales have been used [5,22,41,42], and according to a literature review by Pedrero and colleagues [42], the 'gold standard' scale is the Mobile Phone Problem Use Scale (MPPUS [3]). Unfortunately, the MPPUS is a unidimensional scale, which is problematic given the hypothesized multi-dimensional nature of PMPU. Moreover, the structural validity of the MPPUS was only tested with exploratory factor analysis (EFA), and needs to be confirmed in further studies using confirmatory factor analysis (CFA) and measurement invariance (MI). Another contemporary instrument to assess PMPU is the Problematic Mobile Phone Use Questionnaire (PMPUQ; [4]), which allows the measurement of the multi-dimensional nature of PMPU and was validated through the conjoint use of EFAs and CFAs. The scale assesses the three aforementioned specific types of PMPU. It was initially developed with a four-factor solution, but was recently reduced to a shorter version with three factors (dangerous use, prohibited use, and dependence) and updated to contemporary smartphone use (PMPUQ-SV; [33,35,43]). The fourth factor, related to the occurrence of financial problems, was removed due to the evolution of smartphones (i.e., smartphones being relatively cheap to use compared to when they were first introduced).

Subsequent studies—including some cross-cultural ones [33,35]—have evaluated the factor structure of the PMPUQ in its long or short versions via exploratory [35,43] and confirmatory [33,43,44] approaches in different populations (e.g., young adults [33,43], adults [35,43,44]), and different European languages, especially English [33,35,43,44]. However, psychometric results have been contradictory because some studies have reported adequate properties [33,35], while others have not [43,44]. Finally, to the best of the authors' knowledge, no previous study has tested MI to establish the cross-validity of any of the PMPU scales (i.e., unidimensional or multidimensional) simultaneously across different languages using confirmatory approaches. This is a necessary step to move the field forward in order to establish cross-cultural MI of a scale to guarantee reliable and comparative findings across countries and languages.

The aim of the present study was to test the psychometric properties and measurement invariance of eight versions of the PMPUQ-SV. The languages selected were German, French, English, Finnish, Spanish, Italian, Polish, and Hungarian. A number of non-European countries using the same languages agreed to join the data collection in this first study. In addition to being able to perform future cross-cultural studies, there are a number of reasons for carrying out the present study to validate the PMPUQ-SV in several languages. Firstly, there is little empirical evidence regarding PMPU as a multidimensional construct, especially in adulthood. Secondly, PMPU has almost exclusively been investigated in relation to its addictive use rather than considering other potential problems (such as dangerous or prohibited use). Thirdly, the PMPUQ has been previously tested mostly using exploratory and confirmatory approaches, with no consistent results across different languages (e.g., English), but its MI across different languages remains to be investigated. Consequently, the present study investigated the multidimensional construct of PMPU across specific types of problematic mobile phone use described via the multi-group validation of the PMPUQ-SV across languages. Thus, the objectives were to (i) determine an optimal factor structure for the PMPUQ–SV among university populations using eight languages; and (ii) simultaneously examine the MI of the PMPUQ–SV across all languages in order to assess the linguistic comparability across the eight versions of the scale independently.

Therefore, the main purpose of the present study was to ascertain if the PMPUQ-SV is an appropriate psychometric tool for cross-cultural research. To the best of the authors' knowledge, this is the first study to investigate the three-factor model in a multinational sample and the first to conduct MI on a multidimensional model of the PMPU across multiple linguistic scale versions. Thus, the present study will help fill an important gap in the field of PMPU and make a contribution to the research area because it comprises robust cross-cultural research examining mobile phone use and its associated problems.

2. Materials and Methods

2.1. Participants and Procedure

A total of 5209 respondents participated in the study, which builds upon the Tech Use Disorders (TUD; [45]) project. The items examined in the present study were part of a longer online survey including other questions (e.g., other scales concerning use of technology or personality traits). Participants were not forced to answer questions (because the survey was completely voluntarily). After cleaning the dataset (e.g., removing missing values), a sample of 3038 participants remained. The sample included adults engaged in higher education environments in 2015. The ethics committee of the Psychological Science Research Institute of the Université Catholique de Louvain (Belgium) approved the study protocol in 2014. Participants provided informed consent and voluntarily participated following an assurance of confidentiality and anonymity. The invitation to participate in the online survey (hosted on Qualtrics) used two recruitment strategies: (i) the present authors inviting undergraduates to participate via their respective universities during their 2015 and 2016 lectures; and (ii) via electronic invitations in academic online environments (e.g., university emails, university research participant pools, university social networks, and university virtual learning environments). Missing data were treated with pairwise deletion to maximise the statistical power, and cases were considered to be missing at random (MAR). This left a total sample size of 3038 participants with some not included for several reasons (e.g., young participants were not yet drivers, etc.). The sample breakdown by each respective language is shown in Table 1, alongside key socio-demographic data and reliability estimations.

Consequently, a total of eight languages were included in the present study (see Table 1), which were provided by 14 countries participating in the present study via their respective academic environments: German (i.e., Germany: 12.61% of the sample), French (i.e., Belgium: 16.06%; France: 10.60%; Switzerland: 3.39%; Canada: 5.13%; others who filled in the French adaptation chose not to report their country: 0.23%), English (i.e., United Kingdom (UK): 1.81%, Norway: 1.71%; US: 0.13%; Indonesia: 0.20%), Finnish (i.e., Finland: 14.78%), Spanish (i.e., Spain: 5.13%), Italian (i.e., Italy: 9.48%), Polish (i.e., Poland: 8.49%), and Hungarian (Hungary: 10.24%).

2.2. Instrument

To assess potential PMPU, the 15-item PMPUQ-SV [33,35,43] was adapted from English into the other seven languages using a standard translation and back-translation method [46], except for French (as it was the original language [4]) (see Appendix A). Each subscale comprised five items, which were scored from 1 ('I strongly agree') to 4 ('I strongly disagree'), except for the items that were reverse scored [35] (see Table 1 for descriptive item scores). Overall scores ranged from 15 to 60, with higher scores indicating more potential problems due to mobile phone use. The Cronbach's alphas of the PMPUQ-SV across all languages ranged from 0.56 (English version; prohibited use) to 0.90 in the present study (German version: dependence; French version: dependence; English version: dangerous use).

Table 1. Demographic Information and item scores across all eight adaptations of the PMPUQ-SV.

	All	German	French	English	Finnish	Spanish	Italian	Polish	Hungarian
N	3038	383	1076	117	449	156	288	258	311
Women (*N* (%))	2193 (72%)	262 (68%)	829 (77%)	89 (76%)	308 (69%)	123 (79%)	190 (66%)	187 (72%)	205 (66%)
Age in (Yrs; mean (*SD*))	26.505 (9.395)	25.204 (6.602)	25.22 (10.034)	27.735 (11.319)	28.296 (9.06)	28.045 (11.642)	28.576 (9.555)	25.279 (6.965)	27.833 (9.032)
PMPUQ Score (mean (*SD*))	27.156 (6.869)	26.976 (6.547)	27.005 (7.177)	28.145 (5.564)	26.938 (6.584)	28.5 (6.696)	29.59 (6.533)	28.961 (6.399)	23.539 (6.159)
Cronbach's α									
Dangerous use	0.84	0.88	0.81	0.90	0.87	0.86	0.86	0.77	0.89
Prohibited use	0.69	0.65	0.74	0.56	0.62	0.66	0.68	0.66	0.75
Dependent use	0.88	0.90	0.90	0.83	0.85	0.85	0.84	0.82	0.89
Item Scores (mean (*SD*))									
1. Easy not to use mobile	2.359 (1.034)	2.460 (0.991)	2.352 (1.095)	2.462 (0.915)	2.341 (1.030)	2.455 (0.905)	2.581 (0.921)	2.391 (1.009)	1.971 (0.991)
2. Use mobile when driving ^{R}	1.569 (0.869)	1.525 (0.789)	1.402 (0.784)	1.393 (0.754)	1.933 (0.968)	1.549 (0.930)	1.874 (0.904)	1.721 (0.917)	1.354 (0.773)
3. Don't use when forbidden	1.737 (0.943)	1.593 (0.866)	1.830 (0.981)	1.632 (0.826)	1.506 (0.869)	1.821 (0.962)	1.839 (0.936)	2.128 (1.003)	1.508 (0.823)
4. Difficult not to use mobile ^{R}	2.179 (0.965)	2.285 (0.892)	2.214 (1.039)	2.308 (0.825)	2.049 (0.941)	2.393 (0.884)	2.345 (0.889)	2.236 (0.926)	1.772 (0.867)
5. Avoid using on motorway	1.417 (0.866)	1.324 (0.752)	1.467 (0.931)	1.308 (0.760)	1.379 (0.755)	1.319 (0.825)	1.360 (0.751)	1.721 (1.123)	1.296 (0.755)
6. Using mobile in library	2.354 (1.037)	2.590 (0.988)	2.423 (1.086)	2.624 (0.953)	2.336 (0.998)	2.757 (0.918)	2.205 (0.986)	2.031 (0.994)	1.965 (0.927)
7. Easy to live without mobile	2.473 (0.995)	2.527 (0.900)	2.377 (1.052)	2.778 (0.862)	2.715 (0.993)	2.653 (0.855)	2.784 (0.854)	2.442 (0.937)	1.929 (0.902)
8. Dangerous situations ^{R}	1.721 (0.965)	1.574 (0.782)	1.828 (1.102)	1.410 (0.697)	1.686 (0.846)	1.730 (0.996)	1.839 (0.929)	2.070 (1.011)	1.302 (0.636)
9. Use mobile when forbidden ^{R}	1.581 (0.806)	1.655 (0.753)	1.440 (0.754)	1.701 (0.757)	1.579 (0.804)	1.822 (0.900)	1.909 (0.954)	1.694 (0.829)	1.431 (0.701)
10. Lost without mobile ^{R}	2.205 (0.986)	2.016 (0.859)	2.213 (1.056)	2.607 (0.909)	2.245 (0.953)	2.160 (0.980)	2.345 (0.932)	2.395 (0.945)	1.939 (0.930)
11. Driving danger mobile ^{R}	1.379 (0.762)	1.295 (0.600)	1.437 (0.886)	1.248 (0.642)	1.425 (0.738)	1.299 (0.638)	1.529 (0.827)	1.434 (0.715)	1.125 (0.448)
12. Public transport	1.462 (0.711)	1.496 (0.647)	1.369 (0.709)	1.530 (0.566)	1.410 (0.751)	1.542 (0.668)	1.586 (0.684)	1.659 (0.869)	1.476 (0.621)
13. Hard to turn off mobile ^{R}	1.999 (0.980)	1.984 (0.880)	1.911 (1.023)	2.325 (0.936)	1.795 (0.920)	2.278 (0.978)	2.399 (0.948)	2.019 (0.906)	1.994 (0.987)
14. Driving concentration ^{R}	1.247 (0.597)	1.188 (0.459)	1.171 (0.505)	1.231 (0.578)	1.332 (0.703)	1.306 (0.651)	1.482 (0.787)	1.415 (0.755)	1.087 (0.353)
15. Use mobile in silent place	1.496 (0.749)	1.457 (0.620)	1.578 (0.851)	1.590 (0.697)	1.287 (0.594)	1.558 (0.769)	1.538 (0.672)	1.605 (0.813)	1.373 (0.659)

Note: ^{R} = reverse coded; PMPUQ-SV: Problematic mobile phone use questionnaire short version.

2.3. Analysis

2.3.1. Understanding Measurement Invariance

To investigate whether the PMPUQ-SV is psychometrically valid for use across different languages, an analysis of MI was conducted using multigroup confirmatory factor analysis (MGCFA). MI establishes whether various aspects of the latent structure of a model remain stable across multiple groups, being run in an iterative manner with a set of increasingly constrained confirmatory factor analyses (CFAs). Comparison tests are then undertaken to determine if reliable differences exist between these models [47], which would suggest groups have reliable variations at those specific levels. The first step in this procedure was to conduct individual CFAs in each language group and investigate model fit. Following this, a set of constrained and planned models were implemented as follows. Constraints are given in square brackets and are cumulative throughout:

- A test of configural (or 'pattern', (groups)) invariance that investigates whether the same number of factors and their respective items are the same across groups (i.e., does the specified CFA structure replicate across the groups tested?). Support for configural invariance would suggest that the three-factor solution of the PMPUQ-SV and respective items per factor are valid across groups.
- A test of metric (or 'weak', (loadings)) invariance that estimates whether the factor loading strengths are equivalent across groups. Metric invariance suggests that participants understand and respond to items in the same way across groups.
- A test of scalar invariance (or strong invariance, (thresholds)) investigates if group differences in factor means are unbiased [48], meaning latent scores can be compared across groups.
- A test of strict invariance (residuals) estimates whether observed items have the same residuals, meaning that items have the same measurement error terms across groups.
- An additional fifth model of strict invariance and equally constrained (means) tests if the entire mean structure is invariant. If supported, this suggests that the means of both the latent variables and observed variables are invariant across groups.

2.3.2. Ordinal Data Analysis

Given that all PMPUQ-SV items are assessed on ordinal scales, models that could support non-continuous item analyses were employed [49]. Therefore, multigroup analyses were run using the R program with RStudio [50] using the Lavaan [51], Psych [52], and SemTools [53] packages, all of which have options for assessing ordinal data in a CFA framework (see [54] for a tutorial). Accordingly, thresholds rather than intercepts were constrained. In all CFA models, correlation matrices used were polychoric and model fit statistics were estimated using diagonally weighted least squares scale-shifted (DWLSSS). DWLSSS has been found to be a more effective estimation method for ordinal data than maximum likelihood [55,56], the default estimation for most statistical software.

3. Results

3.1. Factor Structures

As the PMPUQ-SV is a relatively new scale and its first aim was to assess its psychometric properties to ascertain potential reasons concerning previous contradictory results [33,35,43,44], cut-off points for all indices were taken prudently. Firstly, individual CFAs were performed for the overall sample and individually in each language. The correlation matrix and factor loadings of PMPUQ-SV items across all linguistic versions can be found in Tables 2 and 3, respectively. Cut-off values for fit indices were applied as follows (although caution must be taken as these cut-off values appear to result in lower Type II error rates (with acceptable costs of Type-I error rates [57–61])): a Comparative Fit Index (CFI) between 0.90 and 0.95 is indicative of acceptable fit relative to the independent model, and from 0.95 is considered a good fit; Tucker Lewis Index (TLI) values greater than 0.90 have been used as acceptable fit models in the past, but since 2000, this has been increased to approximately 0.95 indicating good fit; Root Mean Square Error of Approximation (RMSEA) values less than or equal to 0.05 can be considered as good fit, values between 0.05 and 0.08 are acceptable, and greater than 0.08 could be considered a mediocre fit, but higher than 0.10 are considered a poor fit (ideally, if the RMSEA is greater than 0.05, the fit of the model is 'close' (i.e., such a model has a specification error, but this is not large; then sample size is a critical factor)); Standardised Root Mean Square Residual (SRMR) values must be less than 0.08 or close to 0.09 (or 0.10), as it is the most sensitive index to models with misspecified factor loadings, and a combination rule has been suggested (i.e., if the RMSEA is greater than 0.05 or close to 0.06, then the SRMR should be greater than 0.06, or close to 0.09–0.10, as it is usually acceptable for sample sizes that are equal to or less than 250 [59]). However, although reported here for the sake of transparency, χ^2 was not used to assess model fit, as it has been found to artificially inflate with an increasing sample size [62].

As can be seen in Table 4, the PMPUQ-SV in German, French, and English yielded good model fit statistics, whilst Hungarian and Finnish versions yielded adequate model fit statistics using less conservative cut-off scores. The Spanish version was almost acceptable, while the Italian and Polish versions had poor fit and were thus not carried forward for MI testing. Including the Italian and Polish versions would likely lead to the multigroup analysis immediately failing because the models already differ between the validated and non-validated languages in terms of factor numbers or structure. The English version was also removed because the highest response score for Item 12 (public transport) was not endorsed by any participants (with Lavaan requiring at least one response per level as a prerequisite).

Table 2. Correlation matrix of the 15 PMPUQ-SV items across all language adaptations.

	Item Number														
	1	*2*	*3*	*4*	*5*	*6*	*7*	*8*	*9*	*10*	*11*	*12*	*13*	*14*	*15*
1	-														
2	0.19 ***	-													
3	0.25 ***	0.22 ***	-												
4	0.58 ***	0.19 ***	0.30 ***	-											
5	0.09 ***	0.51 ***	0.33 ***	0.05 ***	-										
6	0.36 ***	0.07 ***	0.33 ***	0.33 ***	0.14 ***	-									
7	0.78 ***	0.22 ***	0.28 ***	0.56 ***	0.11 ***	0.39 ***	-								
8	0.19 ***	0.45 ***	0.31 ***	0.32 ***	0.35 ***	0.14 ***	0.20 ***	-							
9	0.23 ***	0.44 ***	0.51 ***	0.42 ***	0.31 ***	0.29 ***	0.29 ***	0.52 ***	-						
10	0.56 ***	0.13 ***	0.22 ***	0.56 ***	0.02	0.25 ***	0.63 ***	0.25 ***	0.32 ***	-					
11	0.12 ***	0.57 ***	0.24 ***	0.22 ***	0.41 ***	0.02	0.13 ***	0.53 ***	0.40 ***	0.19 ***	-				
12	0.08 ***	0.06 ***	0.18 ***	0.04 *	0.19 ***	0.12 ***	0.10 ***	0.08 ***	0.12 ***	0.01	0.08 ***	-			
13	0.53 ***	0.15 ***	0.27 ***	0.56 ***	0.05 *	0.23 ***	0.54 ***	0.25 ***	0.37 ***	0.62 ***	0.20 ***	0.09 ***	-		
14	0.15 ***	0.67 ***	0.32 ***	0.23 ***	0.49 ***	0.06 *	0.18 ***	0.56 ***	0.53 ***	0.20 ***	0.68 ***	0.20 ***	0.30 ***	-	
15	0.25 ***	0.08 ***	0.41 ***	0.25 ***	0.22 ***	0.41 ***	0.28 ***	0.22 ***	0.33 ***	0.20 ***	0.15 ***	0.38 ***	0.27 ***	0.21 ***	-

Note: All correlations are polychoric and p-values are FDR corrected for multiple comparisons, * $p < 0.05$, *** $p < 0.001$; PMPUQ-SV: Problematic mobile phone use questionnaire short version.

Table 3. Factor loadings of the PMPUQ-SV items across all languages.

	Factor Loadings		
	Dangerous Use	**Prohibited Use**	**Dependent Use**
14. Driving concentration	0.88		
11. Driving danger mobile	0.74		
2. Use mobile when driving	0.73		
8. Dangerous situations	0.73		
5. Avoid using on motorway	0.57		
9. Use mobile when forbidden		0.80	
3. Don't use when forbidden		0.63	
15. Use mobile in silent place		0.55	
6. Using mobile in library		0.52	
12. Public transport		0.26	
7. Easy to live without mobile			0.87
1. Easy not to use mobile			0.84
10. Lost without mobile			0.74
4. Difficult not to use mobile			0.73
13. Hard to turn off mobile			0.72

Note: PMPUQ-SV: Problematic mobile phone use questionnaire short version.

Table 4. Individual confirmatory factor analyses across all samples and for each language of the PMPUQ-SV 3.2. Measurement invariance.

Version	n	df	$\chi 2$	p	CFI	TLI	RMSEA	RMSEA 90% CI	pClose	SRMR
All languages	3038	87	1858.371	<0.001	0.947	0.936	0.082	0.079–0.085	<0.000	0.074
German	383	87	231.642	<0.001	0.972	0.966	0.066	0.056–0.076	<0.000	0.069
French	1076	87	589.755	<0.001	0.973	0.968	0.073	0.068–0.079	<0.000	0.075
English	117	87	108.124	=0.062	0.976	0.972	0.046	0.000–0.072	<0.000	0.106
Finnish	449	87	383.929	<0.001	0.924	0.908	0.087	0.078–0.096	<0.000	0.092
Spanish	156	87	186.553	<0.001	0.940	0.928	0.086	0.069–0.103	=0.001	0.115
Italian	288	87	337.184	<0.001	0.915	0.897	0.100	0.089–0.111	<0.000	0.108
Polish	258	87	303.358	<0.001	0.880	0.856	0.098	0.086–0.111	<0.000	0.112
Hungarian	311	87	236.557	<0.001	0.954	0.945	0.074	0.063–0.086	<0.000	0.103

Note: PMPUQ-SV = Problematic mobile phone use questionnaire short version; χ^2 = Chi-square value, CFI = comparative fit index, TLI = Tucker–Lewis index, RMSEA = root mean squared error of approximation, pClose = provides a one-sided test of the null hypothesis that the RMSEA is equal to 0.05 in the population, SRMR = standardized root mean square residual.

To test for MI across languages, a series of MGCFAs with increasing constraints were conducted. The degree of difference (Δ) between the pairs of nested models was assessed using ΔCFI, ΔRMSEA, and ΔSRMR, as recommended by Chen [63], with respective cut-off values of $\leq$0.01, $\leq$0.015, and $\leq$0.03 for metric invariance and $\leq$0.01 for scalar invariance [64]. Satorra-Bentler χ^2 difference tests were also calculated between the nested models, although again, these have been found to produce unreliable estimates for large sample sizes, and therefore need to be interpreted with caution. Successive models were only calculated if the previous less constrained invariance in the hierarchy was at least partially supported.

Configural invariance was supported because the majority of fit indices were adequate, and so the next levels of constraint were investigated. Satorra-Bentler tests between all subsequent MI models were reliably different ($p < 0.001$; see Table 5), suggesting that each successive model had a poorer fit than the previous one (this provides evidence against invariance). At each stage, the changes in fit indices were inspected to assess whether this conclusion could be supported. Across all models, the changes in all fit indices were well below the pre-specified cut-off Δ-values, with the exception of ΔCFI for the scalar, strict, and mean models, which exceeded the $\leq$0.01 threshold at 0.012, 0.014, and 0.012, respectively. Considering the excellent values for the remaining fit indices at each stage, overall, the MI results provide evidence for metric invariance and partial evidence for scalar, strict, and mean invariance.

Table 5. Measurement invariance procedure conducted between German, French, Finnish, Spanish, and Hungarian for PMPUQ-SV.

Invariance	df	χ^2	p	CFI	TLI	RMSEA	RMSEA 90% CI	pClose	SRMR	$\Delta\chi^2$	Δdf	ΔRMSEA	ΔCFI	ΔSRMR
Configural	435	1567	<0.001	0.964	0.956	0.074	0.070–0.078	<0.001	0.084					
Config vs. metric ***										133	24	0.002	0.001	0.009
Metric	483	1654	<0.001	0.962	0.959	0.072	0.068–0.075	<0.001	0.093					
Metric vs. Scalar ***										165	54	0.003	0.012	0.006
Scalar	591	2154	<0.001	0.950	0.955	0.075	0.071–0.078	<0.001	0.087					
Scalar vs. Strict ***										272	30	0.005	0.014	0.008
Strict	651	2643	<0.001	0.936	0.948	0.080	0.077–0.084	<0.001	0.095					
Strict vs. Means ***										345	6	0.007	0.012	0.002
Means	663	3038	<0.001	0.924	0.939	0.087	0.084–0.090	<0.001	0.097					

Note: PMPUQ-SV: Problematic mobile phone use questionnaire short version; Satorra-Bentler $\Delta\chi^2$ Tests: Config vs. Metric: $X^2(5) = 23.930$, $p < 0.001$, Metric vs Scalar: $X^2(11) = 29.202$, $p < 0.001$, Scalar vs Strict: $X^2(7) = 69.105$, $p < 0.001$, Strict vs Means: $X^2(1) = 28.590$, *** $p < 0.001$.

3.2. Response Rates

Finally, in order to explore potential reasons for the violated ΔCFI for the fifth (means) model and if any items could be flagged up for refinement in future iterations, the response rates for each item were investigated across all different linguistic versions. As can be seen in Table 6, across all items for all languages, 75.2% of responses were in the 'disagree' or 'strongly disagree' categories. However, as items were reversed (i.e., 2, 4, 8, 9, 10, 11, 13, and 14), it should be interpreted that, for example, strongly agreeing on Item 14 (i.e., 'I use my mobile phone while driving, even in situations that require a lot of concentration') is more indicative of PMPU (i.e., dangerous use), but strongly agreeing on Item 12 (i.e., 'When using my mobile phone on public transport, I try not to talk too loud') is less indicative of PMPU (i.e., prohibited use). Moreover, a pair of items (i.e., Item 12 and Item 14) showed a particularly skewed response pattern with less than 7.6% and 4.9%, respectively, of respondents endorsing PMPU, to some degree, with both statements. Response patterns such as these may suggest that participants were not able to identify with these behaviours and that particular items that comprise the PMPUQ-SV may not be able to adequately discriminate respondents into distinct groups (i.e., as the vast majority of respondents reject these items). Taken together, the MI results illustrate that some of the items of the PMPUQ-SV with more skewed response patterns may prove consistently difficult to identify with across French, German, Hungarian, Finnish, and Spanish respondents.

Table 6. Item response rates per response category across all languages for PMPUQ-SV.

		All Languages			
Factor	**Item**	**Strongly Agree (%)**	**Agree (%)**	**Disagree (%)**	**Strongly Disagree (%)**
Dangerous	14. Driving concentration R	1.8	3.1	13.1	82.0
	11. Driving danger mobile R	4.2	4.5	16.1	75.1
	2. Use mobile when driving R	4.7	11.4	20.0	63.9
	8. Dangerous situations R	8.1	12.3	23.3	56.3
	5. Avoid using on motorway	77.3	10.2	6.1	6.4
Prohibited	9. Use mobile where forbidden R	3.4	10.2	27.7	58.8
	3. Don't use when forbidden	55.1	22.2	16.7	6.0
	15. Use mobile in silent place	63.0	27.2	6.9	2.9
	6. Using mobile in library	27.7	23.7	34.0	14.6
	12. Public transport	64.1	28.3	5.0	2.6
Dependent	7. Easy to live without mobile	21.4	25.6	37.5	15.6
	1. Easy not to use mobile	26.6	26.1	31.9	15.3
	10. Lost without mobile R	11.2	27.3	32.3	29.2
	4. Difficult not to use mobile R	9.7	28.0	32.8	29.5
	13. Hard to turn off mobile R	9.4	19.8	32.1	38.7
	Mean *	*51.2*	*24.0*	*17.0*	*7.7*

Note: PMPUQ-SV: Problematic mobile phone use questionnaire short version; R = reverse coded for questionnaire validations, but actual score given here. * = mean when items 14, 11, 2, 8, 9, 10, 4, and 13 are reversed.

4. Discussion

The objectives of the present study were to determine an optimal factor structure for the PMPUQ–SV among university populations using eight different language versions, and to examine the MI of the PMPUQ–SV across all linguistic versions and across the eight versions. Taken together, the findings suggest that the PMPUQ-SV is a potentially appropriate psychometric tool to screen for prohibited, dangerous, and dependent mobile phone use in adults from countries using these languages (e.g., Europe and America). However, its psychometric properties can be nuanced depending on the respective language. Despite this, there are several potential reasons for adopting this tool. First, it is a very good psychometric tool for French and German mobile phone users, as its fit indices showed good fit in comparison with the other languages (i.e., $CFI_{French\ and\ German} = 0.97$; $TLI_{French\ and\ German} = 0.97$; $RMSEA_{French\ and\ German} = 0.07$; $SRMR_{French} = 0.07$, $SRMR_{German} = 0.08$). Second, in psychometric terms, the English and Hungarian versions were considered robust and the Finnish and Spanish versions were considered acceptable. However, the Italian and Polish versions did not meet the psychometric requirements.

4.1. Main Findings

Using the overall sample, the PMPUQ-SV performed well across its three theoretical factors in terms of their internal consistency, factor loadings, and CFA results. Only two elements were close to the limit of being psychometrically acceptable [59]. More specifically, the Cronbach's alpha for prohibited use was low because it is a coefficient sensitive to the number of items (i.e., the subscale only had five items), and it demonstrated that there was no poor interrelatedness between items or heterogeneous constructs, except for Item 6 (i.e., '*I don't use my mobile phone in a library*') relating to prohibited use. This item does not load strongly enough and could perhaps be dropped in future PMPUQ versions. Furthermore, as evidence for metric invariance demonstrated, factor loadings were similar across the MI groups, so Item 6 was poor across countries. In fact, reliabilities were quite poor for prohibited mobile phone use [65].

It was also demonstrated that the RMSEA value was not ideal [59,60]. The French and German versions were psychometrically excellent, although the latter had the same issue with an acceptable reliability for prohibited use. Consequently, more in-depth research is needed to explore the phenomenology of this specific aspect of PMPU. Similarly, the English and Hungarian versions fitted the model well, although the reliability of the English version using Cronbach's alpha was only acceptable [66]. In addition, the Finnish and Spanish versions can be argued to be acceptable with restrictions due to their mediocre TLI, RMSEA, and SRMS values.

In sum, half of the linguistic versions tested in the present study fitted the proposed model well (i.e., ordered by their respective goodness of fit: French, German, Hungarian, and English). However, two require future testing (i.e., Finnish and Spanish) and the other two (i.e., Polish and Italian) require further review in relation to potentially different mobile phone use patterns in some countries and/or to methodological aspects (e.g., the translation and back-translation method applied), because both had large enough sample sizes to test its factor structure using a confirmatory approach. Regarding the English version, previous studies using this tool have also shown other psychometric weaknesses in relation to its reliability (e.g., in other studies, the α for prohibited use was 0.59 [35], and the α for the dangerous subscale was 0.67 [35], or 0.42 [43]), which are in line with the present findings regarding the limited internal consistency for prohibited use. Consequently, this needs to be cautiously interpreted due to the short length of this particular subscale. However, in relation to previous studies using the full PMPUQ or its short version, its factor structure usually corroborated the underlying theoretical model, except for the English studies, which reported a two-factor solution [43,44].

The lower loadings achieved in the overall sample for one of the items on the dangerous use subscale (i.e., '*I try to avoid using my mobile phone when driving on the motorway*'), and four items on the prohibited use subscale (e.g., '*When using my mobile phone on public transport, I try not to talk too loud*'), could be due to two reasons. First, mobile phone users are probably in a more pre-contemplative stage (i.e., they may not consider their mobile phone behaviour as problematic when asked about it, possibly denying or resisting this possibility), which is in line with research on compulsive internet use and other addictive behaviours [67,68]. Second, some items may not have been appropriate in the present day and age. For instance, '*I don't use my mobile phone in a library*', which appeared appropriate in the present study because university samples were used, but (i) not all respondents may use libraries given the ease in which reading materials can be accessed remotely; and (ii) those who are library users can access their mobile phones using silent option modes (e.g., for checking the time or *Facebook* notifications, using IM, navigating, or listening to music through headphones).

When looking at the descriptive findings, the study also demonstrated that there appeared to be common usage patterns and preferences in relation to mobile phones for dangerous and prohibited use in the respective language versions. Items related to dangerous and prohibited use were very extreme (either strongly agree or disagree on the Likert scales), whereas items related to dependent use were more evenly spread across the response categories (from strongly agree to strongly disagree on the Likert scales). This may explain why a lower internal consistency was found for prohibited use, as well as the fact that the subscale only had five items. Underlying cultural differences may also explain this (e.g., English-speaking participants may have the most diversity in response to Item 12 (the item with lowest loadings on the prohibited use factor); for example, some cultures do not appear to mind talking on a bus, whereas other cultures do not like it at all). The study also demonstrated, in reference to the response patterns in Table 6, that participants endorsed much fewer PMPU behaviors when responding to items on the dangerous (M = 5.025) and prohibited factors (M = 5.9), relative to the dependency factor (M = 12.24). Furthermore, in a previous study [43], Item 12 was excluded from the analysis, because it did not share variance in the body of items.

The results in the present study also demonstrated that cultural differences in self-reported mobile phone usage patterns and contextual factors must be investigated in greater depth (e.g., driving regulations in the countries where the study was conducted). The PMPUQ-SV may be a good tool to initially screen for potential PMPU among adult mobile phone users in some languages (French, German, Hungarian, English, Finnish, and Spanish) if tested independently per country. However, the scale is only useful for mobile phone users who drive vehicles. Consequently, dangerous use is only associated with driving behaviour, instead of other dangers when using mobile phones (e.g., crossing the road). In future developments of the PMPU, it is recommended that the construct should not only be assessed with items related to driving behaviour [3,4,25] because other dangerous behaviours also exist and have been reported in recent research regarding safety, which can be included in future iterations (e.g., collisions or injuries when cycling or walking [24], such as '*I use my mobile phone whilst crossing the road*' [43]), especially if new scales are going to be tested using children, adolescents, or adults that are not drivers. Conversely, in countries like the US where driving is permitted during mid-adolescence [25], items addressing driving behaviours when using mobile phones are recommended (e.g., PMPUQ-SV). In other words, while almost all European countries allow driving individuals to drive from the age of 17 or 18 years old, in other countries, the ages at which individuals can drive are lower, such as 14–16 years old.

Regarding the strongest MI results, only the French, German, Hungarian, Finnish, and Spanish results can be compared when using these versions in a cross-cultural study [68] because it was only in these countries that configural and metric invariances resulted in obtaining the expected value [63,64]. However, the other types of invariance (i.e., scalar, strict, and mean models) slightly exceeded the threshold, but the model can still be considered tenable. In sum, the present findings suggest that the factor structure, loadings and intercepts, and residuals of the PMPUQ-SV are invariant across the French, German, Hungarian, Finnish, and Spanish language versions. Therefore, the present study provides evidence for the equality of meaning of the problematic mobile phone use construct in five out of eight languages, further providing confidence in future use of the PMPUQ-SV in cross-cultural research on PMPU. For instance, among some of the countries where the TUD project [45] was developed (i.e., Belgium, France, Switzerland, Canada, Germany, Hungary, Finland, and Spain), it appears that cross-cultural data can be reliably compared.

4.2. Limitations

The main potential limitations of the present study are the sampling and characteristics of the participants (i.e., convenience community-based self-selected samples), who were adults studying or working in universities (or who were related to those who study or work in higher education institutions). Nevertheless, a large sample was collected during the same period (i.e., in 2015), using similar strategies, and the same online survey, in order to guarantee the standardization of the procedures for collecting reliable data from the three specific aspects of PMPU. The data were also self-reported and are therefore subject to well-known biases and limitations that are inherent within such a methodology. The purpose of the present study was to evaluate the cross-cultural robustness of the PMPUQ-SV to facilitate the development of future epidemiological studies across different cultures as these studies can help better ascertain the potential problems on the phenomenology of maladaptive mobile phone use from a psycho-sociological perspective. The PMPUQ-SV is appropriate for specific types of mobile phone users and has partially been cross-validated, but still presents some weaknesses which need to be studied in future research (i.e., reliability and language adaptability).

4.3. Future Research Directions

PMPU is still open to debate in relation to its potential health and educational harms in individuals' daily lives. For instance, it is not clear if PMPU results from a contemporary psychosocial problem (facilitated through this technology and the online behaviours associated with it in individuals' daily lives) or from other potentially addictive technological behaviours [37]). PMPU is situated somewhere on the continuum between the absence of problems to severe problems, ranging from a normal daily behaviour to potentially dysfunctional behaviours (or as the consequence of an existing disorder [38]). Furthermore, mobile phones are being increasingly used by adolescents and young adults worldwide [1], and given that they are usually utilized mainly for communicative purposes (i.e., information and maintenance of social relationships), some degree of constant use is expected in Eastern and Western societies.

In recent years, research on PMPU has bloomed in East Asian countries, where the condition is often viewed and classified as an addictive behaviour. Recent studies conducted in this region have indicated the moderating and mediating roles of several sociodemographic factors (e.g., gender), usage patterns (e.g., history of mobile phone use), and psychological variables (e.g., personality traits, emotion regulation skills) [69–73]. Most of this research focused on addictive usage patterns (with scales such as the Mobile Phone Addiction Index [74]), and it would thus be relevant to adapt the PMPUQ-SV to these contexts to provide a tool able to measure different types of PMPU. Such adaptation would also allow for interesting cross-cultural studies to be conducted between, for example, Asian and European countries.

One of the first literature reviews that examined both problematic internet and mobile phone use between 1991 and 2005 using five scientific databases determined, at that time, that mobile phone addiction symptoms were less consistently reported than internet addiction symptoms [75]. Sanchez-Carbonell and colleagues stated in 2008 that the use of synchronous apps (such as chatting apps and online games) might increase the likelihood of developing an addictive behaviour, due to the time lapse between engaging in the act and receiving a reward. A recent longitudinal study [30] partially confirmed this hypothesis (especially in relation to social networking apps and messaging services such as *WhatsApp* and *Facebook*). However, recent research on PMPU has not provided this evidence yet [35], in comparison with other internet-related problems, such as gaming, which appears to be the most prevalent because of the immediate rewards [76]. A recent review of cell-phone addiction [77] concluded (irrespective of whether or not it is a genuine addiction) that mobile phones give rise to problems that increasingly affect daily life. For instance, even with the risk of unlimited use (due to the affordability of contracts), the conceptualisation of this problematic mobile phone behaviour is still debated. In general, there is still an overlap in definitions of problematic behaviours related to online activities, including PMPU, Internet addiction, gaming disorder, social network use disorder, and others. Future studies are needed to gather evidence on how a nomenclature can progress and improve.

5. Conclusions

The present study is the first to ascertain that the PMPUQ-SV is an appropriate psychometric tool for cross-cultural comparisons to determine future prevalence estimates of multidimensional PMPU (i.e., dangerous, prohibited, and dependent mobile phone use). This is the first study that has investigated PMPU in an international sample, conducting MGCFAs and MI on a multidimensional model of the PMPUQ-SV across multiple language versions. An optimal factor structure (i.e., three-factor model) was found for the PMPUQ–SV among different university populations using six language versions (French, German, Hungarian, English, Finnish, and Spanish), and the MI of the PMPUQ–SV was examined across eight linguistic versions. The results indicate that five of the language variants (i.e., French, German, Finnish, Spanish, and Hungarian) are comparable for future cross-cultural studies. The PMPUQ-SV has been validated for almost all languages tested in order to be used independently in countries using these languages, and parts of these versions can be used for cross-cultural comparisons. The present study contributes to the behavioural addictions field by cross-validating results that can be used for future cross-cultural research on PMPU.

Author Contributions: Conceptualization: O.L.-F. Data curation: O.L.-F.; Formal analysis: C.D. and L.V.J. Funding acquisition: O.L.-F., J.B. Investigation: O.L.-F., D.J.K., H.M.P., M.F.G., N.M., M.K., H.-J.R., A.B., A.-K.G., L.R., L.K., Y.M., A.R., P.G., Z.D., O.K., A.S., A.P., B.L.-K., J.C., M.C., J.J.Z., E.S. (Emilia Serra), M.D., L.R., D.Z., S.A., N.I.L., E.S. (Eva Suryani), J.M.H., J.P.T., and J.B.; Methodology, O.L.-F.; Project administration: O.L.-F.. Resources: O.L.-F. Supervision: O.L.-F. and J.B. Writing—original draft: O.L.-F. Writing, review and editing: O.L.-F., D.J.K., H.M.P., M.D.G., C.D., L.V.J., N.M., M.K., H.-J.R., A.B., A.-K.G., L.R., L.K., Y.M., A.R., P.G., Z.D., O.K., A.S., A.P., B.L.-K., J.C., M.C., J.J.Z., E.S. (Emilia Serra), M.D., L.R., D.Z., S.A., N.I.L., E.S. (Eva Suryani), J.M.H., J.P.T., and J.B.

Acknowledgments: The present study was supported, first, by the European Commission ("Tech Use Disorders"; FP7-PEOPLE-2013- IEF-627999) through a Marie Curie postdoctoral grant awarded to O.L.-F. (supervisor: J.B.). Second, by the Psychology Department QR Funding at Nottingham Trent University, through a Kickstarter bid grant (2017) awarded to O.L.-F. to develop studies on 'Internet and mobile phone addiction: Cross-cultural epidemiological studies'. O.L.-F. also acknowledges the support of Kim Hoffman from the International Center for Advanced Research and Applied Science (INCAAS), Peru; Carmen Margarita Ilizarbe Pizarro, Universidad Antonio Ruiz de Montoya, Peru; and Katarzyna Gajewska from the Polish Foundation for Humanitarian Aid 'Res Humanae', Poland. The Hungarian part of the study was supported by the Hungarian Scientific Research Fund (grant number: K111938; KKP126835). O.K. acknowledges the support of the ÚNKP-17-4 New National Excellence Program of the Ministry of Human Capacities.

Conflicts of Interest: The authors declare no conflict of interest.

Appendix A

English

In relation with your mobile phone/smartphone, please answer these questions on a scale from 1 to 4, the numbers corresponding to: 1 "Strongly agree", 2 "Agree", 3 "Disagree", 4 "Strongly disagree"

The statement suits you:

1. It is easy for me to spend all day not using my mobile phone.
2. I use my mobile phone while driving.
3. I don't use my mobile phone when it is completely forbidden to use it.
4. Is it hard for me not to use my mobile phone when I feel like it.
5. I try to avoid using my mobile phone when driving on the motorway.
6. I don't use my mobile phone in a library.
7. I can easily live without my mobile phone.
8. I use my mobile phone in situations that would qualify as dangerous.
9. I use my mobile phone where it is forbidden to do so.
10. I feel lost without my mobile phone.
11. While driving, I find myself in dangerous situations because of my mobile phone use.
12. When using my mobile phone on public transport, I try not to talk too loud.
13. It is hard for me to turn my mobile phone off.

14. I use my mobile phone while driving, even in situations that require a lot of concentration.
15. I try to avoid using mobile phone where people need silence.

French

Concernant votre téléphone portable/smartphone, veuillez répondre à ces questions selon une échelle allant de 1 à 4, ces chiffres correspondant à : 1 "Tout à fait ", 2 "Plutôt bien ", 3 "Plutôt mal", 4"Pas du tout "

L'énoncé vous correspond:

1. Il est facile pour moi de passer toute une journée sans utiliser mon téléphone portable.
2. Je téléphone en conduisant.
3. Je n'utilise pas mon téléphone portable dans des lieux où il est formellement interdit de le faire.
4. Il m'est difficile de ne pas utiliser mon téléphone portable lorsque j'en ai envie.
5. J'évite d'utiliser mon téléphone portable quand je conduis sur l'autoroute.
6. Je n'utilise pas mon téléphone portable quand je suis dans une bibliothèque.
7. Je peux facilement me passer de mon téléphone portable.
8. Je me sers de mon téléphone portable dans des situations que je peux qualifier de «dangereuses».
9. Je me sers de mon téléphone portable dans des lieux où la loi l'interdit.
10. Je me sens perdu quand je n'ai pas mon téléphone portable.
11. En conduisant, je me retrouve en situation délicate alors que j'utilise mon téléphone portable.
12. Quand je téléphone dans les transports publics, je fais attention à ne pas parler trop fort.
13. Il est pénible pour moi d'éteindre mon téléphone portable.
14. En conduisant, j'utilise mon téléphone portable dans des situations qui demandent une concentration importante.
15. J'évite d'utiliser mon téléphone portable dans des endroits où il faut être silencieux.

German

Bitte beantworten Sie die nächsten Fragen in Bezug auf Ihr Smartphone/Handy auf einer Skala von 1 bis 4, die Zahlen entsprechend: 1 "Stimme sehr zu", 2 "Stimme zu", 3 "Stimme nicht zu", 4 " Stimme überhaupt nicht zu"

Wählen Sie die Aussage, die am besten zu Ihnen passt:

1. Es fällt mir leicht, den Tag zu verbringen, ohne mein Handy zu benutzen.
2. Ich benutze mein Handy, während ich Auto fahre.
3. Ich gebrauche mein Handy nicht, wenn es absolut verboten ist es zu benutzen.
4. Es fällt mir schwer mein Handy nicht zu verwenden, wenn mir danach ist.
5. Ich versuche zu vermeiden, mein Handy während der Autobahnfahrt zu benutzen.
6. Ich benutze mein Handy nicht in einer Bibliothek.
7. Ohne mein Handy kann ich problemlos leben.
8. Ich benutze mein Handy in Situationen, die als gefährlich gelten würden.
9. Ich verwende mein Handy an Orten, an denen die Nutzung verboten ist.
10. Ohne mein Handy fühle ich mich verloren.
11. Während ich Auto fahre, bringe ich mich in gefährliche Situationen, weil ich gleichzeitig mein Handy benutze.
12. Bei der Verwendung meines Mobiltelefons in öffentlichen Verkehrsmitteln versuche ich nicht zu laut zu sprechen.
13. Es fällt mir schwer, mein Handy auszuschalten.
14. Ich benutze mein Handy während der Autofahrt, selbst in Situationen, die viel Aufmerksamkeit erfordern.

15. Ich versuche meine Handynutzung in Situationen zu vermeiden, in denen Menschen Ruhe brauchen.

Hungarian

Kérjük, válaszold meg az alábbi kérdéseket a mobiltelefonod használatával kapcsolatban egy egytől négyig terjedő skálán, ahol: 1 "Teljesen egyetértek", 2 "Egyetértek", 3 "Nem értek egyet", 4 "Egyáltalán nem értek egyet"

Mennyire értesz egyet az alábbi állításokkal?

1. Könnyű számomra egy egész napot eltölteni anélkül, hogy használnám a mobilomat.
2. Használom a mobilomat vezetés közben.
3. Nem használom a mobilomat olyankor, amikor ez egyértelműen tilos.
4. Nehezemre esik, hogy ne használjam a mobilomat bármikor, amikor csak kedvem van hozzá.
5. Igyekszem nem használni a mobilomat, amikor autópályán vezetek.
6. Nem használom a mobilomat, ha könyvtárban vagyok.
7. Jól elvagyok a mobilom nélkül.
8. Olyan helyzetekben is használom a mobilomat, amikor az mások szerint veszélyes lehet.
9. Olyan helyen is használom a mobilomat, ahol ez tilos.
10. Elveszettnek érzem magam a mobilom nélkül.
11. Sokszor veszélyes helyzetekben találom magam vezetés közben a mobilhasználatom miatt.
12. Amikor tömegközlekedési eszközön használom a mobilomat, igyekszem nem túl hangosan beszélni.
13. Nehezemre esik kikapcsolni a mobilomat.
14. Még olyan helyzetekben is használom a mobilomat vezetés közben, amik nagy koncentrációt igényelnek.
15. Igyekszem nem használni a mobilomat olyan helyeken, ahol másoknak csendre van szükségük.

Finnish

Alla väittämiä matka-/älypuhelimenne käytöstä. Vastatkaa seuraaviin väittämiin valitsemalla itseänne parhaiten kuvaava vaihtoehto asteikolla 1–4: 1 "Täysin samaa mieltä", 2 "Jokseenkin samaa mieltä", 3 "Jokseenkin eri mieltä", 4 "Täysin eri mieltä"

1. Minulle on helppoa olla koko päivä käyttämättä matkapuhelintani.
2. Käytän matkapuhelinta autolla ajaessani.
3. En käytä matkapuhelintani silloin, kun sen käyttö on ehdottomasti kielletty.
4. On vaikeaa olla käyttämättä matkapuhelinta silloin kun haluaisin käyttää sitä.
5. Yritän välttää matkapuhelimen käyttöä moottoritiellä ajaessani.
6. En käytä matkapuhelinta kirjastossa.
7. Voin helposti elää ilman matkapuhelinta.
8. Käytän matkapuhelintani tilanteissa, jotka voidaan luokitella vaarallisiksi.
9. Käytän matkapuhelintani silloinkin, kun sen käyttö on kielletty.
10. Tunnen olevani hukassa ilman matkapuhelintani.
11. Ajaessani huomaan olevani vaarallisissa tilanteissa matkapuhelimen käyttöni seurauksena.
12. Käyttäessäni matkapuhelinta joukkoliikennevälineissä yritän olla puhumatta kovalla äänellä.
13. Minulle tuottaa vaikeuksia laittaa matkapuhelimeni pois päältä.
14. Käytän matkapuhelinta ajaessani autolla myös tilanteissa, jotka vaativat erityistä tarkkaavaisuutta.
15. Yritän välttää matkapuhelimen käyttöä tilanteissa, joissa tarvitaan hiljaisuutta.

Italian

Per favore, indica quanto ciascuna affermazione è adatta a descrivere l'uso che fai del tuo cellulare/ smartphone, su una scala da 1 a 4. I numeri corrispondono a: 1. "Fortemente d'accordo", 2. "D'accordo", 3. "In disaccordo", 4. "Fortemente in disaccordo"

1. E' facile per me trascorrere tutto il giorno senza utilizzare il cellulare.
2. Utilizzo il cellulare mentre guido.
3. Non uso il cellulare quando è assolutamente proibito usarlo.
4. E' difficile per me non utilizzare il cellulare quando mi sento di farlo.
5. Cerco di evitare di utilizzare il cellulare quando guido in autostrada.
6. Non uso il cellular in biblioteca.
7. Posso vivere facilmente senza il cellulare.
8. Uso il cellulare in situazioni che potrebbero essere considerate pericolose.
9. Uso il cellulare dove è proibito farlo.
10. Mi sento perso senza il cellulare.
11. Quando guido mi ritrovo in situazioni pericolose perchè utilizzo il cellulare.
12. Quando uso il cellulare su un mezzo di traporto pubblico, cerco di non parlare troppo forte.
13. E' difficile per me spegnere il cellulare.
14. Uso il cellulare quando guido, anche in situazioni che richiedono molta concentrazione.
15. Cerco di evitare di usare il cellulare nei posti in cui la gente ha bisogno di silenzio.

Spanish

En relación con su móvil/smartphone, responda por favor a estas cuestiones en una escala de valoración del 1 al 4, los números corresponden a: 1 "Muy de acuerdo", 2 "De acuerdo", 3 "En desacuerdo", 4 "Muy en desacuerdo"

La declaración que más le convenga:

1. Es fácil para mí pasar todo el día sin usar el móvil.
2. Uso el móvil mientras conduzco.
3. Yo no uso el móvil cuando está completamente prohibido.
4. Es difícil para mí no usar el móvil cuando quiero usarlo.
5. Trato de evitar el uso del móvil cuando conduzco por la autopista.
6. No uso el móvil en una biblioteca.
7. Puedo vivir fácilmente sin mi móvil.
8. Uso el móvil en situaciones que podrían considerarse peligrosas.
9. Uso el móvil donde está prohibido utilizarlo.
10. Me siento perdido sin el móvil.
11. Me he encontrado en situaciones peligrosas debido al uso del móvil mientras conducía.
12. Cuando uso el móvil en el transporte público, trato de no hablar demasiado alto.
13. Es difícil para mí apagar el móvil.
14. Uso el móvil durante la conducción, incluso en situaciones que requieren mucha concentración.
15. Trato de evitar el uso del móvil en lugares en que la gente necesita silencio (o se ruega silencio).

Polish

Proszę odpowiedzieć na pytania dotyczące Pana(i) telefonu/smartfona, posługując się skalą, w której kolejne cyfry od 1 do 4 znaczą: 1 "Całkowicie", 2 "Raczej tak", 3 "Raczej nie", 4 "Wcale"

Stwierdzenie odpowiadające Panu(i):

1. Łatwo mi spędzić cały dzień bez używania mojego telefonu komórkowego.

2. Rozmawiam przez telefon prowadząc samochód.
3. Nie używam telefonu w miejscach, w których formalnie jest to zakazane.
4. Trudno mi się powstrzymać od używania mojego telefonu komórkowego, kiedy mam na to ochotę.
5. Staram się nie używać telefonu, kiedy prowadzę saochód na autostradzie.
6. Nie używam telefonu kiedy jestem w bibliotece.
7. Z łatwością mogę obejść się bez mojego telefonu.
8. Używam mojego telefonu w sytuacjach, które uważam za « niebezpieczne ».
9. Używam mojego telefonu w miejscach, w których prawo tego zakazuje.
10. Czuję się zagubiony, gdy nie mam ze sobą mojego telefonu.
11. Prowadząc, zdarza mi się znaleźć w trudnej sytuacji podczas gdy używam telefonu komórkowego.
12. Kiedy rozmawiam przez telefon w środkach komunikacji publicznej, zwracam uwagę by nie rozmawiać zbyt głośno.
13. Źle się czuję, kiedy wyłączam mój telefon, nie lubię tego.
14. Prowadząc samochód używam telefonu w sytuacjach, które wymagają szczególnej koncentracji.
15. Unikam używania telefonu w miejscach, w których należy zachować ciszę.

References

1. International Telecommunication Union (ITU). ITU Committed to Connecting the World: ICT Facts and Figures 2017—Global ICT Developments. Available online: https://www.itu.int/en/ITU-D/Statistics/Pages/stat/default.aspx (accessed on 7 March 2018).
2. ProQuest. Search—All Databases. Available online: https://search.proquest.com/results/3E9BFDEC5154401PQ/1?accountid=14693 (accessed on 26 January 2018).
3. Bianchi, A.; Phillips, J.G. Psychological predictors of problem mobile phone use. *Cyberpsychol. Behav.* **2005**, *8*, 39–51. [CrossRef] [PubMed]
4. Billieux, J.; Van der Linden, L.; Rochat, L. The role of impulsivity in actual and problematic use of the mobile phone. *Appl. Cogn. Psychol.* **2008**, *26*, 1195–1210. [CrossRef]
5. Billieux, J. Problematic use of the mobile phone: A literature review and a pathways model. *Curr. Psychiatry Rev.* **2012**, *8*, 299–307. [CrossRef]
6. Lopez-Fernandez, O.; Kuss, D.J.; Griffiths, M.D.; Billieux, J. The conceptualization and assessment of problematic mobile phone use. In *Encyclopedia of Mobile Phone Behavior (Volumes 1, 2, & 3)*; Yan, Z., Ed.; IGI Global: Hershey, PA, USA, 2015; pp. 591–606. ISBN 9781466682399.
7. Billieux, J.; Maurage, P.; Lopez- Fernandez, O.; Kuss, D.J.; Griffiths, M.D. Can disordered mobile phone use be considered a behavioral addiction? An update on current evidence and a comprehensive model for future research. *Curr. Addict. Rep.* **2015**, *2*, 156–162. [CrossRef]
8. Martinotti, G.; Villella, C.; Di Thiene, D.; Di Nicola, M.; Bria, P.; Conte, G.; Cassano, M.; Petruccelli, F.; Corvasce, N.; Janiri, L.; et al. Problematic mobile phone use in adolescence: A cross-sectional study. *J. Public Health* **2011**, *19*, 545–551. [CrossRef]
9. Chóliz, M. Mobile-phone addiction in adolescence: The test of mobile phone dependence (TMD). *Prog. Health Sci.* **2012**, *2*, 33–44. [CrossRef]
10. López-Fernandez, O.; Honrubia-Serrano, M.L.; Freixa-Blanxart, M. Spanish adaptation of the "mobile phone problem use scale" for adolescent population. *Adicciones* **2012**, *24*, 123–130. [CrossRef] [PubMed]
11. Ezoe, S.; Toda, M.; Yoshimura, K.; Naritomi, A.; Den, R.; Morimoto, K. Relationships of personality and lifestyle with mobile phone dependence among female nursing students. *Soc. Behav. Personal.* **2009**, *37*, 231–238. [CrossRef]
12. Gallimberti, L.; Buja, A.; Chindamo, S.; Terraneo, A.; Marini, E.; Rabensteiner, A.; Vinelli, A.; Gomez Perez, L.J.; Baldo, V. Problematic cell phone use for text messaging and substance abuse in early adolescence (11- to 13-year-olds). *Eur. J. Pediatr.* **2016**, *175*, 355–364. [CrossRef] [PubMed]

13. Yang, Y.; Yen, J.; Ko, C.; Cheng, C.; Yen, C. The association between problematic cellular phone use and risky behaviors and low self-esteem among Taiwanese adolescents. *BMC Public Health* **2010**, *10*. [CrossRef] [PubMed]
14. Thomée, S.; Dellve, L.; Härenstam, A.; Hagberg, M. Perceived connections between information and communication technology use and mental symptoms among young adults—A qualitative study. *BMC Public Health* **2010**, *10*, 66. [CrossRef] [PubMed]
15. Thomée, S.; Härenstam, A.; Hagberg, M. Mobile phone use and stress, sleep disturbances, and symptoms of depression among young adults—A prospective cohort study. *BMC Public Health* **2011**, *11*, 66. [CrossRef] [PubMed]
16. Elhai, J.D.; Dvorak, R.D.; Levine, J.C.; Hall, B.J. Problematic smartphone use: A conceptual overview and systematic review of relations with anxiety and depression psychopathology. *J. Affect. Disord.* **2017**, *207*, 251–259. [CrossRef] [PubMed]
17. Elhai, J.D.; Levine, J.C.; Dvorak, R.D.; Hall, B.J. Non-social features of smartphone use are most related to depression, anxiety and problematic smartphone use. *Comput. Hum. Behav.* **2017**, *69*, 75–82. [CrossRef]
18. Stothart, C.; Mitchum, A.; Yehnert, C. The attentional cost of receiving a cell phone notification. *J. Exp. Psychol. Hum. Percept. Perform.* **2015**, *41*, 893–897. [CrossRef] [PubMed]
19. McDaniel, B.T.; Coyne, S.M. "Technoference": The interference of technology in couple relationships and implications for women's personal and relational well-being. *Psychol. Pop. Media Cult.* **2016**, *5*, 85–98. [CrossRef]
20. Bickham, D.S.; Hswen, Y.; Rich, M. Media use and depression: Exposure, household rules, and symptoms among young adolescents in the USA. *Int. J. Public Health* **2015**, *60*, 147–155. [CrossRef] [PubMed]
21. Tao, S.; Wu, X.; Wan, Y.; Zhang, S.; Hao, J.; Tao, F. Interactions of problematic mobile phone use and psychopathological symptoms with unintentional injuries: A school-based sample of Chinese adolescents. *BMC Public Health* **2016**, *16*, 88. [CrossRef] [PubMed]
22. Delgado, M.K.; Wanner, K.J.; McDonald, C. Adolescent cellphone use while driving: An overview of the literature and promising future directions for prevention. *Media Commun.* **2016**, *4*, 79–89. [CrossRef] [PubMed]
23. Takao, M.; Takahashi, S.; Kitamura, M. Addictive personality and problematic mobile phone use. *Cyberpsychol. Behav.* **2009**, *12*, 501–507. [CrossRef] [PubMed]
24. Chotpitayasunondh, V.; Douglas, K.M. How "phubbing" becomes the norm: The antecedents and consequences of snubbing via smartphone. *Comput. Hum. Behav.* **2016**, *63*, 9–18. [CrossRef]
25. Roberts, J.A.; David, M.E. My life has become a major distraction from my cell phone: Partner phubbing and relationship satisfaction among romantic partners. *Comput. Hum. Behav.* **2016**, *54*, 134–141. [CrossRef]
26. Kim, Y.; Jeong, J.-E.; Cho, H.; Jung, D.-J.; Kwak, M.; Rho, M.J.; Yu, H.; Kim, D.-J.; Choi, I.Y. Personality factors predicting smartphone addiction predisposition: Behavioral inhibition and activation systems, impulsivity, and self-control. *PLoS ONE* **2016**, *11*, e0159788. [CrossRef] [PubMed]
27. Lin, T.T.C.; Chiang, Y.; Jiang, Q. Sociable people beware? Investigating smartphone versus nonsmartphone dependency symptoms among young Singaporeans. *Soc. Behav. Personal.* **2015**, *43*, 1209–1216. [CrossRef]
28. Foerster, M.; Roser, K.; Schoeni, A.; Röösli, M. Problematic mobile phone use in adolescents: Derivation of a short scale MPPUS-10. *Int. J. Public Health* **2015**, *60*, 277–286. [CrossRef] [PubMed]
29. Roser, K.; Schoeni, A.; Foerster, M.; Röösli, M. Problematic mobile phone use of Swiss adolescents: Is it linked with mental health or behaviour? *Int. J. Public Health* **2016**, *61*, 307–315. [CrossRef] [PubMed]
30. Carbonell, X.; Chamarro, A.; Oberst, U.; Rodrigo, B.; Prades, M. Problematic Use of the Internet and Smartphones in University Students: 2006–2017. *Int. J. Environ. Res. Public Health* **2018**, *15*, 475. [CrossRef] [PubMed]
31. Long, J.; Liu, T.Q.; Liao, Y.H.; Qi, C.; He, H.Y.; Chen, S.B.; Billieux, J. Prevalence and correlates of problematic smartphone use in a large random sample of Chinese undergraduates. *BMC Psychiatry* **2016**, *16*, 408. [CrossRef] [PubMed]
32. Andreassen, C.S.; Billieux, J.; Griffiths, M.D.; Kuss, D.J.; Demetrovics, Z.; Mazzoni, E.; Pallesen, S. The relationship between addictive use of social media and video games and symptoms of psychiatric disorders: A large-scale cross-sectional study. *Psychol. Addict. Behav.* **2016**, *30*, 252–262. [CrossRef] [PubMed]

33. Lopez-Fernandez, O.; Kuss, D.J.; Romo, L.; Morvan, Y.; Kern, L.; Graziani, P.; Rousseau, A.; Rumpf, H.J.; Bischof, A.; Gässler, A.K.; et al. Self-reported dependence on mobile phones in young adults: A European cross-cultural empirical survey. *J. Behav. Addict.* **2017**, *6*, 168–177. [CrossRef] [PubMed]
34. Jiang, Z.; Zhao, X. Self-control and problematic mobile phone use in Chinese college students: The mediating role of mobile phone use patterns. *BMC Psychiatry* **2016**, *16*, 416. [CrossRef] [PubMed]
35. Lopez-Fernandez, O.; Männikkö, N.; Kääriäinen, M.; Griffiths, M.D.; Kuss, D.J. Mobile gaming and problematic smartphone use: A comparative study between Belgium and Finland. *J. Behav. Addict.* **2018**, *9*, 1–12. [CrossRef] [PubMed]
36. Chung, N. Korean adolescent girls' addictive use of mobile phones to maintain interpersonal solidarity. *Soc. Behav. Personality* **2011**, *39*, 1349–1358. [CrossRef]
37. Billieux, J.; Schimmenti, A.; Khazaal, Y.; Maurage, P.; Heeren, A. Are we overpathologizing everyday life? A tenable blueprint for behavioral addiction research. *J. Behav. Addict.* **2015**, *4*, 119–123. [CrossRef] [PubMed]
38. Chóliz, M. Mobile phone addiction: A point of issue. *Addiction* **2010**, *105*, 373–374. [CrossRef] [PubMed]
39. Baron, N.S.; af Segerstad, Y.H. Cross-cultural patterns in mobile-phone use: Public space and reachability in Sweden, the USA and Japan. *New Media Soc.* **2010**, *12*, 13–34. [CrossRef]
40. Lopez-Fernandez, O. Short version of the Smartphone Addiction Scale adapted to Spanish and French: Towards a cross-cultural research in problematic mobile phone use. *Addict. Behav.* **2017**, *64*, 275–280. [CrossRef] [PubMed]
41. Lopez-Fernandez, O.; Honrubia-Serrano, M.L.; Freixa-Blanxart, M.; Gibson, W. Prevalence of problematic mobile phone use in British adolescents. *Cyberpsycho. Behav. Soc. Netw.* **2014**, *17*, 91–98. [CrossRef] [PubMed]
42. Pedrero, E.J.; Rodriguez Monje, M.T.; Ruiz Sanchez De León, J.M. Mobile phone abuse or addiction. A review of the literature. *Adicciones* **2012**, *24*, 139–152. [CrossRef]
43. Kuss, D.J.; Harkin, L.; Kanjo, E.; Billieux, J. Problematic Smartphone Use: Investigating Contemporary Experiences Using a Convergent Design. *Int. J. Environ. Res. Public Health* **2018**, *15*, 142. [CrossRef] [PubMed]
44. Kuss, D.J.; Kanjo, E.; Crook-Rumsey, M.; Kibowski, F.; Wang, Y.W.; Sumich, A. Problematic Mobile Phone Use and Addiction Across Generations: The Roles of Psychopathological Symptoms and Smartphone Use. *J. Technol. Behav. Sci.* **2018**. [CrossRef]
45. Tech Use Disorders. Technological Use Disorders: European Cross-Cultural Longitudinal and Experimental Studies for Internet and Smartphone Problem Uses, 12 July 2017. Available online: http://cordis.europa.eu/project/rcn/189961_en.html (accessed on 14 July 2017).
46. Beaton, D.E.; Bombardier, C.; Guillemin, F.; Ferraz, M.B. Guidelines for the process of cross-cultural adaptation of self-report measures. *Spine* **2000**, *25*, 3186–3191. [CrossRef] [PubMed]
47. Van de Schoot, R.; Lugtig, P.; Hox, J. A checklist for testing measurement invariance. *Eur. J. Dev. Psychol.* **2012**, *9*, 486–492. [CrossRef]
48. Gregorich, S.E. Do self-report instruments allow meaningful comparisons across diverse population groups? Testing measurement invariance using the confirmatory factor analysis framework. *Med. Care* **2006**, *44*, S78–S94. [CrossRef] [PubMed]
49. Kim, E.S.; Yoon, M. Testing measurement invariance: A comparison of multiple-group categorical CFA and IRT. *Struct. Equ. Model.* **2011**, *18*, 212–228. [CrossRef]
50. RStudio Team. *RStudio: Integrated Development for R*; RStudio, Inc.: Boston, MA, USA, 2015; Available online: http://www.rstudio.com/ (accessed on 7 March 2018).
51. Rosseel, Y. Lavaan: An R package for structural equation modeling and more. Version 0.5-12 (BETA). *J. Stat. Softw.* **2012**, *48*, 1–36. [CrossRef]
52. Revelle, W. *Procedures for Personality and Psychological Research*; Northwestern University: Evanston, IL, USA, 2016; Available online: http://www.personality-project.org/revelle.html and https://cran.r-project.org/web/packages/psych/index.html; (accessed on 7 March 2018).
53. SemTools Contributors. SemTools: Useful Tools for Structural Equation Modelling. R Package Version 0.4-14. 2016. Available online: https://cran.r-project.org/web/packages/semTools/index.html and https://cran.r-project.org/web/packages/semTools/semTools.pdf (accessed on 7 March 2018).
54. Hirschfeld, G.; Von Brachel, R. Multiple-Group confirmatory factor analysis in R—A tutorial in measurement invariance with continuous and ordinal. *Pract. Assess. Res. Eval.* **2014**, *19*, 1–11. Available online: http://pareonline.net/getvn.asp?v=19&n=7 (accessed on 7 March 2018).

55. Mîndrilă, D. Maximum Likelihood (ML) and Diagonally Weighted Least Squares (DWLS) Estimation Procedures: A Comparison of Estimation Bias with Ordinal and Multivariate Non-Normal Data. 2010. Available online: http://infonomics-society.org/wp-content/uploads/ijds/published-papers/volume-1-2010/Maximum-Likelihood-ML-and-Diagonally-Weighted-Least-Squares-DWLS-Estimation-Procedures-A-Comparison-of-Estimation-Bias-with-Ordinal-and-Multivariate-Non-Normal-Data.pdf (accessed on 7 March 2018).
56. Muthén, L.K.; Muthén, B.O. *Mplus User's Guide*; Authors: Los Angeles, CA, USA, 2001; Available online: http://research.socialwork.wayne.edu/pdf/mplus-users-guide.pdf (accessed on 7 March 2018).
57. Flora, D.B.; Curran, P.J. An empirical evaluation of alternative methods of estimation for confirmatory factor analysis with ordinal Data. *Psychol. Methods* **2004**, 9, 466–491. [CrossRef] [PubMed]
58. Hooper, D.; Coughlan, J.; Mullen, M. Structural equation modelling: Guidelines for determining model fit. *Electron. J. Bus. Res. Methods* **2008**, 6, 53–60. Available online: http://arrow.dit.ie/buschmanart (accessed on 7 March 2018).
59. Hu, L.; Bentler, P.M. Cutoff criteria for fit indexes in covariance structure analysis: Conventional criteria versus new alternatives. *Struct. Equ. Model.* **1999**, 6, 1–55. [CrossRef]
60. Browne, M.; Cudeck, R. Alternative ways of assessing model fit. In *Testing Structural Equation Models*; Bollen, K.A., Long, J.S., Eds.; Sage Publications: Beverly Hills, CA, USA, 1993; pp. 111–136. ISBN 978-0803945074.
61. Van de Schoot, R.; Lugtig, P.; Hox, J. A checklist for testing measurement invariance, Eur. *J. Dev. Psychol.* **2012**, 9, 486–492. [CrossRef]
62. Marsh, H.W.; Balla, J.R.; McDonald, R.P. Goodness-of-fit indexes in confirmatory factor analysis: The effect of sample size. *Psychol. Bull.* **1988**, 103, 391–410. [CrossRef]
63. Chen, F.F. Sensitivity of goodness of fit indexes to lack of measurement invariance. *Struct. Equ. Model.* **2007**, 14, 464–504. [CrossRef]
64. Cheung, G.W.; Rensvold, R.B. Evaluating goodness-of-fit indexes for testing measurement invariance. *Struct. Equ. Model.* **2002**, 9, 233–255. [CrossRef]
65. Tavakol, M.; Dennick, R. Making sense of Cronbach's alpha. *Int. J. Med. Educ.* **2011**, 2, 53–55. [CrossRef] [PubMed]
66. Schmitt, N. Uses and abuses of coefficient alpha. *Psychol. Assess.* **1996**, 8, 350–353. [CrossRef]
67. Khazaal, Y.; Chatton, A.; Atwi, K.; Zullino, D.; Khan, R.; Billieux, J. Arabic validation of the Compulsive Internet Use Scale (CIUS). *Subst. Abuse Treat. Prev. Policy* **2011**, 6, 32. [CrossRef] [PubMed]
68. Prochaska, J.O.; DiClemente, C.C.; Norcross, J.C. In search of how people change. Applications to addictive behaviors. *Am. Psychol.* **1992**, 47, 1102–1114. [CrossRef] [PubMed]
69. Jiang, Z.; Zhao, X. Brain behavioral systems, self-control and problematic mobile phone use: The moderating role of gender and history of use. *Personal. Individ. Differ.* **2017**, 106, 111–116. [CrossRef]
70. Liu, Q.; Zhou, Z.; Niu, G.; Fan, C. Mobile phone addiction and sleep quality in adolescents: Mediation and moderation analyses. *Acta Psychol. Sin.* **2017**, 49, 1524–1536. [CrossRef]
71. Lian, L. Alienation as mediator and moderator of the relationship between virtues and smartphone addiction among Chinese university students. *Int. J. Ment. Health Addict.* **2017**, 1–11. [CrossRef]
72. Gao, T.; Li, J.; Zhang, H.; Gao, J.; Kong, Y.; Hu, Y.; Mei, S. The influence of alexithymia on mobile phone addiction: The role of depression, anxiety and stress. *J. Affect. Disord.* **2018**, 225, 761–766. [CrossRef] [PubMed]
73. Zhang, Y.; Lu, G.; Liu, Y.; Zhou, Y. Mediating effect of self-identity on relationship between interpersonal adaptation and mobile phone addiction tendency in college students. *Chin. Ment. Health J.* **2017**, 31, 568–572.
74. Leung, L. Linking psychological attributes to addiction and improper use of the mobile phone among adolescents in Hong Kong. *J. Child. Media* **2008**, 2, 93–113. [CrossRef]
75. Sanchez-Carbonell, X.; Beranuy, M.; Castellana, M.; Chamarro, A.; Oberst, U. Internet and cell phone addiction: Passing fad or disorder? *Adicciones* **2008**, 20, 149–159. [CrossRef] [PubMed]

76. Griffiths, M.D.; Nuyens, F. An overview of structural characteristics in problematic videogame playing. *Curr. Addict. Rep.* **2017**, *4*, 272–283. [CrossRef] [PubMed]
77. De-Sola Gutiérrez, J.; Rodríguez de Fonseca, F.; Rubio, G. Cell-Phone Addiction: A Review. *Front. Psychiatry* **2016**, *7*, 175. [CrossRef] [PubMed]

International Journal of
Environmental Research and Public Health

MDPI

Article

Predicting the Time Spent Playing Computer and Mobile Games among Medical Undergraduate Students Using Interpersonal Relations and Social Cognitive Theory: A Cross-Sectional Survey in Chongqing, China

Li Chen [1,2,3,†], Ruiyi Liu [1,2,3,†], Huan Zeng [1,2,3], Xianglong Xu [1,2,3], Rui Zhu [1,2,3], Manoj Sharma [4,5,6] and Yong Zhao [1,2,3,*]

1 School of Public Health and Management, Chongqing Medical University, Chongqing 400016, China; nameclx@foxmail.com (L.C.); lry981118@foxmail.com (R.L.); zenghuan586@aliyun.com (H.Z.); xianglong1989@126.com (X.X.); 2016111041@stu.cqmu.edu.cn (R.Z.)
2 Research Center for Medicine and Social Development, Chongqing Medical University, Chongqing 400016, China
3 The Innovation Center for Social Risk Governance in Health, Chongqing Medical University, Chongqing 400016, China
4 Department of Behavioural and Environmental Health, Jackson State University, Jackson, MS 39213, USA; manoj.sharma@jsums.edu
5 Health for All, Omaha, NE 68144, USA
6 College of Health Sciences, Walden University, Minneapolis, MN 55401, USA
* Correspondence: zhaoyong@cqmu.edu.cn; Tel.: +86-23-6848-5008; Fax: +86-23-6848-5031
† These authors contributed equally to this work.

Received: 10 July 2018; Accepted: 4 August 2018; Published: 6 August 2018

Abstract: *Background*: Computer and mobile games are widely used among undergraduate students worldwide, especially in China. Our objective was to predict the time spent playing computer and mobile games based on interpersonal relations and social cognitive theory constructs (i.e., expectation, self-efficacy, and self-control). *Methods*: The cross-sectional survey was conducted in two medical universities using a sample of 1557 undergraduate students recruited by cluster sampling. The five-point Likert questionnaire was jointly developed by researchers from Chongqing Medical University and Jackson State University. *Results*: Approximately 30% and 70% of the students played computer and mobile games, respectively. The daily times spent by participants on computer games were 25.61 $\pm$ 73.60 min (weekdays) and 49.96 $\pm$ 128.60 min (weekends), and 66.07 $\pm$ 154.65 min (weekdays) and 91.82 $\pm$ 172.94 min (weekends) on mobile games. Students with high scores of interpersonal relations but low scores of self-efficacy spent prolonged time playing computer games on weekdays and weekends ($p < 0.05$ for all). Students with low scores of expectation spent prolonged time playing computer games on weekdays ($p < 0.05$). Students with high scores of interpersonal relations but low scores of self-efficacy and self-control spent prolonged time playing mobile games on weekdays and weekends ($p < 0.05$ for all). *Conclusions*: The prevalence and duration of playing mobile games were higher than those of playing computer games among medical undergraduate students in Chongqing, China. This study determined the interpersonal relations, self-efficacy, self-control, and expectation of the students at the time of playing computer and mobile games. Future studies may consider studying the interaction among game-related behaviours, environments, and personality characteristics.

Keywords: time; gaming disorder; interpersonal relations; self-efficacy; self-control; expectations

1. Introduction

Globally, computer and mobile games are widely played by undergraduate students and have been identified to cause serious public health problems. The use of computer and mobile games is especially prevalent in China. By the end of December 2015, the number of Chinese teenage netizens (18 to 25 years old) reached 287 million, and of these, 66.5% play computer and mobile games [1]. By 2017, China had more than 390 million students playing computer games, accounting for 25.4% of all Internet users [2]. With the development of portable devices, the number of players shifting to mobile phones from computers has increased. Since 2013, 15.6% of users have not played games on their computers [3]. According to a 2015 survey, 90% of teenagers used mobile phones to surf the Internet, and 51.0% of them played mobile games online [1]. Playing computer games for a long period increases the screen viewing time of undergraduate students, causes dry eye syndrome and visual fatigue [4,5] and raises the risk of sedentary behaviour and obesity [6–8]. Smartphone overuse can lead to neck, wrist and back pains [9]. Excessive smartphone use at night can also shorten sleep time and lead to stress and depression [10]. In addition, a positive correlation was noted between the severity of Internet addiction and depression in adolescents, but no such relationship was found between time spent using social networks and depression [11]. Moreover, being in a state of negative tension, such as anxiety and depression, can promote the use of smartphones and the Internet and even lead to addiction [12,13].

In the study on Internet or smartphone addiction (computer and mobile games), several models of Internet and smartphone addiction are developed, like the IPACE model of Brand or the model concerning problematic smartphone use of Bilieux [12,14,15]. These models suggest that personality characteristics or psychopathological phenomena (depression and anxiety) significantly influence Internet and smartphone addiction. According to Montag's research in 2017, the personality characteristics of Internet and smartphone addiction overlap, and the relationship between personality and Internet addiction is closer, with self-efficacy or self-control as potential precursors of addiction [16]. Empathy and life satisfaction are also linked to Internet and smartphone addiction [17]. In addition, a 2015 study in South Korea found that psychopathological phenomenon, such as anxiety, is a risk factor for Internet and smartphone addiction [13]. Preoccupation and conflict are risk factors for smartphone addiction [18]. Depression and attention deficit both play a critical role in Internet addiction [19]. Internet addiction can also predict stress, depression, anxiety, and loneliness [20]. Personality characteristics, such as self-esteem and resilience, also regulate depression and Internet and smartphone addiction [21]. Moreover, a considerable overlap exists between Internet and smartphone addiction [16,22]. This overlap may explain why studying the factors that influence Internet-use disorders (e.g., computer and mobile games) is crucial. Game disorder has progressed and is now officially included in the ICD-11. However, the inclusion of game disorder in the manual is controversial. The theoretical link between game disorders and personality traits and their role as factors of resilience or vulnerability (e.g., self-directedness, extraversion, impulsivity, and empathy) needs clarification.

Time spent on computer and mobile games is affected by many factors. In terms of gender, males are more interested in playing computer and mobile games than females [23]. A Chinese study found that teenage females spend more time on other extra-curricular and leisure activities, whereas teenage males spend more time playing video games [24]. Previous research showed that the length of time spent playing video games varies with age [25]. Smoking and drinking are associated with video games [26,27]. Family situation is also linked to playing computer and mobile games [6], whereby game behavior can enhance family cohesion to a certain extent [28]. Although game behavior is exhibited separately, most players prefer to share activities with friends or parents [28]. This preference may relatively extend game time. Many studies identified factors that influence the amount of time students spend playing computer and mobile games, but few studies focused on interpersonal relations and social cognitive theory constructs at the time of playing computer and mobile games, especially among Chinese medical undergraduate students.

Interpersonal relations and social cognitive theory constructs (i.e., expectation, self-efficacy, and self-control) are associated with game-playing behaviour [28–30]. Interpersonal relations measure the relationship between a person and those around him/her. A previous study found that people who play video games possess a good friendship network [28]. The social nature of certain video games can extend playing time whilst allowing players to gain new and other relationships [31]. Social cognition theory is a dynamic and reciprocal model based on the interaction among behaviour, personal factors, and environmental influences [32]. In this theory, expectation, self-efficacy and self-control are the core determinants for achieving a goal [33]. Social cognition theory was applied to study Chinese people for smoking cessation [34] and obesity prevention [35,36]. Expectations include the anticipation of the outcome of a particular action and the importance of the value of these results [37]. Self-efficacy describes a person's confidence in exhibiting a particular behaviour at a given moment [38]. Self-control describes a person's capability to regulate behaviour and includes strategies that encourage proximal and distal goal setting and self-rewards [37,38]. Another study found that self-efficacy is a comparatively robust predictor of involvement in massively multiplayer online role-playing game community [29]. In addition, self-control is negatively correlated with online game addiction [30], and students with high self-control spend lesser playing time on video games than those with low self-control. A study on Iranian students revealed that self-control when playing video games differs between male and female students [39].

Despite these efforts, previous studies paid little attention to medical undergraduate students, especially in China. Medical undergraduate students have more academic pressure and longer education years than other college students. Computer and mobile games provide a way for medical undergraduate students to relax. The video game time of medical students and the factors influencing this time may differ from those influencing students of other majors. To the best of our knowledge, this study is the first to use social cognitive theory in predicting the time spent playing computer and mobile games among medical undergraduate students in China. The objectives of this study are to examine the factors influencing students' behaviours in playing computer and mobile games and the effects of interpersonal relations and social cognitive theory constructs (i.e., expectation, self-efficacy, and self-control) on the time spent by students playing computer and mobile games.

2. Materials and Methods

2.1. Study Design

A cross-sectional study on medical undergraduate students was conducted in Chongqing in March 2018. Each medical university/college in Chongqing consists of a comparable number of students. A total of 1557 students were chosen from Chongqing Medical University and Chongqing Medical and Pharmaceutical College by cluster sampling. Cluster sampling was used to select the classes in school, and 27 classes (6 classes from Chongqing Medical and Pharmaceutical College and 19 from Chongqing Medical University) were selected from the two medical schools in Chongqing. The 27 classes were from Grades 1 to 3 (10 classes from Grade 1, 9 classes from Grade 2, and 8 classes from Grade 3). Prior to the investigation, we conducted a pilot survey in March 2018, involving 50 students from Chongqing Medical University and Chongqing Medical and Pharmaceutical College. With the pilot survey as basis, the questionnaire was distributed by a trained student helper to students who gave written consent to participate. For the questionnaire distribution, the student helper was trained on the purpose and method of the research and was the one who explained the research objectives to the students. The students who signed the informed consent filled out the questionnaire 'Study on the video games behaviour of undergraduate students in Chongqing, China', and the participants did not receive incentives. The students were asked to answer the questionnaire in the classroom within 15–20 min. This study was approved by Chongqing Medical University (Reference Number: 2016001), and the ethical approval includes allowing anonymous surveys of minors and adults with the consent of the class counsellor. Written informed consent for processing personal data was obtained from each participant.

2.2. Instruments

Questionnaire refers specifically to the measurement instrument used to obtain games related information in this study. The questionnaire was designed by Chongqing Medical University and Jackson State University researchers. The part on interpersonal relations was developed by research analysts from Chongqing Medical University. The part on social cognitive theory was developed by research analysts from Jackson State University. The questionnaire was translated from English into Chinese. We also checked the readability of the translated questionnaire in the pilot survey. The internal consistency of the total questionnaire was 0.954 (Cronbach's alpha). The games in this study refer to all multiplayer cooperative or independent games that operate on electronic device platforms, including online and offline variants.

The demographic information in this survey included gender, age (15–18, 19–20, 21–28 years old), ethnic group (Han nationals or Minority), grade level (Grade 1, Grade 2, Grade 3) and without siblings (Yes or No). Questions about living habits were also asked, including smoking (Smoker or Non-smoker) and drinking (Drinker or Non-drinker) preferences. This study measured the daily time spent playing computer and mobile games (on weekdays or weekends) with four questions. The participants were asked to answer the questions in min. We considered the score as a game time value which was zero, instead of not playing games when the self-reported game time was zero. The questions about interpersonal relations and social cognitive theory constructs were as follows: 6 items were used to measure interpersonal relations, 4 items were used to measure self-efficacy, 4 items were used to measure the self-control, and 10 items were used to measure the expectation (see Table 1). The internal consistency of the interpersonal relations was 0.929 (see Table 2). The internal consistencies of the self-efficacy, self-control, and expectation subscales were 0.887, 0.915, and 0.944, respectively.

Table 1. Specific problems in interpersonal relationships and social cognitive theory constructs.

Constructs	Specific Problems
Interpersonal relations [1]	How much do you agree with the following statement? I have good relationship with …
	(1) my classmates, (2) my roommates, (3) everyone around me, (4) my parents, (5) my teachers, (6) anyone. I don't often have conflicts with people.
Self-efficacy [2]	How sure are you that you will …
	(1) play computer or mobile games for less than 3 h daily? (2) reduce the time spent playing computer or mobile games, even if you enjoy playing games? (3) reduce the time spent playing computer or mobile games if you have to hand in your homework? (4) reduce the time spent playing computer or mobile games if you have to do something important?
Self-control [3]	How sure are you that you will …
	(1) set a goal to play computer or mobile games for less than 3 h daily? (2) reward yourself for insisting on reducing the time spent playing computer or mobile games daily? (3) remind yourself to insist on playing computer or mobile games for less than 3 h daily? (4) constantly check progress to make sure you play computer or mobile games for less than 3 h daily?
Expectation	If I play computer games or mobile games for less than 3 h daily, I will … [4]
	(1) have additional friends, (2) have more spare time, (3) enjoy more, (4) feel more relaxed, (5) be able to study well.
	Which of the following changes are important to you? [5]
	(6) have additional friends, (7) have more spare time, (8) enjoy more, (9) feel more relaxed, (10) be able to study well

[1] Response options were "Not At All Agree", "Slightly Agree", "Moderately Agree", "Very Agree"," Completely Agree". [2] Response options were "Not At All Sure", "Slightly Sure", "Moderately Sure", "Very Sure", "Completely Sure". [3] Response options were "Not At All Sure", "Slightly Sure", "Moderately Sure", "Very Sure", "Completely Sure". [4] Response options were "Never", "Hardly Ever", "Sometimes", "Almost Always", "Always". [5] Response options were "Not At All Important", "Slightly Important", "Moderately Important", "Very Important", "Extremely Important".

Table 2. Times spent playing games by undergraduate students in Chongqing, China (N, %).

Total Time (min)	Time Spent Playing Computer Game on Weekdays	Time Spent Playing Computer Game on Weekends	Time Spent Playing Mobile Game on Weekdays	Time Spent Playing Mobile Game on Weekends
Mean ± SD	25.61 ± 73.60	49.96 ± 128.60	66.07 ± 154.65	91.82 ± 172.94
0 [1]	898 (72.4)	853 (68.7)	387 (31.2)	400 (32.3)
(0–30] [2]	98 (7.9)	52 (4.2)	282 (22.7)	159 (12.8)
(30–60]	108 (8.7)	92 (7.4)	264 (21.3)	201 (16.2)
(60–90]	21 (1.7)	19 (1.5)	21 (1.7)	36 (2.9)
(90–120]	72 (5.8)	101 (8.1)	181 (14.6)	208 (16.8)
(120–150]	2 (0.2)	3 (0.2)	2 (0.2)	14 (1.1)
(150–180]	11 (0.9)	36 (2.9)	38 (3.1)	71 (5.7)
>180	31 (2.5)	85 (6.8)	66 (5.3)	151 (12.2)

[1] 0 min; [2] greater than 0 min is less than or equal to 30 min.

2.3. Data Analysis

Frequencies and percentages were calculated to summarize the distributions of the categorical variables. A *t*-test was employed to compare the differences in the continuous variables between males and females. Generalized linear models were developed using social cognitive theory constructs (i.e., expectations, self-efficacy and self-control), interpersonal relations, healthy habits (smoking and drinking status), gender, age group, grade level, lack of siblings, and nationality as independent variables and time spent playing computer and mobile games on weekdays and weekends as dependent variables. Statistical tests included a two-sided test, and statistical significance was at $p < 0.05$. All data were analyzed using SPSS22.0 for Windows (SPSS Inc., Chicago, IL, USA).

3. Results

3.1. Characteristics of the Sample

This survey involved 1557 undergraduate students. Of the 1241 persons who answered all the questions, 458 (36.9%) were males and 783 (63.0%) were females. All the participants were 15–28 years old, and the average age of the participants was 19.76 ± 1.30 years old. A total of 89.2% were Han nationals; 10.8%, minorities. Of the 1241 participants, 35.1% were Grade 1, 27.0% were Grade 2, and 38.0% were Grade 3. A total of 43.4% had no siblings, and 56.6% had siblings. Moreover, 6% and 11% were smokers and drinkers, respectively, and 94.0% and 89% did not smoke and drink (see Table 3).

Table 3. Demographic characteristics of undergraduate students in Chongqing, China.

Variables	Daily Time Playing Computer Games				Daily Time Playing Mobile Games				Total
	On Weekdays		On Weekends		On Weekdays		On Weekends		
	0 min	>0 min	0 min	>0 min	0 min	>0 min	0 min	>0 min	
	N (%)	N (%)	N (%)	N (%)	N (%)	N (%)	N (%)	N (%)	N (%)
Gender									
Male	109 (8.8)	349(28.1)	114 (9.2)	344 (27.7)	202 (16.3)	256 (20.6)	172 (13.9)	286 (23.0)	458 (36.9)
Female	278 (22.4)	505 (40.7)	287 (23.1)	496 (40.0)	696 (56.1)	87 (7.0)	681 (54.9)	102 (8.2)	783 (63.0)
Age									
15–18 years old	63 (5.1)	460 (12.9)	67 (5.4)	156 (12.6)	158 (12.7)	65 (5.2)	140 (11.3)	83 (6.7)	223 (18.0)
19–20 years old	207 (16.7)	458 (36.9)	215 (17.3)	450 (36.3)	493 (39.7)	172 (13.9)	467 (37.6)	198 (16.0)	665 (53.6)
21–28 years old	117 (9.4)	236 (19.0)	119 (9.6)	234 (18.9)	247 (19.9)	106 (8.5)	246 (19.8)	107 (8.6)	353 (28.4)
Nationality									
Han nationals	334 (23.9)	773 (62.3)	348 (28.0)	759 (61.2)	793 (63.9)	314 (25.3)	753 (60.7)	354 (28.5)	1107 (89.2)
Minority	53 (4.3)	81 (6.5)	53 (4.3)	81 (6.5)	105 (8.5)	29 (2.3)	100 (8.1)	34 (2.7)	134 (10.8)

Table 3. *Cont.*

Variables	Daily Time Playing Computer Games				Daily Time Playing Mobile Games				Total
	On Weekdays		On Weekends		On Weekdays		On Weekends		
	0 min	>0 min	0 min	>0 min	0 min	>0 min	0 min	>0 min	
	N (%)	N (%)	N (%)	N (%)	N (%)	N (%)	N (%)	N (%)	N (%)
Grade levels									
Grade 1	139 (11.2)	332 (26.8)	144 (11.6)	327 (26.3)	349 (28.1)	122 (9.8)	327 (26.3)	144 (11.6)	435 (35.1)
Grade 2	109 (8.8)	226 (18.2)	111 (8.9)	224 (18.0)	245 (19.7)	90 (7.3)	222 (17.9)	113 (9.1)	335 (27.0)
Grade 3	139 (11.2)	296 (23.9)	146 (11.8)	289 (23.3)	304 (24.5)	131 (10.6)	304 (24.5)	131 (10.6)	471 (38.0)
Without siblings									
Yes	162 (13.1)	376 (30.3)	175 (14.1)	363 (29.3)	364 (29.3)	174 (14.0)	342 (27.6)	196 (15.8)	538 (43.4)
No	225 (18.1)	478 (38.5)	226 (18.2)	477 (38.4)	534 (43.0)	169 (13.6)	511 (41.2)	192 (15.5)	703 (56.6)
Smoking status									
Smoker	21 (1.7)	54 (4.4)	20 (1.6)	55 (4.4)	31 (2.5)	44 (3.5)	27 (2.2)	48 (3.9)	75 (6.0)
Non-smoker	366 (29.5)	800 (64.5)	381 (30.7)	785 (63.3)	867 (69.9)	299 (24.1)	826 (66.6)	340 (27.4)	1166 (94.0)
Drinking status									
Drinker	36 (2.9)	100 (8.1)	41 (3.3)	95 (7.7)	78 (6.3)	58 (4.7)	70 (5.6)	66 (5.3)	136 (11.0)
Non-drinker	351 (28.3)	754 (60.8)	360 (29.0)	745 (60.0)	820 (66.1)	285 (23.0)	783 (63.1)	322 (25.9)	1105 (89.0)

3.2. Daily Time Spent Playing Computer and Mobile Games among Undergraduate Students

Participants reported spending an average of 25.61 ± 73.60 min playing computer games per day on weekdays. A total of 898 (72.4%) reported spending 0 min playing computer games on weekdays, and 31 participants reported spending over 180 min playing computer games on weekdays. A total of 206 participants played computer games from 0 to 60 min on weekdays daily.

Participants reported an average of 49.96 ± 128.60 min playing computer games per day on weekends. A total of 853 (68.7%) reported spending 0 min playing computer games on weekends, and 85 participants reported spending over 180 min playing computer games on weekends. A total of 144 participants played computer games from 0 to 60 min, and 101 participants played computer games from 90 to 120 min on weekends daily.

Participants reported spending an average of 66.07 ± 154.65 min playing mobile games per day on weekdays. A total of 387 (31.2%) reported spending 0 min playing mobile games on weekdays, and 66 participants reported spending over 180 min playing mobile games on weekdays. The majority of participants spent between 0 to 60 min (546 participants) and 90 to 120 min (181 participants) playing mobile games on weekdays daily.

Participants reported spending an average of 91.82 ± 172.94 min playing mobile games per day on weekends. A total of 400 (32.3%) reported spending 0 min playing mobile games on weekends, and 151 participants reported spending over 180 min on playing mobile games on weekends. The majority of participants spent between 0 to 60 min (360 participants) and 90 to 120 min (208 participants) playing mobile games on weekends daily (see Table 2).

3.3. Descriptive Statistics of Interpersonal Relations and Social Cognitive Theory Constructs

Compared with males, females had a significantly higher mean score of interpersonal relations for playing video games (p = 0.041). However, no significant differences between males and females were observed in the mean scores of self-efficacy, self-control, and expectation for playing video games (see Table 4).

Table 4. Interpersonal relations and social cognitive theory constructs among undergraduate students in Chongqing, China.

Interpersonal Relations and Social Cognitive Theory Constructs	Min	Max	Mean (SD)	Standardized Cronbach Alpha	Males	Females	p-Value
Interpersonal relations	0.00	24.00	15.56 (5.40)	0.929	15.98 (5.81)	15.31 (5.14)	0.041 *
Self-efficacy	0.00	16.00	10.36 (4.59)	0.887	10.16 (4.36)	10.49 (4.71)	0.208
Self-control	0.00	16.00	9.13 (4.72)	0.915	8.82 (4.57)	9.31 (4.81)	0.068
Expectation	0.00	40.00	25.00 (9.58)	0.944	24.58 (9.32)	25.24 (9.72)	0.238
Total	-	-	-	0.954			

* Statistically significant ($p < 0.05$).

3.4. Generalised Linear Model Analysis for Factors Affecting the Time Spent Playing Video Games

In multivariable analyses, interpersonal relations, self-efficacy, and expectation were associated with the time spent playing computer games (see Table 4). Male students spent considerable time playing computer games on weekdays ($p < 0.001$) and weekends ($p < 0.001$). Students with a high score in interpersonal relations spent a long time playing computer games on weekdays ($p < 0.001$) and weekends ($p < 0.001$). Students with a high score in self-efficacy spent limited time playing computer games on weekdays ($p = 0.010$) and weekends ($p = 0.011$). Students with a high score in expectation spent limited time playing computer games on weekdays ($p = 0.003$).

3.5. Generalised Linear Model Analysis for Factors Affecting the Time Spent Playing Mobile Games

In multivariable analyses, interpersonal relations, self-efficacy, expectations, and self-control were associated with the time spent playing mobile games (see Tables 5 and 6). Students with a high score in interpersonal relations spent a long time playing mobile games on weekdays ($p < 0.001$) and weekends ($p < 0.001$). Students with a high score in self-efficacy spent limited time playing mobile games on weekdays ($p = 0.002$) and weekends ($p < 0.001$). Students with a high score of self-control spent limited time playing mobile games on weekdays ($p = 0.017$) and weekends ($p = 0.029$).

Table 5. Generalised linear model analysis of factors that affect interpersonal relations and social cognitive theory constructs of time spent playing computer games among all participants in Chongqing, China.

Parameter	Time Spent Playing Computer Games on Weekdays			Time Spent Playing Computer Games on Weekends		
	B	SE	p-Value	B	SE	p-Value
Males vs. Females	39.100	4.186	<0.001 **	79.798	7.214	<0.001 **
19–20 years old vs. 15–18 years old	−15.115	5.728	0.008 *	−27.646	9.872	0.005 *
21–28 years old vs. 15–18 years old	−30.535	7.816	<0.001 **	−56.256	13.471	<0.001 **
Han nationals vs. Minority	5.129	6.293	0.415	−3.666	10.847	0.735
Grade 2 vs. Grade 1	5.68	5.307	0.277	4.514	9.147	0.622
Grade 3 vs. Grade 1	20.917	6.266	0.001 *	17.876	10.801	0.098
Without siblings vs. with siblings	−3.149	3.999	0.431	3.711	6.892	0.590
Smoker vs. Non-smoker	1.288	8.532	0.880	51.145	14.706	0.001 *
Drinker vs. Non-drinker	17.201	6.543	0.009 *	14.158	11.277	0.209
Interpersonal relations and Social Cognitive Theory constructs						
Interpersonal relations	2.358	0.465	<0.001 **	3.211	0.801	<0.001 **
Self-efficacy	−1.619	0.627	0.010 *	−2.756	1.081	0.011 *
Self-control	−0.658	0.560	0.239	−0.878	0.964	0.362
Expectation	−0.892	0.296	0.003 *	−0.881	0.510	0.084

* Statistically significant ($p < 0.05$); ** Statistically significant ($p < 0.001$).

Table 6. Generalised linear model analysis of factors that affect interpersonal relations and social cognitive theory constructs of time spent playing mobile games among all participants in Chongqing, China.

Parameter	Time Spent Playing Computer Games on Weekdays			Time Spent Playing Computer Games on Weekends		
	B	SE	*p*-Value	B	SE	*p*-Value
Males vs. Females	6.569	9.264	0.478	19.573	10.314	0.058
19–20 years old vs. 15–18 years old	−8.581	12.676	0.498	−11.780	14.114	0.404
21–28 years old vs. 15–18 years old	−15.332	17.298	0.375	−12.047	19.260	0.532
Han nationals vs. Minority	17.008	13.928	0.222	16.723	15.507	0.281
Grade 2 vs. Grade 1	−8.537	11.746	0.467	−0.442	13.078	0.973
Grade 3 vs. Grade 1	−1.095	13.869	0.937	−1.762	15.442	0.909
Without siblings vs. with siblings	2.453	8.850	0.782	−4.443	9.854	0.652
Smoker vs. Non-smoker	−12.803	18.883	0.498	−5.613	21.025	0.789
Drinker vs. Non-drinker	21.086	14.480	0.145	9.465	16.123	0.557
Interpersonal relations and Social Cognitive Theory constructs						
Interpersonal relations	3.783	1.028	<0.001 **	4.671	1.145	<0.001 **
Self-efficacy	−4.208	1.388	0.002 *	−6.157	1.546	<0.001 **
Self-control	−2.954	1.238	0.017 *	−3.011	1.379	0.029 *
Expectation	−0.490	0.655	0.454	−0.477	0.729	0.513

* Statistically significant ($p < 0.05$); ** Statistically significant ($p < 0.001$).

4. Discussion

The aim of this study is to determine the interpersonal relations, self-efficacy, self-control, and expectation of the students at the time of playing computer and mobile games among Chinese medical undergraduates. Our results indicate that high interpersonal relations, low self-efficacy, and low expectations lead to longer computer games. We find also that high interpersonal relationship, low self-efficacy, and low self-control lead to longer mobile game time.

Approximately 30% and 70% of the participants played computer and mobile games, respectively. According to a previous study, roughly 63.7% participants are active players of computer and mobile games [40]. Our study reported high gaming rates among medical undergraduate students. In addition, mobile games are more popular than computer games among these undergraduate students. This finding implies that an increasing number of medical undergraduate students are replacing computer games with mobile games. A Chinese study found that 22% of medical undergraduates play mobile games or read online novels in class. The use of mobile phones is not limited by space and time and can even penetrate the classroom. The heavy and tedious courses in medical colleges may lead to high mobile game usage in class. This situation may be one reason why mobile gamers outnumber computer gamers in this setting. Medical schools should strengthen classroom management and enhance interest in the curriculum. Attracting students' attention in class should be done as much as possible to improve quality of teaching and reduce use of mobile games.

Medical undergraduate students with high interpersonal relations and low self-efficacy showed increased times spent playing computer games on weekdays and weekends. A previous study found that gamers who spent more time playing computer games display more prosocial behavior [41] and wider friendship networks [28]. Games have gradually become a way of maintaining interpersonal relationships in real life among college students. Extended game time may be due to games' ability to enhance users' social attributes in real life. Extending game time may help maintain a good relationship and enhance users' social attributes in real life. Good social conditions may also further increase game time. If students' interpersonal communication is low, then they tend to avoid social interaction, and the use of games as a popular social way will be reduced. However, people with social difficulties play games for social comfort [42]. Students with poor relationships may also immerse themselves in a game, which is related to the type of game. People with good relationships may be inclined to choose social games, such as massively multiplayer games. People with poor relationships may be inclined to choose games that make them comfortable to ease social difficulties.

A previous study found that self-efficacy in the real world is negatively related to game addiction [43], which is similar to our finding. Self-efficacy may be enhanced or amplified by environmental encouragement. However, our study marks the first time the relationship between social cognition theory and playing time is tested among a specific segment of the Chinese population. Health education workers can enhance self-efficacy for playing a computer game to reduce screen time among undergraduate students, especially students from higher grade levels. A previous study showed that motivational interviewing to enhance self-efficacy through repetition, reinforcement, and encouragement can strengthen the intrinsic motivations of subjects [44]. Short-term goals and rewards can also improve self-efficacy [45].

Students with low expectation reported spending long times spent playing computer games on weekdays. A previous study found that adolescents' expectation for health behaviour can influence the establishment of individual trajectories of health [46]. Therefore, teenagers can achieve healthy goals by raising expectations and controlling playing time. Students with high expectations are more health conscious and willing to take the initiative to reduce their playing time and gain health. Their parents or peers can encourage healthy behaviours, and such an action can lead to higher expectations. Building high expectations for playing video games for less than 3 h a day is important for adolescents. Health educators can enhance the expectation for playing video games to less than 3 h a day to decrease the time spent playing computer games. Medical undergraduate students who are also drinkers and smokers reported spending long time playing computer games on weekdays. This finding is supported by previous studies [27,47,48]. Previous studies have linked interpersonal relations with social behaviours such as smoking and drinking, and we have found that interpersonal relations are also linked to gaming behaviours, which may suggest that games are a social tool in some way [49,50].

For mobile games, medical undergraduate students with high interpersonal relations, low self-efficacy, and low self-control reported spending long time playing mobile games on weekdays and weekends. Playing mobile games can enhance friendships among peers. A study based on massively multiplayer online role-playing games showed that social interaction is essential in gaming experience. A mobile phone-based social network is a major tool for college students to socialize with others. This feature may extend the time spent by undergraduate students on mobile games. A previous study found that mental health self-efficacy influences the symptom outcomes of a mobile phone user. This study also linked self-efficacy with reduced depression, anxiety, and stress symptoms and reported that this improved work and social functioning [51]. We found that students with high self-control spent lesser time on mobile games than on computer games. Self-control is negatively correlated with mobile phone use [52], which supports our finding. The current study determined that self-control is negatively associated with mobile games. Students with low self-control are unable to control themselves from playing mobile games. Health education workers can enhance the self-efficacy and self-control of undergraduate students for playing mobile games to reduce time spent playing mobile games.

However, gender, age, ethic group, and grade levels are not significantly associated with the time spent playing mobile games. No space or time limit is found for mobile games, and mobile games are more popular than computer ones for males and females. A 2014 research found that when surfing and playing games become the main functions of mobile phones, males are more likely to use mobile phones than females [24]. In recent years, an increasing number of mobile games aimed at female groups have been developed, such as 'Love and Producer', which is popular among young Chinese women. This activity may be the reason why no significant difference was observed between genders in mobile games in this study. Moreover, age is not significantly associated with the time spent playing mobile games. The participants in this study were mostly 18–22 years old. Age is insignificant, possibly due to the concentration of participants in the age group. More age groups should be included to determine the relationship between age and mobile games.

This study has notable limitations. Firstly, the cross-sectional survey data cannot determine the direction of causality. We cannot clarify if constructs precede behaviour. Secondly, the time spent playing games was self-reported by the participants, which may not be accurate for game time. Self-reported game time of zero is probably the average time of the last month of zero, and this finding does not mean the participant has never played a game. This study also does not distinguish between excessive games because no clear standard for excessive game time is available. The self-reported game times may have introduced information and measurement bias. Future studies can collect the time spent playing games with Internet technologies. Thirdly, the questionnaire on interpersonal relations was developed by researchers in Chongqing Medical University. The interpersonal relationship scale was only tested for internal consistency and readability. It lacks wide recognition and use. Fourthly, our study was conducted on medical universities, which do not accurately represent all the undergraduates in China. Future research requires a larger sample size to cover more students with different majors. Fifthly, self-efficacy, self-control, and expectation are strictly focused on computer and mobile games, and this focus may have inflated correlations with self-reported gaming. Future studies may consider studying the relationship between self-efficacy, which is not strictly focused on computer and mobile games, and the time spent playing computer and mobile games. Finally, the types of games are not distinguished in this study, and different types of games may have different playing times.

5. Conclusions

The prevalence and duration of playing mobile games are higher than those of playing computer games among medical undergraduate students in Chongqing, China. This study determined the interpersonal relations, self-efficacy, self-control, and expectation on the time spent playing computer and mobile games. This study reported on students' gender, age, grade level, and drinking and smoking status at the time of playing computer and mobile games. Future studies may consider studying the interaction among game-related behaviours, environments, and personality characteristics.

Author Contributions: All authors contributed to the design of the survey. L.C. designed the study, interpreted and analysed the data and drafted the manuscript. R.L. contributed to data analysis and data interpretation as well as drafted the manuscript. H.Z., X.X., R.Z., M.S. and Y.Z. helped draft the manuscript. All authors have read and approved the final manuscript. All authors have seen and approved the final version of the manuscript.

Funding: This research was funded by the Humanities and Social Science Project Fund of the Ministry of Education of the People's Republic of China grant number 15YJA860020.

Acknowledgments: We wish to thank the research students from Chongqing Medical University for their help in the data collection.

Conflicts of Interest: The authors declare no conflict of interest.

References

1. China Internet Network Information Center. Research Report on Internet Behavior of Chinese Teenagers in 2015. Available online: http://www.cnnic.cn/hlwfzyj/hlwxzbg/qsnbg/201608/P020160812393489128332.pdf (accessed on 10 July 2018).
2. China Culture and Entertainment Industry Association. China Game Industry Development Report in 2017. Available online: http://www.cnccea.com/index.php?m=newscon&id=408&aid=770 (accessed on 10 July 2018).
3. China Internet Network Information Center. 2013 China Mobile Game User Research Report. Available online: http://www.cnnic.cn/hlwfzyj/hlwxzbg/wybg/201409/P020140911538720279158.pdf (accessed on 10 July 2018).
4. Dumery, B.; Grounauer, P.A.; van Toi, V. *Eyestrain, Blink Rate and Dry Eye Syndromes of Video Display Terminal Users*; Springer: Berlin/Heidelberg, Germany, 2010; pp. 270–273.
5. Rajeev, A.; Gupta, A.; Sharma, M. Visual Fatigue and Computer Use Among College Students. *Indian J. Community Med.* **2006**, *31*, 192–193.

6. He, M.Z.; Piché, L.; Beynon, C.; Harris, S. Screen-related sedentary behaviors: Children's and parents' attitudes, motivations, and practices. *J. Nutr. Educ. Behav.* **2010**, *42*, 17–25. [CrossRef] [PubMed]
7. Bhadoria, A.S.; Kapil, U.; Kaur, S. Association of Duration of Time Spent on Television, Computer and Video Games with Obesity amongst Children in National Capital Territory of Delhi. *Int. J. Prev. Med.* **2015**, *6*, 80. [CrossRef] [PubMed]
8. Xu, X.; Pu, Y.; Sharma, M.; Rao, Y.; Cai, Y.; Zhao, Y. Predicting Physical Activity and Healthy Nutrition Behaviors Using Social Cognitive Theory: Cross-Sectional Survey among Undergraduate Students in Chongqing, China. *Int. J. Environ. Res. Public Health* **2017**, *14*. [CrossRef] [PubMed]
9. Cha, S.S.; Seo, B.K. Smartphone use and smartphone addiction in middle school students in Korea: Prevalence, social networking service, and game use. *Health Psychol. Open* **2018**, *5*. [CrossRef] [PubMed]
10. Lemola, S.; Perkinson-Gloor, N.; Brand, S.; Dewald-Kaufmann, J.F.; Grob, A. Adolescents' electronic media use at night, sleep disturbance, and depressive symptoms in the smartphone age. *J. Youth Adolesc.* **2015**, *44*, 405–418. [CrossRef] [PubMed]
11. Banjanin, N.; Banjanin, N.; Dimitrijevic, I.; Panticd, I. Relationship between internet use and depression: Focus on physiological mood oscillations, social networking and online addictive behavior. *Comput. Hum. Behav.* **2015**, *43*, 308–312. [CrossRef]
12. Billieux, J. Problematic Use of the Mobile Phone: A Literature Review and a Pathways Model. *Curr. Psychiatry Rev.* **2012**, *8*, 299–306. [CrossRef]
13. Choi, S.W.; Kim, D.J.; Choi, J.S.; Ahn, H.; Choi, E.J.; Song, W.Y.; Kim, S.; Youn, H. Comparison of risk and protective factors associated with smartphone addiction and Internet addiction. *J. Behav. Addict.* **2015**, *4*, 308–314. [CrossRef] [PubMed]
14. Brand, M. *Theoretical Models of the Development and Maintenance of Internet Addiction*; Springer International Publishing: Cham, Switzerland, 2017; pp. 19–34.
15. Brand, M.; Young, K.S.; Laier, C.; Wölfling, K.; Potenza, M.N. Integrating psychological and neurobiological considerations regarding the development and maintenance of specific Internet-use disorders: An Interaction of Person-Affect-Cognition-Execution (I-PACE) model. *Neurosci. Biobehav. Rev.* **2016**, *71*, 252–266. [CrossRef] [PubMed]
16. Lachmann, B.; Duke, É.; Sariyska, R.; Montag, C. Who's Addicted to the Smartphone and/or the Internet? *Psychol. Pop. Media Cult.* **2017**. [CrossRef]
17. Lachmann, B.; Sindermann, C.; Sariyska, R.Y.; Luo, R.; Melchers, M.C.; Becker, B.; Cooper, A.J.; Montag, C. The Role of Empathy and Life Satisfaction in Internet and Smartphone Use Disorder. *Front. Psychol.* **2018**, *9*, 398. [CrossRef] [PubMed]
18. Lee, H.; Kim, J.W.; Choi, T.Y. Risk Factors for Smartphone Addiction in Korean Adolescents: Smartphone Use Patterns. *J. Korean Med. Sci.* **2017**, *32*, 1674–1679. [CrossRef] [PubMed]
19. Sariyska, R.; Reuter, M.; Lachmann, B.; Montag, C. Attention Deficit/Hyperactivity Disorder is a Better Predictor for Problematic Internet use than Depression: Evidence from Germany. *J. Addict. Res. Ther.* **2015**, *6*, 209. [CrossRef]
20. Ostovar, S.; Allahyar, N.; Aminpoor, H.; Moafian, F.; Nor, M.B.M.; Griffiths, M.D. Internet Addiction and its Psychosocial Risks (Depression, Anxiety, Stress and Loneliness) among Iranian Adolescents and Young Adults: A Structural Equation Model in a Cross-Sectional Study. *Int. J. Mental Health Addict.* **2016**, *14*, 257–267. [CrossRef]
21. Park, H.; Choi, E. Smartphone Addiction and Depression: The Mediating Effects of Self-esteem and Resilience among Middle School Students. *J. Korean Acad. Community Health Nurs.* **2017**, *28*, 280–290. [CrossRef]
22. Min, K.; Kim, D.-J.; Cho, H.; Yang, S. The Smartphone Addiction Scale: Development and Validation of a Short Version for Adolescents. *PLoS ONE* **2013**, *8*, e83558. [CrossRef]
23. Yang, D.-J.; Chiu, J.-Z.; Chen, Y.-K. Examining the Social Influence on College Students for Playing Online Game: Gender Differences and Implications. *Turk. Online J. Educ. Technol.* **2011**, *10*, 115–122.
24. Jiang, X.X.; Hardy, L.L.; Ding, D.; Baur, L.A.; Shi, H.-J. Recreational screen-time among Chinese adolescents: A cross-sectional study. *J. Epidemiol.* **2014**, *24*, 397–403. [CrossRef] [PubMed]
25. Greenberg, B.S.; Sherry, J.; Lachlan, K.; Lucas, K.; Holmstrom, A. Orientations to Video Games Among Gender and Age Groups. *Simul. Gaming* **2010**, *41*, 238–259. [CrossRef]
26. Forsyth, S.R.; Malone, R.E. Smoking in Video Games: A Systematic Review. *Nicotine Tob. Res. Off. J. Soc. Res. Nicotine Tob.* **2016**, *18*, 1390–1398. [CrossRef] [PubMed]

27. Bőthe, B.; Tóth-Király, I.; Orosz, G. Clarifying the Links among Online Gaming, Internet Use, Drinking Motives, and Online Pornography Use. *Games Health J.* **2015**, *4*, 107–112. [CrossRef] [PubMed]
28. Durkin, K.; Barber, B. Not so doomed: Computer game play and positive adolescent development. *J. Appl. Dev. Psychol.* **2002**, *23*, 373–392. [CrossRef]
29. Hopp, T.; Barker, V.; Schmitz, A.W. Interdependent Self-Construal, Self-Efficacy, and Community Involvement as Predictors of Perceived Knowledge Gain among MMORPG Players. *Cyberpsychol. Behav. Soc. Netw.* **2015**, *18*, 468–473. [CrossRef] [PubMed]
30. Kim, E.J.; Namkoong, K.; Ku, T.; Kim, S.J. The relationship between online game addiction and aggression, self-control and narcissistic personality traits. *Eur. Psychiatry* **2008**, *23*, 212–218. [CrossRef] [PubMed]
31. Smyth, J.M. Beyond self-selection in video game play: An experimental examination of the consequences of massively multiplayer online role-playing game play. *Cyberpsychol. Behav.* **2007**, *10*, 717–721. [CrossRef] [PubMed]
32. Bandura, A. *Social Foundations of Thought and Action: A Social Cognitive Theory/Albert Bandura*; American Psychological Association: Washington, DC, USA, 1986.
33. Glanz, K.; Lewis, F.M.; Rimer, B.K. *Health Behavior and Health Education: Theory, Research, and Practice*; Jossey-Bass: San Francisco, CA, USA, 1997; pp. 344–353.
34. Zheng, P.; Guo, F.; Chen, Y.; Fu, Y.; Ye, T.; Fu, H. A randomized controlled trial of group intervention based on social cognitive theory for smoking cessation in China. *J. Epidemiol.* **2007**, *17*, 147–155. [CrossRef] [PubMed]
35. Murnan, J.; Sharma, M.; Lin, D. Predicting childhood obesity prevention behaviors using social cognitive theory: Children in China. *Int. Q. Community Health Educ.* **2006**, *26*, 73–84. [CrossRef] [PubMed]
36. Chen, Y.; Ma, L.; Ma, Y.; Wang, H.; Luo, J.; Zhang, X.; Luo, C.; Wang, H.; Zhao, H.; Pan, D.; et al. A national school-based health lifestyles interventions among Chinese children and adolescents against obesity: Rationale, design and methodology of a randomized controlled trial in China. *BMC Public Health* **2015**, *15*, 210. [CrossRef] [PubMed]
37. Bandura, A. Health Promotion by Social Cognitive Means. *Health Educ. Behav.* **2004**, *31*, 143–164. [CrossRef] [PubMed]
38. Tougas, M.E.; Hayden, J.A.; McGrath, P.J.; Huguet, A.; Rozario, S. A Systematic Review Exploring the Social Cognitive Theory of Self-Regulation as a Framework for Chronic Health Condition Interventions. *PLoS ONE* **2015**, *10*, e0134977. [CrossRef] [PubMed]
39. Haghbin, M.; Shaterian, F.; Hosseinzadeh, D.; Griffiths, M.D. A brief report on the relationship between self-control, video game addiction and academic achievement in normal and ADHD students. *J. Behav. Addict.* **2013**, *2*, 239–243. [CrossRef] [PubMed]
40. Mohamud, S. Effect of Video Games on Medical Students' Academic Performance: A Two-Institutions, Cross-Sectional Study. *Int. J. Sci. Eng. Res.* **2017**, *8*, 653–658.
41. Mengel, F. Computer games and prosocial behavior. *PLoS ONE* **2014**, *9*, e94099. [CrossRef] [PubMed]
42. Kowert, R.; Oldmeadow, J.A. Playing for social comfort: Online video game play as a social accommodator for the insecurely attached. *Comput. Hum. Behav.* **2015**, *53*, 556–566. [CrossRef]
43. Jeong, E.J.; Kim, D.H. Social activities, self-efficacy, game attitudes, and game addiction. *Cyberpsychol. Behav. Soc. Netw.* **2011**, *14*, 213–221. [CrossRef] [PubMed]
44. Walpole, B.; Dettmer, E.; Morrongiello, B.A.; McCrindle, B.W.; Hamilton, J. Motivational interviewing to enhance self-efficacy and promote weight loss in overweight and obese adolescents: A randomized controlled trial. *J. Pediatr. Psychol.* **2013**, *38*, 944–953. [CrossRef] [PubMed]
45. Schunk, D.H. Enhancing Self-Efficacy and Achievement through Rewards and Goals: Motivational and Informational Effects. *J. Educ. Res.* **1984**, *78*, 29–34. [CrossRef]
46. Mcdade, T.W.; Chyu, L.; Duncan, G.J.; Hoyt, L.T.; Doane, L.D.; Adam, E.K. Adolescents' expectations for the future predict health behaviors in early adulthood. *Soc. Sci. Med.* **2011**, *73*, 391–398. [CrossRef] [PubMed]
47. Frangos, C.C.; Frangos, C.C.; Sotiropoulos, I. Problematic Internet Use among Greek university students: An ordinal logistic regression with risk factors of negative psychological beliefs, pornographic sites, and online games. *Cyberpsychol. Behav. Soc. Netw.* **2011**, *14*, 51. [CrossRef] [PubMed]
48. Ko, C.H.; Yen, J.Y.; Yen, C.F.; Chen, C.S.; Weng, C.C.; Chen, C.C. The association between Internet addiction and problematic alcohol use in adolescents: The problem behavior model. *Cyberpsychol. Behav.* **2008**, *11*, 571–576. [CrossRef] [PubMed]

49. Liang, Y.; Zheng, X.; Zeng, D.D.; Zhou, X.; Leischow, S.J.; Chung, W. Characterizing Social Interaction in Tobacco-Oriented Social Networks: An Empirical Analysis. *Sci. Rep.* **2015**, *5*, 1–11. [CrossRef] [PubMed]
50. Monahan, J.L.; Lannutti, P.J. Alcohol as social lubricant. *Hum. Commun. Res.* **2000**, *26*, 175–202. [CrossRef]
51. Jiang, Z.; Zhao, X. Self-control and problematic mobile phone use in Chinese college students: The mediating role of mobile phone use patterns. *BMC Psychiatry* **2016**, *16*, 416. [CrossRef] [PubMed]
52. Clarke, J.; Proudfoot, J.; Birch, M.R.; Whitton, A.E.; Parker, G.; Manicavasagar, V.; Harrison, V.; Christensen, H.; Hadzi-Pavlovic, D. Effects of mental health self-efficacy on outcomes of a mobile phone and web intervention for mild-to-moderate depression, anxiety and stress: Secondary analysis of a randomised controlled trial. *BMC Psychiatry* **2014**, *14*, 272. [CrossRef] [PubMed]

International Journal of
Environmental Research and Public Health

MDPI

Article

Fear of Missing Out as a Predictor of Problematic Social Media Use and Phubbing Behavior among Flemish Adolescents

Vittoria Franchina [1,*], Mariek Vanden Abeele [2,*], Antonius J. van Rooij [3,4], Gianluca Lo Coco [1] and Lieven De Marez [3]

1 Department of Psychology and Educational Sciences, University of Palermo, 90133 Palermo, Italy; gianluca.lococo@unipa.it
2 Tilburg School of Humanities and Digital Sciences, Tilburg University, 5037AB Tilburg, The Netherlands
3 imec-mict-UGent, Department of Communication Sciences, Ghent University, 9000 Ghent, Belgium; trooij@trimbos.nl (A.J.v.R.); lieven.demarez@ugent.be (L.D.M.)
4 Department of Youth & Risky Behavior, Trimbos Institute, 3521 VS Utrecht, The Netherlands
* Correspondence: vittoria.franchina923005@gmail.com (V.F.); m.m.p.vandenabeele@tilburguniversity.edu (M.V.A.)

Received: 6 September 2018; Accepted: 24 September 2018; Published: 22 October 2018

Abstract: Fear-of-missing-out (FOMO) refers to feelings of anxiety that arise from the realization that you may be missing out on rewarding experiences that others are having. FOMO can be identified as an intra-personal trait that drives people to stay up to date of what other people are doing, among others on social media platforms. Drawing from the findings of a large-scale survey study among 2663 Flemish teenagers, this study explores the relationships between FOMO, social media use, problematic social media use (PSMU) and phubbing behavior. In line with our expectations, FOMO was a positive predictor of both how frequently teenagers use several social media platforms and of how many platforms they actively use. FOMO was a stronger predictor of the use of social media platforms that are more private (e.g., Facebook, Snapchat) than platforms that are more public in nature (e.g., Twitter, Youtube). FOMO predicted phubbing behavior both directly and indirectly via its relationship with PSMU. These findings support extant research that points towards FOMO as a factor explaining teenagers' social media use.

Keywords: fear of missing out (FOMO); social media; problematic social media use (PSMU); phubbing; teenagers; adolescents; addiction

---

1. Introduction

Behavioral addiction researchers argue that the psychological processes that explain problematic behavior require greater attention [1–5]. Understanding underlying processes is particularly relevant when examining problematic forms of (digital) media use. Digital devices, such as mobile phones, can be used and misused in a variety of different ways. It is likely that the way in which problem use manifests itself depends on the particular underlying psychological mechanism [6,7].

One psychological process that may underlie problematic digital media use is the Fear-Of-Missing-Out (FOMO). FOMO refers to the "pervasive apprehension that others might be having rewarding experiences from which one is absent" [8]. Persons who have a greater FOMO are assumed to have a greater desire to stay continually up-to-date of what others are doing, for example via the use of social media. According to Przybylski et al. [8], FOMO originates from psychological deficits in people's competence and relatedness needs [9]. One way of satisfying these needs, the authors claim, is through the use of social media applications, because the dynamic nature of social media applications provides users with a consistent stream of social and informational rewards [10].

The purpose of the current study is to contribute to the extant body of research on FOMO in relation to social media use. The study has four aims. First, it investigates whether teenagers with higher levels of FOMO have more social media accounts (i.e., the breadth of social media use) and whether they access these accounts more frequently (i.e., the depth of social media use [11]) than teenagers with lower levels of FOMO. Second, assuming that teenagers with higher levels of FOMO are mostly motivated to check up on people in their personal social networks, we examine whether FOMO is a stronger predictor of the use of platforms that connect teenagers to their offline networks (e.g., Facebook, Snapchat) than of the use of platforms that connect to a largely unknown audience (e.g., Youtube, Twitter). Third, the study examines if teenagers with greater FOMO report higher levels of problematic social media use (PSMU) and, four, are more likely to report one particular form of problematic social media use, which is the use of social media during conversations with co-present others (cf. "phubbing"). These research aims are addressed with data from a large-scale cross-sectional survey that was administered to 2663 Flemish adolescents.

2. Theoretical Framework

People have always had a tendency to think about what others are thinking and doing [12]. In the 1970 and 1980 scholars already identified that some people developed anxieties around missing out, contemplating on the rewarding experiences that others might be having (e.g., in the context of romantic relationships) [13]. FOMO is thus not an entirely new concept.

FOMO can be understood as such an anxiety around missing out on rewarding experiences that results from people's desire for interpersonal attachments [12]. This desire, which is grounded in people's need to belong, is an innate and fundamental motivation which humans have [14]. People gratify this need by seeking belongingness to social groups. Social groups nowadays exist in both physical and virtual shapes and people have access to their social groups in both ways, online and offline. Social media, which can be defined as "Internet-based applications that build on the ideological and technological foundations of Web 2.0, and that allow the creation and exchange of User Generated Content" [15,16], offer a place where users can keep in touch with their social circles. Social network sites (SNS) such as Facebook or Instagram, for example, offer users an online connection to persons in their personal networks, facilitating the practice of keeping in touch.

Nowadays the digital world is considered an extension of the Self [17,18]. In addition to the personal mind and physical body, the Self can be thought to include people, places, physical possessions, as well as affiliation groups to which a person feels attached [19]. Social media platforms are a part of this: they are the digital portals to affiliation groups. In contemporary society we thus manage the affiliation network both offline and online; virtual groups are as real and important as the physical ones. Not being able to connect with those affiliation groups on social media may cause feelings of being out of touch with "real" life [17]. After all, losing or missing a person one is attached to, can bring feelings of loss and grieving as much as if a part of the Self was damaged [19].

The fear of being socially excluded plays a role in experiencing a FOMO [12]. Social exclusion produces a loss of belongingness, and therefore causes anxiety. Thus, when people cannot access their social media accounts, they might feel anxiety because of a fear that they are being socially excluded [12].

Social exclusion also elicits feelings of worthlessness [12]. These feelings lead people to the act of comparing themselves to others on social media [20] with the purpose of deciding upon their own personal value [21]. Social networks offer a place where consumers, particularly young generations, can continuously keep up with what peers are doing and checking on what they are missing out on (e.g., social events, life experiences, life opportunities, and so on and so forth). FOMO can thus drive social media use, as checking up on other people may lead to a temporary relief of one's anxiety.

2.1. FOMO and the Use of Different Social Media Platforms

Users may use different media to gratify different needs [22–24]. This can explain differences in the popularity of certain social media platforms. In January 2015, the Global Web Index summary showed that the most popular social media platform was the social network site Facebook, immediately followed by YouTube, Twitter and Instagram. The same year Instagram's popularity outperformed Twitter. According to some, this is because pictures are more effective than words in achieving self-presentational objectives, which are central motivations for social media use [25]. Given that each social media platform is characterized by its own features and affordances, it is relevant to differentiate between social media platforms in research on social media use: Different platforms may connect users to different persons and networks, and give access to different forms of information of which users may wish to stay up-to-date.

Current research findings reveal that FOMO is a predictor of the use of SNS with which users connect to people in their personal networks, such as Instagram [25] and Facebook [26]. Instagram use, for example, is found to be motivated by the desire to keep in touch with others [25] and to "to keep up with or gain knowledge about what others (i.e., friends, family and strangers) are doing" [27]. FOMO has also been found to predict Facebook use [28,29] and Instagram use [30,31]. The above study findings thus indicate that FOMO predicts the use of at least these social network sites, but potentially also the use of other social media platforms.

With respect to social media use, a difference can be made between the "depth" and the "breadth" of one's use, where the depth of use refers to aspects such as the frequency and duration of social media use, and the breadth of use refers to the variety of social media platforms that are actively used. For teens in particular, not only frequent use, but also the use of a broad variety of social media platforms may serve the purpose of relieving anxieties with regard to not knowing what others are thinking and doing, as the differences in the relational affordances of different platforms [32] mean that they may give access to at least partially different networks and contents. Hence, for the current study we expect that not only the depth, operationalized as the frequency of social media use, but also the breadth, operationalized as the number of active social media platforms teenagers use, are predicted by FOMO:

Hypothesis 1. *Teens who experience greater FOMO, use social media more frequently (Hypothesis 1a;* i.e., *the depth of social media use), and use more different social media accounts (Hypothesis 1b;* i.e., *the breadth of social media use).*

2.2. FOMO and the Use of More Private versus More Public Social Media Platforms

As mentioned above, the use of different social media platforms may gratify different underlying needs. For example, a recent study shows that users significantly differ in the gratifications they derive from using Facebook, Instagram, Twitter and Snapchat [33]. One affordance in which platforms may differ is in the extent to which content is restricted to a (sub-)set of contacts, or fully public—in other words, whether content is shared with a mostly known versus a mostly unknown audience. On platforms such as Facebook, Instagram or Snapchat, people's online social network usually overlaps with their offline affiliation group (e.g., Facebook/Snapchat contacts need to know each other's names or phone numbers to see each other's posts and profiles). Platforms such as Twitter or Youtube, on the other hand, usually make content accessible to a wider audience of mainly unknown individuals, and resemble a broadcasting platform rather than a platform in which content is restricted to people who have been accepted as "contacts", "followers" or "friends".

Given this difference in the public accessibility of platform content, it seems logical to assume that SNSs such as Facebook or Snapchat are more apt at providing relief from FOMO than, for example, video sharing platforms such as Youtube or microblogging services such as Twitter because the former provide greater relief from anxieties surrounding what friends and family are doing.

Indeed, platforms such as Facebook or Instagram are more personal SNS that enable teens to limit content accessibility to the desired public (e.g., friends or friends-of-friends). As a result, these SNS may be especially attractive venues for teens with a high FOMO because it lets them know what people in their primary affiliative groups are thinking and doing. We explore this assumption by asking the following research question:

Research Question. *Is FOMO a stronger predictor of the use of more private social media platforms (that connect mostly to offline networks) than more publicly accessible social media platforms (that connect mostly to online networks)?*

2.3. FOMO and Problematic Social Media Use (PSMU)

When social media use is excessive, it can become problematic. Several studies have addressed problematic social media use (PSMU) [34–36]. There is an ongoing debate in the literature around the differences between problematic social media use and a possible social media behavioral addiction [4,37]. An in-depth discussion of this debate goes beyond the scope of this work. In the current study, however, we use the term problematic social media use, which we define as an unhealthy excessive form of social media use, characterized by a lack of control over the behavior, and continued behavior despite adverse life consequences. Our focus is on revealing the factors that predict to problematic social media use in a general population of teenagers (i.e., the aim is not to diagnose or identify pathological cases).

As mentioned before, one of the aims of this study is to explore if teenagers with a greater FOMO report a higher levels of PSMU. Previous studies suggest that this is the case [38–43], and suggest that those who experience FOMO may try to relieve their anxiety by checking up on other people on social media. Ironically, however, the more people check their social media accounts, however, the more they may find events they are missing out on. Using social media to reduce the anxiety may end up to be another source of FOMO. Therefore this vicious circle may reinforce itself, gradually turning social media use into problematic social media use. Hence we expect:

Hypothesis 2. *Teens who experience greater FOMO, report higher PSMU.*

2.4. FOMO and Phubbing

A final study purpose is to focus on FOMO as a predictor of one particular form of problematic social media use, which is the use of social media during conversations with co-present others. This practice is termed "phubbing" (derived from phone + snubbing), which refers to "the act of snubbing someone in a social setting by concentrating on one's phone instead of talking to the person directly" [44,45]. Experimental studies show that phubbing negatively impacts relational outcomes such as impression formation [46].

Recent findings show that people prefer to use smartphones when going online [47]. Smartphones allows us to be in contact with our affiliation groups on social media, everywhere we are. Therefore, we assume that if people experience anxiety, they may temporary try to reduce it by accessing their social media accounts on their smartphones [48]. It is likely that those high in FOMO, who use social media to address their anxiety, may overuse social media on their smartphones in such a way that it intersects with their offline social interactions, leading them to phub their offline interaction partners.

Hence, the last aim of this study is to explore whether teens with a greater FOMO report to engage in phubbing behavior more frequently, and whether the latter relationship is mediated by PSMU:

Hypothesis 3a. *Teens with a greater FOMO, are more likely to use social media during conversations with co-present others (cf. "phubbing").*

Hypothesis 3b. *PSMU mediates the former relationship.*

3. Method

3.1. Sample and Procedure

In Flanders, a consortium of non-profit organizations collaborates bi-annually on a large-scale survey project that examines the state of affairs of digital media ownership and use of Flemish youths. Apart from a large set of recurring questions, every edition of the survey includes a number of questions on topics that are considered relevant at the time.

The data gathered for the current study were part of the 2016 research project [49]. An omnibus survey was administered to the high school pupils of 11 geographically dispersed high schools in Flanders, Belgium. Using the information made available by the ministry of education, quota sampling was used to select schools, and—within the schools—years and classrooms. This procedure resulted in a final sample that is representative for the population in terms of gender, age and school track (see Van Waeg, D'Hanens, Dooms & Naesens [49] for further details).

Within each school, a local collaborator (e.g., the school's information and communications technology coordinator) organized the survey administration process, according to a set of instructions provided by the project leaders. The survey was administered online, but to avoid self-selection bias, the data collection took part during school hours, in the computer rooms of the schools. Unfortunately, no response rates were registered. However, the local collaborators stated that few pupils did not receive permission to participate. In total, the responses of 3291 pupils were gathered. The project leaders subjected these responses to a rigorous data cleaning procedure, leading to removal of 452 responses that were either substantially incomplete, either contained multiple invalid responses to validation screening items. This procedure resulted in a final sample of 2663 pupils (57.1% girls; M_{age} = 14.87, SD = 1.67). This final dataset was distributed by the consortium to the collaborating researchers for further analysis.

Informed consent was collected from both the participating teenagers and their parents. Given the large sample-size, an opt-out procedure was used for collecting consent from parents. The university's institutional review board approved the study.

3.2. Measures

3.2.1. Breadth of Social Media Platforms Used

Based on interviews with young persons, a list of 25 frequently used social media applications was constructed. We adhered to Kaplan and Haenlein's [16] definition of social media, which includes all platforms in which users can generate content that is (semi-)publicly available to others. The latter definition excludes platforms that focus exclusively on instant messaging (e.g., Whatsapp, Facebook Messenger). The list of included platforms can be consulted in Table 1. The breadth of social media use was assessed by asking for each platform whether the teenager had an active account (1 = yes, 0 = no), and then summing the total number of active accounts per participant.

Table 1. Scale items, means and standard deviations.

	Fear-of-Missing-Out (FOMO)		
	Items	**M**	**SD**
1	I fear my friends have more rewarding experiences than me	2.33	1.11
2	It is important that I understand my friends' "inside jokes"	3.09	1.05
3	It bothers me when I miss an opportunity to meet up with friends	4.16	0.90
4	When I go on summer camp or vacation, I continue to keep tabs on what my friends are doing	2.66	1.14
	Problematic Social Media Use (PSMU)		
	Items	**M**	**SD**
1	How frequently do you find it difficult to quit using social media?	2.89	1.19
2	How frequently do others (e.g., your parents or friends) tell you that you should spend less time on social media?	2.72	1.28
3	How frequently do you prefer using social media over spending time with others (e.g., with friends or family)?	1.89	0.97
4	How frequently do you feel restless, frustrated or irritated when you can't access social media?	2.33	1.16
5	How frequently do you do your homework poorly because you prefer being on social media?	2.51	1.17
6	How frequently do you use social media because you feel unhappy?	2.31	1.23
7	How frequently do you lack sleep because you spent the night using social media?	2.44	1.35
	Phubbing		
	Items	**M**	**SD**
1	How frequently do you use your mobile phone during a conversation in a bar or restaurant?	2.39	0.99
2	How frequently are you engaged with your phone during a conversation?	2.13	0.96
3	How frequently do you check social media on your phone during a personal conversation?	1.89	0.92

3.2.2. Depth of Social Media Platforms Used

The depth of social media use was measured by assessing for each active platform how frequently it was used (1 = less than once per week, 5 = multiple times per day). Questions with respect to the usage frequency of a platform were only answered by participants who had an active account for the platform. As visible in Table 2, a substantial number of platforms was used by (very) few participants. We opted to only include those platforms who were used by at least 5% of the sample. The analyses for Hypothesis 1a, concerning FOMO as a predictor of platform usage frequency (see Tables 3 and 4) are performed on the basis of the subsample of users of each platform.

Table 2. Percentage of teenagers with an account on various social media platforms and average frequency of use among account holders (1 = less than once/week, 5 = multiple times/day).

Social Media Platform	N	% of Total Sample	Average Usage Frequency		Less than Once per Week	Once per Week	Multiple Times per Week	Daily	Multiple Times per Day
			M	*SD*	%	%	%	%	%
Facebook	2360	89%	4.49	0.87	1.4	3.6	6.4	22.2	66.5
Snapchat	1937	73%	4.36	1.00	2.4	4.5	10.1	20.5	62.5
Instagram	1680	63%	4.36	0.97	1.5	4.4	15.7	31.8	46.6
YouTube	1596	60%	4.17	0.95	2.1	4.4	9.7	22.3	61.4
Google+	925	35%	2.67	1.40	28.2	21.5	19.8	16.4	14.1
Twitter	582	22%	3.63	1.35	8.8	14.9	19.4	18.6	38.3
Swarm	510	19%	4.26	1.09	3.7	5.5	10.8	20.6	59.4
We Heart It	329	12%	3.42	1.34	10.6	16.4	21.9	22.2	28.9
Pinterest	324	12%	2.6	1.34	28.1	21.6	24.4	14.5	11.4
Tumblr	298	11%	3.52	1.35	9.4	16.4	21.1	18.8	34.2
Vine	232	9%	3.12	1.38	15.9	19.8	23.3	18.5	22.4
Foursquare	172	6%	3.03	1.72	34.3	8.7	9.9	13.4	33.7
Tinder	120	5%	2.52	1.51	36.7	20	15.8	9.2	18.3
Kiwi	117	4%	3	1.58	25.6	18.8	13.7	13.7	28.2
Ask.fm	92	3%	3.6	1.60	20.7	6.5	12	14.1	46.7
Runkeeper	72	3%	2.17	1.10	31.9	34.7	23.6	4.2	5.6
Reddit	60	2%	3.07	1.48	20	20	18.3	16.7	25
Happening	48	2%	2.65	1.42	29.2	20.8	20.8	14.6	14.6
Vimeo	32	1%	2.91	1.61	34.4	6.3	15.6	21.9	21.9
Strava	25	1%	2.6	1.44	28	28	16	12	16
LinkedIn	24	1%	2.08	1.38	50	16.7	20.8	12.5	/
Periscope	15	1%	3.07	1.28	13.3	20	26.7	26.7	13.3
Endomondo	11	0%	2.73	1.74	36.4	18.2	9.1	9.1	27.3
Ello	8	0%	1.5	1.41	87.5	/	/	/	12.5
Meerkat	6	0%	2.33	2.07	66.7	/	/	/	33.3

Table 3. Gender, age, school track and Fear-of-Missing-Out (FOMO) as predictors of the frequency of Facebook, Snapchat, Instagram, Youtube, Google Plus and Twitter.

	Facebook			Snapchat			Instagram			Youtube			Google Plus			Twitter		
	PE	SE	Wald	PE	SE	Wald	PE	SE	Wald	PE	SE	Wald	PE	SE	Wald	PE	SE	Wald
Gender (boy)	−0.50	0.09	32.55 ***	−0.68	0.09	52.79 ***	−0.61	0.10	35.85 ***	0.72	0.10	56.21 ***	0.18	0.12	2.23	−0.25	0.15	2.58
Gender (girl)	a																	
Age	0.15	0.03	28.1 ***	0.02	0.03	0.30	0.12	0.03	14.36 ***	−0.01	0.03	0.14	−0.09	0.04	6.42 *	0.02	0.05	0.15
School track (voc)	0.41	0.14	8.72 **	0.40	0.14	8.00 **	−0.34	0.14	5.75 *	0.54	0.15	13.58 ***	0.51	0.17	8.79 **	0.36	0.25	2.18
School track (s-voc)	0.28	0.12	5.67 *	0.35	0.12	8.88 **	−0.01	0.12	0.00	−0.01	0.12	0.00	0.40	0.15	6.84 **	0.76	0.18	19.06 ***
School track (ac)	a																	
FOMO	0.48	0.07	55.09 ***	0.28	0.07	16.89 ***	0.34	0.07	22.48 ***	0.18	0.07	7.08 **	0.04	0.09	0.22	0.15	0.11	2.02
Pearson GOF	$X^2(1821) = 1635.34$, $p = 1.00$			$X^2(1731) = 1842.31$, $p = 0.031$			$X^2(1643) = 1716.37$, $p = 0.102$			$X^2(1623) = 1621.94$, $p = 0.503$			$X^2(1335) = 1370.00$, $p = 0.247$			$X^2(1083) = 1131.58$, $p = 0.148$		
−2LL GOF	$X^2(5) = 150.10$, $p < 0.001$			$X^2(5) = 88.07$, $p < 0.001$			$X^2(5) = 79.19$, $p < 0.001$			$X^2(5) = 75.47$, $p < 0.001$			$X^2(5) = 17.20$, $p = 0.004$			$X^2(5) = 29.41$, $p < 0.001$		
Nagelkerke R^2	0.07			0.05			0.05			0.05			0.02			0.05		

* $p < 0.05$, ** $p < 0.01$, *** $p < 0.001$, Note 1: PE = parameter estimate, SE = standard error, Wald = Wald test statistic, voc = vocational, s-voc = semi-vocational, ac = academic, GOF = goodness-of-fit, a = reference category.

Table 4. Gender, age, school track and Fear-of-Missing-Out (FOMO) as predictors of the frequency of Swarm, We Heart It, Pinterest, Tumlbr, Vine, Foursquare and Tinder.

	Swarm			We Heart It			Pinterest			Tumblr			Vine			Foursquare			Tinder		
	PE	SE	Wald	PE	SE	Wald	PE	SE	Wald	PE	SE	Wald	PE	SE	Wald	PE	SE	Wald	PE	SE	Wald
Gender (boy)	−0.48	0.19	6.58 **	−1.41	0.58	5.87 *	−0.59	0.29	4.18 *	−0.41	0.29	2.01	0.19	0.24	0.64	−0.04	0.29	0.02	−0.11	0.34	0.11
Gender (girl)	a																				
Age	0.21	0.06	11.25 *	−0.12	0.07	2.69	−0.03	0.07	0.16	−0.03	0.07	0.16	−0.13	0.08	2.48	0.03	0.11	0.10	−0.17	0.11	2.36
School track (voc)	−0.58	0.25	5.34 *	0.65	0.31	4.36 *	−0.10	0.27	0.15	−0.50	0.33	2.30	−0.21	0.36	0.33	0.25	0.37	0.46	0.75	0.48	2.45
School track (s-voc)	0.36	0.22	2.78	0.01	0.24	0.00	−0.17	0.25	0.43	0.07	0.25	0.08	0.05	0.29	0.03	0.33	0.34	0.93	0.48	0.42	1.33
School track (ac)	a																				
FOMO	0.16	0.13	1.49	0.24	0.14	2.86	0.19	0.17	1.28	0.29	0.15	3.77 *	0.37	0.18	4.22*	0.47	0.21	5.26 *	0.04	0.23	0.03
Pearson GOF	$X^2(915) = 856.28$, $p = 0.917$			$X^2(571) = 607.49$, $p = 0.141$			$X^2(619) = 627.27$, $p = 0.400$			$X^2(619) = 620.67$, $p = 0.474$			$X^2(599) = 616.26$, $p = 0.304$			$X^2(475) = 488.52$, $p = 0.324$			$X^2(371) = 373.34$, $p = 0.456$		
−2LL GOF	$X^2(5) = 28.53$, $p < 0.001$			$X^2(5) = 12.53$, $p = 0.028$			$X^2(5) = 6.24$, $p = 0.283$			$X^2(5) = 689.76$, $p = 0.123$			$X^2(5) = 8.29$, $p = 0.141$			$X^2(5) = 7.47$, $p = 0.188$			$X^2(5) = 4.26$, $p = 0.513$		
Nagelkerke R2	0.06			0.04			0.02			0.03			0.04			0.05			0.04		

* $p < 0.05$, ** $p < 0.01$, *** $p < 0.001$, Note 1: PE = parameter estimate, SE = standard error, Wald = Wald test statistic, voc = vocational, s-voc = semi-vocational, ac = academic, GOF = goodness-of-fit, a = reference category. Note 2: The results for We Heart It, Pinterest and Tinder have to be interpreted with caution as the assumption of proportional odds was violated.

3.2.3. Private Versus Public Accessibility of Social Media Platforms Used

With respect to the private versus public accessibility of platforms, we consider *Facebook* and *Snapchat* as social media platforms on which content is generally less publicly accessible (i.e., content is oftentimes shielded off to a public of "approved" contacts), and *Youtube* and *Twitter* as social media platforms on which content is generally publicly accessible (i.e., content is usually accessible to everyone who visits the platform).

3.2.4. Fear of Missing Out (FOMO)

The omnibus format of the survey implied a constraint on the number of items we could use. We chose to select four items from Przybylski et al.'s [8] 10-item FOMO-scale, as this scale had been pre-validated by the authors. In Przybylski et al.'s study, the ten scale items loaded on one factor, and were internally consistent. In the absence of information on the factor loadings of the individual items in the original study, we chose to select a subset of four items that reflect the diversity of the original scale items well. Those items were: "It bothers me when I miss an opportunity to meet up with friends", "I fear my friends have more rewarding experiences than me", "When I go on vacation or summer camp, I continue to keep tabs on what my friends are doing", and "It is important that I understand my friends "inside jokes"". The items were measured on a 5-point Likert-scale (1 = completely disagree; 5 = completely agree).

A well-known risk of using a diverse set items to construct a short scale, is that internal consistency may be jeopardized [50]. Indeed, although an exploratory factor analysis revealed that the four items loaded onto one factor, with all factor loadings above 0.55, the total variance explained by the factor (42%) was below the advised 60% threshold, and the overall Kaiser-Meier-Olkin measure (0.65) indicated mediocre sampling adequacy. A further examination of the scale's reliability, confirmed that the internal consistency of the scale was weak ($\alpha = 0.56$), and revealed that it could not be further improved via item selection, as the highest inter-item correlation was 0.36 ($p < 0.001$). Appendix A shows the inter-item correlation matrix. Means and standard deviations can be consulted in Table 1.

A solution to the issue of low internal consistency, is to perform analyses with individual scale items, rather than with a scale variable. For the current study, however, such procedure would imply an inflation of results—particularly when answering Hypothesis 1a, which reduces the comprehensibility of the study findings. The alternative is to continue with a sub-optimal measure, knowing that the main risk of using a scale-measure that suffers from a weak Cronbach alpha, is underestimation of the real relationship [51]. In the context of the current study, we opted for the latter solution for reasons of comprehensibility, and thus work with the scale variable to answer Hypothesis 1. The risk of an underestimation of the real relationship should be kept in mind, however, when interpreting the results. For Hypotheses 2, 3 and 4, we report both the results using the FOMO scale variable and the individual scale items.

3.2.5. Problematic Social Media Use (PSMU)

Problematic social media use was assessed using an adapted version of the C-VAT instrument [52], which is a scale based on the CIUS-scale that was developed and validated by Meerkerk, Van den Eijnden et al. [53] and Meerkerk [54]. The adapted version addresses social media use rather than videogaming. The items were measured on a 5 point Likert-scale ($\alpha = 0.82$). The scale items, their means and standard deviations, and their factor loadings can be consulted in Table 1.

3.2.6. Phubbing Behavior

At the time of constructing the questionnaire, we were unaware of scales measuring phubbing behavior. Hence, we measured phubbing behavior with three self-constructed items: "How frequently do you use your mobile phone during a conversation in a bar or restaurant?", "How frequently are you engaged with your phone during a conversation?", and "How frequently do you check social media on your phone during a personal conversation?". The items were measured on a 5-point Likert scale, ranging from 1 (never) to 5 (almost all the time; $\alpha = 0.77$). The scale items, their means and standard deviations can be consulted in Table 1.

3.2.7. Control variables

Gender (1 = boy, 2 = girl), age and school track (1 = vocational, 2 = semi-vocational, 3 = academic) were included as control variables in the linear regression analyses.

3.3. *Analyses*

Hypothesis 1a states that teens who have a greater FOMO use social media platforms more frequently. We used ordinal regression analysis in to test this hypothesis because the frequency of use-measures have ordinal response scales. In the analyses with the frequency of use of Snapchat, We Heart it, Pinterest and Tinder as the dependent variables, the assumption of proportional odds was violated; this occurs more frequently in large samples, because minor violations of the parallel lines test may already be statistically significant [55]. Nonetheless, we advise to interpret these results with caution.

Hypothesis 1b states that FOMO positively predicts the breadth of social media use. The "breadth of social media use" variable was operationalized by counting the number of active social media accounts that teens have (min = 0, max = 25). We used multiple linear regression analysis to test the hypothesis, after assessing that the standardized residuals of the variable were normally distributed (and thus that the assumption of normality was not violated: because measurements gathered in large samples typically have very small standard errors, it is advised to assess normality on the basis of the absolute values of skewness and kurtosis rather than on Z-scores of skewness and kurtoses. Advised critical values for skewness, respectively kurtosis in large samples are 2, respectively 7 [56]. Using these guidelines, the skewness (1.22) and kurtosis (5.95) values of the residuals indicated that the assumption of normality for regression analysis was sufficiently met).

To explore our research question on the comparative strength of the correlations between FOMO and private platforms on the one hand, and between FOMO and public platforms on the other, we first calculated the correlations, and next compared the strength of these correlations using Steiger's Z-test [57] in Lee and Preacher's [58] web-based software. The Steiger Z-test operates on the basis of Pearson correlations between two dependent correlations with one variable in common. Because the correlations have to be drawn from the same sample, we first recoded the 'frequency of platform use' variables so that persons without an active account received the lowest usage score (rather than a missing value). This recoding procedure ensured that there were 2663 responses for each variable. Next, we calculated the Pearson correlation coefficients, which form the input for the Steiger's Z-test. The reader may notice that the Pearson correlation test is a parametric test, whereas the "frequency of platform use" variables are ordinal. However, the variables met the standard guidelines for skewness and kurtosis in large samples [56], and the Pearson correlation coefficients and the Spearman rho correlation coefficients were highly similar (i.e., for only two out of ten correlations the difference between the Pearson and the Spearman correlation coefficient exceeded a value of 0.03).

In social science research, scale variables are oftentimes treated as interval variables, based on the idea that the sets of items that comprise each scale form an index that represents an underlying latent factor [59]. The required assumptions for parametric testing were met. Hence, to examine Hypotheses 2 and 3, we fitted a mediation model using model 4 of Hayes' [60]. Process macro for SPSS with FOMO as the independent variable, PSMU as the mediator and phubbing behavior as the dependent variable. The indirect effect was estimated for 5000 bootstrap samples with a 95% bias-corrected confidence interval.

4. Results

4.1. Descriptives

Before addressing the hypotheses and research questions, we briefly report some descriptive statistics for the media use variables. The means and standard deviations for the items of the FOMO-scale, the PSMU-scale and the phubbing scale can be consulted in Table 1.

For 25 social media platforms, we asked the teenagers in our sample whether they had an active account, and if so, how frequently they use it. In terms of active account ownership, Facebook (89% of teens with an active account), Snapchat (73%), Instagram (63%) and Youtube (60%) are the most popular social media platforms. The teenagers in our sample had on average 4.35 (Median = 4, Mode = 4, Min = 0, Max = 25) active social media accounts. The standard deviation (SD = 2.29) reveals that there is substantial variability between teenagers. Most teens who have an active account on Facebook, Snapchat, Instagram and Youtube, reported using the platform more than once per day (see Table 2). There are other social media platforms that are frequently used, such as the location-sharing platform Swarm (M = 4.26, SD = 1.09), but these oftentimes have small user bases (e.g., only 19% of teens has an active Swarm account).

The teenagers in our sample on average had a fairly neutral FOMO score (M = 3.06, SD = 0.69). Using a multiple regression analysis, we explored whether FOMO was predicted by gender, age and school track. Our analysis showed that girls ($ß = 0.08$, $p < 0.001$) reported a higher FOMO, whereas age ($ß = 0.03$, $p = 0.109$) and school track ($ß = -0.01$, $p = 0.800$) were no significant predictors, $R^2 = 0.01$, $F(3.2659) = 5.96$, $p < 0.001$).

4.2. FOMO as a Predictor of the Depth and Breadth of Social Media Use

Hypotheses 1a and 1b concern FOMO as a predictor of the depth and breadth of social media use. We tested Hypothesis 1a using ordinal regression analysis, with gender, age, school track and FOMO as the predictor variables, and the frequency of use of each respective social media platform that was used by 5% of the sample or more as the dependent variable.

Tables 3 and 4 show the results. After controlling for gender, age and school track, FOMO positively predicted the use of the four most popular social media platforms: Facebook, Snapchat, Instagram and Youtube, as well as the frequency of using foursquare, Tumblr and Vine. These findings lend partial support to our hypothesis. While FOMO appears to be a modest predictor of the most common social media platforms, as well as of platforms that are used more rarely, it was not a consistent predictor of social media platform use.

With respect to the breadth of social media use (Hypothesis 1b), a multiple linear regression analysis with gender, age, school track and FOMO as predictors, revealed that gender ($ß = 0.08$, $p < 0.001$), age ($ß = 0.20$, $p < 0.001$), school track ($ß = -0.05$, $p = 0.006$) and FOMO ($ß = 0.14$, $p < 0.001$; $R^2 = 0.08$, $F(4.2658) = 57.27$, $p < 0.001$) were significant, positive predictors of the number of active social media accounts that teenagers have (H1b supported).

4.3. FOMO in Relation to the Public Accessibility of Platforms

We posited that FOMO would be a stronger predictor of social media use when the social media platform examined is a more private platform than when it as a more public platform (RQ1), because more private platforms such as Facebook or Snapchat connect teenagers more strongly to their offline ties, which can be considered the dominant affiliative group on which they want to keep tabs. We calculated Steiner's Z to statistically compare the correlations between FOMO and Facebook, respectively Snapchat use on the one hand, and between FOMO and YouTube, respectively, and Twitter on the other hand (see Tables 5 and 6).

The findings show that the correlations between FOMO and Facebook ($r = 0.16$, $p < 0.001$), respectively Snapchat use ($r = 0.17$, $p < 0.001$), are significantly stronger than the correlations between

FOMO and YouTube ($r = 0.00$, $p = 0.921$), respectively Twitter use ($r = 0.06$, $p = 0.002$), thus lending support to our research question.

Table 5. Correlations between FOMO and frequency of use of less publicly accessible platforms versus more publicly accessible platforms.

Pearson's *r*	Facebook	Snapchat	YouTube	Twitter
Snapchat	0.48 ***	1		
Youtube	0.13 ***	0.04 *	1	
Twitter	0.15 ***	0.19 ***	0.18 ***	1
FOMO	0.16 ***	0.17 ***	0.00	0.06 ***

* $p < 0.05$, *** $p < 0.001$, Note 1: For these analyses, participants with no active account on the platform were assigned a value of '0' on usage frequency to ensure that the analyses were performed on the same sample size. Note 2: We used Pearson's correlation coefficient, to enable the calculation of a Steiger Z value (see the subsequent set of analyses in Table 6). Note 3: The normality assumption was not violated.

Table 6. Comparison of Pearson correlation strength between FOMO and more private platforms versus FOMO and more public platforms.

Steiner's Z ($r_{FOMO,\ column\ var}$ vs. $r_{FOMO,\ row\ var}$)	Facebook	Snapchat	YouTube	Twitter
Snapchat	−0.41			
Youtube	6.37 ***	6.38 ***		
Twitter	4.18 ***	4.62 ***	−2.30 *	

* $p < 0.05$, *** $p < 0.001$.

4.4. FOMO as a Predictor of PSMU and Phubbing Behavior

Hypotheses 2 stated that FOMO positively predicts PSMU and Hypothesis 3a stated that FOMO positively predicts phubbing behavior. Hypothesis 3b stated that PSMU mediates the relationship between FOMO and phubbing behavior. We tested these hypotheses by estimating a mediation model. The results are depicted in Figure 1. All hypotheses were supported. FOMO has a direct, positive predictor of both PSMU ($ß = 0.40$, $p < 0.001$) and phubbing behavior ($ß = 0.20$, $p < 0.001$). When PSMU is accounted for, the relationship between FOMO and phubbing behavior weakens considerably. The indirect effect is significant ($ß = 0.16$, $p < 0.001$).

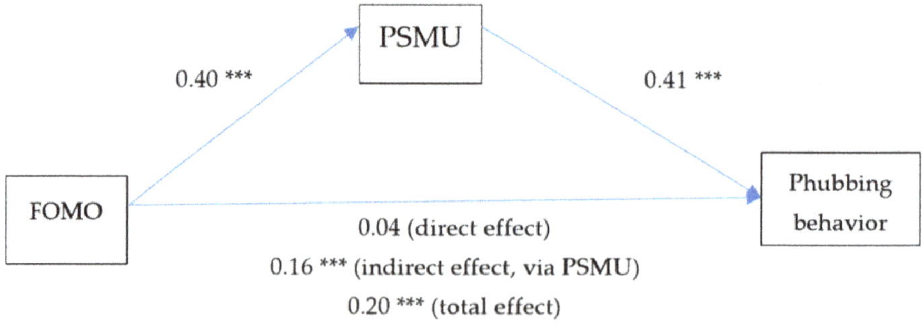

Figure 1. Mediation model of the relationship between FOMO, PSMU and phubbing behavior (*** $p < 0.001$). FOMO: Fear-of-missing-out; PSMU: problematic social media use.

Given that the internal consistency of the FOMO scale was unsatisfactory, we performed the mediation analysis on the individual items of the FOMO scale (see Table 7). For three items, the mediation analysis resulted in a similar results pattern. For the item "It bothers me when I miss an opportunity to meet with friends", however, the direct relationship with PSMU was much weaker (albeit still significant; $ß = 0.05$, $p = 0.008$), and the direct relationship with phubbing behavior was negative ($ß = -0.04$, $p = 0.011$).

Table 7. Mediation analysis results for the individual items of the FOMO scale and the FOMO scale variable.

	a	b	c	c′	Indirect Effect Estimate
FOMO (4-item scale)	0.40 ***	0.41 ***	0.04	0.20 ***	0.16 [0.14; 0.19]
Individual items					
I fear my friends have more rewarding experiences than me	0.21 ***	0.42 ***	0.00	0.09 ***	0.09 [0.07; 0.10]
It is important that I understand my friends' "inside jokes"	0.15 ***	0.43 ***	−0.01	0.05 ***	0.06 [0.05; 0.08]
It bothers me when I miss an opportunity to meet up with friends	0.05 **	0.43***	−0.04 *	−0.02	0.02 [0.00; 0.04]
When I go on summer camp or vacation, I continue to keep tabs on what my friends are doing	0.22 ***	0.38 ***	0.09 ***	0.18***	0.09 [0.07; 0.10]

* $p < 0.05$, ** $p < 0.01$, *** $p < 0.001$. Notes: a = effect of FOMO on PSMU. b = effect of PSMU on phubbing. c = effect of FOMO on phubbing. c′= total effect of FOMO on phubbing.

5. Discussion

Drawing from the results of a survey study among 2663 Flemish teenagers, the aims of this study were fourfold: (1) to examine FOMO as a predictor of the depth and breadth of social media use; (2) to examine whether FOMO relates more strongly to the use of more private social media platforms than more public platforms; (3) to test whether FOMO predicts PSMU and (4) whether FOMO predicts, both directly and indirectly via PSMU, phubbing behavior.

With respect to the breadth of social media use (Hypothesis 1b), we found support for the hypothesis that teens who have a greater FOMO use a wider variety of social media platforms. With respect to the depth of social media use, the study findings partially support the hypothesis that teenagers with a higher FOMO use social media more frequently (Hypothesis 1a), as FOMO was identified as a predictor for the frequency of use of some, but not all social media platforms examined. We did find a consistent relationship with the usage frequency of the four most used platforms: Facebook, Snapchat, Instagram and YouTube. This finding suggests that the relationship between FOMO and the frequency of use of these popular platforms is generalizable to the population of at least Flemish teenagers. These findings also support the findings of earlier work that characterize FOMO as an intrapersonal characteristic that predicts the use of Facebook [28,29] and Instagram [30,31].

The relationships found between FOMO and the frequency of using less popular social media platforms such as Foursquare, Tumblr and Vine are more difficult to interpret, as these social media platforms differ considerably from each other in what they afford the user to do: Foursquare is a location-based social network site, Tumblr is a blogging service, and Vine let users share short videoclips. Future research may explore the relationship between FOMO and the use of these particular platforms more in-depth via the use of interviews with teenagers, as greater insight into their personal experiences with these platforms and their anxieties concerning missing out may shed new light on what makes these particular platforms attractive.

As mentioned above, a pertinent question is whether the social gratification provided by social media use sooths or aggravates the anxieties of teenagers; after all, studies on social comparison on social media platforms [61] suggest that exposure to other people's social media accounts may make the experience of missing out on rewarding experiences even greater. A limitation of our study is its correlational nature, which prevents from making claims concerning the potential bi-directional causality of the relationship between FOMO and social media use. Future research may address this question via the use of longitudinal study designs that enable the modelling of cross-lagged path models.

It is essential for our understanding of FOMO to unravel how it resemblances, but also differs from other, associated personality factors. One factor that has been identified in the extant literature and that appears relevant to consider, is sociotropy [62], which refers to an insatiable need for belongingness to others, visible in an over-reliance on approval from others, which can go at the expense of personal autonomy. In recent work, sociotropy is linked to the ritualistic monitoring of, oftentimes multiple social media platforms [63]. Future research may explore how the fear of negative feedback or rejection that is typical for sociotropic individuals aligns with a fear of missing out.

We questioned whether FOMO would be a stronger predictor of more private social media platforms such as Facebook or Snapchat than of more public platforms such as YouTube and Twitter (RQ1). Our exploratory analysis suggests this is indeed the case. This is unsurprising, given that FOMO itself has been conceptualized and operationalized in the current study as a fear to miss out on what friends are thinking and doing, and information about these friends can be found mostly on platforms that connect to people who are part of one's offline network. Nonetheless the stronger relationship between FOMO and these private platforms, we need to remark that FOMO still remains a weak, yet significant predictor of some more publicly accessible platforms (e.g., Tumblr). A pertinent question is what still drives those higher on FOMO to use the latter platforms, then. As mentioned above, qualitative research seems relevant to further expand on the affordances in which various social media platforms resemble or differ from one another, and how these affordances are perceived, valued and acted upon by those with a higher versus lower FOMO.

FOMO was found to predict PSMU (Hypothesis 2), a result that aligns with the findings from other recent studies which showed that greater FOMO is associated with more problematic internet and smartphone use [40,64,65]. Moreover, FOMO was also associated with phubbing behavior (Hypothesis 3a), consistently with previous research that showed a similar pattern of associations [44]. Interestingly, our results showed that the relationship between FOMO and phubbing behavior was mediated by PMSU (cf. Hypothesis 3b), accordingly to the proposed conceptual model by Chotpitayasunondh & Douglas [44], which found that FOMO was a positive predictor of smartphone addiction, and that smartphone addiction predicted smartphone behavior. Thus, adolescents who are high in their fear of missing out are more likely to overuse the social media and smartphones, which in turns leads them to phub their offline interaction partners [66]. Scholars in the field of problematic media use research advocate to invest greater effort into the study of the pathways that lead to problem behavior [6]. Our study findings represent such an effort, as they reveal that FOMO is an intra-personal characteristic that leads to phubbing behavior by inducing excessive, uncontrolled social media use.

The findings reported in this study are generalizable as they stem from a large-scale survey study that was administered to representative sample of Flemish teenagers. There are a number of limitations to the current study, however. We used a shortened, four item version of the FOMO-scale developed by Przybylski et al. [8] to assess teenagers' FOMO. The internal consistency of the scale was unsatisfactory, which increases the risk of underestimating the real relationship (see Schmitt [51]). The low internal consistency for our scale illustrates the importance for research in this field to use complete and validated scales—for the reason of making reliable claims about one's study, and for enabling valid comparisons between studies.

In light of the unsatisfactory internal consistency of our FOMO-scale, we tested Hypotheses 2 and 3 using both the scale variable and the individual items. This analysis revealed that for one item the relationship with phubbing behavior was reversed: The more teens agreed to feel bothered when missing an opportunity to meet with friends (i.e., indicative of a greater fear-of-missing-out), the less they report phubbing their interaction partner during a face-to-face interaction. This finding indicates that some teenagers attach great importance to face-to-face interactions with friends, leading them to prioritize these interactions over smartphone interactions. This finding suggests that it is relevant to further investigate how FOMO relates to relational behavior, not only online (in the form of social media use), but also offline.

Second, our study used a narrow definition of social media [16], which excludes mobile messaging applications. It is likely that those with a greater FOMO also heavily rely on the use of these messengers to soothe their anxieties about what others in their social networks are doing. The latter applications are different from social media, however, in that they are oftentimes used for small-group communication, and therefore are more dialogical in nature [67]. It is more difficult—if not impossible—to "lurk" in dyadic and small-group conversations, as they generally rely on a certain degree of reciprocity. With respect to FOMO, this difference in the interactional affordances of social media platforms and (mobile) instant messengers raises interesting questions. It may be the case that people with a high FOMO are particularly attracted to social media because they can lurk on these platforms without risking a label of "voyeur", and without having to engage in reciprocal disclosures about themselves. As mentioned above, it seems relevant to explore the difference between FOMO and sociotropy [62] in this context, as active disclosures on social media platforms and messaging platforms may provide sociotropic individuals with a means to gather the social approval they long for, while those with a high FOMO may seek information about other people's experiences without necessarily wanting to engage in interactions with them. Future research may explore this.

In short, this study is valuable because it provides generalizable findings on the relationships between FOMO, social media use, PSMU and phubbing behavior among teenagers. As such, it can serve as a starting ground for future research. This research needs to look into the pathways via which FOMO leads to particular forms of (problematic) media use, and how these pathways are similar to or different from other pathways.

6. Conclusions

To conclude, and on the basis of the findings presented in this article, FOMO is an important factor explaining teenagers' social media use. The present study found support for the hypothesis that teens who have a greater FOMO use a wider variety of social media platforms. Also, the present study found partially support for the hypothesis that teenagers with a higher FOMO use social media more frequently: FOMO was identified as a predictor for the frequency of use of some, but not all social media platforms examined. In particular, there was a consistent relationship between FOMO and the usage frequency of Facebook, Snapchat, Instagram and YouTube. Moreover, FOMO was found to predict PSMU. This result aligns with the findings from other recent studies which showed that greater FOMO is associated with more problematic internet and smartphone use [40,64,65]. Finally, FOMO was associated with phubbing behavior. Our results additionally showed that the relationship between FOMO and phubbing behavior was mediated by PSMU. Teens who are high in their fear of missing out are more likely to overuse the social media and smartphones, which in turns may lead them to phub their offline interaction partners [66].

Author Contributions: V.F. and M.V.A. performed analyses and wrote initial drafts of the manuscript. A.J.v.R., G.L.C., and L.D.M. proofread drafts, contributed ideas and edited the various versions.

Funding: This research received no external funding.

Conflicts of Interest: The authors declare no conflict of interest.

Appendix A

Table A1. Intercorrelation matrix of items measuring the Fear-of-Missing-Out (FOMO).

Items FOMO Scale	1	2	3	4
I fear my friends have more rewarding experiences than me	1	0.36 ***	0.17 ***	0.26 ***
It is important that I understand my friends' "inside jokes"		1	0.25 ***	0.21 ***
It bothers me when I miss an opportunity to meet up with friends			1	0.15 ***
When I go on summer camp or vacation, I continue to keep tabs on what my friends are doing				1

*** $p < 0.001$.

References

1. Moreno, M.A.; Jelenchick, L.; Koff, R.; Eikoff, J.; Diermyer, C.; Christakis, D.A. Internet use and multitasking among older adolescents: An experience sampling approach. *Comput. Hum. Behav.* **2012**, *28*, 1097–1102. [CrossRef]
2. Billieux, J.; Schimmenti, A.; Khazaal, Y.; Maurage, P. Are we overpathologizing everyday life? A tenable blueprint for behavioral addiction research. *J. Behav. Addict.* **2015**, *4*, 119–123. [CrossRef] [PubMed]
3. Bean, A.M.; Nielsen, R.K.; Van Rooij, A.J.; Ferguson, C.J. Video game addiction: The push to pathologize video games. *Prof. Psychol. Res. Pract.* **2017**, *48*, 378–389. [CrossRef]
4. Kardefelt-Winther, D.; Heeren, A.; Schimmenti, A.; Van Rooij, A.; Maurage, P.; Carras, M.; Edman, J.; Blaszczynski, A.; Khazaal, Y.; Billieux, J. How can we conceptualize behavioral addiction without pathologizing common behaviours? *Addiction* **2017**, *112*, 1709–1715. [CrossRef] [PubMed]
5. Van Rooij, A.J.; Daneels, R.; Liu, S.; Anrijs, S.; Van Looy, J. Children's motives to start, continue, and stop playing video games: Confronting popular theories with real-world observations. *Curr. Addict. Rep.* **2017**, *4*, 323–332. [CrossRef]
6. Billieux, J. Problematic use of the mobile phone: A literature review and a pathways model. *Curr. Psychiatry Rev.* **2012**, *8*, 299–307. [CrossRef]
7. Billieux, J.; Maurage, P.; Lopez-Fernandez, O.; Kuss, D.J.; Griffiths, M.D. Can disordered mobile phone use be considered a behavioral addiction? An update on current evidence and a comprehensive model for future research. *Curr. Addict. Rep.* **2015**, *2*, 156–162. [CrossRef]
8. Przybylski, A.K.; Murayama, K.; DeHaan, C.R.; Gladwell, V. Motivational, emotional, and behavioral correlates of fear of missing out. *Comput. Hum. Behav.* **2013**, *29*, 1841–1848. [CrossRef]
9. Deci, E.L.; Ryan, R.M. *Intrinsic Motivation and Self-Determination in Human Behavior*; Plenum Press: New York, NY, USA, 1985.
10. Oulasvirta, A.; Rattenbury, T.; Ma, L.; Raita, E. Habits make smartphone use more pervasive. *Personal Ubiquitous Comput.* **2012**, *16*, 105–114. [CrossRef]
11. LaPlante, D.A.; Nelson, S.E.; Gray, H.M. Breadth and depth involvement: Understanding Internet gambling involvement and its relationship to gambling problems. *Psychol. Addict. Behav.* **2014**, *28*, 396. [CrossRef] [PubMed]
12. Abel, P.J.; Buff, C.L.; Burr, S.A. Social media and fear of missing out: Scale development and assessment. *J. Bus. Econ. Res.* **2016**, *14*, 34–44. [CrossRef]
13. Simon, J. Love: Addiction or road to self-realization, a second look. *Am. J. Psychoanal.* **1982**, *42*, 253–263. [CrossRef] [PubMed]
14. Baumeister, R.F.; Leary, M.R. The need to belong: Desire for interpersonal attachments as a fundamental human motivations. *Psychol. Bull.* **1995**, *117*, 497–529. [CrossRef] [PubMed]
15. Boyd, D.; Ellison, N.B. Social network sites: Definition, history and scholarship. *J. Comput. Mediat. Commun.* **2008**, *13*, 210–230. [CrossRef]
16. Kaplan, A.M.; Haenlein, M. Users of the world, unite! The challenges and opportunities of Social Media. *Bus. Horiz.* **2010**, *53*, 59–68. [CrossRef]
17. Clayton, R.B.; Leshner, G.; Almond, A. The extended iself: The impact of iphone separation on cognition, emotion and physiology. *J. Comput. Mediat. Commun.* **2015**, *20*, 119–135. [CrossRef]
18. Kruger, D.J.; Djerf, J.M. High ringxiety: Attachment anxiety predicts experiences of phantom cell phone ringing. *Cyberpsychol. Behav. Soc. Netw.* **2016**, *19*, 56–59. [CrossRef] [PubMed]
19. Belk, R.W. Possession and the extended self. *J. Consum. Res.* **1988**, *15*, 139–168. [CrossRef]
20. Tandor, E.C.; Ferrucci, P.; Duffy, M. Facebook use, envy, and depression among college students: Is facebooking depressing? *Comput. Hum. Behav.* **2014**, *43*, 139–146. [CrossRef]
21. Casale, S.; Rugai, L.; Fioravanti, G. Exploring the role of positive metacognitions in explaining the association between the fear of missing out and social media addiction. *Addict. Behav.* **2018**, *85*, 83–87. [CrossRef] [PubMed]
22. Katz, E. Mass communications research and the study of popular culture: An editorial note on a possible future for this journal. *Stud. Public Commun.* **1959**, *2*, 1–6.
23. Katz, E.; Blumler, J.G. *The Uses of Mass Communications: Current Perspectives on Gratifications Research*; Sage Publications: Beverly Hills, CA, USA, 1974.

24. Ajzen, I. The theory of planned behavior. *Organ. Behav. Hum. Decis. Process.* **1991**, *50*, 179–211. [CrossRef]
25. Lee, E.; Lee, J.A.; Moon, J.H.; Sung, Y. Pictures speak louder than words: Motivations for using Instagram. *Cyberpsychol. Behav. Soc. Netw.* **2015**, *18*, 552–556. [CrossRef] [PubMed]
26. Sheldon, K.M.; Abad, N.; Hirsch, C. A two-process view of Facebook use and relatedness need-satisfaction: Disconnection drives us and connection rewards it. *J. Personal. Soc. Psychol.* **2011**, *100*, 766–775. [CrossRef] [PubMed]
27. Sheldon, P.; Bryant, K. Instagram: Motives for its use and relationship to narcissism and contextual age. *Comput. Hum. Behav.* **2015**, *58*, 89–97. [CrossRef]
28. Beyens, I.; Frison, E.; Eggermont, S. "I don't want to miss a thing": Adolescents' fear of missing out and its relationship to adolescents' social needs, Facebook use, and Facebook related stress. *Comput. Hum. Behav.* **2016**, *64*, 1–8. [CrossRef]
29. Błachnio, A.; Przepiórka, A. Facebook Intrusion, fear of missing out, narcissism, and life satisfaction: A cross-sectional study. *Psychiatry Res.* **2018**, *259*, 514–519. [CrossRef] [PubMed]
30. Barry, C.T.; Reiter, S.R.; Anderson, A.C.; Schoessler, M.L.; Sidoti, C.L. "Let me take another selfie": Further examination of the relation between narcissism, self-perception, and Instagram posts. *Psychol. Popul. Media Cult.* **2017**. [CrossRef]
31. Salim, F.; Rahardjo, W.; Tanaya, T.; Qurani, R. Are self-presentation of instagram users influenced by friendship-contingent self-esteem and fear of missing out? *Makara Hubs Asia* **2017**, *21*, 70–82. [CrossRef]
32. Bayer, J.B.; Ellison, N.B.; Schoenebeck, S.Y.; Falk, E.B. Sharing the small moments: Ephemeral social interaction on Snapchat. *Inf. Commun. Soc.* **2016**, *19*, 956–977. [CrossRef]
33. Alhabash, S.; Ma, M. A tale of four platforms: Motivations and uses of Facebook, Twitter, Instagram, and Snapchat among college students? *Soc. Media Soc.* **2017**, *3*, 1–13. [CrossRef]
34. Caplan, S. Relations among loneliness, social anxiety, and problematic internet use. *Cyberpsychol. Behav.* **2006**, *10*, 234–242. [CrossRef] [PubMed]
35. Milani, L.; Osualdella, D.; Di Blasio, P. Quality of interpersonal relationships and problematic internet use in adolescents. *CyberPsychol. Behav.* **2009**, *12*, 681–684. [CrossRef] [PubMed]
36. Chakraborty, K.; Basu, D.; Vijaya, K. Internet Addiction: Consensus, Controversies and the Way Ahead. *East Asian Arch Psychiatry.* **2010**, *20*, 123–132. [PubMed]
37. Griffiths, M.D.; Kuss, D.J. Adolescent Social Media Addiction (revised). *Educ. Health* **2017**, *35*, 49–52.
38. Buglass, S.L.; Binder, J.F.; Betts, L.R.; Underwood, J.D. Motivators of online vulnerability: The impact of social network sites use and FOMO. *Comput. Hum. Behav.* **2017**, *66*, 248–255. [CrossRef]
39. Fuster, H.; Chamarro, A.; Oberst, U. Fear of missing out, online social networking and mobile phone addiction: A latent profile approach. *Revista de Psicologia Ciències de l'Educació i de l'Esport* **2017**, *35*, 23–30.
40. Oberst, U.; Wegmann, E.; Stodt, B.; Brand, M.; Chamarro, A. Negative consequences from heavy social networking in adolescents: The mediating role of fear of missing out. *J. Adolesc.* **2017**, *55*, 51–60. [CrossRef] [PubMed]
41. Wegmann, E.; Oberst, U.; Stodt, B.; Brand, M. Online-specific fear of missing out and Internet-use expectancies contribute to symptoms of Internet-communication disorder. *Addict. Behav. Rep.* **2017**, *5*, 33–42. [CrossRef] [PubMed]
42. Elhai, J.D.; Levine, J.C.; Alghraibeh, A.M.; Alafnan, A.A.; Aldraiweesh, A.A.; Hall, B.J. Fear of missing out: Testing relationships with negative affectivity, online social engagement, and problematic smartphone use. *Comput. Hum. Behav.* **2018**, *89*, 289–298. [CrossRef]
43. Scott, H.; Woods, H.C. Fear of missing out and sleep: Cognitive behavioural factors in adolescents' nighttime social media use. *J. Adolesc.* **2018**, *68*, 61–65. [CrossRef] [PubMed]
44. Chotpitayasunondh, V.; Douglas, K.M. How "phubbing" becomes the norm: The antecedents and consequences of snubbing via smartphone. *Comput. Hum. Behav.* **2016**, *63*, 9–18. [CrossRef]
45. Kenny, M. Cell Phones, College Students, and Conversations: Exploring Mobile Technology Use during Face-to-Face Interactions. Bachelor's Thesis, The Florida State University, Tallahassee, FL, USA, 2016.
46. Vanden Abeele, M.M.V.; Antheunis, M.L.; Schouten, A.P. The effect of mobile messaging during a conversation on impression formation and interaction quality. *Comput. Hum. Behav.* **2016**, *62*, 562–569. [CrossRef]
47. Ofcom. *The Communications Market Reports*; Ofcom: Warrington, UK, 2015.

48. Kuss, D.J.; Kanjo, E.; Rumsey-Crook, M.; Kibowski, F.; Wang, G.Y.; Sumich, A. Problematic mobile phone use and addiction across generations: The roles of psychopathological symptoms and smartphone use. *J. Technol. Behav. Sci.* **2018**, *3*, 141–149. [CrossRef] [PubMed]
49. Van Waeg, S.; D'hanens, K.; Dooms, V.; Naesens, J. Onderzoeksrapport Apestaartjaren 6/Research Report Apestaartjaren 6. Available online: www.apestaartjaren.be/onderzoek/apestaartjaren-6 (accessed on 17 August 2018).
50. Tavakol, M.; Dennick, R. Making sense of Cronbach's alpha. *Int. J. Med. Educ.* **2011**, *2*, 53. [CrossRef] [PubMed]
51. Schmitt, N. Uses and Abuses of Coefficient Alpha. *Psychol. Assess.* **1996**, *4*, 350–353. [CrossRef]
52. Van Rooij, A.J.; Ferguson, C.J.; Van de Mheen, D.; Schoenmakers, T.M. Time to abandon Internet Addiction? Predicting problematic Internet, game, and social media use from psychosocial well-being and application use. *Clin. Neuropsychiatry* **2017**, *14*, 113–121.
53. Meerkerk, G.; Van den Eijnden, R.J.J.M; Vermulst, A.A.; Garretsen, H.F.L. The Compulsive Internet Use Scale (CIUS): Some psychometric properties. *Cyberpsychol. Behav.* **2009**, *1*, 1–6. [CrossRef] [PubMed]
54. Meerkerk, G.; Pwned by the Internet. Explorative research into the causes and consequences of compulsive internet use. Ph.D. Thesis, Erasmus University, Rotterdam, The Netherlands, 2007.
55. Williams, R. Generalized ordered logit/partial proportional odds models for ordinal dependent variables. *Stata J.* **2006**, *6*, 58.
56. Kim, H. Statistical notes for clinical researchers: Assessing normal distribution (2) using skewness and kurtosis. *Restor. Dent. Endod.* **2013**, *38*, 52–54. [CrossRef] [PubMed]
57. Steiger, J.H. Tests for comparing elements of a correlation matrix. *Psychol. Bull.* **1980**, *87*, 245–251. [CrossRef]
58. Lee, I.A.; Preacher, K.J. Calculation for the Test of the Difference between Two Dependent Correlations with One Variable in Common (Computer Software). Available online: http://quantpsy.org (accessed on 2 October 2018).
59. Allen, I.E.; Seaman, C.A. Likert scales and data analyses. *Qual. Prog.* **2007**, *40*, 64–65.
60. Hayes, A.F. *Introduction to Mediation, Moderation, and Conditional Process. Analysis*; Guilford Press: New York, NY, USA, 2013.
61. Fox, J.; Moreland, J.J. The dark side of social networking sites: An exploration of the relational and psychological stressors associated with Facebook use and affordances. *Comput. Hum. Behav.* **2015**, *45*, 168–176. [CrossRef]
62. Beck, A.T. Cognitive therapy of depression: New perspectives. In *Treatment of Depression: Old Controversies and New Approaches*; Clayton, P.J., Barrett, J.E., Eds.; Raven Press: New York, NY, USA, 1983; pp. 265–290.
63. Orchard, L.J.; Fullwood, C.; Galbraith, N.; Morris, N. Individual differences as predictors of social networking. *J. Comput. Mediat. Commun.* **2014**, *19*, 388–402. [CrossRef]
64. Elhai, J.D.; Levine, J.C.; Dvorak, R.D.; Hall, B.J. Fear of missing out, need for touch, anxiety and depression are related to problematic smartphone use. *Comput. Hum. Behav.* **2016**, *63*, 509–516. [CrossRef]
65. Stead, H.; Bibby, P.A. Personality, fear of missing out and problematic internet use and their relationship to subjective well-being. *Comput. Hum. Behav.* **2017**, *76*, 534–540. [CrossRef]
66. Roberts, J.A.; David, M.E. My life has become a major distraction from my cell phone: Partner phubbing and relationship satisfaction among romantic partners. *Comput. Hum. Behav.* **2016**, *54*, 134–141. [CrossRef]
67. Ling, R.; Lai, C. Microcoordination 2.0: Social Coordination in the Age of Smartphones and Messaging Apps. *J. Commun.* **2016**, *66*, 834–856. [CrossRef]

International Journal of
Environmental Research and Public Health

MDPI

Article

Exploring the Opportunities and Challenges of the Digital World for Early Childhood Services with Vulnerable Children

Leona Harris [1,2], Niki Davis [1,2,*], Una Cunningham [3], Lia de Vocht [1], Sonja Macfarlane [4], Nikita Gregory [2], Saili Aukuso [1,2], Tufulasifa'atafatafa Ova Taleni [2] and Jan Dobson [5]

1 E-Learning Research Lab, University of Canterbury, Christchurch 8041, New Zealand; leona.harris@canterbury.ac.nz (L.H.); lia.devocht@canterbury.ac.nz (L.d.V.); saili.aukuso@pg.canterbury.ac.nz (S.A.)
2 Child Well-being Research Institute, University of Canterbury, Christchurch 8041, New Zealand; nikita.gregory@pg.canterbury.ac.nz (N.G.); tufulasi.taleni@canterbury.ac.nz (T.O.T.)
3 Department of Education, Uppsala University, 75236 Uppsala, Sweden; una.cunningham@edu.uu.se
4 Te Rū Rangahau: The Māori Research Laboratory, University of Canterbury, Christchurch 8041, New Zealand; sonja.macfarlane@canterbury.ac.nz
5 Sector Enablement and Support, Ministry of Education, Christchurch 8013, New Zealand; jan.dobson@education.govt.nz
* Correspondence: niki.davis@canterbury.ac.nz; Tel.: +64-3-369-3650

Received: 31 July 2018; Accepted: 25 October 2018; Published: 30 October 2018

Abstract: Potentially addictive behaviours supported by the internet and mobile phones raise concerns in education services for early childhood. Although there is evidence that screen media can distract the attention of young children, there was a massive uptake of digital devices by early childhood centres (ECCs). We investigated practices of families (n = 85) and of six ECCs serving vulnerable children in New Zealand, many of whom are emergent bilinguals. Descriptions of the limited and exemplary choice of screen media of the ECCs include digital portfolios containing children's learning stories in multiple languages illustrated with digital photos. This was facilitated by increasing partnership with the families and the inclusion of their languages in the physical and digital landscapes of the ECCs. However, these families and the ECCs are seeking additional guidance to face the complex challenges of the digital world. These early findings from our national research programme, A Better Start, E Tipu E Rea, already informed significant changes in the ECCs; we also identified the potential for young children to act as agents of change.

Keywords: internet addiction; mobile phone (or smartphone) use; young children; early childhood education; parenting; emergent bilinguals; intergenerational language transmission

1. Introduction

Childhood today is increasingly technologically constructed [1] with digital technologies such as computers, tablets, and smartphones now commonplace in many education systems and homes worldwide [2]. While the ubiquity of these technologies has beneficial outcomes for youngsters' learning and socialisation, it contributes also, among other detrimental outcomes, to the loss of minority languages and cultures [3]. This loss is of concern given the increasing body of evidence that bilingualism and multilingualism can have lifelong cognitive, health, and economic benefits (see, for example, Reference [4]).

Addictive behaviours supported by technologies, such as the internet and mobile phones, raise concerns, particularly in health services for early childhood. The bold claims of the American Academy of Paediatrics Committee on Education in 2001 included recommendations reducing the exposure of children under six years of age to screen media. These arose from concerns about overuse of digital devices, resulting in young children missing developmentally appropriate activities and/or not getting enough exercise; in other words, typical activities in childhood are displaced by the adoption of digital media. Evidence that media can distract the attention of young children increased since those recommendations were published [5]). However, directives to simply reduce time spent with digital devices are unhelpful and impractical in a world in which digital devices have become ubiquitous, and when a range of devices with apps and digital environments now marketed include some supporting developmentally appropriate activities that were adopted by early childhood educators and other experts in the field [2].

In addition to the concerns about the exposure of young children to digital media, concerns about the influence of digital media on the behaviour of family members emerged over the last decade. Given the complexity of effective parent and sibling interaction with young children [6], it is possible that increasing the use of mobile phones and other devices can adversely impact child development, particularly in families suffering from some form of media addiction.

The study documented in this paper was partly prompted by these concerns. It describes the linguistic landscapes, as well as the deployment of digital technologies within them, of six early childhood education (ECE) centres (ECCs) in New Zealand to explore how such centres contribute to young children's language development. The provision of high-quality early childhood education was shown to have a beneficial impact on the development of young children, particularly those from low-income families [7].

We set our study within these centres because the children attending them are of an age (four to six years) when they are rapidly developing their language and social skills. We also wanted to gain a better understanding of how digital technologies might further support the evolution of linguistic landscapes in ways that provide a better start within education systems for young children, particularly those who acquire more than one language (emergent bilinguals; cf. [8]), and for whom an enriched linguistic landscape, visibly valuing their home language, has the potential to make the most impact [9].

New Zealand is, for several reasons, a particularly interesting location in which to explore the interface between digital technology use in ECE and the effort to support emergent bilinguals attending ECCs.

1. Opportunities and challenges linked with the digital world recently grew in New Zealand with the rollout of ultrafast broadband nationwide and the consequent expanded internet access in schools, a great many homes, community facilities, and workplaces [10]; from 2015 to 2017, the "frequency, range of activities, and the number of devices people use to access the internet all increased" [11].
2. New Zealand's ECE curriculum, Te Whāriki [12], is grounded in the country's founding document from 1840, te Tiriti o Waitangi (Treaty of Waitangi), which is highly respectful of cultural and linguistic diversity. Written in both Māori and English, Te Whāriki is based on four principles (empowerment, holistic development, family and community, and relationships) woven together with five strands (wellbeing, belonging, contribution, communication, and exploration). The term "oral language" used within Te Whāriki refers to any method of communication that "the child uses as a first language" (p. 42). The culturally safe development of children's language skills is essential (e.g., Reference [13]).
3. New Zealand's awareness of the lifelong benefits of bilingualism and multilingualism is relatively recent [14]. Māori did not become one of the country's three official national languages until 1987 (the other two are English and New Zealand sign language), and despite New Zealand's increasingly multilingual population, migrants are generally encouraged to adopt English at the expense of their heritage languages [15]. English remains the language most commonly spoken in New Zealand and is the de facto language of nearly all educational provision in the country.

4. New Zealand is recognised as a linguistically "superdiverse" country with over 160 languages, in addition to English and Māori, with Samoan, Hindi, and Northern Chinese more commonly spoken [16]. Te reo Māori words are included in this article in line with common practice.

We aimed to gather evidence in a way that spread good practice in regard to enhancing the linguistic landscapes in the ECCs and minimising the risks that come with the digital world.

2. Growing Up in a Digital World

Hsin, Li, and Tsai's comprehensive review of studies on information and communications technology (ICT) use in ECE [17] found positive associations between children's language and literacy development and their use of digital tools and toys, although the learning outcomes generally depended on how the children used the technology. Chaudron's Europe-wide study investigating the digital technology behaviours of children (0–8 years) [18] showed children typically using digital devices on their own, with parental management restricted to limiting length and frequency of use. Parents had little knowledge of what their children were actually doing in the digital world, however. Aubrey and Dahl [19] found ECE educators likewise lacked detailed knowledge of children's digital experiences, and that parents rarely shared information with educators about how their children were using digital technologies at home. Marklind and Dunkels examined the introduction of tablets by early childhood teachers in Sweden [20], sometimes in the face of opposition from colleagues or parents, to develop children's literacy and school-oriented language development, which is especially relevant for multilingual children.

Adult and adolescent addictive behaviours relating to problematic mobile phone use (PMPU) [21], have been found to be "positively associated with stress, depression, sleep disturbances, extraversion, female gender, young age, and poor academic or professional competence or performance" (p. 1213) and to affect learning [22]. However, it remains important to be cautious when applying the label addiction using the connection with substance abuse as a way of categorising online behaviour as disordered [23]. In the context of early childhood, the poor competence or performance of particular concern is around parenting and the contribution of siblings to the healthy development of young children. Kildare and Middlemiss's comprehensive literature review [24] identified that there was a range of both parenting benefits and complications with the integration of mobile devices in their day-to-day lives, suggesting that the impact on any particular child was wide-ranging and complex. The common themes that emerged included "parent's level of absorption with their mobile devices, child safety in the presence of parents' mobile distractions and parents conflicted attitudes regarding device use, and decreased parental responsiveness and sensitivity towards children while distracted." (p. 590). Perhaps most interestingly, they note that children's screen time increases with parents' screen time [25].

The findings from these studies highlight the importance of adults (most notably parents and educators) as mediators of children's learning via ICT, as well as role models. Research by McPake, Plowman, and Stephen [26] found that young children's use of digital technology at home tended to expand children's early communicative and creative experiences. They also observed that maximisation of this learning once the children were in more formal educational settings relied on educators' knowledge of these home-based learning experiences (including digital), and on their ability to carefully and creatively help the children build on those experiences.

However, adults cannot be effective mediators if they do not know how children are using these technologies at home and in formal educational settings, and if they do not understand how that use might be influencing the children's holistic (including linguistic) development.

3. Emergent Bilinguals

Recent findings in neuroscience and from longitudinal cohort studies indicate that young children who acquire more than one language and retain them into adulthood have significantly better lifetime outcomes [27,28]. Research also shows that children learn a new language more effectively when they continue to use and develop their heritage language [29], and that pre-schoolers develop more positive identities as learners and are more engaged in literacy activities when their ECE settings include their home cultures and experiences [30].

Increasing opportunities in ECE settings for young children to engage in their family's languages and cultures is important not only for reasons such as these, but also because the children's respective languages and cultures encapsulate their heritage and enhance their mana (a word used in Māori to denote power, prestige, acumen, and efficacy). However, while children's ability to develop and retain more than one language is challenging in educational contexts that are largely monolingual [31], proficiency in their parents' language is important for the children of migrants [32].

Dressler [33] found that, as children move around their educational settings, they draw conclusions about the relative importance of the languages evident in the signs and artefacts around them, because it is these that indicate what is socially supported within those places. Thus, the presence of signs in more than one language and of culturally based artefacts in ECCs prompts and supports the use of multiple languages, enabling children and adults to co-build multilingual language and social skills [9,34].

Early childhood educators can value and support children's use of their heritage languages by working in partnership with the children's whānau (families, including extended family members) to ensure the authenticity of these inclusions in the linguistic landscapes of the centres. Bridging language practices between home and educational settings through parental participation, such as being expert partners in bi-literacy development through the co-creation of dual-language texts, reinforces children's bilingual language development, as well as building their self-confidence, and strengthening their bilingual identities and their English language competencies [35–38].

However, there is very little research into the duration of internet and/or mobile phone use in educational contexts that aim to provide holistic support for young children, including ECCs. The benefits for emergent bilinguals are particularly relevant for vulnerable children.

4. Materials and Methods

Our main research question was "How are ECCs contributing to the development of the languages of young bilingual children who are growing up in a digital world?" In addition, for whānau with children who have lower levels of oral language ability in English, (i) How long do children spend on digital media? (ii) What are whānau views on the importance of their child's multilingualism?

The study presented in this article conformed to the provisions of the Declaration of Helsinki in 1995, revised in Edinburgh in 2000 [39]. Ethical approval was granted by the University's committee for human ethics (number 2016/21/ERHEC). For this paper, we bring together two complementary datasets of vulnerable emergent bilinguals who grew up in one locality, at a time when the digital world blended into their physical world: (1) qualitative ethnographic case studies of the linguistic landscapes of six ECCs in 2016 and in 2017, and (2) a multidisciplinary survey of families that included some of the same children during their first year of primary school in the same locality (n = 85). It may also be important to note that this multidisciplinary mixed-methods study was conducted within our national research programme's Vision Matauranga, which braids western scientific methods with indigenous kaupapa Māori principles [40]. The research was also associated with a multi-agency initiative called "You Matter to Us" that emerged to support young children in an area with some of the most complex challenges faced in this country [41].

These children were situated within three suburbs of a large city purposefully selected because of its cultural diversity and over-representation of low-income families. Around 22% of people in this area speak more than one language compared with 15.8% across the city [42]. The languages other than English most commonly spoken in this area are Māori and Samoan. The details of our approach and methods are provided in other publications [43,44]; thus, in this paper, we provide a brief summary along with details of the participants. We begin with the methodology used with the ECC before describing the survey methodology.

4.1. Case Studies of Six Early Childhood Centres

The first set of evidence was linguistic ethnographies [45] gathered in collaboration with six mainstream English medium ECCs. Reports from the national government department responsible for reviewing the quality of ECCs in New Zealand, called the Education Review Office, considered five of the ECCs "well placed" and the sixth ECC "very well placed" to promote positive learning outcomes for children (the highest two levels) [46]. To acknowledge the value of their participation, each ECC received a report of their linguistic landscape to use for their own purposes (plus advising edits). In addition, some development time was gifted on request. Figure 1 is derived from an image that was created as koha (a gift).

Figure 1. An example of innovative practice co-constructed with the pilot Māori early childhood centre (ECC) in response to parents' need for support to sing along in the Māori language with their children at home [44]. The image of the child was retrieved from https://pixabay.com, which is released under the Creative Commons, as is this image CC BY SA3.

The ethnographic data for each linguistic landscape consisted of images in the form of still pictures, videos, and online screenshots. Still pictures and two videos (child height and adult height) captured the displays on the walls within each centre. Videos and screenshots (of publicly accessible pages online) were used to support interviews. Sixteen interviews were conducted with 18 staff (15 with individuals and one with a small group), as well as a few with whānau. In most ECCs, there was one interview with the head teacher and other teachers, and one interview with member(s) of a bilingual child's whānau. In addition, there was one interview with the outreach librarian who provided library services for those with limited ability to visit community libraries. She visited many of these ECCs to promote the library services through storytelling, songs, and digital storytelling.

Linguistic content in the photos was classified by the language(s) visible. Selected artefacts (for example, the e-portfolio) were used to discuss each centre's language and digital technology policies and practices, and to identify each centre's engagement with whānau and the communities that supported the language development of its emergent bilingual children. The interviews were semi-structured and conversational, and questions about what they did to support the language development of their emergent bilinguals stemmed from the participants' narratives and explanations of the physical and digital artefacts. The interviews elicited reflections on adults' practices with emergent bilinguals and prompted ideas for future strategies. Interviews were transcribed and analyses were both inductive and deductive. Firstly, a short report was produced for the ECCs to verify and use for their own purposes, such as an annual review and future planning. Then themes around policies and practices relating to language and digital media were identified through coding references to the use of digital technologies and/or languages other than English. Other themes also emerged, such as teachers prioritizing social competencies, physical play, and the development of relationships over digital technology use, and the challenges faced engaging whānau when whānau were under increased pressures in everyday life. Expert researchers then validated the evidence for each theme, applying their expertise in ECE, linguistics, and/or the digital world.

4.2. The Survey of Families

Our overarching programme of research focuses on vulnerable children because our aim is to give them a better start [47]. As a proxy for vulnerability, we chose to focus on those with lower language levels. These children were in the first year of schooling in the same suburbs as the ECC and included children who previously attended the six participating ECCs. A questionnaire was developed to ask whānau (extended family) about the hauora (health and wellbeing), hononga (reading together at home), hinengaro (reading practices), and harikoa (positive identities) of their children, who were identified with lower levels of oral language ability.

These children were selected from all 247 children in their first year attending the seven participating primary schools (see Figure 2). The screening employed two sub-tests from the Clinical Evaluation of Language Fundamentals Preschool, Second Edition (CELF-P2) [48], and the initial phoneme identity task from the New Zealand Computer-Based Phonological Awareness Database (CBPA) [49]. The 152 children who scored less than eight on the CELF-P2 subtest and/or less than six on the CBPA assessment were identified with lower levels of oral language ability.

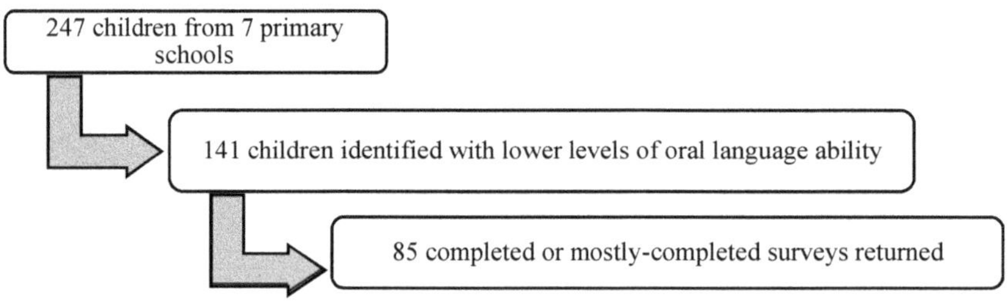

Figure 2. Visual representation of staged recruitment process of the whānau survey.

All 141 whānau of those children with lower levels of oral language ability were invited to complete the questionnaire. Subsequently, 85 (60%) of these whānau questionnaires were returned, and the demographics of participants are shown in Table 1. There was unlikely to be much difference with those who did not complete the survey, because they had a similar demographic makeup to the 141 children identified with lower levels of language ability (see Table 1). Whānau completing a survey or interview were given a $15 New Zealand dollar (NZD) voucher for groceries to acknowledge the gift of their time. Various follow-up methods were employed to reach this number, including community workshops, one-on-one meetings (occasionally with a researcher who spoke their home language), phone calls, and emails. Only items related to multilingualism and the digital world are considered in this paper; another article will report in full on this survey [43].

Table 1. Demographic information of the 85 children whose whānau participated in the survey compared with national demographics for children of the same age. NZ—New Zealand.

Characteristic	Surveys Returned (%) (n = 85)	Children Who Entered School with Lower Levels of Oral Language Ability (n = 141)	Percentage Nationwide of Children Aged 5–6 Years of Age in 2013 [42]
Gender			
Male	50 (58.8%)	49.60%	50.90%
Female	35 (41.2%)	50.40%	49.10%
Ethnicity			
NZ European and other ethnicity	48 (56.5%)	54.50%	62.80%
Pasifika	15 (17.64%)	21%	8.80%
Māori	22 (25.9%)	24.50%	20.10%

5. Findings

5.1. Linguistic Landscapes of Six ECC

An overview of the languages that could be seen in the physical and digital landscapes of the six ECCs is followed by a discussion of themes relating to the perception and use of digital media, ending with a review of challenges. Table 2 provides a summary of the languages in the six ECE centres that was evident in their signs and artefacts. Of the 469 images taken of all wall displays across the six centres, 354 contained linguistic items. These signs and artefacts were categorised as containing only one language (e.g., English only), as having an equal presence of two languages (e.g., English and Māori), or as having the majority of one language and some presence another language (e.g., English with some Māori). This indicated that these centres were reflecting Te Whāriki curriculum guidelines relating to commitment and work within the bilingual (Māori and English) context of Aotearoa New Zealand. Other languages were also represented, with these typically responsive to the languages and cultures of the children attending the centres and purposively chosen to meet centre policies and practices, such as collaboration (see Table 2 and Figures 3 and 4).

Table 2. Pictures containing linguistic items across the six early childhood education centres (ECCs) classified on the basis of the language(s) they contained.

Language(s) in the Items	Number of Items (n = 354)
English only	171
English with some Māori	43
English and Māori	50
Māori with some English	30
Māori only	22
Samoan only	10
Samoan and English	9
English/Samoan/Māori	8
Multilingual	11

Most teachers interviewed said that the main purpose of the linguistic landscape was to engage whānau in their children's learning and development. This engagement facilitated face-to-face conversations with whānau, making it easier to communicate centre aims and plans, discuss children's activities, and promote conversations, collaboration, and language development. It was notable that all ECCs had a multilingual welcome near the entrance and online (see Figure 3).

Figure 3. Screenshot of one early childhood education (ECE) centre website with bilingual welcome.

Crucially, teachers emphasised that centres relied on whānau support and collaboration with respect to children's bilingual language development. Multilingual languages were most commonly seen in basic greetings used in displays, e-portfolios, and written communication to welcome, respect, and celebrate culture, identity, and language. A Whānau Aspirations Tree was observed in most ECCs. The one illustrated in Figure 4 was particularly rich in languages and culture. All the children's photos taken with a digital camera were place on the tree, an image that echoes Māori culture, and beside each photo was the family's aspirations in their home languages that focused mainly on the social aspects. A copy was placed within each child's e-portfolio. Pictures of the co-construction surrounded this large display.

Figure 4. The Whānau Aspirations Tree echoes Māori culture. It was co-constructed by centre staff and whānau and set within the ECC's main learning area. The photos were taken with a digital camera.

After English, the Māori language, along with cultural symbols and references, was most prevalent with evidence in artefacts such as signs, labels, songs, commands, books, portfolios, and communications to the children's whānau. The writing on the walls also served as reminders and prompts for the teachers to use languages with their emergent bilingual children. Pasifika languages were also used in songs, instructions, resources, and books that also contained cultural symbols and references; some were obtained through digital technologies such as iPads and computers.

The outreach librarian stressed the importance of supporting emergent bilinguals by creating an environment which welcomed the use of all children's languages beyond the context of the child's home by stretching across the contexts of the home and ECC through the child's digital world. "It's really interesting that [bilingual children] actually can't translate for you and they usually won't speak that language when they're at preschool. It is out of context, this isn't where that happens, [the bilingual children] are just not old enough to put that in place. [The digital world] is a huge part of who [the child] is and who their family is and it doesn't just belong in that house, it can actually exist everywhere else." (Outreach librarian interview, 2016).

5.1.1. Blended and Digital Resources Were Used Purposefully

Table 3 provides an overview of the way in which physical and digital resources were commonly used in all ECCs. The key digital resource was the e-portfolio, often used alongside paper portfolios. The e-portfolio was used to collate children's learning stories and compile observational records of children's competencies and learning dispositions, as recommended in Te Whāriki. Portfolio content, which included digital photos, videos, and text in English, Māori, and the children's home languages, was shared with centre colleagues, homes, and whānau residing elsewhere, and it provided an opportunities for whānau to contribute in their home languages. The portfolios prompted children to reflect on, share, and talk about their activities across contexts.

From one centre's perspective, the e-portfolio had both benefits and disadvantages. One advantage to using the e-portfolio was the ability to connect, engage, and collaborate with family/whānau and a platform to use home languages. "In our learning stories and children's [portfolio] books we're really encouraging the partnership between whānau, the teaching team, and the kindergarten, so that we are encouraging them to contribute photos of things that are of cultural significance to them, things that are important to their family. So, photos are a really good way of doing that. We're trying to use greetings as we are writing the story and maybe some images." (Teacher interview, 2016). The disadvantage was that some parents chose not to access their computer until after their children went to bed; thus, the child missed the sharing experience. The teacher told us, "So, we've gone back to redoing them, we are still loading them electronically, but we are doing it in the book as well." (Teacher interview, 2016).

Table 3. Representative sample of resources commonly observed in the six ECCs, classified by purpose(s) and an illustrative use. Figures refer to the examples of ECC artefacts.

Resource	Example of Use
Physical resource	
Whiteboards	Multilingual greetings and sentences in Māori and other languages (Figure 3)
Paper portfolio	Multilingual greetings and learning stories with Māori words and concepts
Labels	Prompt for teachers to use languages and welcomes
Cultural protocols	Normal practice in ECC, e.g., songs and prayers; mostly sung in Māori and Samoan (Figure 1)
Displays, national events	Resources supporting Samoan language week, Diwali Festival, etc.
Digital/blended resource	
E-portfolios	Children's learning stories included multilingual greetings, Māori words and concepts, and family languages (e.g., Educa ePortfolio)
iPads	Cloud-based games supporting curriculum carefully managed; most apps in English
E-newsletters	Instead of or complementing paper version, these included multilingual greetings, and Māori words and concepts
Email	Used to send newsletters and photos to family (Figure 1)
Texting	Used occasionally for quick communication with families (e.g., to update a parent on their child)
Facebook	Produced by national organisations for their ECC; one had multilingual greetings
Websites	Produced by national organisations for their ECC; one had multilingual greetings (Figure 3)
Digital cameras	Extensive use by teachers for displays and artefacts to take home (Figures 1 and 4)

All centres had a digital camera and used digital photos extensively, enabling teachers, children, and whānau to share experiences across environments and generate conversations regardless of language. Digital photos were displayed on walls, in portfolios (both types) on iPads, and on a large screen in the mat or seating areas. Each centre had one of these screens, and teachers played videos (recorded on the digital camera) of the children's activities on it and occasionally multicultural stories and celebrations. Whānau could send digital photos via email to the centre or place them in the e-portfolios. Teachers often added text to the photos and asked whānau if they could translate the text into home languages. One teacher interviewed said, "I gave (the family) some photos of their two children, two boys, what they were doing, and then asked them to take them home and talk to the boys about those (photos). So, they've come back with some of their own dialect (written on them)."

Despite the ability to connect language and experiences across the children's environments of home and the ECCs, there were concerns that digital media could be distracting. Teachers in all ECCs expressed concerns about the impact digital media may have that reduces the real-world experiences that children have, which generate the conversations of shared experiences that are necessary for language development. For example, "I don't hear our children talk about those experiences ... So if they are not talking about them to us, then I would guess that the parents aren't talking to the children about doing them ... So they might be doing them but there's not a lot of language going on while they are doing them, (adults) are not mediating for (young children)." (Teacher interview 2016).

The outreach librarian also reflected that parents were not engaged with the literacy activities, as was hoped. However, she noted that many parents approached her asking about the apps she used during the story time session to download themselves. This response was different from that to physical books. She believed it was the instant accessibility and the desire for access to quality apps that was motivating (Researcher journal, 2016).

All the centres had iPads, used for games (in English), music, and accessing content stemming from the children's interests. One ECC also had apps in Samoan and Māori. Centres carefully managed children's access to iPads. For one centre, this entailed the use of a booking system that allowed the child 10 min of iPad use controlled with the input of a passcode. This system was self-managing, and the next child on the booking schedule would prompt the transfer of the iPad. At the time of the study, two of the centres were not actively using the iPads, due to technical issues. Neither prioritised funding for maintaining them, partly due to the difficulties in managing them, as well as the teachers' stated belief that whānau preferred children to play outdoors and socialise with one another rather than spend time on screens.

Because all of the centres deemed (through policy and practice) the children's emotional regulation and social competency a priority, their staff took care to ensure that digital technology use and associated language artefacts were meaningful to the children and likely to engage their whānau in supporting that child's development. The majority of teachers thought that most of the children attending the centres had access to at least one digital device at home. If teachers knew that a whānau did not have home-based access, they typically encouraged them to use the free internet at a public library. However, and crucially, all teachers said they did not know how the children used devices at home and what content they could access on them, one said "I haven't had any experiences [of children's digital worlds]. I have to say it is not something that has been a passion for me or really interested me, so unless somebody is really struggling or there is a child that I haven't spent time with and I think that the only time that I'll be able to spend with them is on an iPad, then for me it's not one of the top priorities."

5.1.2. Issues and Access

There was concern about digital media distracting from young children's social and physical development, particularly mobile devices, and a number of adults mentioned that controlling their use was problematic in all locations, including arguments at home among siblings. However, there was no mention of any "screen addiction". However, fear of being judged, lack of shared information about how much is too much, and limited knowledge of such addictions may be why the term addiction was avoided. Alternatively, it may be less likely among families with low incomes. Indeed, in the previous year during the stakeholder meeting that informed the research design so that ECCs were included, one of the teachers passionately objected to the "Bring Your Own Device" policy that was being introduced in the neighbouring schools, because of the stress that it placed on families that already struggled to feed and clothe their children.

5.2. Survey Findings

As introduced earlier, the whānau survey enabled us to extend our findings relating to digital world and whānau perceptions of their children's multilingualism. Despite the small sample size (n = 85) of the survey, there were some significant findings. The 85 questionnaires returned by whānau described the selected 50 boys and 35 girls with low language abilities attending their first year of schooling (age mean = 68.13 months, SD = 3.46 months) including many of the children that previously attended an ECC described earlier. Table 1 outlines the demographics of the children surveyed, as well as that for all the children with lower levels of oral language ability and the same nationwide demographics from the most recent census [42]. To aid in interpreting our results, we also compared our finds with the Growing Up in New Zealand (GUiNZ) cohort study of four year olds in 2013/2014 [50], while also noting that our findings are limited and not generalizable.

One of the questions in the survey asked, "During a typical week at home, approximately how many hours would your tamaiti/child/tama spend on digital media?". In order to improve the cultural relevance of the survey, the questions inter-languaged English between Māori and Samoan words, e.g., "tamaiti/child/tama". As shown in Table 4, we found that 6.8% of children did not engage with any digital media. When compared to the GUiNZ findings that 5% of four year olds did not engage with digital media, this suggests the proportion of cautious parents does not appear to be increasing, despite the increasing access to digital devices reported by the World Internet Project for New Zealand [11]. We also found that nearly one-third (32%) of children engaged with digital media for more than 5 h per week, equivalent to an average of more than 43 min per day. When comparing this result to GUiNZ data (41% of four year olds spending one or more hours a day on digital devices), our findings suggest that there is less use reported by whānau of these children than for the four year olds of the GUiNZ study. A limitation of our survey data was the way in which the question was asked and answered, and its variation with respect to the categories in GUiNZ survey, which also allowed open ended answers. These differences in survey design may be the cause for the differences in findings between our questionnaire and that in the GUiNZ study. As discussed later, any simple measure of time is problematic given the blending of physical and digital worlds.

Table 4. Frequency table of hours spent of digital media and of the importance of bilingualism.

		Frequency	Percent	Valid Percent
Hours on digital media per week	Never	5	5.9	6.8
	Less than an hour	11	12.9	14.6
	1–3 h	18	21.2	24.3
	3–5 h	16	18.8	21.6
	More than 5 h	24	28.2	32.4
	Sub total	74	87.1	100
Missing		11	12.9	
Total		85	100	
Importance of bi- and multilingualism	Not at all important	17	20	20.5
	Not that important	24	28.2	29.9
	Quite important	22	25.9	26.5
	Very important	20	23.5	24.1
	Sub total	83	97.6	100
Missing		2	2.4	
Total		85	100	

Our statistical analysis to investigate potential relationships with this use of digital devices identified a two negative associations that were nearing significance: the number of hours on digital media and the whānau report of whether the child slept well through the night ($r = -0.184$, $p = 0.122$)

and also with the whānau report of amount of sleep ($r = -0.196$, $p = 0.09$). Most of these children appear to be having the recommended 10–12 h of sleep (mean = 10 h 57 min, SD = 57 min), with only 4.9% having less than 10 h sleep per night. This result was similar to the hours of sleep per night (mean = 10 h 45 min) for four year olds in the GUiNZ study.

Whānau indicated that there were 54 multilingual children in this sample who spoke one or more languages in addition to English. In order to look into whānau perceptions, the importance of bi- or multilingualism was scaled from 0, not important at all, through to 3, very important. Counts are shown in Table 4; further analysis showed that whānau of multilingual children were the most affirmative. When looked at with reading practices, a positive correlation was found with children's ability to read words that were pointed to ($r = 0.315$, $p = 0.012$). We also found that speaking more than one language was positively associated with children's ability to read words that were pointed to ($r = 0.308$, $p = 0.014$), children's ability to point and say words without being asked ($r = 0.279$, $p = 0.027$), and children's ability to write their own surname ($r = 0.471$, $p < 0.01$). These findings may indicate the value for multilingualism in a child's literacy development, such as richer language exchanges with whānau.

6. Discussion

Our descriptions of the linguistic landscapes of six mainstream ECE centres are reassuring in the way that they indicate purposeful integration of the digital world in the experiences of the children and the adults who support them, including the use of multilingual e-portfolios to share with whānau and to develop the children's learning stories across time and space. Evidence from our survey also supported previous research indicating parental care of children. All the ECCs respected the children's rights to access digital media and also controlled that access to ensure a healthy balance of activities in the knowledge that most children had a lot of exposure at home. This is exemplary, including the recognition of the need for further organisational and professional development.

Of particular value to multilingual children and their whānau, the teachers used languages other than English in the physical and digitally mediated linguistic landscape to welcome, respect, and celebrate the cultures and identities of all the children attending. The diverse languages presented in signs and artefacts also reminded and prompted teachers to use diverse languages with their emergent bilingual children. Teachers created most of the displays in the centres, especially the bilingual or multilingual ones, and this effort reflected their motivation for supporting emergent bilingual children.

Teachers and whānau prioritised social competencies, interaction, and emotional regulation as foundational for the development of their children's languages. Centres carefully managed digital resources as an additional means of strengthening connections between the centre and home. Our later research showed that whānau engagement with the e-portfolio grew significantly from 2016 to 2017, along with the teachers' skills and confidence in engaging with the software to collaborate on their children's linguistic and other experiences. The linguistic landscapes nurtured reciprocal relationships with families and their communities to support children's language development, including intergenerational transmission of language that is likely to result in a better start [36], particularly for the more vulnerable children who grow up with whānau speaking more than one language.

Incorporation of the digital world in the centres was not without concern, however, with teachers expressing interest in professional development and guidance in this area, particularly in terms of strategies that enable adults to share and shape children's experiences in the digital world across both home and centre environments. Although early exemplars of learning stories to guide assessment practice in New Zealand ECE included resources such as digital cameras, email, and home languages [51], our study suggests that, as the complexity of the digital world increases, teachers and children's families are likely to need more guidance to mediate that world for young children in ways that support both the children's linguistic and holistic development.

While centres found it relatively easy to share children's learning stories with the children's whānau by encouraging them to add images and translations to the children's e-portfolios, development of these portfolios created challenges for teachers, such as increased workload. The portfolios also highlighted issues relating to digital equity. For example, lack of technology or data to access e-portfolios at home reduced the opportunity to work with ECC staff to support their child. This lack of access could also potentially compromise the cultural safety interface between an ECC and homes [52].

Of most concern is the widespread knowledge that many young children are overexposed to digital media outside ECCs. Paudel et al.'s systematic literature review [53] suggests that children growing up in families with addictions such as those related to mobile screen media are at most risk and identified the following correlates: "older children, children better skilled in using mobile screen media devices, and those having greater access to such devices at home and whose parents had high mobile screen media use were more likely to have higher use of mobile screen media devices." According to Lindenberg et al. [54], who identified prevalence rates of over 5% among adolescents in Germany, the World Health Organisation recently listed gaming disorder in addition to internet use disorder, indicating that the rising rates of various types of overuse of screen media are an increasing cause of health concerns. Our study updates our understanding of the protective factor of quality ECE [7]. Our study also raises the possibility that ECCs can guide and support families coping with these addictive behaviours to reduce that impact on these young children.

The writing of this article prompted us to consider ways in which to guide ECC. Concerns about the influence of digital media on the behaviour of family members emerged over the last decade. For example, Raman et al., whose detailed study [55] of screen exposure during young children's daily routines explored the risk of having social-emotional delay, showed that children were exposed to screens throughout daily routines that would typically involve face-to-face interaction and conversational turn-taking, such as eating and sleeping routines. This persistent exposure to digital screens has the potential to distract attention away from the daily conversational interactions between adults and young children, which are necessary for healthy early development. Several teachers were concerned that parents were not mediating the use of digital media for their children, with one teacher saying that there was "not a lot of language going on".

Digital distraction may cause these and other conversations to become shallow, thus reducing young children's opportunities. Given the complexity of effective parent and sibling interaction with young children [6], it is possible that the increasing use of mobile phones and other devices can adversely impact child development, particularly in families suffering from some form of media addiction. For example, the Tizard and Hughes study [56] of the interactions of four-year-old girls with their mother at home in the 1980s contradicted the Piagetian thinking, which was prevalent at the time of their research, that the young child is illogical or whimsical (p. xiv). Instead, they concluded that children have an intense need to understand the world, which is reflected in the many "why" questions they asked at home (averaging 26 questions per hour). The analysis also showed that the turn-taking of the children and their mother supporting them to ask questions in a persistent and logical way in order to extend their understanding.

This unexplored relationship between digital distraction and young children's need for interaction with parents indicates the need for additional research. It is also worth noting that Raman et al.'s methodology [55] clarified our understanding of how children experience the impact of the digital world for themselves and their whānau. This impact is spread across the entire day, from waking up until going to sleep, so that it becomes misleading to describe it in a unit of time, such as the 2.1 h from the GuiNZ study [50]. It will be important to apply this to the interpretation of the literature and in the design of future research, policy, and guidance that we share on our web sites (see Supplementary Materials).

7. Recommendations and Conclusions

The invitation to write this article in response to the IJERPH special issue on "Internet and Mobile Phone Addiction: Health and Educational Effects" led us to reflect deeply on that topic, and we now plan to apply this to our mission. The mission of our national programme of multidisciplinary research includes a strand on mental health within which Merry and her colleagues are developing e-health interventions, e.g., Reference [57]. "To predict, prevent, and treat vulnerability in obesity, poor literacy, and mental health through research excellence that will achieve healthy, well-adjusted, and well-educated tamariki and young people. We aim to achieve our mission by taking both a life-course and a "braided river" approach to integrate themes, research disciplines, and both western and indigenous models of knowledge and practice, as well as incorporating the use of digital technology into our proposed solutions" [47].

We identified the need for additional research into the effects of addictions related to the internet and mobile devices. In particular, there is a need to provide more nuanced guidance for parents and siblings so that these young children have more opportunities for deep learning of languages, concepts, and social behaviour. Such research could identify mitigating strategies to include policy guidance for health and educational services, an action that we previously recommended within our national "Better Start" programme of research.

Drawing on the findings presented, this policy guidance is recommended to include strategies that enable adults to share children's experiences in the digital world. Vaughan and Beers' account of iPad-related professional development for ECE teachers [58] could be adapted to integrate strategies that benefit emergent bilinguals.

Improvements in software and interface design are also needed. Miller and Kocurek's five principles for the design and development of educational games for children under five years of age [59] could apply equally well to the development of digitally based multilingual activities and artefacts: (1) developmentally appropriate content; (2) reference to theoretical frameworks from the learning science field; (3) embedding learning in socially rich contexts; (4) ensuring diversity of content; and (5) creating a balance between play and real-world learning opportunities. To these five principles, we would add (6) opportunities for adults to engage alongside the children in their digital world(s) and to use all their languages.

We also have a recommendation with regards to multilingualism and the intergenerational transfer of languages. In Aotearoa New Zealand, Māori is recognised as a national language, and the ECE curriculum promotes its use as a living language. We found that it was often present in the six ECC landscapes as a language additional to English. Similar approaches could be extended to support the over 160 other languages in this linguistically "superdiverse" country [16], particularly for the more commonly spoken Samoan, Hindi, and Northern Chinese. Because the wider ecosystems influence the practices within ECE centres and homes [60], we recommend that policy documents, curricula, and related web resources be improved by weaving into them relevant languages and cultural images to support the enrichment of linguistic landscapes. Increased guidance and professional development is also recommended to strengthen language practices in ECE so that they are inclusive of linguistically diverse children. This approach was one that whānau and the community members who participated in our study and related workshops identified and endorsed.

Our future work includes the following commitments: (1) to expand our collection of evidence to describe changes in the landscapes in the first year of school in our target area; (2) to disseminate research-informed guidance for policy-makers locally nationally and globally that addresses the complexity of the digital world in which young children are growing up, including the support of intergenerational languages transmission. We also plan to further align our research with the "You Matter To Us" (YMTU) consortium which is exploring how inter-agency action combined with community development can influence the evolution of organisational policies and practices (system change) to enhance wellbeing of children aged 0–5 years.

Conclusions

This research indicates that vulnerable children in quality ECC do have a better start in the digital world. ECE is improving multilingualism in Aotearoa New Zealand through its Te Whāriki curriculum, which ECCs are applying to evaluate and develop such practice. This practice limits "screen time" while also promoting strategies that deliberately deploy the digital world to serve the purposes of ECE in appropriate sociocultural ways. In addition to improving the quality of engagement, ECC staff deployed the digital world to increase the range and strength of their partnerships with whānau. As Taylor et al. [36] found, such purposeful strategies leverage the diverse languages of whānau with cultural support for their children, so it is also likely to have a positive impact on these families and their minority communities, which may result in a virtuous circle back to support all the children in the ECCs. In our research, the Whānau Aspirations Tree increased the sense of belonging, participation, and safety; mostly, it gave a strong presence in the ECC of the whānau voices. Given the fact that the area of the city in which the ECCs were located included one of the three most deprived suburbs in the country [61], this could have increased participants' feelings of safety. Multiple sources of evidence evaluated for the related community project "You matter to Us" indicate that such improvements are possible [41].

Therefore, there is evidence that ECCs can contribute to the resilience of vulnerable children and given them a better start in the digital world. The increased thoughtfulness of the adults and siblings of young children is also likely to lead to increases in resilience to potential internet and mobile phone addictions. For example, United States guidance [62] (p. 77) recommends media mentors to guide children and states, "As the primary role models for technology and media use, adults should be aware of and set limits on their own technology and media use when children are present, and focus on children having well-rounded experiences, including moderate, healthy media use."

The actions of these ECCs to deploy the digital world purposefully with developmentally appropriate sociocultural strategies included partnership with families. Projects by the World Organisation for Early Childhood Education (OMEP) [63] provide evidence that young children can act agents of change in their families to promote health and wellbeing. In the future, young children and their ECCs may be recognised as change agents redressing the challenges of addiction that are associated with growing up in a digital world.

Supplementary Materials: The Better Start research programme and EBinDW strand have web sites: http://www.abetterstart.nz/ and https://ebdwwebsite.wixsite.com/ebdw.

Author Contributions: L.H. was the lead research assistant for "emergent bilinguals in a digital world" (EBinDW), and piloted the data collection design, collected all of the linguistic landscapes, analysed most of the EBinDW data, and led the writing of that methodology and findings. N.D. is co-principal investigator for the EBinDW strand of research; she co-conceived, co-designed, and co-analysed the EBinDW study, and she also collaborated on most aspects of the survey study. N.D. also conceptualised this article and led its writing. U.C. was the other co-principal investigator for the EBinDW research (2015–2017) who co-conceived, co-designed, and co-analysed the study, and she contributed to writing of the article. L.d.V. contributed her expertise in early childhood education, including application of literature to the analysis and writing of this article. S.M. was the principal investigator for the survey research, who led its conception, design and analysis. N.G. undertook the statistical analysis of the survey data and drafted that methodology and findings section. S.A. and T.O.T. gathered data volunteered by Pacific participants and also contributed to the analysis in both parts of the study. J.D. supported engagement of the whole project with stakeholders and contributed the fifth future strategy in this article.

Funding: This research was supported by funding from the New Zealand Ministry of Business and Innovation (MBIE) grant number 15-02688. L.H., S.A., and N.G.'s PhD scholarships were funded by the University of Canterbury.

Acknowledgments: The data were collected within the project "Eke pānui, eke tamaiti: Braiding health and education services to ensure early literacy success and healthy well-being for vulnerable children" co-led by Gail Gillon and Angus Macfarlane. We acknowledge the collaborative community of researchers in our National Science Challenge. No conditions were imposed by the financing beyond the Vision Matauranga cited in this article. We thank all the participants in the study, including the leaders of the ECC, community library, and schools. In addition, we acknowledge the support of the Ministry of Education.

Conflicts of Interest: The authors declare no conflict of interest.

References

1. Fleer, M. Technologically constructed childhoods: Moving beyond a reproductive to a productive and critical view of curriculum development. *Aust. J. Early Child.* **2011**, *36*, 16–24.
2. Stephen, C.; Edwards, S. *Young Children and Learning in a Digital Age: A Cultural and Critical Perspective*; Routledge: Abingdon, UK, 2018.
3. UNESCO. *Recognizing the Potential of ICT in Early Childhood Education*; UNESCO Institute for Information Technologies in Education: Moscow, Russia, 2010.
4. Cunningham, U. *Growing up with Two Languages*, 3rd ed.; Routledge: London, UK, 2011.
5. Pempek, T.A.; Kirkorian, H.L.; Richards, J.E.; Anderson, D.R.; Lund, A.F.; Stevens, M. Video comprehensibility and attention in very young children. *Dev. Psychol.* **2010**, *46*, 1283–1293. [CrossRef] [PubMed]
6. Bornstein, M.H. Parenting science and practice. In *Handbook of Child Psychology: Child Psychology in Practice*, 6th ed.; Renniger, K.A., Sigel, I.E., Eds.; Wiley: Hoboken, NJ, USA, 2006; Volume 4, pp. 893–949.
7. Lamb, M.E.; Ahnert, L. Nonparental child care: Context, concepts, correlates, and consequences. In *Handbook of Child Psychology: Child Psychology in Practice*, 6th ed.; Renniger, K.A., Sigel, I.E., Eds.; Wiley: Hoboken, NJ, USA, 2006; Volume 4, pp. 950–1016.
8. García, O.; Kleifgen, J.A.; Falchi, L. *From English Language Learners to Emergent Bilinguals: Equity Matters Research Review No. 1*; Campaign for Educational Equity; Teachers College, Columbia University: New York, NY, USA, 2008.
9. Brown, K.D. The linguistic landscape of educational spaces: Language revitalization and schools in southeastern Estonia. In *Minority Languages in the Linguistic Landscape*; Gorter, D., Marten, H.F., Van Mensel, L., Eds.; Palgrave-Macmillan: Basingstoke, UK, 2012; pp. 281–298.
10. Crothers, C.; Smith, P.; Urale, P.W.B.; Bell, A. *The Internet in New Zealand 2015*; Institute of Culture, Discourse & Communication, Auckland University of Technology: Auckland, New Zealand, 2016.
11. Díaz Andrade, A.; Hedges, M.R.; Karimikia, H.; Techatassanasoontorn, A. *World Internet Project: The Internet in New Zealand 2017*; New Zealand Work Research Institute: Auckland, New Zealand, 2018.
12. Ministry of Education. *Te Whāriki: He Whāriki Mātauranga mō ngā Mokopuna o Aotearoa Early Childhood Curriculum*; Learning Media: Wellington, New Zealand, 2017.
13. Education Council. *Our Code, Our Standards*; Education Council: Wellington, New Zealand, 2017.
14. King, J.; de Vocht, L.; Cunningham, U.; Davis, N.E. *Languages Policy Development: A Guidance Note. He Kupu Āwhina kia Whakaritea He Kaupapa mō Ngā Reo*; Unpublished note to Minister of Education, University of Canterbury: Christchurch, New Zealand, 2017.
15. Office of Ethnic Affairs. Language and Integration in New Zealand. Available online: https://ethniccommunities.govt.nz/story/language-information-and-advice (accessed on 30 July 2018).
16. Royal Society of New Zealand. *Languages in Aotearoa New Zealand*; Royal Society of New Zealand: Wellington, New Zealand, 2013.
17. Hsin, C.-T.; Li, M.-C.; Tsai, C.-C. The influence of young children's use of technology on their learning: A review. *J. Educ. Technol. Soc.* **2014**, *17*, 85–99.
18. Chaudron, S. *Young Children (0–8) and Digital Technology: A Qualitative Exploratory Study across Seven Countries*; Publications Office of the European Union: Luxembourg, 2015.
19. Aubrey, C.; Dahl, S. The confidence and competence in information and communication technologies of practitioners, parents and young children in the early years foundation stage. *Early Years* **2014**, *34*, 94–108. [CrossRef]
20. Marklund, L.; Dunkels, E. Digital play as a means to develop children's literacy and power in the Swedish preschool. *Early Years* **2016**, *36*, 289–304. [CrossRef]
21. Lopez-Fernandez, O.; Kuss, D.; Pontes, H.; Griffiths, M.; Dawes, C.; Justice, L.; Männikkö, N.; Kääriäinen, M.; Rumpf, H.J.; Bischof, A.; et al. Measurement invariance of the short version of the problematic mobile phone use questionnaire (PMPUQ-SV) across eight languages. *Int. J. Environ. Res. Public Health* **2018**, *15*, 1213. [CrossRef] [PubMed]
22. Rozgonjuk, D.; Saal, K.; Tahl, K. Problematic smartphone use, deep and surface approaches to learning, and social media use in lectures. *Int. J. Environ. Res. Public Health* **2018**, *15*, 92. [CrossRef] [PubMed]

23. Van Rooij, A.J.; Ferguson, C.J.; Colder Carras, M.; Kardefelt-Winther, D.; Shi, J.; Aarseth, E.; Bean, A.M.; Bergmark, K.H.; Brus, A.; Coulson, M.; et al. A weak scientific basis for gaming disorder: Let us err on the side of caution. *J. Behav. Addict.* **2018**, *7*, 1–9. [CrossRef] [PubMed]
24. Kildare, C.A.; Middlemiss, W. Impact of parents mobile device use on parent-child interaction: A literature review. *Comput. Hum. Behav.* **2017**, *75*, 579–593. [CrossRef]
25. Lauricella, A.R.; Wartella, E.; Rideout, V.J. Young children's screen time: The complex role of parent and child factors. *J. Appl. Dev. Psychol.* **2015**, *36*, 11–17. [CrossRef]
26. McPake, J.; Plowman, L.; Stephen, C. Pre-school children creating and communicating with digital technologies in the home. *Br. J. Educ. Technol.* **2013**, *44*, 421–431. [CrossRef]
27. Bak, T.H.; Nissan, J.J.; Allerhand, M.M.; Deary, I.J. Does bilingualism influence cognitive aging? *Ann. Neurol.* **2014**, *75*, 959–963. [CrossRef] [PubMed]
28. Olsen, R.K.; Pangelinan, M.M.; Bogulski, C.; Chakravarty, M.M.; Luk, G.; Grady, C.L.; Bialystok, E. The effect of lifelong bilingualism on regional grey and white matter volume. *Brain Res.* **2015**, *1612*, 128–139. [CrossRef] [PubMed]
29. Cummins, J. *Language, Power, and Pedagogy: Bilingual Children Caught in the Crossfire*; Multilingual Matters Ltd.: Clevedon, UK, 2000.
30. Rowe, D.W.; Miller, M.E. Designing for diverse classrooms: Using iPads and digital cameras to compose eBooks with emergent bilingual/biliterate four-year-olds. *J. Early Child. Lit.* **2016**, *16*, 425–472. [CrossRef]
31. King, J.; Cunningham, U. Intergenerational transmission of minority languages in New Zealand: Methodological issues. In *Advances in Understanding Multilingualism: A Global Perspective*; Grucza, S., Olpińska-Szkielko, M., Romanowski, P., Eds.; Peter Lang Verlag: Frankfurt, Germany, 2016; pp. 61–77.
32. Cunningham, U.; King, J. Language, ethnicity, and belonging for the children of migrants in New Zealand. *SAGE Open* **2018**, *8*. [CrossRef]
33. Dressler, R. Sign geist: Promoting bilingualism through the linguistic landscape of school signage. *Int. J. Multiling.* **2015**, *12*, 128–145. [CrossRef]
34. Cenoz, J.; Gorter, D. Linguistic landscape and minority languages. *Int. J. Multiling.* **2006**, *3*, 67–80. [CrossRef]
35. Sneddon, R. Young bilingual children learning to read with dual-language books. *Engl. Teach. Pract. Crit.* **2008**, *7*, 71–84.
36. Taylor, L.K.; Bernhard, J.K.; Garg, S.; Cummins, J. Affirming plural belonging: Building on students' family-based cultural and linguistic capital through multiliteracies pedagogy. *J. Early Child. Lit.* **2008**, *8*, 260–294. [CrossRef]
37. Gorter, D. Linguistic landscapes and trends in the study of schoolscapes. *Linguist. Educ.* **2017**, *44*, 80–85. [CrossRef]
38. Cooper, M.; Hedges, H. Beyond participation: What we learned from Hunter about collaboration with Pasifika children and families. *Contemp. Issues Early Child.* **2014**, *15*, 165–175. [CrossRef]
39. World Medical Association. Declaration of Helsinki. Ethical principles for medical research involving human subjects. *Bull. World Health Organ.* **2001**, *79*, 373–374.
40. A Better Start, E Tipu E Rea. Vision Mātauranga and He Awa Whiria Guide Our Research and Operations. Available online: https://www.abetterstart.nz/the-science/vision-matauranga/ (accessed on 1 September 2018).
41. Dobson, J. (Ed.) *You Matter to Us (YMTU)*, Unpublished Concept Paper. 2017.
42. Statistics New Zealand. *2013 Census QuickStats about a Place*; Statistics New Zealand: Wellington, New Zealand, 2013.
43. Schaughency, E.; Riordan, J.; Reese, E.; Derby, M.; Wilson, L.; Gillon, G. Developing a community-based oral language shared book reading program: Exploring feasibility and social validity. Submitted. 2018.
44. Harris, L. An Ethnographic Case Study of the Linguistic Landscape of an Award-Winning Māori Immersion Early Childhood Centre. Master's Thesis, University of Canterbury, Christchurch, New Zealand, 2016.
45. Copland, F.; Creese, A.; Rock, F.; Shaw, S. *Linguistic Ethnography: Collecting, Analysing and Presenting Data*; SAGE: Los Angeles, CA, USA, 2015.
46. Education Review Office. Early Childhood and School Reports. Available online: http://www.ero.govt.nz/review-reports/ (accessed on 30 July 2018).
47. A Better Start National Science Challenge E Tipu e Rea. Revised Research and Business Plans. Available online: http://www.abetterstart.nz/about/ (accessed on 30 July 2018).

48. Wiig, E.H.; Secord, W.; Semel, E. *Clinical Evaluation of Language Fundamentals—Preschool*, 2nd ed.; Harcourt Assessment: London, UK, 2006.
49. Carson, K.; Gillon, G.; Boustead, T. Computer-administered versus paper-based assessment of school-entry phonological awareness ability. *Asia Pac. J. Speech Lang. Hear.* **2011**, *14*, 85–101. [CrossRef]
50. Morton, S.M.B.; Grant, C.C.; Berry, S.D.; Walker, C.G.; Corkin, M.; Ly, K.; de Castro, T.G.; Atatoa Carr, P.E.; Bandara, D.K.; Mohal, J.; et al. *Growing Up in New Zealand: A Longitudinal Study of New Zealand Children and Their Families. Now We Are Four: Describing the Preschool Years*; Growing Up in New Zealand: Auckland, New Zealand, 2017.
51. Carr, M.; Lee, W.; Jones, C. Information and communication technology (ICT). In *Kei Tua o te Pae: Assessment for Learning: Early Childhood Exemplars*; Ministry of Education; Learning Media: Wellington, New Zealand, 2009.
52. Conteh, J.; Brock, A. 'Safe spaces'? Sites of bilingualism for young learners in home, school and community. *Int. J. Biling. Educ. Biling.* **2011**, *14*, 347–360. [CrossRef]
53. Paudel, S.; Jancey, J.; Subedi, N.; Leavy, J. Correlates of mobile screen media use among children aged 0–8: A systematic review. *BMJ Open* **2017**, *7*, e014585. [CrossRef] [PubMed]
54. Lindenberg, K.; Halasy, K.; Szász-Janocha, C.; Wartberg, L. A phenotype classification of internet use disorder in a large-scale high-school Study. *Int. J. Environ. Res. Public Health* **2018**, *15*, 733. [CrossRef] [PubMed]
55. Raman, S.; Guerrero-Duby, S.; McCullough, J.L.; Brown, M.; Ostrowski-Delahanty, S.; Langkamp, D.; Duby, J.C. Screen exposure during daily routines and a young child's risk for having social-emotional delay. *Clin. Pediatr.* **2017**, *56*, 1244–1253. [CrossRef] [PubMed]
56. Tizard, B.; Hughes, M. *Young Children Learning*, 2nd ed.; Blackwell Publishing: Malden, MA, USA, 2002.
57. Fleming, T.; Bavin, L.; Lucassen, M.; Stasiak, K.; Hopkins, S.; Merry, S. Beyond the trial: Systematic review of real-world uptake and engagement with digital self-help interventions for depression, low mood, or anxiety. *J. Med. Int. Res.* **2018**, *20*, e199. [CrossRef] [PubMed]
58. Vaughan, M.; Beers, C. Using an exploratory professional development initiative to introduce iPads in the early childhood education classroom. *Early Child. Educ. J.* **2017**, *45*, 321–331. [CrossRef]
59. Miller, J.L.; Kocurek, C.A. Principles for educational game development for young children. *J. Child. Media* **2017**, *11*, 314–329. [CrossRef]
60. Davis, N.E. *Digital Technologies and Change in Education: The Arena Framework*; Routledge: New York, NY, USA, 2018.
61. Schluter, P.J.; Audas, R.; Kokaua, J.; McNeill, B.; Taylor, B.; Milne, B.; Gillon, G. The efficacy of preschool developmental indicators as a screen for early primary school-based literacy interventions. *Child. Dev.* **2018**, in press. [CrossRef] [PubMed]
62. Donohue, C.; Shomburg, R. Technology and Interactive Media in Early Childhood Programs. What We've Learned from Five Years of Research, Policy, and Practice. *Young Children* **2017**, *72*, 72–78.
63. Engdahl, I. Early childhood education for sustainability: The OMEP world project. *Int. J. Early Child.* **2015**, *47*, 347–366. [CrossRef]

International Journal of
Environmental Research and Public Health

MDPI

Article

Generalised Versus Specific Internet Use-Related Addiction Problems: A Mixed Methods Study on Internet, Gaming, and Social Networking Behaviours

Olatz Lopez-Fernandez [1,2,3]

[1] Turning Point, Eastern Health Clinical School, Monash University, 110 Church Street, Richmond VIC 2131, Australia; olatz.lopez-fernandez@monash.edu or lopez.olatz@gmail.com; Tel.: +61-8413-8509
[2] International Gaming Research Unit, Psychology Department, Nottingham Trent University, Nottingham NG1 4FQ, UK
[3] Laboratory for Experimental Psychopathology, Psychological Sciences Research Institute, Catholic University of Louvain, 1348 Louvain-la-Neuve, Belgium

Received: 23 June 2018; Accepted: 24 November 2018; Published: 19 December 2018

Abstract: The field of technological behavioural addictions is moving towards specific problems (i.e., gaming disorder). However, more evidence of generalised versus specific Internet use-related addiction problems (generalised pathological Internet use (GPIU) vs. specific pathological Internet use (SPIU)) is still needed. This mixed methods study aimed to disentangle GPIU from SPIU. A partially mixed sequential equal status study design (QUAN→QUAL) was undertaken. First, through an online survey, which adapted the compulsive Internet use scale (CIUS) for three types of problems (i.e., generalised Internet use, and specific online gaming and social networking). Second, potential problem users' perceptions of the evolution of these problems (aetiology, development, consequences, and factors) were ascertained, through semi-structured interviews, together with their opinion on present Internet gaming disorder (IGD) criteria adapted to each problem studied. Findings showed the CIUS remains valid and reliable for GPIU and SPIUs examined; a prevalence between 10.8% and 37.4% was estimated for potential at-risk problem gamers and Internet users, respectively, who reported their preference for maintaining their virtual lives. Half of the sample had a risk of a unique or mixed profile of these problems. Moreover, device patterns, gender, and age issues emerged, such as problem gamers being proportionally equal male and female young or middle-aged adults. GPIU was highly associated with problem social networking use, and weakly with problematic gaming, but both SPIUs were independent. Concerning addictive symptoms, salience, deception, and tolerance required redefinition, especially for SPIUs, while better-valued IGD criteria applied to GPIU and SPIUs were: Risk relationships or opportunities, give up other activities, withdrawal, and continue despite problems. Thus, although problems studied are present as risk behaviours, SPIUs seem to cover the addictive symptomatology in those categorised as potential problem users, online gaming being the most severe behavioural addiction problem.

Keywords: behavioural addictions; generalised versus specific problem Internet uses; Internet addiction; gaming disorder; social networking; mixed methods research

1. Introduction

The field of behavioural addictions related to technological uses (i.e., technological behavioural addictions) has been growing exponentially since 1995 [1–4] and not without scientific, clinical, and social debates. In mid-nineties, the phenomenon was recognised by the umbrella term of 'Internet addiction', a generalised addiction problem covering all online activities together. Almost automatically, this was conceptualised as a clinical disorder [5], initially closely aligned with 'impulse control disorder'. In 2013, it was proposed as a future 'addictive disorder' in the third appendix of the fifth Diagnostic and Statistical Manual of Mental Disorders (DSM-5) by the American Psychiatric Association (APA) [6], and at present it has been recognised as a health disease in the eleventh revision of the International Classification of Diseases (ICD-11) by the World Health Organization (WHO) [7]. However, this international recognition has come about solely for a specific technological addictive problem—problematic gaming—even though other technological use-related addiction problems coexist (e.g., cybersex addiction).

The terminology requires an update, as it covers the emergent health issues related to excessive online uses, which emerged together with the development of technologies at the end of last century, and it has been consolidated in the 21st Century. Thus, although it seems Internet use-related addiction problems were initially mainly studied as a generalised problem [8], there is a scientific and clinical production of other specific problems simultaneously studied, such as problematic video gaming [9] or social networking [10]. Both generalised and specific Internet use-related addiction problems seem to produce addictive symptomatology (i.e., the classic symptoms for substance use or gambling disorders) in a few users, together with functional physical or psychological impairments (i.e., when the online activity(-ies) negatively affects other areas of a user's organism or lives; e.g., sight or academic/work facets), and distress (i.e., when the online activity(-ies) may reflect a maladaptive behaviour, failure in coping or adaptation processes).

Concerning statistics records, the International Telecommunication Union (ITU) has published its Information and Communication Technologies (ICT) Facts and Figures 2017 [11], showing the continuous worldwide expansion of Internet use through fixed or mobile subscriptions. For instance, in 2017, 3578 million individuals were using the Internet (compared to 495 million in 2001), 830 million being young people (i.e., 15–24-year-olds), which represents 80 per cent of the youth population in the 104 countries studied (see Figure 1).

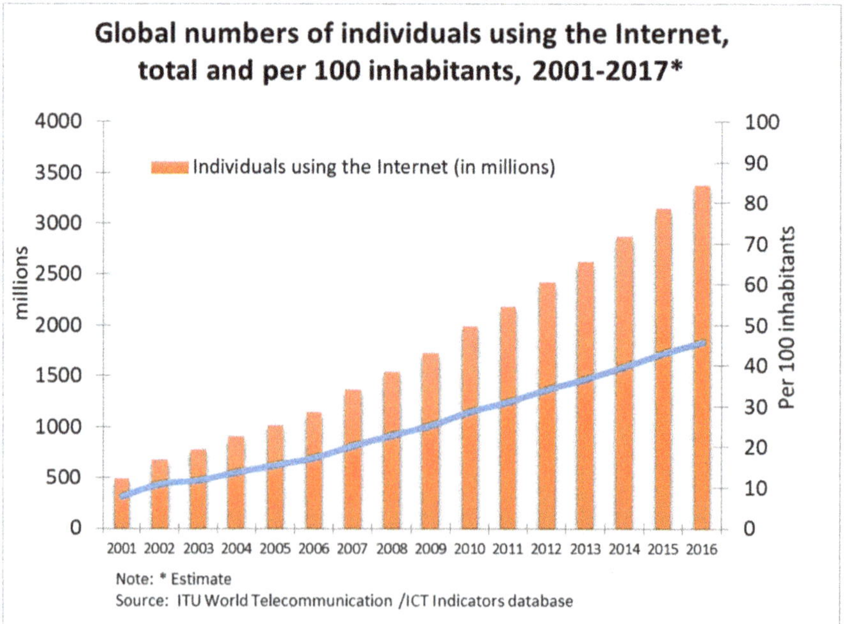

Figure 1. Evolution of global Information and Communication Technologies (ICT) from 2001 to 2017 (according to the International Telecommunication Union [ITU] World Telecommunications and its ICT indicators database).

Similarly, bibliometric evidence extracted from three ProQuest Central scientific databases (i.e., Health and Medical Collection, Psychology Database, and Public Health Database [12]) shows this impactful increase in research on Internet use-related addiction problems during the last two decades (1995–2017), which have especially risen from the beginning of the 21st century [13].

Indeed, an advanced search was conducted to observe the increase of production on these addiction problems. The procedure for the search of Internet addiction production was: It was introduced the terms 'Internet addiction' OR 'problem* Internet use' OR 'pathological Internet use' OR 'excessive Internet use' (to cover the main terminology used in Internet addiction). Simultaneously, the options to refine the search were: 'Peer reviewed' (i.e., to ensure the outputs were articles from official editorial processes), 'Exclude duplicate documents' (i.e., to avoid any article is detected more than once through the databases selected), 'Show additional terms included in the search' (to suggest and add alternative terminology used in relation to the terms selected), and a period from 1 January 1995 to 31 December 2017. A total of 116,455 outputs were obtained, showing a growing tendency in its bibliometric production, where each bar corresponded to the number of articles published in a year, going from 146 records in 1995 to 11,630 in 2017 (see Figure 2, first plot). The same procedure was used with the other two specific online behavioural addiction problems studied. Concerning problematic gaming (i.e., 'gaming addiction' OR 'problem* gaming use' OR 'pathological gaming use' OR 'excessive gaming use' OR 'gaming disorder' OR 'internet gaming disorder'), 7246 outputs were addressing problems within video gaming (e.g., computer gaming, digital or electronic game addiction). These ranged from 35 records in 1995 to 839 in 2017, with a progressive increment from 2010 and especially from 2014 (after the inclusion of 'Internet gaming disorder' (IGD) in the appendix of the DSM-5 by the APA [6]; see Figure 2, second plot). Similarly, problematic social networking obtained the most outputs in the search: A total of 202,045 (using as keywords: 'social network* addiction' OR 'problem* social network* use' OR 'pathological social network* use' OR 'excessive social network* use' OR 'social network* disorder'); from 1474 in 1995 to 18,595 in 2017; with a continuous growth accelerating since 2008 (i.e., the year that Facebook started to be internationally used with the design and functionalities most known; see Figure 2, third plot).

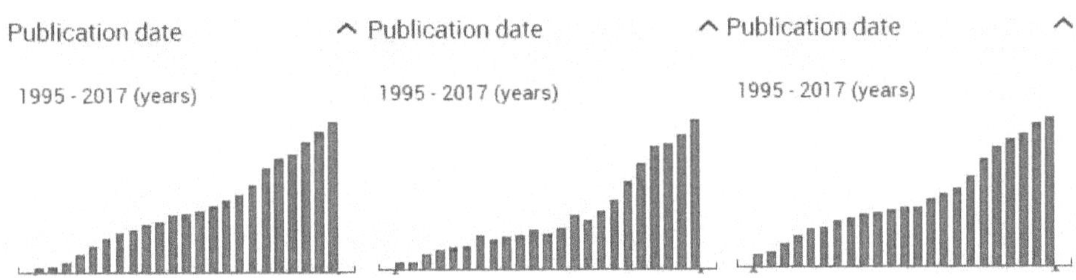

Figure 2. Evolution of Internet use-related addiction problems from 1995 to 2017 (according to the ProQuest Central databases on Medicine, Psychology, and Public Health). Note: The diagrams of bars are ordered from left to right as follow: Internet addiction, problematic gaming, and problematic social networking.

Research into these generalised and specific potential behavioural addictions has provided fundamental advances in clarifying conceptualisation and operationalisation within the spectrum of these problems.

Firstly, Davis in 2001 [14] introduced one of the few existing theoretical models on Internet addiction. His cognitive–behavioural model of pathological Internet use (PIU) proposed dividing the problem into two types: Specific Internet use-related addiction problems (SPIU, referring to the condition in which an individual pathologically uses the Internet for a purpose; e.g., online gaming) and generalised PIU (GPIU, referring to the global set of online behaviours). The basic idea was that cognitive distortions (e.g., thoughts such as 'the Internet is the only place where I am respected') were automatically and unintentionally enacted whenever a stimulus associated with the Internet was available, resulting in emotional or behavioural outputs (i.e., GPIU or SPIU symptoms). Thus, maladaptive cognitions impact SPIU or GPIU, but the latter type was found to be more complicated, as several external factors could be the cause (e.g., social isolation, lack of social support), with negative consequences, such as procrastination and other daily functioning problems (e.g., putting off responsibilities to be online without a directive purpose). To the authors' knowledge, no previous author has asked yet why the same external problems would not be present in generalised and specific Internet-use addition problems. This is a central question, as few works have addressed these factors in SPIUs, which are precisely the ones being treated, studied, and even recognised internationally in the case of gaming disorder.

Between 2002 and 2010, Caplan [15,16] tested and updated this model for GPIU, indicating some users had a preference for online social interactions and used the Internet for mood regulation, which predicted deficient online self-regulation (i.e., compulsive Internet use and a cognitive preoccupation regarding the Internet). Subsequently, Haagsma and colleagues in 2013 [17] tested the model for a SPIU (i.e., online gaming) into an adolescent population and obtained similar findings. However, although these studies showed a few readjustments in Davis' model (e.g., internal factors, such as a preference for online interaction, online mood regulation, and online deficient self-regulation could predict both GPIU or SPIU), there is still a need for more theoretical development on generalised and specific online problems, especially addressing external (i.e., environmental) factors and structural (i.e., video games) factors.

Secondly, Griffiths in 2005 [18] adapted Brown's [19] proposal for gambling and developed the component model of addiction to operationalise the common features among substance and behavioural addictions. The addictive problem is formed in a biopsychosocial framework (i.e., as a consequence of individual, situational, and structural factors) and could be defined by six components (i.e., symptoms; see Table 1). Recently, Griffiths [20] has clarified common components are essential keys to delineate behavioural and substance addictions, as although different addictions have idiosyncrasies, these also have similarities (i.e., components), which are critical to the behaviour being labelled as an addiction. Thus, the six components need to be endorsed to operationally define an addiction (e.g., independently if generalised or specific problems). However, in public and mental health organisations (e.g., APA), and in their respective health manuals (e.g., the DSM), the components are usually articulated as criteria. Thus, to diagnose a disease or a mental health illness, the patient has to endorse a number of criteria (e.g., in the IGD proposed by the DSM-5, it is five out of nine). This quantitative approach has generated concern to estimate the prevalence of potential problematic users, as it has been dependent on several factors (e.g., the criteria, psychometric tools). Moreover, some criteria are more prominent and relevant than others, which have been tackled by few authors.

Thirdly, concerning Internet addiction, Tao and colleagues [21] proposed renowned diagnostic criteria for Internet addiction (see Table 1). This had an operationalisation which introduced specifications: First, the relevance of symptoms (i.e., some being more important than others); second, the timing for the problem course (i.e., the addictive problem requires a period over which to be developed); third, exclusion tactics to diagnose the problem (i.e., the addictive problem should be differentiated from other disorders); fourth, the importance of the functional impairment (i.e., the addictive problem affects users' real-life, impeding at least a facet of a user's functioning). In concrete terms, they stated preoccupation and withdrawal should always be present, as these are the most important symptoms. Subsequently, at least one of the other symptoms should also be present (i.e., tolerance, persistent desire and difficulty to control it, continued use disregarding harmful consequences, loss of interests, or alleviation of negative emotions). The exclusion criteria were psychotic or bipolar I disorders, together with the need for a clinical impairment criterion (in other duties or relationships), and course the criterion of using the Internet excessively for at least three months for six hours per day as entertainment. Tao and colleagues' proposal [21] together with substance use disorder (SUD) and gambling disorder criteria were taken as a source by the APA to propose the IGD criteria [22], which slightly differ in the symptoms, time of development (i.e., a year), and exclusion criteria (other excessive Internet uses). The IGD criteria are: (1) Preoccupation, (2) withdrawal, (3) tolerance, (4) reduction/stop, (5) giving up other activities, (6) continuing despite problems, (7) deceit/cover-up, (8) escaping adverse moods, (9) risk/loss of relationships/opportunities (the wording proposed for these criteria has been introduced in Table 1). Nevertheless, the field has recently been characterised by intense debate on whether and how IGD or Internet addiction (among other addiction problems) should become official disorders. These debates are central to the question of how general versus specific behavioural addiction forms.

Thus, critiques of the component model, Internet addiction (i.e., as GPIU), or IGD (i.e., as SPIU) have emerged [23,24]. For instance, Griffiths' and Tao's proposals were criticised by Van Rooij and Prause [23], as they considered that the evidence base was not enough to support the diagnostic of the generalised Internet addiction as a behavioural addiction. They suggested studying common unpaired dimensions with the support of neuroimaging proofs, and identifying the changes in the rewarding element of using the Internet (e.g., if users are responding to the hedonic reward indistinctly if it is sex, drugs, or online behaviours, and how this response changes their usage pattern) [23]. Similarly, Kardefelt-Winther [24] criticised the IGD criteria for being too adhered to other previous addictive disorders included in the fourth DSM (DSM-IV), rather than capturing the phenomenology of online gaming; he argued some criteria were weak to diagnose SPIU (e.g., preoccupation, withdrawal, loss of interests, and tolerance; especially the later one, which belongs to SUDs). Moreover, it is worth noting how a relevant component, such as salience, which originally covered person's thinking, feelings, and behaviours [18,19] has been reduced to a sole cognitive facet in this century's research and clinical works; even the Davis model was only based on the cognitive approach of the maladaptive behaviour of Internet addiction.

Fourthly, Charlton and Danforth in 2007 [25] stated the distinction between Internet use-related addiction problems and high online engagements through specific components (i.e., criteria). They established the difference between core and peripheral criteria for behavioural addictions, initially performed for the GPIU, and subsequently applied to an SPIU: Online gaming (i.e., massively multiplayer online game playing [MMORPG]; e.g., Asheron's call; see Table 1). The core criteria which defined an addictive problem were: Conflict (with other personal activities or other persons), withdrawal, relapse and reinstatement, and behavioural salience; while peripheral criteria were present in nonproblem online users (i.e., high-engagement users): Cognitive salience (e.g., preoccupation), tolerance, and euphoria. Moreover, time spent on online gaming was positively associated with those who highly scored on the core criteria. A study recently carried out by Lehenbauer and Fohringer [26] has found similar results regarding online gaming. They adapted the version of a previous MMORPG (i.e., World of Warcraft), and found differences between highly-engaged and addicted gamers to the same core versus peripheral criteria, together with more time spent on the core criteria (i.e., addicted gamers were gaming 30 h per week, while highly-engaged players played around 20 h). Furthermore, this study also showed the quality of life for addict gamers was significantly lower than highly-engaged players, especially in physical and psychological health indicators.

For problem Internet and gaming addictions, other research has been undertaken this decade and provided evidence about other potential behavioural addictions which could be classed into this spectrum of online addiction problems [27]. However, the existing theoretical studies (e.g., critiques, reviews) have scarce theoretical development, and only a few of the classic studies have developed attempts of theoretical models (e.g., Davis' model [14]). Studies are usually empirical and use a quantitative approach (e.g., surveys), while qualitative or mixed methods approaches are still scarce, as in other complementary fields, such as behavioural or educational sciences [28,29]. At present, there is a need for knowledge on the phenomenology (i.e., nature) of these problems and more theoretical development. Young developed the latest published study addressing the phenomenon of Internet addiction in 2004 [30], eight years after she coined the term [2,5]. She provided awareness about the nature of Internet use and its potential abuse, as a decade ago this phenomenon had not been identified and defined yet [31,32]. Similarly, recent research has been published, tackling the phenomenology of traditional gaming versus gaming addiction through insights from the gamers' perspective [33], especially in younger generations. Thus, GPIU and SPIU (gaming) might become emerging public health problems as, according to Grant, Schreiber, and Odlaug [34], behavioural addictions are characterised by the inability to resist a drive, resulting in actions that are harmful to oneself and others.

There is a need to cover the gap to start ascertaining the addictive nature of these Internet use-related addiction problems; especially when a public health organisation, such as the WHO, has included 'gaming disorder' as a behavioural addiction into the category for 'Disorders due to substance use or addictive behaviours', subcategory 'Disorders due to addictive behaviours', together with gambling (WHO, 2018 [7]). Furthermore, it seems that while other potential addictive problems connected to technological uses are still under investigation by academic and clinical professionals, there is limited evidence to consider their official recognition yet. This is the case of social media or social networking addiction, where a small number of controversial studies have reported these problems related to the maladaptive use of these media as entertainment and communication tools [35,36]. However, although a few psychometric instruments have been developed to measure this new phenomenon (based on previous addictive criteria, such as SUDs or IGD [37,38]), the evidence is limited and concentrated on adolescents and young adults.

On the other hand, it seems to be established that the existing scales which have been developed in the field of technological (behavioural) addictions have usually been developed through other current addiction criteria and validated using student community samples. An outstanding scale is the compulsive Internet use scale (CIUS [39]), which has been recognised as one of the most psychometrically stable tools (e.g., similar factorial structure among several language adaptations) to measure both generalised and specific (i.e., cybersex) use-related addiction problems. In this study, this scale has been selected to quantitatively measure the phenomena of GPIU (i.e., Internet) and SPIUs (i.e., gaming and social networking), respectively (see Table 1). Indeed, psychometrically, the CIUS has excellent reliability, a unique factor structure demonstrated by exploratory and confirmatory approaches, and has shown measurement invariance [39]. However, as the purpose of this paper is to go further in depth and observe the commonalities and differences among GPIU versus SPIUs, a qualitative measure was performed by interviewing participants potentially classed as problem Internet, gaming, and/or social network users. This strategy required a mixed methods approach of a community sample, accomplishing the following methodological requirements provided below.

Thus, the principal aim is to understand adults' online uses in their personal sphere (i.e., non-academic or professional) to ascertain if these behaviours could be classed as GPIU and/or SPIU and to know about the problems from those who could be categorised as potential problem users. This is articulated in a four-fold specific aim: (i) To validate the CIUS adapted to these three Internet use-related addiction problems (i.e., Internet, gaming, and social networking) to compare them; (ii) to estimate the prevalence of potential problem Internet, gaming, and social networkers users to explore them by sociodemographic variables and addictive symptomatology; (iii) to examine potential problem users' knowledge, experiences, and perceptions with regard to the nature and development of these problems; and (iv) to know their opinion on IGD criteria adapted to the GPIU and SPIUs studied.

Table 1. Comparison of components, criteria, criterions in GPIUs and SPIUs proposals.

Components/Criteria	Subcomponents/Criterions	GPIU (Addiction; Griffiths, 2005 [18])	GPIU (Internet addiction; Tao et al., 2010 [21])	SPIU (IGD; APA, 2013 [22])	SPIU (online gaming; Charlton & Danforth, 2007 [25])	GPIU (CIUS; Meerkerk, Van Den Eijnden, Vermulst, & Garretsen, 2009 [39])
Salience [18,39], Preoccupation [21,22,25]	Cognitive salience [25,39]	When the activity becomes the most important thing and dominates person's thinking, feelings, and behaviours	A strong thinking ongoing online	Do you spend a lot of time thinking about games even when you are not playing, or planning when you can play next?	I rarely think about playing when I am not using a computer	6. Do you think about the Internet, even when not online? 7. Do you look forward to your next Internet session?
	Behavioural salience [25,39]				I often fail to get enough sleep/miss meals because of playing	4. Do you prefer to use the Internet instead of spending time with others (e.g., partner, children, parents)?
Mood modification [18,25], alleviation of negative emotions [21,39], deceit or cover up [22] or escaping adverse moods [22,39]	Manage tension	Subjective experience as a consequence of engaging in the activity to increase or decrease tension to escape, or disconnect	Being online to escape or being relieved	Do you lie to family, friends or others about how much you game, or try to keep your family or friends from knowing how much you game?	I often experience a buzz of excitement while playing	12. Do you go on the Internet when you are feeling down?
	To escape or relieve			Do you game to escape from or forget about personal problems, or to relieve uncomfortable feelings, such as guilt, anxiety, helplessness or depression?		13. Do you use the Internet to escape from your sorrows or get relief from negative feelings?
Tolerance [18,21,22,25]		The need to increase amounts of the activity to achieve the preceding pleasant effects	Marked increase in online use to achieve satisfaction	Do you feel the need to play for increasing amounts of time, play more exciting games, or use more powerful equipment to get the same amount of excitement you used to get?	I tend to want to spend increasing amounts of time playing	
Withdrawal [18,21,25,39]		Unpleasant feeling states or physical effects when the activity is reduced or stopped	Dysphoric mood, anxiety, or boredom after days without online activity	Do you feel restless, irritable, moody, angry, anxious or sad when attempting to cut down or stop gaming, or when you are unable to play?	When I am not playing, I often feel agitated	14. Do you feel restless, frustrated, or irritated when you cannot use the Internet?

Table 1. *Cont.*

Components/Criteria	Subcomponents/Criterions	GPIU (Addiction; Griffiths, 2005 [18])	GPIU (Internet addiction; Tao et al., 2010 [21])	SPIU (IGD; APA, 2013 [22])	SPIU (online gaming; Charlton & Danforth, 2007 [25])	GPIU (CIUS; Meerkerk, Van Den Eijnden, Vermulst, & Garretsen, 2009 [39])
Conflict [18,25,39], loss of interests [21], give up other activities [22]	Intrapersonal [18,25,39], clinical impairment [21]	Conflicts from within the individual themselves		Do you lose interest in or reduce participation in other recreational activities (hobbies, meetings with friends) due to gaming? Do you risk or lose significant relationships, or job, educational or career opportunities because of gaming?	My social life/work has sometimes suffered because of my playing	8. Do you think you should use the Internet less often? 10. Do you rush through your (home) work in order to go on the Internet?
	Interpersonal [18,39], loss of interests and clinical impairment [21]	Conflicts between the addict and those around them	Online use substitutes (e.g., hobbies)		Arguments have sometimes arisen at home because of the time I spend playing	3. Do others (e.g., partner, children, parents) say you should use the Internet less? 11. Do you neglect your daily obligations (work, school, or family life) because you prefer to go on the Internet?
Relapse [18] and relapse reinstatement [25]		Tendency for repeated reversions to earlier patterns of the activity to be quickly restored after time of abstinence or control			I have made unsuccessful attempts to reduce the time I spend playing	
Persistent desire and difficulty to control it [21,39], reduction/stop [22,39]			Not being able to maintain a regular online usage pattern	Do you feel that you should play less, but are unable to cut back on the amount of time you spend playing games?		1. Do you find it difficult to stop using the Internet when you are online? 2. Do you continue to use the Internet despite your intention to stop? 5. Are you short of sleep because of the Internet? 9. Have you unsuccessfully tried to spend less time on the Internet?
Continued use disregarding harmful consequences [21,22]			Being online even causing psychological or physical harm to oneself	Do you continue to play games even though you are aware of negative consequences, such as not getting enough sleep, being late to school/work, spending too much money, having arguments with others, or neglecting important duties?		

Note: GPIU = Generalised Problematic Internet Use; SPIU = Specific Problematic Internet Use; IGD = Internet Gaming Disorder; APA = American Psychiatric Association; CIUS = Compulsive Internet Use Scale.

2. Materials and Methods

Permission to conduct this study was obtained from the ethics committee of the Psychological Science Research Institute (IPSY) at the Catholic University of Louvain (UcL; Belgium) in 2014. Respondents participated in leading research, which builds upon the Tech Use Disorders (TUD; [40]) project, carried out between 2014 and 2016. The qualitative part was undertaken at UcL, which is the main reason only Belgian results have been included in this paper. None of the data analysed in this paper have been used in any other article from this Marie Curie Intra-European Fellowship (FP7-PEOPLE-2013-IEF; ID 627999 [40–44]).

2.1. Design

Following the notation proposed by Morse [45] for the mixed methods designs, and according to the combination of data collection strategy (i.e., sequential implementation/time orientation) and priority (i.e., equal weight/emphasis) [46], a partially mixed sequential equal status ('QUAN→QUAL') design was used. The purpose of this design was to integrate in the discussion of the present study two different types of data [47–49] to clarify and illustrate the results obtained with the quantitative method by applying the qualitative one (i.e., whereby the interviews may help to evaluate and interpret the psychometric results obtained, estimating potential problem users who excessively perform a general or a specific online use(s)). A few of the previous studies on problematic online gaming have recently requested and undertaken this strategy [33] or the qualitative approach [50], but more should be done to cover the methodological gaps of finding what is behind the phenomena usually extracted from a sole psychometric measure. Another complementary purpose was to enable expansion (i.e., seeking to analyse and explore different facets of the GPIU and SPIU phenomena studied; e.g., to know if they seem independent or are interconnected, and in what sense). The mixed methods research in this study attempts to look for the way both methods complement each other to obtain a more productive, realistic, and detailed understanding of Internet addiction (as GPIU), online gaming disorder (as $SPIU_1$), and maladaptive use of social networks (as $SPIU_2$).

2.2. Participants and Procedure

2.2.1. Quantitative

The study online surveyed a convenience sample from Belgian higher education environments with 581 adult participants (85% originally francophone; 25.5% male; age range 18–79 years, mean (*M*) age 26.9 years, standard deviation (*SD*) = 12 years), with student and staff members who voluntarily agreed to participate. Participants had an information sheet and consent form on the first page of the survey and they provided informed consent and voluntarily participated following assurance of confidentiality and anonymity. The invitation to join the online survey used three recruitment strategies: Undergraduates' lectures, master or doctoral supervisions; via electronic requests or pools in online academic environments; and web sites or quick response code advertisements.

2.2.2. Qualitative

Participants who scored equal or over 21 in each CIUS [51] were invited to interviews, as they could potentially be classed as problem users. They were usually university students (except one who was an employee), all speaking fluent French. The invitation to participate in the qualitative part of the study was included at the end of the survey. The author contacted all participants who achieved the inclusion criteria reported and eight (1.4% of the overall sample) agreed to engage in the interview with an economic compensation (20 euros). They received an information sheet, provided informed consent by signing an agreement, and voluntarily participated following assurance of confidentiality and anonymity using pseudonyms. Interviews had a duration between 45 min and an hour, which took place at the IPSY (UcL). Permission was received for audio-recording.

2.3. Instrument and Analytical Strategy

2.3.1. Quantitative

The quantitative method was an online survey developed using Qualtrics which comprised: (i) Sociodemographics (gender, age, civil status, occupation status); (ii) online usage time (i.e., usual minutes per week day, and usual minutes per weekend day); (iii) technologies (i.e., fixed/desktop computer, laptop/notebook, tablet, smartphone, fixed game console or nonportable console, portable game console, television, other devices); (iv) main online activity used (i.e., emails (e.g., Google Mail, Oulook.com), messaging and chat (e.g., Skype, Hangouts), maintaining a blog (e.g., WordPress, Tumblr), online videos or streaming (e.g., YouTube, Netflix), downloads (e.g., music, movies), reading (e.g., newspapers, e-books), search for specific information (e.g., weather forecasts, city maps), casual games (e.g., Candy Crush, Farmville), solo video games (e.g., Dragon age, Assassin's Creed), vehicle simulation games (e.g., Euro Truck Simulator, Flight Simulator), strategy and management games (e.g., Age of Empire, StarCraft, Civilization), sport games (e.g., Pro Evolution Soccer, Virtua Tennis), shooting games online or first-person shooter (FPS; e.g., Call of duty, Planet side 2), multiplayer online battle arena (MOBA; e.g., League of Legends, Dota 2), MMORPG (e.g., World of Warcraft, Guild Wars 2), online gambling (e.g., PlayHugeLottos.com, OnlineBingo.eu), sport bets (e.g., PMU, horseracing, football), poker online (e.g., PokerStars.com), online casino (e.g., Casino Online, Blackjack), online slot machines, dating sites (e.g., Match, Meetic), erotic sites (e.g., Erotica), pornographic sites (e.g., Youporn), online shopping (e.g., eBay, Amazon), social networks (e.g., Facebook, Twitter)); and (v) the psychometric scales assessing generalised and specific Internet use-related addiction problems reported. Other studies have used and analysed other sections of this survey (e.g., other questionnaires in other samples from different countries or group ages), usually related to problematic mobile phone use and gaming [41–44].

Sociodemographics examined gender, age, relationship status (i.e., single, in a relationship, legally cohabitating, married, separated, divorced, or other), profession (i.e., student, employed, unemployed, retired, housewife/husband, self-employed, or other). The CIUS [39] measures general problem or compulsive Internet use (i.e., Internet addiction). It is based on the fourth DSM criterion for SUDs and pathological gambling. It contains 14 items rated from 0 "never" to 4 "very often". Scores ranged from 0–56, with the higher scores meaning to the more top potential generalised compulsive online use. The original scale in Dutch showed factorial, content, and concurrent validity and excellent reliability (Cronbach α between 0.89 and 0.90), as did the French adaptation used [52], which obtained similar outstanding psychometric properties (i.e., Cronbach $\alpha = 91$). In this study, the original item had three options to be answered: General Internet use, online gaming, and social networking. For instance, item 1 (in French: 'A quelle fréquence, trouvez-vous difficile d'arrêter d'utiliser Internet pendant que vous êtes en ligne (c'est à dire s'arrêter, stopper l'activité)'; in English: How often do you find it difficult to stop using the Internet when you are online?) had three responses: (a) For Internet in general (CIUS), (b) for online gaming (CIUS-G), and (c) for social networking (CIUS-SNS). Thus, participants completed each item with these online uses. This innovation allows to assume more than one online use could be present simultaneously for any participant.

All statistical analyses were performed using IBM SPSS (version 23) software (IBM Corp. Released, Armonk, NY, USA) and a significance level of $p < 0.05$. Concerning the psychometric properties, the factor validity of each scale adaptation was assessed by exploratory factor analysis (EFA) using the principal components technique, with the Kaiser–Mayer–Olkin index (KMO) and Bartlett's test of sphericity to confirm the adequacy of the sample and procedure, respectively. The rationale for using the EFA is no study has adapted the CIUS to measure other specific Internet use-related addiction problems until the present, as far as the author is aware. The analysis yielded one factor, which is consistent with the theory and previous research using the CIUS [39]; with eigenvalues above 1 (factor loading > 0.4) to obtain an acceptable factor based on its explained variance. Internal consistency was estimated through the Cronbach's alpha coefficient, and an item analysis was performed to compare all

CIUS forms. Construct and content validity were obtained through associations of the total score with several indicators (i.e., time spent online using the Internet in general, or spent mainly gaming or mainly on social networks; or activities usually used by each of the CIUS studied). Comparisons between CIUS total scores and sociodemographic variables (i.e., gender and age groups) were performed through *t* and *U* Mann–Whitney tests. The proposed cut-off score reported by Guertler and colleagues [51] was chosen (i.e., score of 21 out of 56) to estimate the prevalence of potential problem users. However, caution should be considered, as present cut-offs still do not represent a threshold for clinical relevance and impairment. The sum of all potential problem users was computed to know how many participants could have one (or more) of the problems studied, together with the items and symptoms endorsed. Finally, to compare core versus peripheral symptoms [25,26] in those classed as potential problem users, the three CIUS were divided in two subgroups, according to Meerkerk and colleagues' [39] and Charlton and Danforth's [25] proposals (core symptoms: Items 3, 4, 5, 8, 10, 11, 14; peripheral symptoms: Items 1, 2, 6, 7, 9, 12, 13). Only an item (5) was modified from Meerkerk's proposal (i.e., from 'loss of control' was moved to 'intrapersonal conflict').

2.3.2. Qualitative

The interviews were undertaken and transcribed. The methods selected for the qualitative data analysis were two [53]: First, thematic analysis [54,55] with an etic coding strategy (i.e., coding from the perspective of the observer based on the literature) was used for examining perceptions and attitudes about potential Internet use-related addiction problems (causes, development, consequences, factors). Second, content analysis [56,57] to focus at a more micro level to provide both counts and opinions on IGD criteria proposed by the APA applied to the three problems studied. Both methods belong to qualitative descriptive phenomenological approaches [53], with a lower level of interpretation, but higher detail on the description of the meanings (i.e., focused on the knowledge that can originate and establish new definitions and substantial findings). At this stage of the development of the field, detail is needed on the nature of these problems. Synthetically, thematic analysis constitutes a purely qualitative approach used for identifying, analysing (codes), reporting patterns (themes) within textual data (i.e., involving the search for and identification of common threads), and interpreting aspects of the research question [56]. On the other hand, content analysis is a systematic coding and categorising approach used to explore text in detail to find patterns of words used, their frequencies (counts), their relationships, and their structures, which will allow answering the research questions [58].

3. Results

3.1. Quantitative

3.1.1. Participants Characteristics: Sociodemographic, Technologies Usage Patterns, and Online Activities

The majority of the sample (71.8%) were students (only 17.3% were employees, and the rest were 2.9% self-employed, 2% retired, 1.1% unemployed, 0.05% housewives or househusbands, among other). Half of the sample (58.2%) were single (36.5% were couples/in legal cohabitation/married, 4.9% separated/divorced, among other), without children (only 15.3% had progeny); their maximum education level was above all between secondary (59.8%) and higher education (38.8%). The participants (99.8%) had an Internet connection where they lived, had at least a computer (desktop or laptop; 97.5%) and a mobile phone (99.3%: without Internet access (25.1%) or a smartphone (74.2%)), and a third had (34.7%) a Tablet (e.g., iPad, Samsung galaxy tab). In all technologies, they usually had monthly contracts (i.e., Internet at home 95.5% of the sample, in the mobile phone 75.1%, and for the Tablet 35.3%).

3.1.2. Psychometric Properties of the French Generalised and Specific CIUS in Belgium

The CIUS as a GPIU psychometric tool was previously validated [39,52], but not the CIUS as SPIUs scales in gaming and social networking. In this study, the three measures showed factorial validity (i.e., CIUS: KMO = 0.91, $\chi^2_{(91)} = 2673.09$, $p < 0.001$; CIUS–G: KMO = 0.93, $\chi^2_{(91)} = 5223.13$, $p < 0.001$; CIUS–SNS: KMO = 0.93, $\chi^2_{(91)} = 3519.31$, $p < 0.001$) and excellent internal consistency (see Table 2; where the descriptive, Cronbach alpha coefficients, and correlations between the three CIUS are reported). It is worth noting the high positive association between CIUS and CIUS–SNS, and the low positive association between CIUS and CIUS–G; however, between the two specific uses (CIUS–G and CIUS–SNS), there is no correlation.

Table 2. Descriptive, reliability, and correlation matrix across all adaptations for the CIUS.

			CIUS Adaptations		
Scale Adaptations	**Descriptive *M*(*SD*); Range**	**Cronbach alpha**	**CIUS**	**CIUS–G**	**CIUS–SNS**
CIUS	14.63(10.10); 0–56	0.90	-		
CIUS–G	4.78(8.97); 0–51	0.95	0.22 **	-	
CIUS–SNS	11.63(10.79); 0–56	0.93	0.59 **	−0.02	-

Note: CIUS: Compulsive Internet use scale; CIUS–G: CIUS for gaming; CIUS–SNS: CIUS for social networking sites; *M*: Mean; *SD*: Standard deviation; ** $p < 0.01$.

The CIUS yielded one factor (i.e., generalised problematic Internet use), which explained 48.86% of the total variance with loads to this sole factor between 0.85 (item 14) to 0.72 (item 10); such as the CIUS–G (i.e., specific problematic (video) gaming) that explained 60.12% with loads to between 0.69 (item 9) to 0.84 (item 7), and the CIUS–SNS (i.e., specific problematic social networking) that explained 50.50% with loads to between 0.58 (item 4) to 0.79 (item 2). The scree plots of the three versions show the one-factor of these adaptations (see Figure 3), which allows to compare items by descriptive, correlations, and factors loadings (see Table 3).

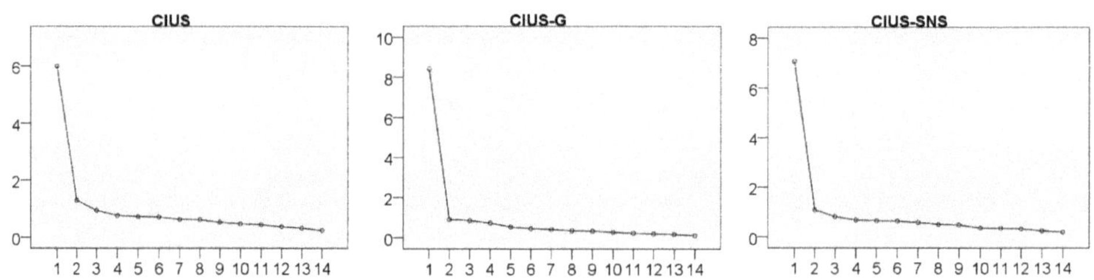

Figure 3. Scree plots from the original CIUS (GPIU), the adaptation for gaming (CIUS–G; $SPIU_1$), and the adaptation for social networking (CIUS–SNS; $SPIU_2$); in the y axis the eigenvalues, and in the x axis the component numbers.

Table 3. Item analysis: descriptive, correlation item per total correlation, and factor loadings across all adaptations for the CIUS.

Items	**Descriptive *M*(*SD*)**			**Correlation Item-Total Correlation Per CIUS Adaptations**			**Factor Loading Per CIUS Adaptations**		
	CIUS	CIUS–G	CIUS–SNS	CIUS	CIUS–G	CIUS–SNS	CIUS	CIUS–G	CIUS–SNS
1	1.98(1.267)	0.66(1.159)	1.84(1.372)	0.60	0.79	0.71	0.67	0.82	0.76
2	1.84(1.354)	0.52(1045)	1.73(1.468)	0.65	0.76	0.74	0.71	0.79	0.79
3	0.80(1.089)	0.27(0.765)	0.78(1.131)	0.46	0.63	0.61	0.53	0.69	0.67
4	0.90(1.003)	0.31(0.768)	0.63(0.945)	0.56	0.75	0.52	0.63	0.78	0.58
5	1.07(1.147)	0.28(0.750)	0.79(1.094)	0.57	0.68	0.64	0.64	0.73	0.69

Table 3. *Cont.*

Items	Descriptive *M*(*SD*)			Correlation Item-Total Correlation Per CIUS Adaptations			Factor Loading Per CIUS Adaptations		
	CIUS	CIUS–G	CIUS–SNS	CIUS	CIUS–G	CIUS–SNS	CIUS	CIUS–G	CIUS–SNS
6	0.99(1.045)	0.29(0.716)	1.06(1.087)	0.56	0.74	0.67	0.63	0.79	0.72
7	1.16(1.107)	0.39(0.863)	1.00(1.103)	0.56	0.80	0.64	0.64	0.84	0.70
8	1.33(1.187)	0.39(0.872)	1.39(1.321)	0.59	0.75	0.65	0.66	0.79	0.71
9	0.78(1.017)	0.22(0.615)	0.82(1.094)	0.63	0.64	0.62	0.70	0.69	0.68
10	1.01(1.090)	0.29(0.739)	0.84(1.107)	0.66	0.78	0.69	0.72	0.83	0.75
11	1.07(1.175)	0.28(0.718)	0.89(1.137)	0.63	0.74	0.64	0.70	0.79	0.69
12	2.05(1.240)	0.57(1.095)	1.69(1.419)	0.60	0.79	0.71	0.67	0.82	0.75
13	1.54(1.304)	0.41(0.943)	1.18(1.318)	0.59	0.72	0.65	0.66	0.76	0.71
14	1.20(1.172)	0.23(0.657)	0.91(1.132)	0.51	0.69	0.67	0.85	0.74	0.73

Note: CIUS: Compulsive Internet use scale; CIUS–G: CIUS for gaming; CIUS–SNS: CIUS for social networking sites; *M*: Mean; *SD*: Standard deviation.

Regarding descriptive results per item in each version, CIUS and CIUS–SNS have higher scores than CIUS–G. The higher punctuations in all of them being for items 1, 2, and 12 (which were associated with loss of control and mood modification symptoms [39]). However, differences between adaptations emerged in the lowest scores: Item 9 on Internet and gaming (loss of control [39]), and item 4 in social networking (preoccupation [39]). Concerning the correlation of each item with the total correlation of each adaptation, all CIUS presented high positive correlations between each item and the total score, especially the gaming version (r between 0.63 and 0.80). Concerning factor loading, all were greater than 0.5; by order from higher to lower scores: Gaming (minimum 0.69), social networking (minimum 0.58), and Internet (minimum 0.53). Consequently, when dividing the CIUS scores by core and peripheral symptoms, descriptive results were higher for peripheral ones (i.e., Internet use in general ($M = 10.34$, $SD = 5.87$), social networking ($M = 9.32$, $SD = 6.77$), and gaming ($M = 3.06$, $SD = 5.25$)) than the core symptoms (i.e., Internet ($M = 7.38$, $SD = 5.32$), social networking ($M = 6.23$, $SD = 5.70$), and gaming ($M = 2.04$, $SD = 4.12$)).

Furthermore, associations between the time indicators and the three CIUS total scores strengthen the construct validity (see Table 4). In this regard, almost all the time variables were significantly associated with the three CIUS, but differently according to the technology. For instance, moderate correlations on the Internet and gaming addiction problems were present in almost all technologies, while social networking addiction problems were present, especially using smartphones. Interestingly, a device pattern emerged as significant associations were quite heterogeneous; for instance, all technologies were associated with time per days (weekday or weekend day) when using Internet in general, but when gaming, only computers and tablets were correlated with time per days, and when social networking, only computers and smartphones were associated with time spent on a day. However, the maximum time spent as leisure was usually on computers.

Similarly, content validity was also established through associations to CIUS total scores and across all the online activities as follows: Firstly, the CIUS (GPIU) with 'messages and chats' ($r = 0.19$, $p < 0.001$), 'online videos and streaming' ($r = 0.17$, $p < 0.001$), 'Twitter' ($r = 0.16$, $p < 0.001$), 'Facebook' ($r = 0.14$, $p < 0.001$), 'Instagram' ($r = 0.12$, $p < 0.05$), 'Hi5' ($r = 0.11$, $p < 0.05$), 'blogs' ($r = 0.10$, $p < 0.05$), and 'porn sites' ($r = 0.09$, $p < 0.05$). Secondly, the CIUS–G ($SPIU_1$) was associated to the following games: 'MOBA' ($r = 0.45$, $p < 0.001$), 'MMORPG' ($r = 0.43$, $p < 0.001$), 'Solo games' ($r = 0.42$, $p < 0.001$), 'FPS' ($r = 0.39$, $p < 0.001$), 'games strategy' ($r = 0.33$, $p < 0.001$), 'casual games' ($r = 0.23$, $p < 0.001$), and 'simulation games' ($r = 0.22$, $p < 0.001$), together with porn sites ($r = 0.24$, $p < 0.001$), and reading ($r = 0.1$, $p < 0.05$). Lastly, the CIUS–SNS ($SPIU_2$) was related to the following social networks: 'Facebook' ($r = 0.36$, $p < 0.001$), 'Instagram' ($r = 0.18$, $p < 0.05$), and 'Twitter' ($r = 0.14$, $p < 0.001$), including messages and chats ($r = 0.23$, $p < 0.01$), downloading ($r = 0.2$, $p < 0.001$), streaming ($r = 0.18$, $p < 0.001$), and blogging ($r = 0.1$, $p < 0.05$).

3.1.3. Prevalence of the French Generalised and Specific CIUS in Belgium

In this study, no statistical difference was found between the original CIUS total score per gender ($t_{(469)} = -0.08$, $p = 0.94$; male: $M = 17.65$, $SD = 11.05$; female: $M = 17.75$, $SD = 10.48$), but differences between gender were found in the specific CIUS: CIUS–G ($U = 13{,}420$, $p < 0.001$; male: $M = 10.05$, $SD = 12.30$; female: $M = 3.68$, $SD = 7.42$), CIUS–SNS ($U = 13{,}945$, $p < 0.001$; male: $M = 11.25$, $SD = 10.55$; female: $M = 16.78$, $SD = 12.03$). With respect to the age groups, young adults (i.e., 18–39 years old) compared to middle-aged adults (i.e., 40–65 years old) presented significant differences in all measures: CIUS ($t_{(459)} = -5.54$, $p < 0.001$; young: $M = 18.58$, $SD = 10.35$; middle-aged: $M = 10.41$, $SD = 8.85$), CIUS–G (*$U = 9160$, $p < 0.05$*; young: $M = 5.34$, $SD = 9.37$; middle-aged: $M = 2.81$, $SD = 6.42$), and CIUS–SNS ($U = 4493.5$, $p < 0.001$; young: $M = 17.05$, $SD = 11.70$; middle-aged: $M = 5.89$, $SD = 7.66$).

Concerning GPIU, taking into account the overall 471 participants as potential problem users, there were 176 potential Internet addicts (37.4%: 8.1% males and 29.3% females, $\chi^2_{(1)} = 23.58$, $p < 0.001$; 39.8% young and 13% middle-aged, $\chi^2_{(1)} = 14.79$, $p < 0.001$), 51 potential problematic online gamers (10.8%: 5.3% males and 5.5% females, $\chi^2_{(1)} = 0.08$, $p = 0.78$; 11.3% young and 3.7% middle-aged, $\chi^2_{(1)} = 2.95$, $p = 0.09$), and 153 problematic social networkers (32.5%: 3.4% males and 29.1% females, $\chi^2_{(1)} = 18.09$, $p < 0.001$; 32.4% young and 7.4% middle-aged, $\chi^2_{(1)} = 23.58$, $p < 0.001$). Finally, when computing the potential problem users, 49.7% of the sample has not classed in any user problem category, 23.6% had only one problematic online use, 23.1% had two potential online problems, and 3.6% had all the problems.

From these three groups of problem users, descriptive and endorsements to each item (i.e., on the basis of answering 'often' or 'very often' should be considered) were computed (see Table 5). Interestingly, the highest scores and endorsement in all GPIU and SPIUs were in items 1 and 2 for the loss of control symptom, and item 12 for the mood modification symptom; while the lower scores and endorsements were more heterogeneous. For instance, item 3 (conflict) being lower in the problematic Internet (CIUS), but the less endorsed were items 4 (preoccupation) and 9 (loss of control); while in gaming (CIUS–G) item 9 (loss of control) was the one with lower score and the less endorsed by none potential problem gamer; lastly, item 4 (preoccupation) obtained the lower score and endorsement in problematic social networking.

Table 4. Descriptive and correlation across all validated adaptations for the CIUS in relation to time.

Scales and Descriptive	Days Per Week (Weekly Frequency) *r*			Minutes Per Day in a Weekly Day *r*			Minutes Per Day in a Weekend Day *r*		
	Computers	Tablets	Phones	Computers	Tablets	Phones	Computers	Tablets	Phones
CIUS	0.18 **	0.08	0.09	0.37 **	0.25 **	0.24 **	0.39 **	0.28 **	0.21 **
CIUS–G	0.14 **	0.2 7 **	0.07	0.27 **	0.26 **	0.05	0.22 **	0.21 **	0.02
CIUS–SNS	0.09	−0.02	0.21 **	0.11 *	0.08	0.35 **	0.22 **	0.13	0.37 *
M(SD)	6.17(1.62)	3.87(2.80)	4.62(2.98)	118.71(93.45)	46.10(64.99)	82.70(116.08)	183.91(200.53)	68.16(100.13)	90.99(124.45)

Note: Weekly Day: from Monday to Friday; Weekend Day: from Saturday to Sunday; CIUS: Compulsive Internet use scale; CIUS–G: CIUS for gaming; CIUS–SNS: CIUS for social networking sites; *r*: r Pearson (correlation); *M*: Mean; *SD*: Standard deviation; * $p < 0.05$; ** $p < 0.01$.

Table 5. Item and symptoms from potential problem users: descriptive, and frequency of endorsement across all adaptations for the CIUS.

Item Number, Symptom According to Meerkerk et al., 2009 [39] and Charlton & Danforth, 2007 [25]	Descriptive *M*(*SD*)			Endorsement *f* i (%)		
	CIUS	CIUS–G	CIUS–SNS	CIUS (*n* = 176)	CIUS–G (*n* = 51)	CIUS–SNS (*n* = 153)
1—Loss of control–Peripheral	2.86(0.942)	2.86(1.02)	3.08(0.924)	114(64.8)	34(66.7)	112(73.2)
2—Loss of control–Peripheral	2.91(0.987)	2.59(1.134)	3.07(0.926)	124(70.5)	28(54.9)	113(73.9)
3—Conflict–Core	1.48(1.214)	1.55(1.301)	1.73(1.262)	31(17.6)	12(23.5)	38(24.8)
4—Preoccupation–Core	1.53(1.008)	1.82(0.974)	1.25(1.079)	26(14.8)	11(21.6)	18(11.8)
5—Loss of control–Core	1.85(1.162)	1.59(1.236)	1.67(1.261)	43(24.4)	13(25.5)	41(26.8)
6—Preoccupation–Peripheral	1.70(1.039)	1.71(1.082)	2.04(0.973)	39(22.2)	9(17.6)	46(30.1)
7—Preoccupation–Peripheral	1.9(1.057)	2.18(1.072)	1.94(1.04)	52(29.5)	19(37.3)	41(26.8)
8—Conflict–Core	2.19(1.082)	2.02(0.99)	2.46(1.088)	68(38.6)	14(27.5)	77(50.3)
9—Loss of control–Peripheral	1.56(1.057)	1.20(1.077)	1.70(1.181)	29(16.5)	7(13.7)	37(24.2)
10—Conflict–Core	1.86(1.073)	1.80(1.114)	1.85(1.146)	52(29.5)	13(25.5)	43(28.1)
11—Conflict–Core	1.97(1.146)	1.67(1.089)	1.82(1.227)	53(30.1)	12(23.5)	40(26.1)
12—Mood modification–Peripheral	2.92(0.977)	2.59(1.268)	2.95(1.022)	124(70.5)	30(58.8)	113(73.9)
13—Mood modification–Peripheral	2.48(1.181)	2.12(1.291)	2.30(1.22)	97(55.1)	20(39.2)	67(43.8)
14—Withdrawal–Core	1.89(1.175)	1.43(1.171)	1.90(1.134)	55(31.3)	11(21.6)	47(30.7)

Note: CIUS: Compulsive Internet use scale; CIUS–G: CIUS for gaming; CIUS–SNS: CIUS for social networking sites; *M*: Mean; *SD*: Standard deviation; *f* i: frequency; % is valid percentage; *n*: Subsample size; and item 5 was moved from 'Loss of control' [39] to Conflict by the author, for this reason has been classed as Core symptom [25].

Finally, when comparing core and peripheral symptoms in each CIUS peripheral ones used to have higher punctuation than core ones in the potential problem users in GPIU and SPIUs (see Table 5). Furthermore, for those categorised as problem Internet users, moderate associations between time variables and core or peripheral items were significant in the time spent using Internet on a weekday (core: $r = 0.19$, $p < 0.05$; peripheral: $r = 0.21$, $p < 0.05$) or a weekend day (core: $r = 0.23$, $p < 0.01$; peripheral: $r = 0.22$, $p < 0.05$); while in problem social networkers there existed only an association between the core symptoms and weekend days ($r = 0.25$, $p < 0.01$), and none association existed in problem gamers (e.g., core: Time spent gaming in a weekday ($r = 0.21$, $p = 0.19$)).

3.2. Qualitative

3.2.1. Participants Characteristics: Sociodemographic, Technologies Usage Patterns, and Online Activities

A total of 165 (out of 581) provided their contact details to participate in the qualitative part of this study; eight of them (4.85% of those who left this information) accepted the invitation to participate in an individual interview. Participants' characteristics are reported in Table 6.

Table 6. Participant sociodemographic characteristics and potential problem online uses.

Pseudonyms	Variables					
	Gender	Age	Civil Status	CIUS	CIUS–G	CIUS–SNS
Leia	Female	31	Partner	**24**	**29**	17
Moira	Female	20	Single	**26**	5	**28**
Aneka	Female	20	Partner	**37**	14	**37**
Victor	Male	20	Single	**36**	**31**	**37**
Elektra	Female	35	Divorced	**23**	5	8
Carol	Female	21	Partner	13	**25**	19
Scarlet	Female	18	Single	8	0	**21**
Martin	Male	19	Partner	9	**22**	4

Note: CIUS: Compulsive Internet use scale; CIUS–G: CIUS for Gaming; CIUS–SNS: CIUS for Social networking sites; numbers in bold are the scores above the cut-off in each CIUS adaptation as potential problems.

Almost all participants were female Belgian young adults who scored as potential problem users of Internet in general ($n = 5$), gaming ($n = 4$), and/or social networking ($n = 4$). Therefore, half of them had mixed problem profiles, usually as excessive Internet in general and social network users, except one presenting all problems studied (i.e., GPIU, $SPIU_1$, and $SPIU_2$), who was a young male.

3.2.2. Themes Related to Generalised Internet Problem, and Specific Gaming and Social Networking Problems

Themes Related to Its Evolution: Causes, Development, and Consequences

Individuals' experiences and reflections mapped into the central themes which are presented in Tables 7–10 below by types of GPIU vs. SPIU.

The causes of excessively using specific online activities were diverse (see Table 7), and they could be summarised in three types of aetiological facets: the individual, the social, and the contextual ones. First, the individual aspect relates to avoiding boredom and to escaping reality (as a daily routine or because of a trauma). In the case of gamers, it seems there is a profile: they tend to be tech-savvy (i.e., study or work in informatics or the technologies sector) or really enjoy technologies (to be daily connected, multitasking, and managing studies/work and leisure), together with being a homebody person with an introvert personality, and to have a specific desire or mood to only play video games. By contrast, social networkers would like to have consistent updates in all areas and use SNS obsessively to scrutinise others' lives and compare them with their own life. Second, the social facet seems to be related to sharing a hobby with others who have similar interests in technologies and online activities (e.g., relatives, boyfriend/girlfriend, friends from real life, virtual colleagues); gamers usually are highly committed to their guild and tend to supply (or complement) their social life through virtual social friendships developed (or maintained) through the game. It seems online users tend to build (or strengthen) bonds with those who share the same online activities and they are constantly connected through games and networks sharing information (e.g., about competitions, gossips). Third, the contextual facet is usually linked to having plenty of free time (e.g., because they are studying at secondary school or university, unemployed, in convalescence, or they are housewives/househusbands); a perceived 'poor' or 'hostile' external context (e.g., difficult relationships (or not fulfilling ones) with family, partner, colleagues), an environment where everybody is connected to the Internet (e.g., gamer guilds, Facebook profiles), being in a post-traumatic recovery (e.g., having lost a loved one(s), having lost a job), or feeling the need to consume technology when something new and appealing appears in the market (e.g., new video game, new social network, latest version). Interestingly, no quote was extracted from GPIU, and almost all were obtained from gaming.

Table 7. Superordinate and subordinate themes about the aetiology of the SPIUs studied, together with a few quotations (adapted from French to English) as illustration.

Theme	Subthemes		
		CIUS–G	CIUS–SNS
Causes	Individual facet	'I started to be a gamer at the same time I started at the university. It is quite usual in people who like technologies to have online hobbies, as we have difficulties communicating with other people; games could be an escape' (Leia) 'I think the pathology in gaming appears when you cannot avoid craving; when the negative feelings emerge from not being able to play, it is the lack of something in you' (Victor)	'In the game, when you win you feel you are valorised; on Facebook, you only see positive things and this makes people feel positive, as it is easier to see someone through Facebook than call him or her' (Elektra)
	Social facet	'What makes me play more and more is playing online with those I know, to compete among us' (Victor) 'I play League of Legends, a MMORPG, not a MOBA, as it needs an objective and it is a network; you play with friends, and this is what I really like' (Martin)	'It is the wish to share; as sharing your emotions. Above all when there is good news you share it through Facebook' (Moira)
	Contextual facet	'Gaming can increase if there is a lot of free time, studying at university or being unemployed. Also because external relations are difficult, at school, friendships and above all in the family' (Leia)	'With all technologies around us, you feel the obligation to be connected all the time, to know what is happening' (Moira)

Note: CIUS–G: Compulsive Internet use scale for gaming; CIUS–SNS: CIUS for social networking sites; MMORPG: massively multiplayer online role-playing games; MOBA: multiplayer online battle arena.

The development of these Internet problems is described as a process, which starts during adolescence with an acquired enjoyable habit that slowly and unconsciously takes progressively more and more from the user (i.e., not only more time; by coming above all other life aspects of the user, such as feelings, emotions, cognitions, behaviours, relationships, activities). This gradual mining of the problematic online uses could be observed through the addictive symptomatology described by excessive users (see Table 8).

Regarding salience, it is usually described covering emotional and cognitive facets (i.e., craving or preoccupation respectively; e.g., as a personal drive or desire to be connected or worrying, even being obsessive about personal or social online duties, such as updating your profile, tending to notifications in the SNS, or accomplishing guild expectations in a role-play game). Sometimes, it is associated with an unconscious need to be connected even when not possible, or an alleviation of negative emotions, such as wanting to escape or being 'anesthetised'. It seems to be quite a usual symptom in the three problems studied.

With regard to mood modification, salience is present in the GPIU and SPIUs studied, but it seems more prevalent in gaming. Gamers report using various kinds of games to produce different types of moods, such as MMORPG or FPS to induce tension, and casual games to relax; they state they balance their spirit in each moment through the games they choose to play. Similarly, social networkers, for instance, use Facebook to channel emotions (e.g., to share good news vs. to look for someone to cheer up). Networkers state they use SNS to share positive things (e.g., positive messages, images, videos). SNS contents seem to be positively biased, but networkers are aware and happy with it. However, sometimes, images or other information observed through the SNS could affect a user's humour negatively (e.g., as the user could compare his or her life with others; especially in adolescence), and feel upset caused by negative feelings (e.g., jealousy, sadness). This could also happen with the online series that people watch in streaming too, usually used to relax (e.g., users could compare their lives, and this could affect them, especially when being an adolescent).

Tolerance is a symptom which only emerged in both SPIUs studied. In gaming, it seems to be linked to all types of continuous increments: Time, expertise, the sensation of progress, achievement, and advancement in the game to produce similar satisfaction or rewards. Therefore, the 'dose' is not only a quantity of time playing, it is also an increment of intensity and development of the gamer's skills in the game, as well as the level of personal achievement in the game (e.g., score, level) or the level of social recognition by other gamers (e.g., status in the guild). Similarly, in social networking, the tolerance symptom could be understood for the continuous need and increment of publishing information (e.g., posts, pictures) about one's own (or another's) life, to observe social reactions (e.g., how many likes and/or comments the post received). This reinforces specific online behaviours, as the user self-perceives his or her life as a gamer or networker good enough to be valued by those who share the same online activity (e.g., through score comparisons, others' life achievements).

The other classic addictive symptom studied was withdrawal, which has usually been associated with negative feelings (i.e., frustration, irritability) when not being connected for gaming. This sense is directly linked to the need of connection after interruption or a forced stop. It seems being disconnected is being in severe discomfort, above all when the user is younger (e.g., an adolescent), alone at home (i.e., with plenty of time to play games), and a gamer (i.e., playing 'big video games', which usually are MMORPG, FPS, or MOBA). Thus, in the GPIU and SPIU related to SNS, withdrawal has not emerged as a subtheme.

Concerning conflicts, intrapersonal and interpersonal facets have emerged for all Internet use-related addiction problems. Gamers are the users who appeal more often to intrapersonal conflicts, as they play everywhere when a computer is available to them (e.g., university, work, or home), and when they supposedly should do other activities (e.g., attending a session in university, working in a workplace, or sleeping during night). However, social networkers have reported similar problems with depression, anxiety, and stress, as well as sleep or meal deprivation, which could affect their duties or relationships. Subsequently, there are problems with family, friends, or partners because of excessive online behaviours, which cause disputes, possible loss of contacts (e.g., boyfriend, girlfriend) or less of other relevant activities (e.g., studies, job).

Concerning relapse, this only appeared once in a gamer interview, who expressed how rigid rules created by the guild could constrain the gamer and obligate him or her to play. If the player leaves the game, social pressure appears by the guild, as well as the loss of the feelings of being in a 'virtual world' created by the gamers during an extended period. Thus, the gamer continually returns to the game.

The persistent desire and difficulty to control, reduce, or stop online behaviour is a common difficulty for all excessive Internet users. For instance, gaming is an extended hobby with an industry which supports periodic updates to play better and differently (e.g., characters, collections, the system of dropping). Moreover, the fact that some games reproduce 'persistent worlds' facilitates continuous playing (e.g., to compete). Similarly, the continued use with disregard of harmful consequences (e.g., physical problems or unhealthy attitudes or behaviours) is usually due to the habit already created to manage the user's mood, user's time, social relationships, etc. It seems essential to be continuously connected to the Internet, games, or SNS to maintain users' personal and social life actively.

Lastly, deceit or cover-up has only appeared in problematic gaming, as gamers usually report they live a 'secret double life', the one in the 'real world' and the one in the 'virtual world' (i.e., the game(s)). Gamers are quite aware they invert time and effort in the games; contrary to other established addictive behaviours, gamers work hard and actively to maintain their gaming behaviour. One reason is other gamers (e.g., the guild) when playing in the 'big games' share their own online spaces and codes to constantly communicate (even if they know themselves or not in 'real life', as sometimes gamers are from other countries or cultures). Thus, deception is not only hiding what is consumed through technology to others (e.g., hiding playing games at night to their parents), it is a more complex issue (i.e., managing languages, identities, relationships, and duties in different 'worlds'). In this sense, they express that those who do not play games usually cannot understand them; they directly maintain this 'other (online) life' in secret for many reasons: To avoid conflicts, to be judged, along with other reasons.

Concerning the consequences (see Table 9), the positive outcomes are usually associated with enriching users' social lives, as they could experience other 'virtual (online) worlds' and relationships that they consider improving their lives. For instance, they can connect differently with others who are more related to them (i.e., transmitting the online 'virtual self' through SNS or through an avatar in a game, creating and maintaining friendships or affective relationships through private online spaces in SNS or through games (e.g., 'feeling together'), dating and finding partners). They have other types of connections parallel to their 'real life' ones (i.e., relatives, partner, peers, colleagues, friends, or other persons with whom they interact face-to-face). Furthermore, gamers report an improvement in their cognitive processes (e.g., memory, attention, and learning).

However, the adverse self-perceived outcomes by heavy users are mainly those problems which have been unconscious developed until they became aware of the addiction problem, as something (or someone) alerted them. This discovery usually requires them to start behaving differently to naturally recover from a maladaptive to an adaptive pattern of usage. Only in gamers has natural recovery emerged as a process of maturity (i.e., need to develop other life challenges), such as to enter the workplace (which requires responsibility, time, effort, energy, among other skills before invested above all in the game); although, sometimes, they cannot do it by themselves. For instance, it has been reported heavy gamers are living in 'another world' which is alternative to real life, and not all of them know how to manage both or do not see the benefit in renouncing 'virtual life'. A few of them have lost studies/jobs, friends/partners, even some aspects of their health; a few have experienced chronic sleeping and eating problems, and discomfort in some bones, muscles, and organs, probably due to the maintained tension during extended periods (e.g., cervical, hands, or eyes, respectively). Those who mix gaming with drug consumption (e.g., cannabis) experience other problems (e.g., see images of the game when the player was not gaming).

With regard to neutral consequences, users appeal to curiosity and relationships reinforced by the virtual connections. It is observed that female gamers seem to have boyfriend gamers, as they have met them and/or matched with them better as a couple because of sharing this hobby or passion for games, which ensures their common understanding as gamers and lets them share their virtual lives. Similarly, social networkers report the relevance of being intellectually fed by being updated continuously. Thus, related to these consequences, detail has been provided only for SPIUs.

Finally, few risk factors were reported and directly connected with the symptomatology in all Internet use-related addiction problems (see Table 10), where developmental and contextual factors seem the most impactful ones. For instance, some of them are: Parent styles, such as authoritarian, partner already 'addicted' to gaming, drug consumption together with gaming, the tendency to be indoors with online hobbies, lost someone(s) loved, adolescent crisis, games substituting 'virtual' prices with money. Only a few protective factors have also emerged (i.e., family limits, activities outdoors, and transforming the online activity to a healthy activity or profession: 'eSport').

Table 8. Superordinate and subordinate themes about the development (i.e., addictive symptomatology) of the GPIU and SPIUs studied, together with a few quotations (adapted from French to English) as illustration.

Theme	Subthemes	Quotes		
		CIUS	CIUS–G	CIUS–SNS
Development (addictive symptoms)	Salience	'If you have a need, such as looking for a job, and you always have your smartphone with you, you could get obsessive about checking the Internet continuously. You are in your own world. If I am not checking it, I think I am losing opportunities' (Aneka)	'Gaming is excessive when it is a priority. Years ago, I felt I needed to play when I was going out with others, and I realised I had a problem. It is not a question of time playing. It is about when you cannot be without gaming, and there is no Internet. For instance, a gamer will take public transport for hours until finding a place with Internet to play video games. The gamer substitutes going out with friends for video gaming, and gamers are always connected even through a smartphone' (Carol) 'There are guild obligations, as we agree to be there to do something, and if you do not do it, you put people in trouble' (Leia) 'The 'vagaries' of life, there are periods when I wish to play. I do not think craving is in myself or my brain; I think it is outside' (Carol)	'In the SNS we think of other things, to free our minds of negative feelings' (Moira)
	Mood modification	'If I am sad, I watch an online series or films to cheer up, which sometimes is better than gaming' (Carol)	'I integrate myself in the story and into a character of an MMORPG to disconnect with reality. When you have had a stressed day, your reward is gaming. As I made an effort, I have the right to escape; it is very relaxing' (Carol) 'When gaming (RPG or FPS) we can quickly become annoyed or nervous, but sometimes it's the contrary, playing casual games helps you to relax. I have all types of different games I play depending on my mood, and how I want to balance my emotions' (Leia) 'When I am alone and upset, I play to calm myself, but I do not regulate myself as I feel tense or nervous after stopping' (Martin)	'You go to Facebook to look for something to cheer you up' (Moira)
	Tolerance		'The feelings of success and gratification could be a stimulus that makes us think of the game and makes us feel well, and produces the wish to play' (Carol) 'I need to have my little dose every day to feel like I'm advancing in the game. Before I could play for 15 hours daily, but now only 2 h. For example, if you stop for a day everything is reset, so sometimes we have to play to keep the game and the gamers together, and when you win you are happy' (Leia)	'It is the need to look into other's lives. It is about being jealous, to posting pictures, messages, and to observing reactions through the numbers of likes and comments we receive, as these reinforce you' (Elektra)
	Withdrawal		'Ten years ago, I had to play games on my computer. If I couldn't, I was frustrated; it was emotionally automatic. I had only one desire: to enter the game and play' (Leia) 'Once we were on vacation for weeks and I could not access the Internet, I felt the craving. I was thinking all the time about it. I learnt I could not put gaming as a priority, and I started to control the periods of gaming or not gaming' (Carol)	
	Conflict (intrapersonal and interpersonal)	'Problematic use is when the use of the Internet is affecting the family or the couple. Or if it is affecting sleep, work, or social life' (Elektra)	'I was gaming in class sessions, when I came back home I spent whole nights on my computer gaming. I could play 15 or 16 hours per day, but other people left or hindered their studies or jobs, or had conflicts with their partners' (Leia) 'You neglect other activities, your course, your family, other social contacts, even your health. Gamers have a lack of vitamin D because they do not go out, they eat poorly and quickly, so they can return to the game' (Carol) 'Excessive gaming is evident as lack of sleep shows in your eyes; hygiene, as you are not taking showers for gaming; the body, as you skip meals. These have consequences for your family, life, and work. It is the same problem during adolescence or adulthood, but the consequences are worst in adulthood. It causes tiredness. It is a loop' (Martin) 'I think the more you play, the more it affects your vision; as you start to seeing images from your thoughts; and these are engaged with the game' (Martin)	'When I am online too much, without sleeping and with troubles in my daily life, I have observed others like me have real problems with their studies, with both games and SNS' (Moira)
	Relapse		'If you want to leave a game such as a FPS, the group require you to return to maintain the same number of gamers in the teams as before. If not, they need to look for other gamers who are not so good. It is like a team sport, we have microphones, it is not simply a game, and it is another dimension. It is about speaking, planning strategies, indeed it is a world that we develop for a long time. Thus, when dates are fixed it is too restrictive' (Victor)	
	Difficulty to control it	'When I come home, I connect myself to Facebook from 5 to 10 or 11 pm, and if I cannot sleep, I use the Internet' (Aneka)	'Each new version of an RPG causes an increment of gamers and game play again. These are persistent worlds and being in one country or another does not change anything' (Leia) 'You make a false plan; it's almost a fake virtual plan' (Victor) 'One of my relatives was online gaming in an RPG with a guild, and we told him he needed to do something else, but he did not stop' (Victor)	'When I play with my partner, he cannot stop gaming if he is not winning. It is a fact of being successful, to have a goal. When a gamer develops an addiction, I think it is because they are attached to a world which is not real life. This is different from SNS users who are overly connected with real life. Gaming and SNS are really different' (Moira)
	Continued use disregard		'I passed hours gaming even when my eyes hurt' (Martin)	
	Deceive		'I had an alternative life; it is like private groups in Facebook, but in the game, and we are constantly in communication through software. We have the impression of living a double life, we do not speak about our real lives in the game, because people ask questions and judge us. It is a secret double life, it is a habit, like a drug' (Leia)	

Note: CIUS: Compulsive Internet use scale; CIUS–G: CIUS for gaming; CIUS–SNS: CIUS for social networking sites; MMORPG: massively multiplayer online role-playing games; RPG: role-playing games; FPS: first-person shooter.

Table 9. Superordinate and subordinate themes about the consequences of the GPIU and SPIUs studied, together with a few quotations (adapted from French to English) as illustration.

Theme	Subthemes	CIUS–G	CIUS–SNS
Consequences	Positive	'I play for the strategy, the research, the challenge, the learning through connections of things. We get into to a story and this promotes your memory, intelligence, and the capacity to maintain attention and quickly answer, to plan and foresee consequences, to adapt yourself. We are less confined in a virtual world and we could see and live more and differently than in real life' (Leia)	'Facebook could be a place to meet with my partner, to communicate or play games. I need to maintain this bond daily to maintain our news' (Moira)
	Negative	'Gaming is addictive without knowing exactly why. The people around you or your financial situation will not stop you, only the circumstances such as the professional world, physical problems, and a partner if he or she is not a gamer. Gamers can be confined and isolated; some of them have lost courses, jobs, partners; it can be dramatic' (Leia) 'Online games sometimes cause negative consequences at a social level, because these are simultaneously promoting isolation of the gamer; maybe there were previous social problems which promoted this isolation through online games. In any case, there is a progressive reclusion, I have observed this in a close friend' (Victor)	'Playing games excessively is a step out of reality; they should go outdoors more, as SNS users usually do' (Moira)
	Neutral	'Gaming is for curiosity, for enjoyment; gaming makes you happy; it does not always affect your real life' (Carol)	'I am also very curious about newsfeeds' (Moira)

Note: CIUS–G: Compulsive Internet use scale for gaming; CIUS–SNS: CIUS for social networking sites.

Table 10. Superordinate and subordinate themes about the prevention of the GPIU and SPIUs studied, together with a few quotations (adapted from French to English) as illustration.

Theme	Subthemes	Quotes		
		CIUS	CIUS–G	CIUS–SNS
Prevention	Risk factors	'A close friend lost a relative, and he was gaming between 5-6 hours per day, plus watching online series, which affected his studies. It was to compensate for the loss. Now he has reduced his gaming and we do other things' (Carol)	'I had a growing crisis in my adolescence; I had to detach from my family, and I could not do it physically, only through the games. I discovered another world and friends' (Carol) 'When I was an adolescent, I spent a lot of time at home alone gaming. In University you live alone, and you do what you want' (Martin) 'I had a friend who had difficulties with his father, then he played games to prove his value, as he needed to valorise himself, to obtain recognition' (Carol) 'I had a boyfriend who introduced me to gaming; it became a vicious circle. There were a lot of external thoughts that made us think about the game, remembering our wellbeing when gaming or the feeling to do something which challenged us was hard; but I started to look for real accomplishments' (Carol) 'I know someone who was an extreme gamer, and he smoked cannabis and played video games simultaneously which became a habit. The cannabis was only reinforcing the habit. The cause was that he was living alone with a parent who worked a lot. He only stopped to help his parent, but he spent the money to buy things for the MMORPG, or to buy cannabis' (Martin) 'There is a system that makes you play more, as you win and you are repaid receiving points. These points let you buy characters without real money; although you could buy things for the game with real money. It is a vicious cycle. The virtual money is the points you accumulate, and there is a ranking; then reputation also makes you play more' (Martin)	'A trauma could encourage you to stay behind the computer' (Moira) 'I had a close friend who was using the SNS all day until she found a partner. Problematic use could appear in a period of solitude to replace the lack of relationships with others' (Elektra) 'I was on Facebook a lot when I broke up with my partner; to avoid thinking' (Scarlet)
	Protective factors		'I had a friend who could not control his time online. I recommended dancing to him, as I had other hobbies apart of the Internet, like dancing 4 hours per week which diminished the hours of my gaming' (Carol) 'There is eSport at a global level, which is a way to win money, where you have to have real teams to compete. They are famous, and win a lot of money with sponsors. I think these gamers are not addicts, as their rewards are real, they are professional gamers; this is a career' (Martin)	'My parents were against the technologies. We did not receive education about them, and we were too connected at home. Thus, they started to switch off the Wi-Fi in the evening, encouraged us to go out, to start doing other activities: dance, music' (Moira)

Note: CIUS: Compulsive Internet use scale; CIUS–G: CIUS for gaming; CIUS–SNS: CIUS for social networking sites; MMORPG: massively multiplayer online role-playing games.

3.2.3. Themes Related to IGD Criteria Adapted for the Three Problems

The nine IGD criteria adapted (to excessive general Internet use, video game use (original IGD), or social networks use) were first quantitatively analysed by codes: by frequency (i.e., 2 = essential criterion without a problem detected in its wording; 1 = good criterion with a problem detected; 0 = not an adequate criterion with problems detected; thus, the minimum score was 0, and maximum was 18); with some quotes as illustrations presented in Table 11 below.

The criteria were ranked by order (i.e., from higher to lower frequency of agreement) and the most positively valued and important ones were: Risk or loss of relationships or opportunities, giving up other activities, withdrawal, and continuing despite problems. Following the first set of crucial symptoms which emerged, a second rank order appeared: Escape adverse moods, deceit or cover-up, preoccupation, and being unable to reduce or stop. Moreover, users stated that some of these criteria should not be stated like they are proposed in the criteria set, and in the scales, as usually users are not aware of them and probably select that they do not fulfil these criteria when they do, but are unaware of it (i.e., false negatives). This was usually happening in the following criteria: Preoccupation, reduction or stop, continuing despite problems, deceit or cover-up, and risk or loss of relationships or opportunities. Thus, clinical external evaluation is needed to explore if criteria are met or not by the user. Almost all criteria were well valued except tolerance, as users requested that it be reworded or that it cover not only quantitative time online, but also frequency, money spent on the game, accumulating points/accomplishments/likes, along with other similar elements.

Table 11. Internet, gaming, and social networking criteria encoded through the frequency of agreement and a few quotations (adapted from French to English) as illustration.

Criteria	Frequency of Agreement	Quotes
1: Preoccupation	Σ = 8	'This is a good criterion, if we are only thinking about what is happening through Facebook, it's as if we are addicted to it' (Moira, as a social networker) 'It remains in your head, we are permanently thinking, but it is also constantly in your feelings. Thus, I think this criterion should include feelings, and the cause of the suffering' (Victor, as a gamer) 'I would eliminate the phrase 'a lot of time', as it requires reflection' (Elektra, as a social networker) 'The problem is when you recognise your strange behaviour when not gaming: the gamer is thinking of the game outside the game, such as reflecting on strategies for playing. I have also observed this in girls who excessively shop online' (Carol, as a gamer)
2: Withdrawal	Σ = 10	'This is a good criterion, it is difficult to stop using Facebook. I am a bit addicted as I am on it 3-4 hours a day. I have started to tell myself I should do another thing in these hours, but I have not stopped. It is a need, but it is different from other online activities because it's about being in touch' (Moira, as a social networker) 'This is the most important criterion because the diagnostic includes suffering, what the user feels' (Victor, as a gamer) 'I think it should be divided into two criterions as there are two different verbs which refer to two different things' (Elektra, as a social networker) 'These excessive gamers can be with you to a degree, but they are so nervous and anxious to return to the games' (Martin, as a gamer)
3: Tolerance	Σ = 1	'Only if the game is not on computers. I think it is not that relevant, as I did not have this need as stated, but I was addicted to games' (Leia, as a gamer) 'To have a hobby which you would like to do more or improve is not a problem, as long as there are no financial problems. Everybody is online' (Victor, as a gamer) 'I would eliminate the words 'excitation' and 'pleasure' as these are intimate words, the word 'satisfaction' is better. Should add the notion of frequency, as in the SNS we are not spending much time but a lot of times' (Elektra, as a social networker) 'When one is addicted to a game there is no need to play more for the same state of excitement; the problem is that we do not want to leave the game. This behaviour is not like drugs; we do not need to increment the dose, above all for MMORPG. For those who have problems, to invest money in improving their materials to play could be a sign. If the gamer has this problem is because he or she has trouble with the excitement. Gaming addicts want to acquire more and more, as the feeling is not to wait, because the game continues without the gamer. I only accumulate accomplishments for avoiding the fear of losing, for not losing events' (Carol, as a gamer) 'To have new materials improves your quality of playing games, then more time and more things, the question is complex' (Martin, as a gamer)

Table 11. *Cont.*

Criteria	Frequency of Agreement	Quotes
4: Reduce or stop	Σ = 7	'Some gamers are not aware of their problem, but when they realise, they have already achieved a step towards recovery' (Leia, as a gamer) 'I think this criterion is above all for those who spend too much time alone' (Moira, as a social networker) 'I think it is for users who are aware of their excessive use, but there are periods in life you are not aware of it; it is a global unhappiness' (Elektra, as a social networker) 'You never take the pauses even when thinking of them' (Martin, as a gamer)
5: Give up other activities	Σ = 10	'This is a good criterion, but it is important to be aware this affects real life' (Leia, as a gamer) 'This is a fundamental criterion. Others stop their activities to be in the SNS' (Moira, as a social networker) 'The problem is users could consider their online activity as a sole hobby' (Elektra, as a social networker) 'I know I had an excessive gaming behaviour, playing a lot with craving, but nobody knew about it, as I was doing 'normal' life in school, and with other activities. You could be an addict and not accomplish this criterion' (Carol, as a gamer) 'This is a discriminative criterion, the impact on your life, when will everything be affected. I have lived it. It lacks the notion of time, the frequency when gaming' (Martin, as a gamer)
6: Continue despite problems	Σ = 10	'It is a good criterion if SNS are affecting our relationships negatively; as when you are going out to dinner, and everybody is always on their smartphones chatting by the SNS instead of with those who are at the dinner' (Moira, as a social networker) 'It is important, but maybe the user is not aware and continues gaming' (Victor, as a gamer) 'As the previous one, you could be an addict gamer and not accomplish this criterion, as others will not observe it' (Carol, as a gamer) 'This is only useful if the person is conscious' (Martin, as a gamer)
7: Deceive or cover-up	Σ = 9	'This is a good criterion. Gamers are a bit ashamed as we live in persistent worlds and we communicate with friends who are gamers' (Leia, as a gamer) 'This is a good criterion, if the user is conscious he or she is addicted to SNS' (Moira, as a social networker) 'This depends on the context, such as a strict family which puts pressure on you, maybe you will secretly game '(Victor, as a gamer) 'In the moment the gamer lies he or she is conscious' (Martin, as a gamer)
8: Escape adverse moods	Σ = 9	'This is a good criterion, as there are users who use the games as a shelter' (Leia, as a gamer) 'This is a criterion for those who play video games, but it should differentiate those who play for coping with a traumatic experience and those who play without any excuse' (Moira, as a social networker) 'Some gamers play to compensate, but others for the adrenaline, some to be alone, some to be happy' (Carol, as a gamer) 'It is needs to include the notion of frequency to achieve the concept of habit' (Martin, as a gamer)
9: Risk or loss of relationships or opportunities	Σ = 13	'This is a good criterion too, as it is influencing real-life' (Leia, as a gamer) 'This is a good criterion, it is too extreme but exists' (Moira, as a social networker) 'It is a subjective criterion because maybe the gamer treats this activity as work and he is fine like this, but if he is not aware and it affects him negatively, it is a problem (Victor, as a gamer) 'If you say yes, you could be an addict gamer; but if you say no, maybe you are accomplishing the criterion without being aware of it, and you could pass as healthy when you are not' (Carol, as a gamer) 'This is the most discriminative, as we lost something for playing and the addiction takes something from you' (Martin, as a gamer)

Note: Σ means to sum all quantitative values (codes). SNS: social networking sites; MMORPG: massively multiplayer online role-playing games.

4. Discussion

This mixed methods study had as a primary aim to understand potential technological behavioural addictions in young adults based on their self-reported online uses (i.e., Internet, gaming, and social networking) to compare generalised versus specific addiction problems and verify if these could be classed as public health outcomes. This approach has added an innovative research study type into the literature of this field, as the individuals were targeted based on unique trends of Internet use, gaming, and social networking simultaneously in a mixed methods study. The specific aims were: To validate the generalised and specific CIUS to compare these problems, to estimate their prevalence, and to examine problem users' perceptions regarding the phenomenology of these problems, and their opinion about the IGD criteria for the GPIU and SPIUs studied.

Concerning the first specific aim, the main psychometric properties were outstanding, showing the commonalities of the three CIUS adaptations tested. Findings showed excellent reliability in all versions (Cronbach alpha coefficients: From α_{CIUS} = 0.90 to $\alpha_{CIUS–G}$ = 0.95), which were even higher than the previous French version [52]. The scales' factor validity was consistent with its unidimensional model [39], achieving greater explained variance than the original version (from CIUS: 49% to CIUS–G: 60%), and with factor loadings being high enough (from CIUS with a minimum of 0.53 to CIUS–SNS with 0.58).

Regarding the construct validity, positive associations between the overall scores of each problem and time variables supported the degree to which the adaptations measure what they claim. Nevertheless, differences among the three problems started to appear. For instance, a device profile emerged for the different problems: The GPIU is mainly associated with computer use, although it was also moderately related to time spent per day with other devices (e.g., tablets and smartphones). Gaming only was modestly associated with daily time in computers and tablets, and social networks above all in smartphones (a part of computers on a weak basis). Thus, gaming is not a usual problem when playing on the mobile phone [44,59], as probably modern smartphones and mobile game apps are not yet developed enough for gamers, at least in Europe. However, computers seem to act as an object of these addiction problems and the Internet can provide a vehicle for facilitating some of the addictive symptomatology, as reported in other behavioural addictions (e.g., gambling [60]).

Concerning content validity, the three forms measured different facets of the 'Internet addiction' umbrella construct. For instance, general use was weakly associated with a set of common online behaviours, with messaging, viewing, social networking, and sex consumption standing out. An issue which emerged is the taste for online TV shows/series and film consumption, a recent research line developed in the field [61–63], which is usually linked to Internet addiction and gaming as a sedentary behaviour affecting mental health [64,65]. Regarding gaming, game genres with a strong association to this problematic use included by order: MOBA [66,67], MMORPG [50,68–71], FPS [70,72,73], and solo games [71] (i.e., almost all big games). It is unusual that MOBA and especially 'solo games' have emerged, as the literature around problem gaming tends to include only role-playing games. The 'solo games' genre makes sense based on the qualitative findings associated with the aetiology about excessively play to avoid or prevent loneliness and boredom, reported in problematic gaming research [71,74–77]. On the other hand, regarding problem SNS usage, only Facebook emerged as the leading network with which users present addiction symptoms similar to Internet addiction, as previous studies reported [10,78]. As far as the author knows, it has scarcely been clarified what activities are really under the labels of GPIU and these SPIUs, such as gaming [79,80].

Interestingly, another set of differences was found between the three problems at a psychometric level. The general Internet and social networking overall scores were strongly positively correlated, which probably means that their constructs are very close, as current SNS have plenty of resources (e.g., Facebook have options to communicate, to share files, to play games, to consume information, even online gambling) that are starting to emulate a generalised online use [10,78,81]. Moreover, the findings of the content validity supported why the general purpose of the Internet and the specific use of social networks were so close in their interpretation. This could explain why when analysing the scales by items, the CIUS and CIUS–SNS have a similar pattern with moderate scores (i.e., CIUS: a M from 0.78 in item 9 to 2.05 in item 12 out of 4; CIUS–SNS: An M from 0.63 in item 4 to 1.84 in item 1) compared to the CIUS–G (i.e., an M from 0.22 in item 9 to 0.66 in item 1). However, the association between general Internet use and gaming was weak, and it did not exist between gaming and social networking, meaning both SPIUs were independent between them. Moreover, qualitative findings support this different phenomenology of both constructs, as users state many differences between problem gaming and problem networking (e.g., the first being detached from real life, while the second overly attached to real life). This independence observed could be interpreted as those who mainly use

SNS are not heavy gamers, and inversely [82]. This is a crucial finding as, from a phenomenological perspective, this indicates gaming could not be classed as a social networking problem and vice versa. These findings also suggest that while from an adult's perspective, their potential addictive symptomatology is higher when using the Internet in general or when networking, only a few users play games in an excessive and problematic form, and these are all types of adults (i.e., males and females, young and middle-aged ones). In a recent study testing the spectrum hypothesis using the network approach, the umbrella construct of Internet addiction was also highlighted; this study recommended the focus on the specific Internet- and technology-mediated addictive behaviours: Gaming, smartphone, and cybersex [83]. Present quantitative and qualitative findings support the specificity of the SPIU of gaming. Regarding gender issues, the differences were detected on SPIUs at a descriptive level, as while descriptively males tend to be gamers and view online porn, females tend to use social networks, as Andreassen and colleagues also reported [82]. However, when analysing only those classed as problem users, findings change.

Concerning the second specific aim, the estimated prevalence of potential at-risk addiction problem users, findings were (in descending order): 37.4% Internet addicts (generally women and young adults), 32.5% problematic social networkers (usually women and young adults too), and 10.8% problematic online gamers (both genders equally, as well as all group ages in adulthood). Thus, this suggests women are starting to be problem users, especially in addictive gaming, which traditionally was a problem in males [84]. Moreover, while young adulthood has a higher prevalence in general Internet and social networking uses, in the case of gaming, all adults have a similar level of addiction problems, as previously reported [85]. These findings provide evidence for IGD proposal and gaming disorder [6,7], as the problem is extended to both genders and adult group ages. The prevalence estimated was higher than usual for a European study, as reported previously by Laconi and colleagues [80,86]. This overestimation is probably due to a set of methodological factors, the different scales selected in similar studies (e.g., the Internet addiction test; IAT [5] vs. the CIUS [39]), the different interpretation of the symptoms measured as addiction criteria (e.g., see Table 1), and the cut-off score applied to the CIUS, which could be considered a threshold for risk instead of an addiction problem (i.e., as the theoretical median of the CIUS is 28, and 21 was used based on Guertler and colleagues' study [47], which it was found in the sole self-reported measures from German problem and pathological gamblers, but without clinical support as it is usual in the field). Furthermore, the cut-off was not performed by gender and by type of problem use, and the present sample was mainly composed of women, as it was based on Belgian academic environments of social sciences in higher education. However, almost no research has provided validated cut-off scores using the CIUS, except for the short CIUS [87], which implies a future research line. For this reason, prevalence should be taken with caution as a proposed proportion of users in risk of potential GPIU or SPIU.

Moreover, as a usual practice in the field, potential problem users were chosen by a psychometric measure, without explicit attention to the symptoms accomplished and their relevance [18,21,22,28] or clinical supervision. Other problems were the results obtained by endorsement per items, as there is no agreement on how to operationalise it (e.g., only extracting those who state 'often' and 'very often' or only 'very often' [88]). Interestingly, a clear difference between core and peripheral symptoms has not clearly emerged by the GPIU and SPIUs studied, contrary to other studies [89], as even the core ones were not related to time spent playing, unlike in previous research [25,26]. It is probably because the quantitative sample contained fewer gamers and problem gamers (the lower prevalence), and usually only players of the 'big games' seem to be the ones with more addictive problems [25,26,66–70,72]. Nevertheless, in the qualitative findings, the main addiction problem was gaming through 'big games', covering all symptomatology and causing more harm compared to the other two problems studied.

Thus, only half of the sample could be classed as potential or at risk of being a problem user, a quarter as having a potential Internet use-related addiction problem, and another quarter more than one problem (i.e., mixed profile: Usually by problem general Internet and networking uses), such as in another European study [80], which was closer in its phenomenology, as reported. It is worth noting that most prevalent symptoms in the quantitative evidence on GPIU and SPIUs studied were not those classed as core criteria according to Charlton and Danforth [25] (e.g., conflict, withdrawal) and Tao and colleagues [21] (i.e., preoccupation, withdrawal). Indeed, the commonality between the GPIU and SPIUs studied at a symptomatologic level was with items addressing loss of control and mood modification. This could highlight the fact that potential problematic users are being classed by criteria which are not the core ones, at least in problem gaming; in other words, the overestimation of these potential problem users could be due to a proportion of them being high enthusiasts of using the Internet in general, gaming and/or social networking. In this study, quantitative evidence has highlighted the loss of control was the symptom with high scores, and according to Griffiths, this is not central to addiction [90] (i.e., it is not a necessary component or a consequence of addictive behaviour). However, a few studies reported a difference in the conceptual and nosological entities of these problems using quantitative approaches, especially between GPIU versus SPIU (i.e., Internet addiction vs. problem online gaming [80]), although the distinction was mainly based on gender, which it seems to be changing based on present findings. Furthermore, the female gender seems at risk of these potential behavioural addictions, even gaming, which is a novel finding against the current literature [91]; this highlights that more research in female gamer in behavioural addictions is needed.

Regarding the third specific aim, the aetiology observed through the qualitative evidence shows more causes have emerged than in previous qualitative studies (e.g., usually only focused on boredom, mood feeling, stress, and escapism [58]); for instance, being an indoors person, with an introvert personality, real and virtual friendship with the guild, and being tech-savvy, feeling the need to be updated, connected, and obsessively informed by others in the case of social networkers. Curiously, environmental factors, such as weak or perceived difficult interpersonal context, have emerged as facilitating problematic gaming, but not other PIUs. Thus, maybe not all external factors facilitate same GPIU or SPIU. Similarly, Griffiths' component model was effective only for problem gaming [18], as a phenomenon caused by a biopsychosocial perspective; followed by problematic social networking; and finally, GPIU with less symptoms, which do not facilitate covering the need for more evidence to promote theoretical development on GPIUs [14] (e.g., only relapse was present as a symptom in problematic gaming).

According to addiction symptoms, salience seems not to only be a cognitive component, as its emotional facet has emerged, as it was stated by Griffiths [18], and it appears to include craving; however, it was not involved in other Internet addiction criteria or IGD [21,22]. Moreover, this addictive symptom is usually used to diagnose SUDs and gambling disorder, but not included in, for example, IGD [6,92]). Davis' model [15] did not cover this affective aspect of the nature of GPIU or SPIUs. Indeed, Caplan statements about the preference for online interaction, mood regulation, and deficient self-regulation [15,16] were supported by all interviewees. However, tolerance is more complicated than stated in the most contemporary proposals to diagnose behavioural addictions [21,22], especially in gaming [33]; as it is the increment of not only time (i.e., intensity, expertise, or advancement in the game), and not only for excitement or desire (wording usually used in IGD [22], which has been highly criticised in the interviews). Concerning deception or cover-up, like tolerance, it is more complicated than initially described in addictive symptomatology [18–21,28,33], as well as in the existing scales (e.g., the IAT [87]). It seems gamers usually maintain a 'secret double life' in the game, as they have explained, and as it has scanty been reported as a shared virtual environment among MMORPG players [93,94]. They have 'game friends' (the partner, guild or other teams) and their codes to communicate (own language), dedicating daily time to set goals and enjoy within the virtual world of the games; this essential element of gaming behaviour has not been studied in-depth yet [95]. If quantitative and qualitative findings are observed simultaneously, according to Tao and colleagues [21], preoccupation and withdrawal are weakly and moderately present, respectively, in the main sample. However, they are only present in the problematic gamers subsample, and persistent desire and difficult to control are strongly present in the sample for all (GPIU and SPIUs) studied, as well as in all problem users studied. Thus, based on Tao's criteria, problems studied could be classed as Internet use-related addiction problems with caution; only gaming seems to fit in their proposal. Nevertheless, for IGD [6,22] criteria, only gaming could also be categorised as SPIU based on all findings. Remarkably, the core components [25] could just be checked on the qualitative findings extracted from potential problem users, which concurrently stated conflict was the main addiction symptom (i.e., the one with a higher frequency of agreement), followed by withdrawal, giving up other activities, and continuing despite problems, as probable indicators of behavioural salience. In other words, matching all Charlton and Danforth core criteria for addictive gaming [25,26], qualitative findings support the health problem, together with IGD opinions, which are considered.

Regarding the consequences of excessive online uses, it should be highlighted not all of them are negative. Almost all of the users agreed they have experienced (or observed in others close to them) the risk of being addicted to the Internet or an online activity (usually gaming). They reported developmental and contextual factors as those supporting how a habit such as gaming could slowly and unconsciously develop into an addictive behaviour. For instance, to be a gamer and play video games during adolescence indoors during a long time, avoiding loneliness, establishing an identity in the game, and virtual social life may relate to real life (e.g., real friends are also gamers of the same role-playing games). The risk factors highlighted compared to the symptomatology reported [18–25], which were connected to both developmental and contextual factors (e.g., adolescent crises, parent styles) were found to be more important than individual factors (e.g., personality traits). This assertion goes against our current research in the field (i.e., focused on personality traits [73,96]), while few contextual or structural risk factors [97] or protective factors have been reported. Nevertheless, it should be argued that maybe these participants are not so aware of their characteristics that may help to develop extreme online behaviours, as in other addictive problems (e.g., SUDs). However, none of the interviewees recognised they might be a problem user at present, which questions the cut-off score selected in the adapted CIUS or supports that they probably had (or have had) an addiction problem (i.e., only one gamer recognised to have had this problem in the past, and another gamer described the game transfer phenomena [98]).

Finally, with respect to the fourth specific aim about IGD criteria opinions for diagnosing these problems, qualitative findings highlighted as the most critical addiction symptoms: Risk or loss of relationships or opportunities (conflict), giving up other activities, withdrawal, and continuing despite problems, which coin the core components previously pointed out [25] and match research findings [33]. Moreover, users clearly state 'intrapersonal or interpersonal conflict' [18] is equally important, sometimes associated with a 'functional impairment' [21]. The qualitative evidence seems to agree with critiques considering GPIU with not enough entity to be as a behavioural addiction [23,80], as a few subthemes did not provide evidence (e.g., not emerging in all addiction symptomatology analysed). In this sense, other excessive online behaviours, such as watching series online, have emerged and caused public health concerns with sedentary lifestyle and time in front of the screens [61–63]. Other criteria, such as escaping adverse moods, deceit or cover-up, preoccupation or not being able to reduce or stop online behaviour, are quite ordinary and excessive users agreed with them, the only criterion where users disagree being tolerance, one of which Kardefelt-Winther [24] criticised in the current IGD [22], and about which other authors have found similar evidence [33]. Tolerance, giving up other activities, or escaping adverse moods require a part of rewording them and covering their complexity, an indicator of frequency, as Ko [99] also stated when IGD was published in the DSM-5 appendix [6], as the need for using the intensity and frequency criteria to distinguish subjects with IGD from casual online gamers.

Some limitations of this cross-sectional and self-report study have been already discussed, but it could be added that it was performed with a non-random community and academic sample, with an online survey and interviewing only those who accepted the invitation to participate in the second part of this study (e.g., economic motivation). However, similar quantitative studies have been performed with smaller samples than the quantitative [80] and qualitative [50] parts of the study. The study's strengths include a survey with psychometric and epidemiological validated techniques, with a considerable and sufficient sample size based on Nunnally's recommendations for psychometrics [100], and Smith and Osborn's assertion [101] for the qualitative part of the study. The provision of a Belgian French adaptation of the CIUS and IGD for specific Internet use-related addiction problems was supported by previously validated adaptations [22,52]. The qualitative participants were psychometrically and theoretically selected after validating the CIUS scales to have the potential problem online users from all profiles studied to shed light on the phenomenology of these new addiction problems from adults.

In summary, technology uses are growing within contemporary societies and bibliographic productivity is attending these phenomena quantitatively. However, confirmatory approaches could have been chosen as a method for factor validity, which is a future research study under development. Finally, although this is not the first mixed methods study done in the field [33,92], it is one of the few that have been undertaken covering mixed user profiles of these problems. Therefore, it is an original contribution looking for mixed evidence on what seems to constitute the phenomenology of these problems through potential or at-risk adult problem users, although no functional impairment or other possible mental disorders were measured to disentangle the problem users with a clinical approach.

5. Conclusions

The present study is one of the first to ascertain what seems to be behind current psychometric measures and to start to disentangle generalised and specific Internet addiction problems, addressing their phenomenology with quantitative and qualitative evidence.

The findings show that around half of Belgian (European) adults seem to present at least a risk of Internet use-related addiction problems. However, a lower proportion could have a mixed profile, covering the risk of more than one problem. The CIUS for generalised problematic internet use and specific problematic gaming and social networking has been validated, and the potential prevalence in at-risk users have been estimated. Young or middle-aged adults in both genders have been shown to be equally vulnerable to problematic gaming. Furthermore, the addiction symptoms are present in the problems studied, but not with the same weight or level of importance. The higher scores are in general Internet addiction and problem social networking, which seem to have a common phenomenology, while problem gaming seems different. However, all of them have loss of control and mood modification as main prevalent symptoms. Nevertheless, when problem gaming is analysed in-depth, it is difficult to confirm all addictive components, symptoms, and criteria, in quantitative and qualitative findings. Furthermore, the classic core components have been tested but were only ratified in the qualitative results, where the relevant symptoms were conflicting, giving up other activities, withdrawal, and continuing despite problems.

An in-depth theoretical and empirical review of the public health approach usually used in behavioural addictions literature applied to generalised and specific addiction problems has been provided. This strategy together with analysing quantitative and qualitative evidence of these problems has shown what seems to be behind each addictive criterion usually used to screen users and classify them as a potential problem user (i.e., through DSM criteria, component addiction symptoms, and core vs. peripheral symptoms). Moreover, it has been documented that at-risk users' perceptions about the phenomenology of these problems, which start in adolescence and remain during adulthood, with a potential of increased harm. The aetiology is more diverse than previously described, and its development is supported by psychosocial, environmental, and technological factors, leading to a few cases of extreme problems. Potential at-risk problem users' opinions about IGD criteria have highlighted that some criteria (e.g., risk or loss of relationships or opportunities) are more important and better developed than others (e.g., tolerance), but the need to apply them with clinical resources is essential in the evaluation and diagnosis of these mental health problems. Nevertheless, natural recovery has emerged as a finding for some potential problem users, although mental health systems to support those users who have developed the problem(s) and cannot naturally recover are required.

Other preventive actions in education, social, and family relationships could promote the maintenance of users' health and wellbeing.

Thus, it seems these problems exist, but probably in a lower proportion than reported. However, when they do appear they present addictive symptomatology, especially in problem online gaming. These findings seem to provide evidence for the IGD proposed by the APA (with the restrictions and improvements suggested) and the recent recognition of gaming disorder by the WHO (especially for specific online gaming behaviours).

Funding: This research was funded by the European Commission grant number FP7-PEOPLE-2013-IEF-627999—Tech Use Disorders research project.

Acknowledgments: The author would like to thank Joel Billieux for supervising the project, to Jory Deleuze, Gaetan Devos, Aline Wery, and William De Brueger for supporting the survey design in French, and to Aurelien Cornil and Ophelie Devrient for supporting the interviews in French. Furthermore, the author would like to thank the ITU and the Proquest for their permission to publish the figures.

Conflicts of Interest: The author declares no conflict of interest.

References

1. OReilly, M. Internet addiction: A new disorder enters the medical lexicon. *Can. Med. Assoc. J.* **1996**, *154*, 1882–1883.
2. Young, K.S. Psychology of computer use: XL. Addictive use of the Internet: A case that breaks the stereotype. *Psychol. Rep.* **1996**, *79*, 899–902. [CrossRef] [PubMed]
3. Stein, D.J. Internet addiction, Internet psychotherapy letter; comment. *J. Am. J. Psychiatry* **1997**, *154*, 890. [PubMed]
4. Griffiths, M. Internet addiction: Fact or fiction? *Psychologist* **1999**, *12*, 246–250. [CrossRef]
5. Young, K.S. Internet addiction: The emergence of a new clinical disorder. *CyberPsychol. Behav.* **1998**, *1*, 237–244. [CrossRef]
6. American Psychiatric Association (APA). *Diagnostic and Statistical Manual of Mental Disorders*, 5th ed.; American Psychiatric Association: Arlington, VA, USA, 2013.
7. World Health Organization (WHO). Gaming Disorder. WHO, 2018. Available online: https://icd.who.int/browse11/l-m/en#/http%3a%2f%2fid.who.int%2ficd%2fentity%2f1448597234 (accessed on 4 December 2018).
8. Shaw, M.; Black, D.W. Internet addiction: Definition, assessment, epidemiology and clinical management. *CNS Drugs* **2008**, *22*, 353–365. [CrossRef] [PubMed]
9. Ng, B.D.; Wiemer-Hastings, P. Addiction to the Internet and Online Gaming. *Cyberpsychol. Behav.* **2005**, *8*, 110–113. [CrossRef]
10. Kuss, D.J.; Griffiths, M.D. Social Networking Sites and Addiction: Ten Lessons Learned. *Int. J. Environ. Res. Public Health* **2017**, *14*, 311. [CrossRef]
11. International Telecommunication Union (ITU). ITU Committed to Connecting the World: ICT Facts and Figures 2017—Global ICT Developments. Available online: https://www.itu.int/en/ITU-D/Statistics/Pages/stat/default.aspx (accessed on 27 June 2018).
12. ProQuest. Search—All Databases. Available online: https://search.proquest.com/results/3E9BFDEC5154401PQ/1?accountid=14693 (accessed on 26 January 2018).
13. Becoña, E. Conductas adictivas: El problema del siglo XXI? *Psicol. Contemp.* **1998**, *5*, 4–15.
14. Davis, R.A. A cognitive–behavioral model of pathological Internet use. *Comput. Hum. Behav.* **2001**, *17*, 187–195. [CrossRef]
15. Caplan, S.E. Theory and measurement of generalized problematic internet use: A two-step approach. *Comput. Hum. Behav.* **2010**, *26*, 1089–1097. [CrossRef]
16. Caplan, S.E. Problematic Internet use and psychosocial well-being: Development of a theory-based cognitive-behavioral measurement instrument. *Comput. Hum. Behav.* **2002**, *18*, 553–575. [CrossRef]
17. Haagsma, M.C.; Caplan, S.E.; Peters, O.; Pieterse, M.E. A cognitive-behavioral model of problematic online gaming in adolescents aged 12–22 years. *Comput. Hum. Behav.* **2013**, *29*, 202–209. [CrossRef]

18. Griffiths, M.D. A "components" model of addiction within a biopsychosocial framework. *J. Subst. Use* **2005**, *10*, 191–197. [CrossRef]
19. Brown, R.I.F. Some contributions of the study of gambling to the study of other addictions. In *Gambling Behavior and Problem Gambling*; Eadington, W.R., Cornelius, J., Eds.; University of Nevada Press: Reno, NV, USA, 1993.
20. Griffiths, M.D. Behavioural addiction and substance addiction should be defined by their similarities not their dissimilarities. *Addiction* **2017**, *112*, 1718–1720. [CrossRef]
21. Tao, R.; Huang, X.; Wang, J.; Zhang, H.; Zhang, Y.; Li, M. Proposed diagnostic criteria for internet addiction. *Addiction* **2010**, *105*, 556–564. [CrossRef]
22. Petry, N.M.; Rehbein, F.; Gentile, D.A.; Lemmens, J.S.; Rumpf, H.; Mößle, T.; Bischof, G.; Tao, R.; Fung, D.S.; Borges, G.; et al. An international consensus for assessing internet gaming disorder using the new DSM-5 approach. *Addiction* **2014**, *109*, 1399–1406. [CrossRef]
23. van Rooij, A.J.; Prause, N. A critical review of "Internet addiction" criteria with suggestions for the future. *J. Behav. Addict.* **2014**, *3*, 203–213. [CrossRef]
24. Kardefelt-Winther, D. A critical account of DSM-5 criteria for Internet gaming disorder. *Addict. Res. Theory* **2015**, *23*, 93–98. [CrossRef]
25. Charlton, J.P.; Danforth, I.D.W. Distinguishing addiction and high engagement in the context of online game playing. *Comput. Hum. Behav.* **2007**, *23*, 1531–1548. [CrossRef]
26. Lehenbauer-Baum, M.; Fohringer, M. Towards classification criteria for internet gaming disorder: Debunking differences between addiction and high engagement in a German sample of World of Warcraft players. *Comput. Hum. Behav.* **2015**, *45*, 345–351. [CrossRef]
27. Lopez-Fernandez, O. How has internet addiction research evolved since the advent of internet gaming disorder? An overview of cyberaddictions from a psychological perspective. *Curr. Addict. Rep.* **2015**, *2*, 263–271. [CrossRef]
28. Lopez-Fernandez, O.; Molina-Azorin, J. The use of mixed methods research in the field of behavioural sciences. *Qual. Quant. Int. J. Methodol.* **2011**, *45*, 1459–1472. [CrossRef]
29. Lopez-Fernandez, O.; Molina-Azorin, J. The use of mixed methods research in interdisciplinary educational journals. *Int. J. Mult. Res. Approaches* **2011**, *5*, 269–283. [CrossRef]
30. Young, KS. Internet Addiction: A New Clinical Phenomenon and Its Consequences. *Am. Behav. Sci.* **2004**, *48*, 402–415. [CrossRef]
31. Stepien, K. Internet Addiction. The Phenomenon of Pathological Internet Use—Problems of Interpretation in the Definition and Diagnosis. *Intern. Secur.* **2014**, *6*, 79–90. [CrossRef]
32. Liu, T.C. Phenomenology and epidemiology of problematic Internet use. In *The Oxford Handbook of Impulse Control Disorders*; Grant, J.E., Potenza, M.N., Eds.; The Oxford Handbook of Impulse Control Disorders; Oxford University Press: New York, NY, USA, 2012; pp. 176–185, Chapter xvii.
33. Colder Carras, M.; Porter, A.M.; Van Rooij, A.J.; King, D.; Lange, A.; Carras, M.; Labrique, A. Gamers' insights into the phenomenology of normal gaming and game "addiction": A mixed methods study. *Comput. Hum. Behav.* **2018**, *79*, 238–246. [CrossRef]
34. Grant, J.E.; Schreiber, L.R.; Odlaug, B.L. Phenomenology and Treatment of Behavioural Addictions. *Can. J. Psychiatry* **2013**, *58*, 252–259. [CrossRef]
35. Andreassen, C.S. Online social network site addiction: A comprehensive review. *Curr. Addict. Rep.* **2015**, *2*, 175–184. [CrossRef]
36. Lee, E.W.J.; Ho, S.S.; Lwin, M.O. Explicating problematic social network sites use: A review of concepts, theoretical frameworks, and future directions for communication theorizing. *New Media Soc.* **2017**, *19*, 308–326. [CrossRef]
37. van den Eijnden, R.J.J.M.; Lemmens, J.S.; Valkenburg, P.M. The Social Media Disorder Scale. *Comput. Hum. Behav.* **2016**, *61*, 478–487. [CrossRef]
38. Hormes, J.M.; Kearns, B.; Timko, C.A. Craving Facebook? Behavioral addiction to online social networking and its association with emotion regulation deficits. *Addiction* **2014**, *109*, 2079–2088. [CrossRef] [PubMed]
39. Meerkerk, G.J.; van Den Eijnden, R.J.; Vermulst, A.A.; Garretsen, H.F. The compulsive internet use scale (CIUS): Some psychometric properties. *Cyberpsychol. Behav.* **2009**, *12*, 1–6. [CrossRef]

40. Tech Use Disorders. Technological Use Disorders: European Cross-Cultural Longitudinal and Experimental Studies for Internet and Smartphone Problem Uses. 12 July 2017. Available online: http://cordis.europa.eu/project/rcn/189961_en.html (accessed on 14 July 2017).
41. Lopez-Fernandez, O. Short version of the Smartphone Addiction Scale adapted to Spanish and French: Towards a cross-cultural research in problematic mobile phone use. *Addict. Behav.* **2017**, *64*, 275–280. [CrossRef] [PubMed]
42. Lopez-Fernandez, O.; Kuss, D.J.; Romo, L.; Morvan, Y.; Kern, L.; Graziani, P.; Rousseau, A.; Rumpf, H.-J.; Bischof, A.; Gässler, A.-K.; et al. Self-reported dependence on mobile phones in young adults: A European cross-cultural empirical survey. *J. Behav. Addict.* **2017**, *6*, 168–177. [CrossRef] [PubMed]
43. Lopez-Fernandez, O.; Männikkö, N.; Kääriäinen, M.; Griffiths, M.D.; Kuss, D.J. Mobile gaming and problematic smartphone use: A comparative study between Belgium and Finland. *J. Behav. Addict.* **2018**, *9*, 1–12. [CrossRef]
44. Lopez-Fernandez, O.; Kuss, D.J.; Pontes, H.M.; Griffiths, M.D.; Dawes, C.; Justice, L.V.; Männikkö, N.; Kääriäinen, M.; Rumpf, H.-J.; Bischof, A.; et al. Measurement Invariance of the Short Version of the Problematic Mobile Phone Use Questionnaire (PMPUQ-SV) across Eight Languages. *Int. J. Environ. Res. Public Health* **2018**, *15*, 1213. [CrossRef]
45. Morse, J. Approaches to qualitative-quantitative methodological triangulation. *Nurs. Res.* **1991**, *40*, 120–123. [CrossRef]
46. Johnson, B.; Onwuegbuzie, A. Mixed methods research: A research paradigm whose time has come. *Educ. Res.* **2004**, *33*, 14–26. [CrossRef]
47. Greene, J.; Caracelli, V.; Graham, W. Toward a conceptual framework for mixed-method evaluation designs. *Educ. Eval. Policy Anal.* **1989**, *11*, 255–274. [CrossRef]
48. Creswell, J. *Research Design: Qualitative, Quantitative and Mixed Methods Approaches*, 2nd ed.; Sage: Thousand Oaks, CA, USA, 2003.
49. Leech, N.; Onwuegbuzie, A. A typology of mixed methods research designs. *Qual. Quant. Int. J. Methodol.* **2009**, *43*, 265–275. [CrossRef]
50. Beranuy, M.; Carbonell, X.; Griffiths, M.D. A Qualitative Analysis of Online Gaming Addicts in Treatment. *Int. J. Ment. Health Addict.* **2013**, *11*, 149–161. [CrossRef]
51. Guertler, D.; Rumpf, H.; Bischof, A.; Kastirke, N.; Petersen, K.U.; John, U.; Meyer, C. Assessment of problematic internet use by the Compulsive Internet Use Scale and the Internet Addiction Test: A sample of problematic and pathological gamblers. *Eur. Addict. Res.* **2014**, *20*, 75–81. [CrossRef] [PubMed]
52. Khazaal, Y.; Chatton, A.; Horn, A.; Achab, S.; Thorens, G.; Zullino, D.; Billieux, J. French Validation of the Compulsive Internet Use Scale (CIUS). *Psychiatr. Q.* **2012**, *83*, 397–405. [CrossRef] [PubMed]
53. Vaismoradi, M.; Turunen, H.; Bondas, T. Content analysis and thematic analysis: Implications for conducting a qualitative descriptive study. *Nurs. Health Sci.* **2013**, *15*, 398–405. [CrossRef] [PubMed]
54. Braun, V.; Clarke, V. Using thematic analysis in psychology. *Qual. Res. Psychol.* **2006**, *3*, 77–101. [CrossRef]
55. Braun, V.; Clarke, V.; Terry, G. Thematic analysis. *Qual. Res. Clin. Health Psychol.* **2014**, *24*, 95–114.
56. Wilkinson, S. Women with Breast Cancer Talking Causes: Comparing Content, Biographical and Discursive Analyses. *Fem. Psychol.* **2000**, *10*, 431–460. [CrossRef]
57. Krippendorff, K. *Content Analysis: An Introduction to Its Methodology*, 2nd ed.; Sage Publications: London, UK, 2004.
58. Li, W.; O'Brien, J.E.; Snyder, SM.; Howard, M.O. Characteristics of internet addiction/pathological internet use in US university students: A qualitative-method investigation. *PLoS ONE* **2015**, *10*, e0117372. [CrossRef]
59. Roberts, J.A.; Petnji Yaya, L.H.; Manolis, C. The invisible addiction: Cell-phone activities and addiction among male and female college students. *J. Behav. Addict.* **2014**, *3*, 254–265. [CrossRef]
60. Shaffer, H.J. Understanding the means and objects of addiction: Technology, the internet and gambling. *J. Gambl. Stud.* **1996**, *12*, 461–469. [CrossRef] [PubMed]
61. Sussman, S.; Moran, MB. Hidden addiction: Television. *J. Behav. Addict.* **2013**, *2*, 125–132. [CrossRef] [PubMed]
62. Lopez-Fernandez, O. Chapter One: Online TV shows and series addiction: An exploratory francophone cross-country comparison. In *Combining Aesthetic and Psychological Approaches to TV Series Addiction*; Camart, N., Lefait, S., Paquet-Deyris, A.-M., Romo, L., Eds.; Cambridge Scholars Publishing: Newcastle upon Tyne, UK, 2018; pp. 1–16, ISBN 978-1-5275-0914-6.

63. Flayelle, M.; Maurage, P.; Billieux, J. Toward a qualitative understanding of binge-watching behaviors: A focus group approach. *J. Behav. Addict.* **2017**, *6*, 457–471. [CrossRef] [PubMed]
64. Hoare, E.; Milton, K.; Foster, C.; Allender, S. The associations between sedentary behaviour and mental health among adolescents: A systematic review. *Int. J. Behav. Nutr. Phys. Act.* **2016**, *13*. [CrossRef]
65. Gentile, D.A.; Berch, O.N.; Choo, H.; Khoo, A.; Walsh, D.A. Bedroom media: One risk factor for development. *Dev. Psychol.* **2017**, *12*, 2340–2355. [CrossRef]
66. Nuyens, F.; Deleuze, J.; Maurage, P.; Griffiths, M.D.; Kuss, D.J.; Billieux, J. Impulsivity in Multiplayer Online Battle Arena gamers: Preliminary results on experimental and self-report measures. *J. Behav. Addict.* **2016**, *5*, 351–356. [CrossRef]
67. Bertran, E.; Chamarro, A. Videojugadores del League of Legends: El papel de la pasión en el uso abusivo y en el rendimiento. *Adicciones* **2016**, *28*, 28–34. [CrossRef]
68. Achab, S.; Nicolier, M.; Mauny, F.; Monnin, J.; Trojak, B.; Vandel, P.; Sechter, D.; Gorwood, P.; Haffen, E. Massively multiplayer online role-playing games: Comparing characteristics of addict vs non-addict online recruited gamers in a French adult population. *BMC Psychiatry* **2011**, *11*, 12. [CrossRef]
69. Bergmark, K.H.; Bergmark, A. The diffusion of addiction to the field of MMORPGs. *Nordic Stud. Alcohol Drugs* **2009**, *26*, 415–426. [CrossRef]
70. Metcalf, O.; Pammer, K. Physiological arousal deficits in addicted gamers differ based on preferred game genre. *Eur. Addict. Res.* **2013**, *12*, 23–32. [CrossRef]
71. Männikkö, N.; Ruotsalainen, H.; Demetrovics, Z.; Lopez-Fernandez, O.; Myllymäki, L.; Miettunen, J.; Kääriäinen, M. Problematic gaming behavior among Finnish junior high school students: Relation to socio-demographics and gaming behavior characteristics. *Behav. Med.* **2018**, *44*, 324–334. [CrossRef] [PubMed]
72. Na, E.; Choi, I.; Lee, T.; Lee, H.; Rho, M.J.; Cho, H.; Jung, D.J.; Kim, D.J. The influence of game genre on Internet gaming disorder. *J. Behav. Addict.* **2017**, *6*, 248–255. [CrossRef] [PubMed]
73. Metcalf, O.; Pammer, K. Impulsivity and related neuropsychological features in regular and addictive first person shooter gaming. *Cyberpsychol. Behav. Soc. Netw.* **2014**, *17*, 147–152. [CrossRef] [PubMed]
74. Bergmark, K.H.; Bergmark, A.; Findahl, O. Extensive Internet Involvement-Addiction or Emerging Lifestyle? *Int. J. Environ. Res. Public Health* **2011**, *8*, 4488–4501. [CrossRef]
75. Myrseth, H.; Olsen, O.K.; Strand, L.Å.; Borud, E.K. Gaming behavior among conscripts: The role of lower psychosocial well-being factors in explaining gaming addiction. *Mil. Psychol.* **2017**, *29*, 128–142. [CrossRef]
76. Carras, M.C.; Van Rooij, A.J.; Van de Mheen, D.; Musci, R.; Xue, Q.; Mendelson, T. Video gaming in a hyperconnected world: A cross-sectional study of heavy gaming, problematic gaming symptoms, and online socializing in adolescents. *Comput. Hum. Behav.* **2017**, *68*, 472–479. [CrossRef] [PubMed]
77. Lemmens, J.S.; Valkenburg, P.M.; Gentile, D.A. The Internet Gaming Disorder Scale. *Psychol. Assess* **2015**, *27*, 567–582. [CrossRef] [PubMed]
78. Enrique, E.; De Corral, P. Addiction to new technologies and to online social networking in young people: A new challenge. *Adicciones* **2010**, *22*, 91–96. [CrossRef]
79. Király, O.; Griffiths, M.D.; Urbán, R.; Farkas, J.; Kökönyei, G.; Elekes, Z.; Tamás, D.; Demetrovics, Z. Problematic Internet use and problematic online gaming are not the same: Findings from a large nationally representative adolescent sample. *Cyberpsychol. Behav. Soc. Netw.* **2014**, *17*, 749–754. [CrossRef] [PubMed]
80. Laconi, S.; Tricard, N.; Chabrol, H. Differences between specific and generalized problematic internet uses according to gender, age, time spent online and psychopathological symptoms. *Comput. Hum. Behav.* **2015**, *48*, 236–244. [CrossRef]
81. Griffiths, M.D. Facebook addiction: Concerns, criticism, and recommendations—A response to Andreassen and colleagues. *Psychol. Rep.* **2012**, *110*, 518–520. [CrossRef] [PubMed]
82. Andreassen, C.S.; Billieux, J.; Griffiths, M.D.; Kuss, D.J.; Demetrovics, Z.; Mazzoni, E.; Pallesen, S. The relationship between addictive use of social media and video games and symptoms of psychiatric disorders: A large-scale cross-sectional study. *Psychol. Addict. Behav.* **2016**, *30*, 252–262. [CrossRef] [PubMed]
83. Baggio, S.; Starcevic, V.; Studer, J.; Simon, O.; Gainsbury, S.M.; Gmel, G.; Billieux, J. Technology-mediated addictive behaviors constitute a spectrum of related yet distinct conditions: A network perspective. *Psychol. Addict. Behav.* **2018**, *32*, 564–572. [CrossRef] [PubMed]
84. Ko, C.; Yen, J.; Chen, C.; Chen, S.; Yen, C. Gender Differences and Related Factors Affecting Online Gaming Addiction among Taiwanese Adolescents. *J. Nerv. Ment. Dis.* **2005**, *193*, 273–277. [CrossRef] [PubMed]

85. Scharkow, M.; Festl, R.; Quandt, T. Longitudinal patterns of problematic computer game use among adolescents and adults—A 2-year panel study. *Addiction* **2014**, *109*, 1910–1917. [CrossRef]
86. Laconi, S.; Kaliszewska-Czeremska, K.; Gnisci, A.; Sergi, I.; Barke, A.; Jeromin, F.; Groth, J.; Gamez-Guadix, M.; Ozcan, N.K.; Demetrovics, Z.; et al. Cross-cultural study of Problematic Internet Use in nine European countries. *Comput. Hum. Behav.* **2018**, *84*, 430–440. [CrossRef]
87. Besser, B.; Rumpf, H.; Bischof, A.; Meerkerk, G.; Higuchi, S.; Bischof, G. Internet-related disorders: Development of the Short Compulsive Internet Use Scale. *Cyberpsychol. Behav. Soc. Netw.* **2017**, *20*, 709–717. [CrossRef]
88. Stavropoulos, V.; Beard, C.; Griffiths, M.D.; Buleigh, T.; Gomez, R.; Pontes, H.M. Measurement invariance of the internet gaming disorder scale–short-form (igds9-sf) between Australia, the USA, and the UK. *Int. J. Ment. Health Addict.* **2018**, *16*, 377–392. [CrossRef]
89. Charlton, J.P.; Danforth, I.D. Validating the distinction between computer addiction and engagement: Online game playing and personality. *Behav. Inf. Technol.* **2010**, *29*, 601–613. [CrossRef]
90. Griffiths, M.D. Is "loss of control" always a consequence of addiction? *Front. Psychiatry* **2013**, *4*, 36. [CrossRef]
91. Paaßen, B.; Morgenroth, T.; Stratemeyer, M. What is a True Gamer? The Male Gamer Stereotype and the Marginalization of Women in Video Game Culture. *Sex Roles* **2017**, *76*, 421–435. [CrossRef]
92. Cornil, A.; Lopez-Fernandez, O.; Devos, G.; de Timary, P.; Goudriaan, A.E.; Billieux, J. Exploring gambling craving through the elaborated intrusion theory of desire: A mixed methods approach. *Int. Gambl. Stud.* **2018**, *18*, 1–21. [CrossRef]
93. Ramirez, F.A. From good associates to true friends: An exploration of friendship practices in massively multiplayer online games. In *Social Interactions in Virtual Worlds: An Interdisciplinary Perspective*; Lakkaraju, K., Sukthankar, G., Wigand, R.T., Eds.; Cambridge University Press: New York, NY, USA, 2018; pp. 62–79, Chapter xii; ISBN 9781107128828 (Hardcover).
94. Yee, N. The Psychology of Massively Multi-User Online Role-Playing Games: Motivations, Emotional Investment, Relationships and Problematic Usage. In *Avatars at Work and Play. Computer Supported Cooperative Work*; Schroeder, R., Axelsson, A.S., Eds.; Springer: Dordrecht, The Netherlands, 2006; Volume 34, Chapter 9, pp. 187–207, ISBN 978-1-4020-3883-9 (Printed). [CrossRef]
95. Farrow, R.; Iacovides, I. Gaming and the limits of digital embodiment. *Philos. Technol.* **2014**, *27*, 221–233. [CrossRef]
96. Ko, C.H.; Yen, J.Y.; Cheng-Chung, C.; Chen, S.; Wu, K.; Yen, C.F. Tridimensional Personality of Adolescents With Internet Addiction and Substance Use Experience. *Can. J. Psychiatry* **2006**, *51*, 887–894. [CrossRef] [PubMed]
97. Hull, D.C.; Williams, G.A.; Griffiths, M.D. Video game characteristics, happiness and flow as predictors of addiction among video game players: A pilot study. *J. Behav. Addict.* **2013**, *2*, 145–152. [CrossRef] [PubMed]
98. Ortiz de Gortari, A.B.; Aronsson, K.; Griffiths, M. Game Transfer Phenomena in video game playing: A qualitative interview study. *Int. J. Cyber Behav. Psychol. Learn. (IJCBPL)* **2011**, *1*, 15–33. [CrossRef]
99. Ko, C.H. Internet Gaming Disorder. *Curr. Addict. Rep.* **2014**, *1*, 177–185. [CrossRef]
100. Nunnally, J.C. *Psychometric Theory*, 2nd ed.; McGraw-Hill: New York, NY, USA, 1978.
101. Smith, J.A.; Osborn, M. Interpretive Phenomenological Analysis. In *Qualitative Psychology: A Practical Guide to Research Methods*; Smith, J.A., Ed.; SAGE Publications Ltd.: London, UK; Thousand Oaks, CA, USA; New Delhi, India; Singapore, 2007; pp. 53–81, ISBN 978-1-4129-3083-3.

MDPI
St. Alban-Anlage 66
4052 Basel
Switzerland
Tel. +41 61 683 77 34
Fax +41 61 302 89 18
www.mdpi.com

International Journal of Environmental Research and Public Health Editorial Office
E-mail: ijerph@mdpi.com
www.mdpi.com/journal/ijerph

www.ingramcontent.com/pod-product-compliance
Lightning Source LLC
Chambersburg PA
CBHW041752220926
51847CB00014B/175
* 9 7 8 3 0 3 8 9 7 6 0 4 2 *